CLIMATE RISK AND SUSTAINABLE WATER MANAGEMENT

Climate change is leading to changing patterns of precipitation and increasingly extreme global weather. There is an urgent need to synthesize our current knowledge on climate risks to water security, which in turn is fundamental for achieving sustainable water management. *Climate Risk and Sustainable Water Management* discusses hydrological extremes, climate variability, climate impact assessment, risk analysis and hydrological modelling. It provides a comprehensive interdisciplinary exploration of climate risks to water security, helping to guide sustainable water management in a changing and uncertain future. The relevant theory is accessibly explained using examples throughout, helping readers to apply the knowledge learned to their own situations and challenges. This textbook is especially valuable to students of hydrology, resource management, climate change and geography, as well as a reference textbook for researchers, civil and environmental engineers and water management professionals concerned with water-related hazards, water cycles and climate change.

QIUHONG TANG is a Professor at the Institute of Geographic Sciences and Natural Resources Research at the Chinese Academy of Sciences. He is Associate Editor of *Wiley Interdisciplinary Reviews: Water*, *Journal of Hydrometeorology* and the *Journal of Geographical Sciences*.

GUOYONG LENG is a Professor at the Institute of Geographic Sciences and Natural Resources Research at the Chinese Academy of Sciences. He is an Editorial Board Member of *Science of The Total Environment*, an Associate Editor of *Geoscience Data Journal*, and an Editorial Board Member of *Water*.

CLIMATE RISK AND SUSTAINABLE WATER MANAGEMENT

Edited by

QIUHONG TANG
Institute of Geographic Sciences and Natural Resources Research, Chinese Academy of Sciences

GUOYONG LENG
Institute of Geographic Sciences and Natural Resources Research, Chinese Academy of Sciences

CAMBRIDGE
UNIVERSITY PRESS

CAMBRIDGE
UNIVERSITY PRESS

University Printing House, Cambridge CB2 8BS, United Kingdom

One Liberty Plaza, 20th Floor, New York, NY 10006, USA

477 Williamstown Road, Port Melbourne, VIC 3207, Australia

314–321, 3rd Floor, Plot 3, Splendor Forum, Jasola District Centre, New Delhi – 110025, India

103 Penang Road, #05–06/07, Visioncrest Commercial, Singapore 238467

Cambridge University Press is part of the University of Cambridge.

It furthers the University's mission by disseminating knowledge in the pursuit of education, learning, and research at the highest international levels of excellence.

www.cambridge.org
Information on this title: www.cambridge.org/9781108479837
DOI: 10.1017/9781108787291

First published 2022

Printed in the United Kingdom by TJ Books Limited, Padstow Cornwall

A catalogue record for this publication is available from the British Library.

Library of Congress Cataloging-in-Publication Data
Names: Tang, Qiuhong, editor. | Leng, Guoyong, editor.
Title: Climate risk and sustainable water management / edited by Qiuhong Tang, Chinese Academy of Sciences, Guoyong Leng, Oxford University Centre for the Environment.
Description: Cambridge, United Kingdom ; New York, NY, USA : Cambridge University Press, [2022] | Includes bibliographical references and index.
Identifiers: LCCN 2021026923 (print) | LCCN 2021026924 (ebook) | ISBN 9781108479837 (hardback) | ISBN 9781108787291 (epub)
Subjects: LCSH: Watershed management. | Floods–Risk assessment. | Droughts–Risk assessment.
Classification: LCC TC413 .C57 2022 (print) | LCC TC413 (ebook) | DDC 333.91–dc23
LC record available at https://lccn.loc.gov/2021026923
LC ebook record available at https://lccn.loc.gov/2021026924

ISBN 978-1-108-47983-7 Hardback

Contents

Contributors

Meysam Alizamir, Department of Civil Engineering, Hamedan Branch, Islamic Azad University, Hamedan, Iran

Parisa Ansari, Department of Natural Resources & Environment, Shiraz University, Shiraz, Iran

Yunyun Ban, Institute of Geographic Sciences and Natural Resources Research, Chinese Academy of Sciences, Beijing, China

Binod Baniya, Land Surface Processes and Global Changes Research Group, Institute of Geographic Sciences and Natural Resources Research, Chinese Academy of Sciences, Beijing, China, University of Chinese Academy of Sciences, Beijing, China and Department of Environmental Science, Patan Multiple Campus, Tribhuvan University, Nepal

Hom Bahadur Baniya, Department of Physics, Tribhuvan University, Kathmandu, Nepal

Ximing Cai, Department of Civil and Environmental Engineering, University of Illinois at Urbana-Champaign, Urbana, IL, USA

Yongqin David Chen, School of Humanities and Social Science, The Chinese University of Hong Kong, Shenzhen, China and Department of Geography and Resource Management, The Chinese University of Hong Kong, Hong Kong, China

Gyan Chhipi-Shrestha, School of Engineering, The University of British Columbia, Kelowna, BC, Canada

Huijuan Cui, Key Laboratory of Land Surface Pattern and Simulation, Institute of Geographic Sciences and Natural Resources Research, Chinese Academy of Sciences, Beijing, China

Seyed Rashid Fallah Shamsi, Department of Natural Resources & Environment, Shiraz University, Shiraz, Iran

Stefano Luigi Gariano, CNR-IRPI Research Institute for Geo-Hydrological Protection, Perugia, Italy

Gebremedhin Gebremeskel Haile, Key Laboratory of Water Cycle and Related Land Surface Processes, Institute of Geographic Sciences and Natural Resources Research, Chinese Academy of Sciences, Beijing, China, University of Chinese Academy of Sciences, Beijing, China and Tigray Agricultural Research Institute, Mekelle, Ethiopia

Seyed-Mohammad Hosseini-Moghari, Key Laboratory of Water Cycle and Related Land Surface Processes, Institute of Geographic Sciences and Natural Resources Research, Chinese Academy of Sciences, Beijing, China

Zhongwei Huang, School of Hydrology and Water Resources, Nanjing University of Information Science and Technology, Nanjing, China

Guoqiang Jia, Institute of Geographical Sciences and Natural Resources Research, Chinese Academy of Sciences, Beijing, China

Marwan Kheimi, Department of Civil Engineering, King Abdulaziz University, Jeddah, Saudi Arabia

Yangchen Lai, Department of Geography, Hong Kong Baptist University, Hong Kong, China and Key Laboratory for Geo-Environmental Monitoring of Coastal Zone of the National Administration of Surveying, Mapping and GeoInformation & Shenzhen Key Laboratory of Spatial Smart Sensing and Services, Shenzhen University, Shenzhen, China

Guoyong Leng, Key Laboratory of Water Cycle and Related Land Surface Processes, Institute of Geographic Sciences and Natural Resources Research, Chinese Academy of Sciences, Beijing, China

Guangdong Li, Institute of Geographical Sciences and Natural Resources Research, Chinese Academy of Sciences, Beijing, China

Haiming Li, College of Marine and Environmental Sciences, Tianjin University of Science & Technology, Tianjin, China

Jianfeng Li, Department of Geography, Hong Kong Baptist University, Hong Kong, China

Xu Li, Key Laboratory of Urban Underground Engineering of Ministry of Education, Beijing Jiaotong University, Beijing, China

Xingcai Liu, Institute of Geographical Sciences and Natural Resources Research, Chinese Academy of Sciences, Beijing, China

Hui Lu, Department of Earth System Science, Tsinghua University, Beijing, China

Masoud Masoudi, Department of Natural Resources & Environment, Shiraz University, Shiraz, Iran

Massimo Melillo, CNR-IRPI Research Institute for Geo-Hydrological Protection, Perugia, Italy

Mengfei Mu, Key Laboratory of Water Cycle and Related Land Surface Processes, Institute of Geographical Sciences and Natural Resources Research, Chinese Academy of Sciences, Beijing, China and University of Chinese Academy of Sciences, Beijing, China

Tolulope Olutimehin, College of Agriculture and Human Sciences, Prairie View A&M University, Prairie View, TX, USA

Hamid Reza Pourghasemi, Department of Natural Resources & Environment, Shiraz University, Shiraz, Iran

Ram L. Ray, College of Agriculture and Human Sciences, Prairie View A&M University, Prairie View, TX, USA

Xiaogang Shi, School of Interdisciplinary Studies, University of Glasgow, Dumfries, UK

Rajendra P. Sishodia, College of Agriculture and Human Sciences, Prairie View A&M University, Prairie View, TX, USA

Siao Sun, Key Laboratory of Regional Sustainable Development Modeling, Institute of Geographical Sciences and Natural Resources Research, Chinese Academy of Sciences, Beijing, China

Qiuhong Tang, Key Laboratory of Water Cycle and Related Land Surface Processes, Institute of Geographic Sciences and Natural Resources Research, Chinese Academy of Sciences, Beijing, China and University of Chinese Academy of Sciences, Beijing, China

Yin Tang, Key Laboratory of Water Cycle and Related Land Surface Processes, Institute of Geographic Sciences and Natural Resources Research, Chinese Academy of Sciences, Beijing, China

Jie Wang, Institute of Geographical Sciences and Natural Resources Research, Chinese Academy of Sciences, Beijing, China

Lei Wang, Key Laboratory of Tibetan Environment Changes and Land Surface Processes, Institute of Tibetan Plateau Research, Chinese Academy of Sciences, Beijing, China, Chinese Academy of Sciences Center for Excellence in Tibetan Plateau Earth Sciences, Beijing, China and University of Chinese Academy of Sciences, Beijing, China

Yueling Wang, Key Laboratory of Water Cycle and Related Land Surface Processes, Institute of Geographic Sciences and Natural Resources Research, Chinese Academy of Sciences, Beijing, China

Zhonggen Wang, Key Laboratory of Water Cycle and Related Land Surface Processes, Institute of Geographic Sciences and Natural Resources Research, Chinese Academy of Sciences, Beijing, China

Nigel Wright, School of Architecture, Design and the Built Environment, Nottingham Trent University, Nottingham, UK

Huan Wu, School of Atmospheric Sciences, Sun Yat-sen University, Guangzhou, Guangdong Province, China

Han Xiao, College of Marine and Environmental Sciences, Tianjin University of Science & Technology, Tianjin, China

Muqi Xiong, Key Laboratory of Water Cycle and Related Land Surface Processes, Institute of Geographic Sciences and Natural Resources Research, Chinese Academy of Sciences, Beijing, China

Ximeng Xu, Key Laboratory of Water Cycle and Related Land Surface Processes, Institute of Geographic Sciences and Natural Resources Research, Chinese Academy of Sciences, Beijing, China

Xiaomeng Yin, Key Laboratory of Water Cycle and Related Land Surface Processes, Institute of Geographic Sciences and Natural Resources Research, Chinese Academy of Sciences, Beijing, China and University of Chinese Academy of Sciences, Beijing, China

Yuanyuan Yin, Key Laboratory of Tibetan Environment Changes and Land Surface Processes, Institute of Tibetan Plateau Research, Chinese Academy of Sciences, Beijing, China and Key Laboratory of Water Cycle and Related Land Surface Processes, Institute of Geographic Sciences and Natural Resources Research, Chinese Academy of Sciences, Beijing, China

Xiaobo Yun, Institute of Geographical Sciences and Natural Resources Research, Chinese Academy of Sciences, Beijing, China

Qinghuan Zhang, Chinese Research Academy of Environmental Sciences, Beijing, China and Key Laboratory of Water Cycle and Related Land Surface Processes, Institute of Geographic Sciences and Natural Resources Research, Chinese Academy of Sciences, Beijing, China

Mohammad Zounemat-Kermani, Water Engineering Department, Shahid Bahonar University of Kerman, Kerman, Iran

Preface

As anthropogenic disturbance has reached unprecedented levels, the terrestrial water cycle is undergoing rapid changes, threatening water scarcity and raising new challenges for sustainable water management. In order to achieve the Sustainable Development Goals (SDGs) set by the United Nations General Assembly, water management will have to be improved by reducing climate risks and building climate resilience. Water stress and water-related hazards such as drought, flood, landslide, storm surge and saltwater intrusion have caused huge impacts to the social-economy and the environment. Such water risks are projected to grow under future socioeconomic scenarios and climate scenarios characterized with rising temperature, changing precipitation patterns and increasing extreme weathers. There is an urgent need to synthesize current state-of-art knowledge on climate risks facing the water sector, especially given that professionals in water management agencies are often not well updated with scientific progress. Therefore, it is time to write a book on Climate Risk and Sustainable Water Management using ordinary-language explanations and examples throughout, such that it can be used as a textbook for training courses as well.

This book covers various topics including hydrological extremes, climate variability and climate, climate impact assessment, risk analysis, hydrological modelling, etc. The inter-disciplinary and multi-disciplinary feature of this book could provide a comprehensive knowledge on climate risks to water security, thus guiding sustainable water management in an uncertain future. This book provides relevant theory step-by-step using ordinary-language explanations and examples throughout, making the theory accessible to readers without specialized training in science. The information and advice are conveyed in such an efficient and realistic way that readers can apply the knowledge learned from this book to their own situation and problems.

This book is organized into three closely connected parts. Part I focuses on the interactions between water-related hazards and climate change. It starts with two chapters on the modelling and analysis of flood risk, followed by droughts, soil erosion and saltwater intrusion. Specifically, Chapter 1 investigates various types of flood risk in a delta area, while Chapter 2 assesses the flood risk in a transboundary river basin impacted by both climate change and reservoir operations. Projection of future drought change patterns in East Africa is presented in Chapter 3. Chapters 4 and 5 focus on the changes in global water erosion vulnerability under climate change and the driving factors behind the changing water erosions, respectively. The last chapter in Part I, Chapter 6 investigates climate change impacts on saltwater intrusion into coastal aquifers.

Part II covers various climate risks to the human and natural systems. Chapter 7 evaluates the impacts of urbanization on local climate in Southeast China, which could further feedback to the spatial-temporal variation of water-related hazards. Chapter 8 investigates the impact of climate change on vegetation dynamics, land cover change and ecological risk in Nepal, followed by Chapters 9 and 10 reviewing the effects of climate change on frozen soil on the Tibetan Plateau and global agriculture production, respectively. Chapter 11 focuses on the impacts of human water use on the terrestrial water cycle, while Chapter 12 assesses the benefits of China's South-to-North Water Diversion project in regulating water scarcity in a water-receiving area under climate change. Chapter 13 focuses on the detection and monitoring of rainfall-induced landslide at global scale, while Chapter 14 introduces the methodology of estimating aquifer depth using satellite remote sensing data.

Part III provides an overview of recent advances in theory and practice towards sustainable water management under future uncertainty. It starts with Chapter 15, on how to manage urban flood risk in a changing climate, followed by Chapter 16, on the application of soft computing techniques for water management. Chapter 17 introduces the benefits of rainwater harvesting for enhancing the resilience to climate risks. Chapter 18 focuses on the variations of runoff coefficient and precipitation elasticity at watershed scale, while the uncertainties from model calibration for future hydrological projections are explored in Chapter 19. Chapters 20 and 21 present two case studies for water risk management under future uncertainty, with one focusing on water scarcity and the other on the shrinking inland lake.

Some of the chapters included in this book are based on discussions with scientists and practitioners at regular conferences organized by the International Association of Chinese Youth in Water Sciences (CYWater). Parts of the chapters are also used for graduate courses at the University of Chinese Academy of Sciences (UCAS) as well as the international training workshop by the Chinese Academy of Sciences (CAS).

We hope this book can facilitate multi-disciplinary and transdisciplinary discussions and exchange between scientists and practitioners for sustainable water management under deep uncertainty, and serve as a textbook for graduate and postgraduate courses in climate change, hydrology, water management, risk analysis and geography.

Acknowledgements

We would like to thank the reviewers for their helpful comments and suggestions that led to substantial improvement of this book. We also acknowledge the Chinese Academy of Sciences, Natural Science Foundation of China (NSFC) and the Royal Society – Newton Advanced Fellowships for providing financial support, and Dr. Andre Francois Python for English language editing.

Part I

Water-Related Risks under Climate Change

1

Pluvial, Fluvial and Coastal Flood Risks and Sustainable Flood Management in the Pearl River Delta under Climate Change

JIANFENG LI, XIAOGANG SHI, YONGQIN DAVID CHEN AND YANGCHEN LAI

1.1 Introduction

The Pearl River Delta (PRD) is a megalopolis located in South China, which includes the Special Administrative Regions (SAR) of Hong Kong and Macau, and nine cities, namely Guangzhou, Shenzhen, Foshan, Zhuhai, Dongguan, Huizhou, Jiangmen, Zhaoqing and Zhongshan. As the largest city region in the world in terms of size and population, the PRD nurtures about 4 per cent of the Chinese population, i.e. 57.15 million, and produces about 20 per cent of national GDP. The Pearl River Basin (PRB) is the second largest basin in China in terms of annual discharge and provides abundant water resources for the socioeconomic development of the PRD cities (Figure 1.1). Around 80 per cent of the water supply for Hong Kong is from the PRB. The PRB is dominated by a sub-tropical and humid monsoon climate, with over 70 per cent of rainfall concentrated in the summer months, making the PRD region highly exposed to both fluvial floods resulting from high upstream river flow and pluvial floods due to localized heavy rainstorms and inadequate drainage capacities. Due to its coastal location by the South China Sea, the PRD is often attacked by typhoons and thus under the threat of coastal floods, such as the severe inundation of low-lying areas in the PRD due to the storm surge and heavy rainfall caused by super typhoon Mangkhut in September 2018. In this connection, multi-flood prevention and mitigation systems are particularly important for the long-term economic prosperity and sustainable development of this region. Furthermore, along with the merging of individual cities, the recent urban expansion has contributed to an increase in the PRD area by ten-times. The PRD has become one of the world's largest megalopolises in size, population and economic output, causing unprecedented stresses on urban

The work described in this paper was supported by research grants from the Research Grants Council of the Hong Kong Special Administrative Region, China (No. HKBU12303517 and No. HKBU12302518), the University of Glasgow CoSS Strategic Research Fund, the National Key R&D Program of China (No. 2019YFC1510400), and Guangdong-Hong Kong Joint Laboratory for Water Security (2020B1212030005).

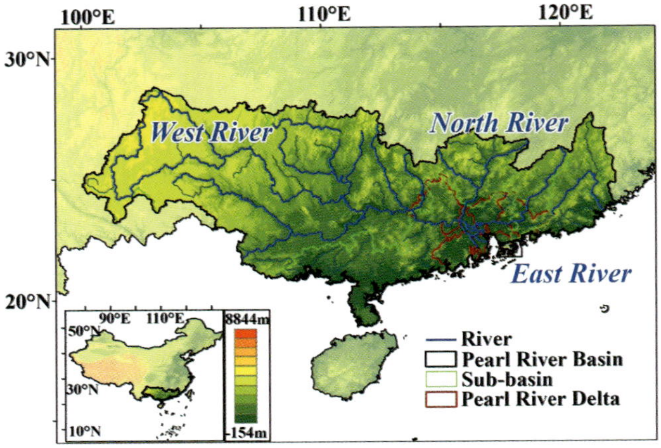

Figure 1.1 Locations of the Pearl River Delta and Pearl River Basin

drainage. Such urban sprawl has substantially altered land–atmosphere interactions in favour of extreme localized rainstorms, as well as increased land surface imperviousness and, thus, enhanced runoff generation, resulting in increases in the risk of both pluvial and fluvial floods.

Due to its unique geographical location and rapid urbanization, the PRD megacity region is affected by various types of floods triggered by different mechanisms, including pluvial, fluvial and coastal floods. Heavy rainstorms caused by various weather systems (e.g., typhoons and convections) usually bring about excessive rainwater inside the cities in a short time. In the PRD, the highly and rapidly urbanized land surface undermines the water storage capacity of the land surface and reduces the capacity of drainage systems due to clogging for various reasons (e.g., trash blockage and back water effect), causing greater threats of pluvial floods. The PRD is located in the estuary of the PRB, which is the convergence of the three major rivers, namely Xijiang (West River), Beijiang (North River) and Dongjiang (East River). Large-scale rainstorms in the upstream of the PRB would cause excessive discharge to the downstream and may trigger fluvial flood hazards in the PRD. In past decades, fluvial floods were the major flood type in the PRD before flood prevention infrastructures such as dikes and dams were built and commissioned. In recent years, fluvial floods have been effectively controlled and prevented, while pluvial floods have become more frequent due to the rapid urbanization; hence, more studies are focused on pluvial floods (Yin et al., 2014; Huang et al., 2018). As the PRD is located on the coasts of the South China Sea, the sea level rises under global climate change and storm surge induced by typhoons are the major causes of coastal floods in the regions (Yu et al., 2018; Hong Kong Observatory, 2019b). Different types of floods may co-occur simultaneously or in sequence because they can be physically caused by the same weather systems. For example, super typhoons can bring excessive rainwater, high storm surge and strong winds and, therefore, lead to pluvial and coastal floods. The co-occurrence of different types of floods have greater impacts on the cities (Ikeuchi et al., 2017; Shen et al., 2019). For instance, coastal floods reduce the city drainage capacity, which makes it more difficult to drain out the flooding water in a city directly resulting from pluvial or fluvial floods.

Under climate change, coastal regions across the world have experienced intensifications of different types of flood hazards. For example, coastal Bangladesh is a delta region affected by multi-type floods such as pluvial, fluvial and coastal floods (Adnan et al., 2019). Among these flood types, pluvial floods are more frequent, while fluvio-tidal floods and coastal floods cause larger inundation areas. Adnan et al. (2019) found that polders in Bangladesh prevent moderate storm surges and fluvio-tidal floods on one hand, but exacerbate the inundation areas of pluvial floods on the other hand. In the Mekong River delta of Vietnam, flood levels depend on the combined effects of upstream flows, sea level rise, storm surges and siltation of the Mekong Estuary. Affected by engineering structures along the Mekong River in Vietnam and nearby countries, the water is deeper in the rivers and channels in these areas because of the increases of flow velocity, bank erosion and siltation, resulting in more frequent floods in the non-protected areas and greater risks of dike failure in the protected areas (Le et al., 2007). The Rhine-Meuse delta in the Netherlands, a well-developed and densely populated delta region in Europe, is threatened by sea level rise and storm surges (De Moel et al., 2011). The long-term efforts in protecting vulnerable areas against rising sea levels, such as mechanic pumping stations and dike-rings, successfully protect the flood-prone areas against floods with pre-defined return periods ranging from 1,250 to 10,000 years in this region. However, combining the effects of population growth, economic development and climate change, the flood damage in the Netherlands is projected to increase ten-fold by 2100 compared to 2000. Thus, flood management in a context of increasing flood risks under climate change is a major challenge for coastal cities across the globe (Vis et al., 2003; Klijn et al., 2012). Therefore, investigating the causes, changes and mitigation measures of multi-type floods in the PRD under climate change carries scientific and practical importance for improving the understanding of climate change impact on flood risks and for formulating mitigation strategies against the increasing multi-type flood risks in a coastal metropolis.

1.2 Localized Precipitation Extremes and Pluvial Flood Risks in the PRD under Climate Change: Observations and Projections

1.2.1 Past Observations and Analyses

Pluvial floods, also called urban waterlogging, are caused by localized precipitation extremes and inadequate drainage capacities of an urban area. The

PRD is controlled by the sub-tropical monsoon systems, with annual precipitation ranging from 1,000 to 2,000 mm, and the summer precipitation accounts for 72–88 per cent of the annual total, with June being the month with the highest precipitation (J. Li et al., 2017). Over the years, the PRD has often suffered from the impacts of heavy rainstorms. Such extreme rainstorms have led to serious waterlogging and resulted in economic damage and even loss of life in the PRD cities. On 29 May 1889, the centurial extreme rainstorm led to a total rainfall of 697.1 mm in 24 hours, which was about one third of the annual total precipitation, resulting in serious floods and landslides in Hong Kong (Lee et al., 2016). As one of the most destructive rainstorms in the history of Hong Kong, this extreme event ruined a number of roads, killed 27 people and resulted in 17 people missing. The estimated cost of the damage to government property was $112,783 at that time, which was about 6 per cent of the annual government expenditure in 1889. On 7 May 2010, extreme rainstorms caused serious water-logging in Guangzhou, leading to inundated streets and transport chaos (Huang et al., 2018). On 7 May 2017, an extreme rainstorm with a return period of 60 years hit the PRD (Zhang et al., 2019). The rain gauge in Zengcheng district in Guangzhou recorded 586 mm precipitation in three hours, causing severe pluvial floods in Zengcheng, Huadu and Huangpu districts.

Precipitation extremes are the major trigger of pluvial floods. Under global climate change, local changes in precipitation extremes can significantly alter the pluvial flood risks in the affected region. Observations showed that the precipitation regimes in the PRD have changed over the past decades (Zhang et al., 2012). Ai and Wu (2018) indicated that the intensity of rainstorms and the contribution of heavy rainstorms to annual precipitation slightly increased in Guangzhou over 1961–2015. Lenderink et al. (2011) analysed the hourly precipitation observations since 1885 in Hong Kong and suggested that the increases in daily precipitation extremes under global climate change generally follow the Clausius-Clapeyron relation (C-C relation). The C-C relation indicates that the maximum moisture content of the atmosphere is expected to increase with about 7 per cent per degree Celsius when temperature increases. They also found that the hourly precipitation extremes in Hong Kong increased by 10–14 per cent per degree Celsius temperature rise, which is faster than the C-C increase rate and called the super-CC relation. The findings of Lenderink et al. (2011) suggested more intensification of hourly and daily precipitation extremes under global climate change, implying greater risks of pluvial floods.

A number of precipitation indices, which represent various characteristics of precipitation regimes, have been applied to analyse the changes in precipitation in the PRD. Although these indices may differ in the PRD, they generally showed that precipitation had become more intense in the past decades. Based on daily precipitation of 1960–2005 from 42 stations across the PRB, Zhang et al. (2012) evaluated the spatial–temporal changes of different precipitation extreme indices, including annual total precipitation amount, annual total rainy days, annual precipitation intensity and annual mean rainy days. The modified Mann–Kendall trend test method was used to detect the trends of these extreme indices. The results showed that the PRB was characterized by increasing precipitation intensity, especially in the middle and eastern parts of the basin where the PRD is located. Frequencies of short-duration wet periods and total precipitation amount have increased in the past decades, raising the risks of floods. Heavy precipitation was associated with wet spells of shorter durations, which suggests that the precipitation process intensified in the PRB, especially in the PRD located in the lower part of the basin. Zhao et al. (2014) analysed the daily precipitation during 1960–2012 and evaluated the changes in another set of precipitation extreme indices across the PRB. They did not find significant trends in the number of heavy precipitation days, however, the simple daily intensity index increased significantly at the 95 per cent level. Other extreme precipitation indices, such as maximum 1-day precipitation, maximum 5-day precipitation and intensity of extreme precipitation with intensity greater than its 99th percentile exhibited no significant trends. They also concluded that Pacific Decadal Oscillation (PDO) and Southern Oscillation Index (SOI) are important factors that affect precipitation changes. Fischer et al. (2012) indicated that extreme precipitation indices in the PRB changed abruptly in 1986 and 1997. Based on the precipitation observations in Hong Kong since 1885, Wong et al. (2011) examined the past trends in precipitation extremes using extreme indices developed by the Expert Team on Climate Change Detection, Monitoring and Indices (ETCCDMI) under the World Meteorological Organization (WMO). They found that the frequency of 1-, 2- and 3-hour precipitation extremes increased significantly. The contribution of precipitation extremes greater than the daily 95th percentile to annual total precipitation increased substantially by 22 mm per decade. The return period of 1-hour precipitation greater than 100 mm decreased significantly from 37 years in 1900 to 18 years in 2000.

Previous studies mostly focused on the changes in the probabilistic behaviour of precipitation extremes. In recent years, increasing concerns have been drawn to the joint behaviours of precipitation extremes. Zhang et al. (2013) investigated spatiotemporal patterns of historical precipitation extremes in China during 1960–2005 based on Copula, an important tool in the bivariate analysis of precipitation extremes. The joint probability analysis of precipitation extremes showed that precipitation extremes were intensifying in South China, specifically the PRB. The intensification of precipitation extremes in the PRB was associated with the decreasing number of rainy days and increasing number of consecutive non-rainy days, suggesting potential increases in the occurrences of floods and droughts.

1.2.2 Future Projections

Under the future global climate change, the frequency and intensity of extreme precipitation in the PRD are expected to continue to change. Because the spatial resolutions of Global Climate Models (GCMs), a major tool for climate projections, are too coarse for projecting precipitation changes in the PRD at the regional scale, the outputs of GCMs were downscaled to a finer resolution in previous studies. The future projections of the Hong Kong Observatory (HKO), based on statistical downscaling, showed that extremely wet years (i.e., annual precipitation greater than 3,168 mm) would reach 12 years, and extremely dry years (i.e. annual precipitation less than 1,289 mm) would be only 2 years in Hong Kong under the future high-emission Representative Concentration Pathways 8.5 scenario (RCP8.5; Hong Kong Observatory, 2014). The annual number of rainy days was projected to decrease from 102 days in 1986–2005 to 97 days in 2091–2100, and in the same two periods the average precipitation intensity should increase from 23.4 mm/day to 26.7 mm/day as rainy days decrease (Hong Kong Observatory, 2014). The annual number of extreme rainy days were projected to increase to 5.1 days in 2091–2100 from 4.2 days in 1986–2005, and the annual maximum daily precipitation should increase considerably from 221 mm in 1986–2005 to 273 mm in 2091–2100 in Hong Kong (Hong Kong Observatory, 2014). More detailed changes in different precipitation extreme indices in Hong Kong can be found in Figure 1.2. Decreases in the number of rainy days under climate change have been widely reported in different regions, such as the PRB, China, and Europe (Qian et al., 2007; Zolina et al., 2010; Zhang et al., 2012; Li et al., 2013). This can be explained by the increases in maximum moisture content of the atmosphere given temperature rises according to the C-C relation (J. Wu et al., 2015). Given a stable level of water vapour content, it is harder for a warmer atmosphere to reach dew-point temperature.

Li et al. (2013) applied a statistical downscaling method on the basis of Quantile–Quantile relationship and transfer functions to downscale precipitation extremes from the GCM outputs of the Coupled Model Intercomparison Project Phase 5 (CMIP5) to the site scale. The Taylor diagrams showed that the normalized standard deviations of most models are close to 1 and the correlations between the simulations and observations are close to 0.8 or larger, indicating that the downscaled extreme indices were reasonably matched with the observations. The projections based on the downscaled indices indicated that precipitation processes are expected to intensify with increased frequency and intensity in the PRD in the future. Future changes in precipitation extremes exhibited larger

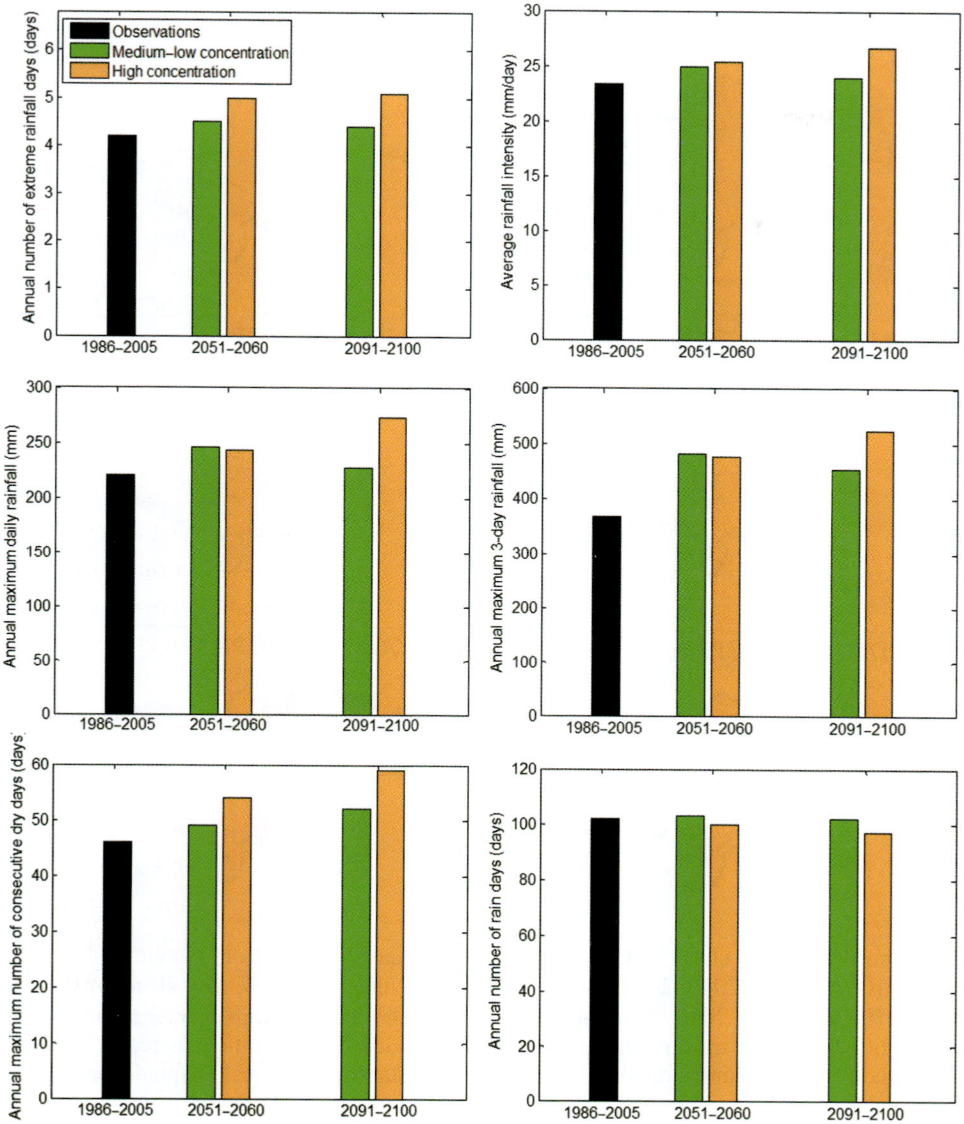

Figure 1.2 Annual number of extreme rainfall days (days), average rainfall intensity (mm/day), annual maximum daily rainfall (mm), annual maximum 3-day rainfall (mm), annual maximum number of consecutive dry days (days) and annual number of rain days (days) in 1986–2005, 2051–2060 and 2091–2100. Black denotes the actual value of the extreme index. Green and orange denote low–medium concentration and high concentration scenarios, respectively (adapted from climate projections of the Hong Kong Observatory, 2014)

magnitudes of change under RCP8.5 than RCP2.6. The future changes in precipitation extremes are gradual processes (Figure 1.3). The change rates of precipitation extremes are expected to be higher in the eastern part of the PRB and lower in the western part during 2010–2039 under RCP2.6 and RCP8.5. In

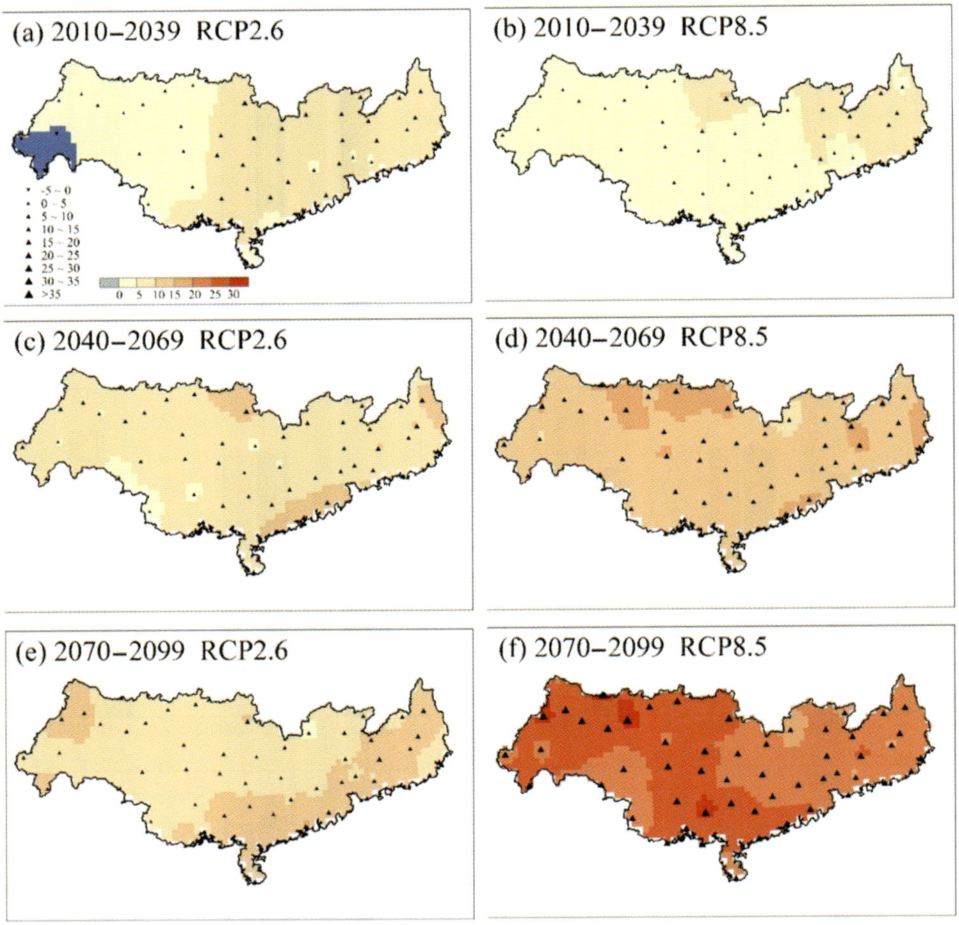

Figure 1.3 The changes (%) in maximum 5-day precipitation amount of (a) 2010–2039, (c) 2040–2069 and (e) 2070–2099 under RCP2.6, as well as of (b) 2010–2039, (d) 2040–2069 and (f) 2070–2099 under RCP8.5, compared to 1960–2005. The upward triangle denotes the index of that station increases and the downward triangle denotes the index of that station decreases (adapted from Li et al., 2013)

2040–2069, the change rates are projected to be similar across the PRB. During 2070–2099, the change rates are projected to be higher in the southern and eastern parts of the PRB under RCP2.6 and the central and northwestern parts under RCP8.5. The directions of changes may alter during the temporal evolution of precipitation extremes. Changes under RCP8.5 become more pronounced over time. Under the peak-and-decline RCP2.6, changes of some indices are not expected to decrease correspondingly during 2070–2099, even though the radiative forcing predicted during 2070–2099 is less than that during 2040–2069. The projected increases in precipitation extremes suggest the pluvial flood risks in the PRD cities are expected to increase in the future.

Future possible changes in joint probabilistic behaviours of precipitation extremes were evaluated by Li et al. (2015). Five joint return periods based on six extreme precipitation indices were defined and analysed in this study, considering joint extreme heavy precipitation events in terms of different combinations of extreme precipitation amount, intensity, fractional contribution to annual precipitation days and consecutive wet periods. The Weather Generator Model (WGEN) was employed to downscale the model outputs. The results of this study showed that, under climate change, the changes in marginal distributions and bivariate relationships between extreme indices might be the causes of joint probability distribution shift. During 2021–2050 and 2071–2100, there should be more joint extreme heavy precipitation events with high intensity and large total amount, implying higher risks of extreme floods in the future. Furthermore, the magnitude of change rates in joint return periods under RCP8.5 is more significant than those under RCP2.6. Even under RCP2.6 with the 2°C warming target achieved, the changes in joint return periods remain substantial.

Moreover, Ou et al. (2013) pointed out that model outputs may underestimate precipitation extremes in South China, which is one of the uncertainty sources of future precipitation projections, implying that the risks of water-logging in the PRD cities may be underestimated. In other words, the pluvial flood risks in the PRD cities may be greater than those projected by model simulations.

1.3 Upstream River Flow and Fluvial Flood Risks in the PRD under Climate Change: Observations and Projections

1.3.1 Past Observations and Analyses

Fluvial floods are basically defined as high streamflow in rivers. Different from pluvial floods, which are mostly affected by localized rainstorms and inadequate drainage capacities, fluvial floods are usually associated with basin-scale precipitation extremes and human activities across the basin. As the PRD is located in the downstream, increases in extreme streamflow from the upstream naturally result in higher risks of fluvial floods in the PRD cities. In June and July 1994, heavy rainstorms occurred widely in Guangdong province and were particularly extreme in some regions (Liang, 1997; Zhang, 1997). The rainstorms were characterized by high intensity, long duration and extensive spatial coverage. These rainstorms in the two months caused two extreme fluvial flood events in Xijiang and Beijiang, which hence converged the PRD and was blocked by tidal backwater, leading to persistent high water levels across the PRD region. The water levels due to the flooding tides of rivers in Shunde and Zhongshan reached or exceeded the 50-year return period and some cities, such as Guangzhou and Zhuhai, were severely affected. Two thirds of Guangdong areas and more than 9.5 million people were affected by those flood events, causing a direct economic loss of 14.6 billion Yuan. In June 2005, the continuous rainstorms in the PRB and extremely intensive precipitation in parts of the region caused a substantial rise in water levels in major rivers (Yi, 2005). Due to the rainstorms, a 100-year extreme fluvial flood, a 10-year fluvial flood and the greatest flood event in the past 20 years hit the mid- and low-stream of the Xijiang, Beijiang and Dongjiang, respectively. Flood water of these events went downstream to converge in the PRD and the hazard was aggravated by the greatest astronomical tide in the past 19 years. The co-occurrence of the fluvial floods in the three sub-basins and the extreme astronomical tide resulted in extreme floods in the PRD.

Previous studies have investigated the changes of fluvial floods based on streamflow observations at hydrological stations in the PRB. Due to climate change, especially changes in precipitation regimes across the basin, the observed annual maximum streamflow in Xijiang and Beijiang showed an increasing trend, and the flood peaks had increased from 1951 to 2010 (Gu et al., 2014). Wu et al. (2013) also analysed the observed streamflow at hydrological stations in the PRB in 1951–2010 and used the Pearson type III distribution for flood frequency analyses. They also found significant increases in extreme fluvial floods across the region since 1980, especially in Xijiang and Beijiang. The flood events at stations in Dongjiang basin showed decreasing trends. The decreases in fluvial floods as represented by decreases in maximum streamflow in Dongjiang can be explained by flood control operations in the reservoirs in Dongjiang (Zhang et al., 2014). Compared to the other two sub-basins, Xijiang and Beijiang, the streamflow in Dongjiang is highly regulated by three major reservoirs (i.e., Xinfengjiang, Fengshuba and Baipanzhu). The observed multi-day maxima at hydrological stations in Dongjiang basin, such as Longchuan, Heyuan, Lingxia and Boluo stations, have dropped after the three reservoirs went into operation. Zhang et al. (2015) also found that the PDO index linearly affected the variance of annual maximum streamflow at Lingxia and Boluo stations and nonlinearly affected Longchuan and Heyuan stations in Dongjiang. They also concluded that the changes in flood discharge in Dongjiang can be non-stationary due to climate change and reservoir operations.

1.3.2 Future Projections

Although previous studies have projected future possible changes in precipitation extremes in the PRD, studies investigating future changes in fluvial floods have been overlooked. Among a few previous studies, changes in precipitation extremes have been used directly to suggest the possible changes in fluvial floods. However, it has been argued that these studies did not sufficiently consider the complex surface hydrological processes. Given the recent development of GCMs and hydrological models, a growing number of studies have employed hydrological models to evaluate future river flood risks in the PRD. Yan et al. (2015) applied the Variable Infiltration Capacity (VIC) model to simulate streamflow in the Pearl River and found that flood discharge would increase more substantially than average streamflow. Yuan et al. (2016) used the Xinanjiang (XAJ) hydrological model for streamflow simulations in the PRB based on historical climatic data and two future precipitation scenarios. Before being used to drive the XAJ model, the precipitation simulations from GCMs were bias-corrected using delta-change methods. The simple delta-change (SDC) method with a constant scaling factor and the quantile–quantile delta-change (QQDC) method with a quantile mapping-based non-uniform delta factor were applied. Their results showed that annual maximum daily streamflow at longer return periods, such as 20 or 50 years, are expected to increase consistently in the future. Other studies using different methods or downscaling methods also predicted more extreme maximum streamflow across the PRB in the future, such as those by Wu et al. (2014) and Li et al. (2016).

The Intergovernmental Panel on Climate Change (IPCC) AR5 report indicates that the confidence of surface hydrological projections is medium (IPCC, 2013). Many studies have analysed the contributions of different sources of uncertainty in hydrological projections in the PRB, such as GCMs, downscaling methods, bias-correction methods and hydrological models. C. H. Wu et al. (2015) used five GCMs (i.e., BCC-CSM1.1, CanESM2, CSIRO-Mk3.6.0, GISS-E2-R and MPI-ESM-LR) to drive the VIC model to simulate future extreme floods in the PRB to evaluate the contributions of GCMs to uncertainties of flood projections. The VIC model demonstrated good performance in simulating extreme floods, with a daily Nash-Sutcliffe efficiency coefficient (NSE) of 0.91. They found that GCMs and future emission scenarios are the major sources of uncertainty in future flood projections in the PRB and the overall uncertainty range for changes in extreme flood magnitudes can be well represented by those five GCMs. Yuan et al. (2016) compared the projections of streamflow changes of the PRB using two delta-change methods (i.e., SDC and QQDC) to correct the biases of GCMs. They found that the two delta-change methods caused significant impacts on streamflow projections. The 20-year daily maximum streamflow simulated by the XAJ hydrological model driven by QQDC precipitation was about 7.0–65.0 per cent higher than that of SDC precipitation. Their results indicated that bias-correction or downscaling is an important source of uncertainty. Liu et al. (2013) concluded that the uncertainty from GCMs outweigh those of downscaling methods. Li et al.

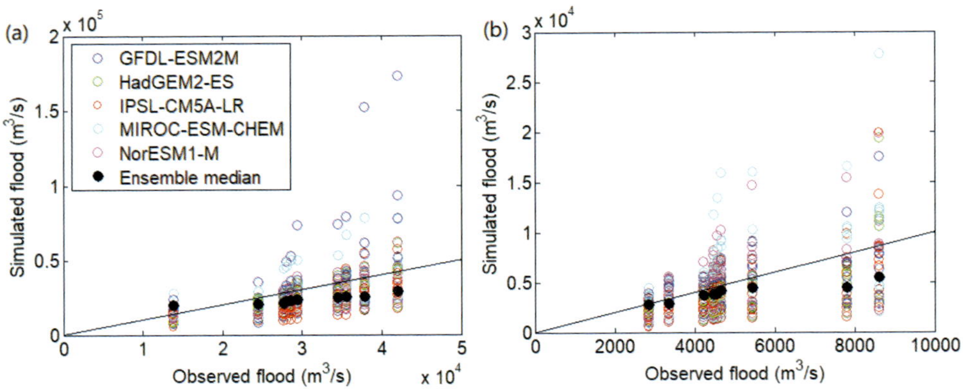

Figure 1.4 Quantile–Quantile plots of observed and simulated annual maximum daily discharge at (a) Wuzhou and (b) Boluo stations during 1971–1980. Wuzhou and Boluo stations are the controlling hydrological stations of Xijiang and Dongjiang, respectively. The colour circles are hydrological models driven by different GCMs and the black circles are the ensemble mean. The black line denotes the 1:1 line (adapted from Li et al. 2016; © American Meteorological Society. Used with permission)

(2016) further evaluated how various hydrological models can affect the uncertainty of future flood projections. Discharge simulations of eight hydrological models driven by five GCMs from the Inter-Sectoral Impact Model Intercomparison Project (ISI-MIP) were analysed to evaluate future changes in fluvial floods under RCP2.6 and RCP8.5 and to assess the uncertainty associated with fluvial flood projections. The results of Li et al. (2016) showed the multi-model median of the ISI-MIP simulations has an acceptable ability in modelling floods based on the comparison with observations in Wuzhou and Boluo, which are the major hydrological stations in Xijiang and Dongjiang, respectively (Figure 1.4). Meanwhile, the uncertainty among different individual combinations of GCMs and hydrological models was considerable, and some combinations exhibited systematic biases relative to the observations. Furthermore, the ensemble median tends to underestimate the upper tail for annual maximum daily discharge. Most of the combinations of GCMs and hydrological models suggested that the magnitude of fluvial floods in the PRD would increase during 2070–2099 under RCP8.5 relative to the period of 1971–2000 (Figure 1.5). In the PRB, 50–60 per cent of the combinations indicated that five-year floods in the north would increase by 10–50 per cent in the future. In the southeastern part of PRB where the PRD is located, the increases of 5-year floods were identified with a higher model agreement. The increasing rates of floods are higher than those of the smaller ones. The high model agreement of the projections suggested that fluvial floods in the PRD would increase in the future. More intensive precipitation extremes are a major driver of

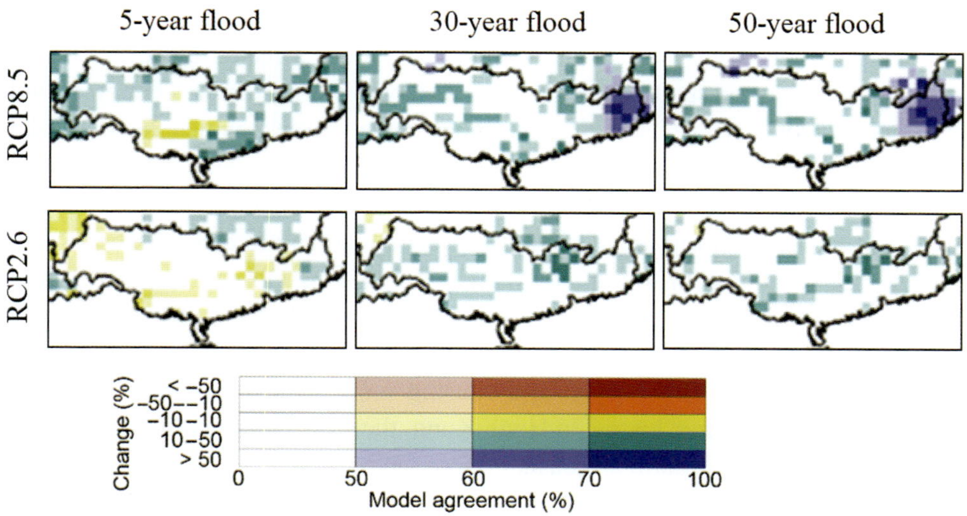

Figure 1.5 Multi-model ensemble median of changes (%) and model agreement (%) of floods and water availability across China over the period of 2070–2099 relative to the period of 1971–2000. Shown are change rates of (a, b) 5-year flood; (c, d) 30-year flood; and (e, f) 50-year flood under (left) RCP8.5 and (right) RCP2.6. Each colour denotes the corresponding change rate. The lighter (darker) colour denotes the lower (higher) model agreement of that category of change rates (from Li et al. 2016; © American Meteorological Society. Used with permission)

more severe fluvial floods in the PRD. Furthermore, Li et al. (2016) also analysed the contributions of GCMs and hydrological models to the uncertainty in hydrological projections. The uncertainties from GCMs outweigh those of projections in fluvial floods in the PRD.

The intensification of precipitation extremes has caused increases in the risk of fluvial floods in the PRD in the past decades (Lenderink et al., 2011; Zhang et al., 2012, 2013, 2014, 2015; Wu et al., 2013; Gu et al., 2014; Zhao et al., 2014; Ai & Wu, 2018). Human activities, such as reservoir regulations in Dongjiang, can reduce fluvial flood discharge. Under the future climate, the fluvial flood risks in the PRD are projected to continue to increase. Although the magnitude of increases in floods is associated with uncertainty, different model simulations and studies consistently show more severe floods in the future.

1.4 Coastal Flood Hazards in the PRD under Climate Change: Observations and Projections

1.4.1 Past Observations and Analyses

Global climate change causes changes in storm surge and sea level rise which can trigger and aggravate coastal floods in the PRD. The occurrence and severity of storm surge are highly associated with typhoons. In recent years, the PRD coastal cities, e.g., Hong Kong, Macau and Zhuhai, have been affected by several super typhoons. In the past decades, typhoons have become more extreme and accompanied by stronger winds and higher storm surges, which have caused destructive coastal floods that resulted in loss of life and damaged economies (Emanuel, 2005; R. C. Li et al., 2017). For example, Super Typhoon Hato, in 2017, led to severe storm surges, causing great threats to the coastal cities in the PRD, especially for Hong Kong and Macao (Hong Kong Observatory, 2019a). In Macao, a maximum sea level rise of 5.58 m was recorded at A-Ma station, the highest since 1925. Extensive areas were flooded in Macao, causing 10 fatalities, wounding more than 240 people, and causing more than 8.3 billion Macanese Patacas (MOP) economic losses. In Hong Kong, the storm surge induced by Super Typhoon Hato caused serious coastal floods and damage in low-lying coastal areas. A number of village houses, shops and carparks were flooded by seawater. Residents in these areas were trapped and required evacuation. A number of vessels ran aground in the southwest of Hong Kong. The storm surge induced by Super Typhoon Mangkhut, in 2018, caused even more damage than Hato (Hong Kong Observatory, 2019b). A number of village houses and coastal structures such as sewage treatment works and sports grounds suffered different levels of damage and hundreds of vessels ran aground or were damaged. According to the statistics of the International Emergency Disasters Database (EM-DAT), typhoon-related hazards in 1960–2018 caused the deaths of 569 people and resulted in economic losses of more than 1 billion US dollars in Hong Kong. Furthermore, sea level rises aggravate coastal flood hazards when storm surges occur. Observations in the Pearl River estuary showed increasing trends in sea level, raising the risks of coastal floods associated with storm surge (Lee et al., 2010; Yu et al., 2018). The mean sea level in Victoria Harbour in Hong Kong increased by 31 mm/decade over 1954–2019 (Hong Kong Observatory, 2019c). In Macao, as shown in Table 1.1, sea level rise has accelerated, as indicated by the increasing rates of 13.5 mm/decade in 1925–2010 and 42 mm/decade in 1970–2010 (Wang et al., 2016).

1.4.2 Future Projections

The future risks of coastal floods will be mainly affected by the changes in storm surge and sea level. Previous studies indicate that the number of typhoons affecting the PRD region is expected to drop down slightly, but the intensity should increase in the future (Murakami et al., 2011, 2012). Such projected intensification of typhoons may lead to stronger storm surge and hence greater risks of coastal floods in the PRD (Church et al., 2013; Walsh et al., 2016). The persistent global temperature increases in the future would further strengthen thermal expansion of

Table 1.1. Observed and projected sea level rise around the PRD

Source	Study area	Observation period	Observed rate of sea level rise (mm/decade)	Height of sea level rise (cm)			
				2050 RCP4.5	2050 RCP8.5	2100 RCP4.5	2100 RCP8.5
Yu et al. (2018)	Hong Kong	1957–2010	21 ± 3	20	25	40	75
He et al. (2016)	Hong Kong–Macao	1954–2013	29	38	43	67	84
Lee et al. (2010)	Hong Kong	1954–2009	26	–	–	–	–
Qu et al. (2019)	Hong Kong–Macao	1980–2016	19 ± 8	–	56 (38–73)	–	92 (53–192)
Wang et al. (2016)	Macao	1970–2010 (1925–2010)	42 (13.5)	28 (6–10)	53 (36–71)	33 (24–42)	74 (53–98)
Johnson et al. (2016)	Hong Kong	1954–2014	30	–	–	62–70	73–91

ocean water and glacier melting, causing continuous a global sea level rise. By 2050, the sea level around the PRD is projected to increase by 38 (29–47) cm and 43 (32–53) cm under RCP4.5 and RCP8.5, respectively. By 2100, these two figures are estimated to be 67 (50–84) cm and 84 (63–107) cm, respectively (He et al., 2016; Johnson et al., 2016; Hong Kong Observatory, 2019d; Qu et al., 2019). The projected intensification of typhoon and sea level rise both raise the risks of coastal floods in the low-lying areas of the PRD. Along with climate change, Hanson et al. (2011) also considered the future possible economic development to estimate the economic exposure of Hong Kong to coastal floods in 2070, which is expected to exceed 1,000 billion US dollars, which exceeds by far the current economic exposure, estimated at 36 billion US dollars.

In the past decades, typhoons have become more intense and accompanied by higher storm surge, leading to greater intensity of coastal floods. Furthermore, a sea level rise under global climate change further raises the risks of coastal floods. The increasing trends of coastal flood risks are projected to continue under climate change, suggesting that coastal cities in the PRD should prepare for more severe coastal floods in the future.

1.5 Urbanization and Sustainable Flood Management in the PRD

1.5.1 Impacts of Urbanization on Flood Risks

As one of the largest megapolises in the world, the PRD has experienced rapid urbanization over the past decades: the population increased from 4.6 million in 1980 to more than 57 million in 2015, and the construction land increased from 2,910 km^2 to 7,487 km^2 over 1990–2015 (Chan et al., 2012; Liu et al., 2019). The intensive urbanization in the low-lying areas raises the exposure of the population and economic assets to flood hazards, resulting in greater flood risks. When taking the highly exposed population and assets into account, Hong Kong, Macau, Shenzhen and Guangzhou are the most vulnerable cities in the PRD (Yang et al., 2014). As estimated by Hanson et al. (2011), the population and asset exposure to flood hazards in Guangzhou are 2.7 million and $84.2 billion, respectively. The exposed population and assets are projected to be 10.3 million and $3,357.8 billion in 2070 under the extreme scenario, respectively. Furthermore, urbanization can affect the geomorphology and thus significantly alter the hydrological processes in cities and basins, such as faster runoff velocity and higher flood peak (Chen et al., 2015). Such changes in hydrological processes due to increases in urban imperviousness are another key factor that raises flood risks, especially pluvial floods. Huang et al. (2018) indicated that the impervious surface in the PRD showed a continuous rising trend from 2009 to 2015 based on the Landsat

Thematic Mapper (TM) data. Furthermore, in the intensively urbanized area, it may be difficult to construct engineering facilities to mitigate pluvial floods and limit the potential for improvement of drainage capacities of cities. Moreover, the flood prevention standards of drainage systems vary in different PRD cities (H. Liu et al., 2014; Drainage Services Department, 2017). For example, the standard of the main trunk of urban drainage in Hong Kong is a 200-year return period, while those in other major PRD cities, such as Shenzhen and Guangzhou, are lower and less than a 1-year return period in some areas. All these factors determine the occurrence and intensity of pluvial floods. By analysing urban floods in 2009–2015, Huang et al. (2018) indicate the urban flood situations are improving in Guangzhou, which is attributed to drainage improvement.

Therefore, precipitation extremes in the PRD cities have become more frequent and intense in the past decades, and the intensification of precipitation extremes is projected to continue under the future climate change scenario. The rapid urbanization is another reason for the increasing risks of floods in the PRD cities by altering urban hydrological processes and increasing exposure to flood hazards. Previous studies also indicated that drainage improvement is an effective way to mitigate urban flood hazards. Under the future climate change scenario, pluvial and fluvial floods are very likely to become more severe and frequent due to intensification of precipitation extremes. The populations and assets in the PRD will continue to increase due to urban growth. Considering the increasing exposure and more weather extreme events under climate change in the future, more holistic and integrated flood risk management strategies and flexible adaptation options are needed to deal with the potential greater threats of future floods.

1.5.2 Sustainable Flood Management

Because the nature and mechanism of different types of flood hazards are different, various measures have to be considered and adopted in the PRD. Different PRD cities have their unique geographical settings, socio-economic development and history of adaptation measures. One major cause of pluvial floods is insufficient drainage capacities in urban areas. To respond to pluvial floods, the PRD cities have adopted different adaptation measures, including engineering and non-engineering measures. For engineering measures, grey infrastructure has been commonly used to deal with flooding blackspots to improve drainage capacities for flood mitigation, such as pipes, pumping stations and deep tunnels (Wu et al., 2016; Huang et al., 2018). However, in highly urbanized areas, such grey infrastructure can sometimes be difficult to implement, given the land surface has already been urbanized and occupied by buildings and the underground has been filled with other infrastructures such as subways (Chen et al., 2015). Hong Kong,

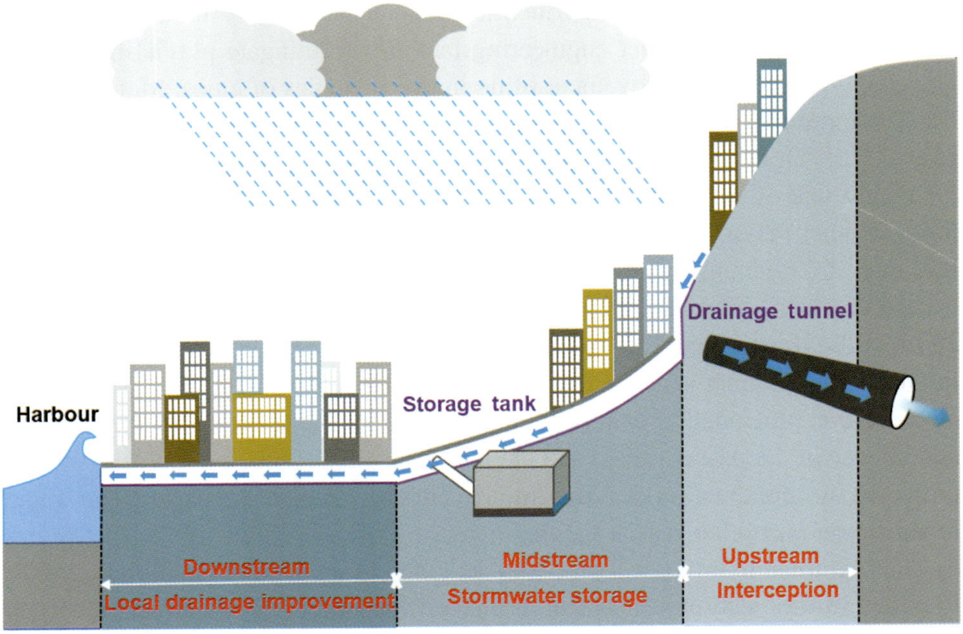

Figure 1.6 Three-pronged flood prevention strategies in Hong Kong (adapted from Drainage Services Department, 2017; Drainage Services Department of Hong Kong SAR. Used with permission)

as an example, has made enormous efforts to improve and enhance flood prevention and management. The design standard for the main trunk of urban drainage in Hong Kong is set for 200 years, which is up to the top international level (Drainage Services Department, 2017). To adapt the diverse geographical settings and socio-economic conditions in different districts of the city, a three-pronged flood prevention strategy has been adopted to enhance stormwater drainage capacities (Figure 1.6; Drainage Services Department, 2017). In the upstream, a rainwater drainage tunnel is adopted to intercept and transfer stormwater from upland areas for discharging directly to the sea or river. In the midstream, underground stormwater storage facilities are used for temporary stormwater retention to reduce the flood peak. In the downstream, river training works are adopted to improve the drainage capacities. Some rivers are trained by straightening, widening, deepening and provision of linings. Furthermore, in recent years, green infrastructures (which are more sustainable flood mitigation measures) have been increasingly adopted, because they exert less of an impact on hydrological processes. Sponge city is a modern and sustainable stormwater management strategy that emphasizes making use of natural river networks in urban regions, increasing green areas and pervious surfaces in cities and constructing and strengthening urban green facilities. The sponge city approach aims to manage and mitigate urban flood problems through natural hydrological processes such as infiltration, storage, purification, reuse and diversion. In Hong Kong, the Drainage Services Department (DSD) is making efforts to revitalize water bodies, construct flood retention lakes and improve green elements in flood prevention facilities, with the aim to enhance infiltration and reduce surface runoff. Shenzhen is a pilot sponge city in China, and rainwater system construction projects with the sponge city concept have been conducted in Guangming district (Ma et al., 2017; Xia et al., 2017). Non-engineering measures, such as risk zoning and impact assessment in urban development planning, monitoring, early warning and scheduling systems of pluvial floods, are also key elements in sustainable flood management. Hong Kong has implemented the Outline Zoning Plans (OZPs) under both the Town Planning Ordinance and the Drainage Impact Assessment. In that case, the development plans that require a change in land use specified in OZPs should seek approval from the Town Planning Board. The projects that might significantly affect the drainage situation are required to adopt stormwater drainage mitigation measures and conduct the Drainage Impact Assessment. The performance of rainstorm warning systems has been significantly improved recently, given the advancement in scientific and technological support. The HKO provides various weather information and warning messages for rainstorm, typhoon and storm surge to the general public via various means such as the Internet, TV, radio and mobile applications. Furthermore, the PRD cities also have further improved policies and legislations of flood management (Wu et al., 2016; Chen et al., 2017). By these efforts, although precipitation extremes have increased under global climate change in the past decades, the number of blackspots of pluvial floods in Hong Kong has been substantially reduced to six in 2019 from more than 90 in 1995, and the urban flood conditions have also been improved in Guangzhou (Huang et al., 2018). However, given more extreme and frequent rainstorms in the future, the PRD cities should further review their existing flood prevention capacities and improve flood management accordingly.

To deal with fluvial floods caused by excessive streamflow from the upstream, the fluvial flood prevention system in the PRD has been substantially improved, including basin-level construction and improvement of reservoirs, dams, dikes, detention basins and drainage networks (Yang et al., 2014). For example, Zhang et al. (2014) found that the observed fluvial floods, i.e. multi-day maximum streamflow, have decreased significantly after the construction of reservoirs in the Dongjiang basin. The reservoirs and dams in the Pearl River play an important role in flood control and mitigation (F. Liu et al., 2014). To prepare for more extreme fluvial floods in the future climate, Yang et al. (2014) pointed out that non-engineering measures of fluvial flood management shall be improved. For example, public awareness and knowledge of flood prevention shall be further enhanced, especially for vulnerable people. More funding, resources, information and facilities shall be available for local communities to prevent and mitigate floods. These resources can enhance the networks and collaborations among local communities and facilitate mutual help in the face of fluvial flood hazards. The PRD cities are different in many aspects, such as rights and obligations of development and conservation for flood control. Therefore, these differences shall be considered to formulate cooperation mechanisms and plans among PRD cities for flood prevention and mitigation.

To adapt to the projected increased risks of coastal floods in coastal regions due to stronger storm surges and sea level rises, the Civil Engineering and Development Department of Hong Kong SAR considers the combined effects of sea level rise and storm surge under climate change in the design and construction of coastal flood prevention facilities such as seawalls and breakwaters (Chan et al., 2013, 2018). The design standards of the coastal prevention facilities have been improved to return periods of 100–200 years. During storm surges, seawater is pushed forward to the coastal areas and, hence, increases the water level of rivers, which may reduce the drainage capacities of stormwater in the cities. Therefore, the DSD of Hong Kong SAR constructs and improves urban drainage systems to meet the design standards of 100–200 return years (Environment Bureau et al., 2015). Apart from these hard measures, soft measures are also important dimensions of flood risk management. Monitoring and early warning systems of rainstorms and typhoons have been well established and implemented in Hong Kong. The warning signals can be effectively propagated to the general public via various means, such as the Internet, TV, radio and mobile applications. Through these systems, the public can be informed of the warning signals of the hazards on time and take actions to avoid and mitigate causality and economic loss induced by the hazards (Chan et al., 2013; Takagi et al., 2018; Hong Kong Observatory, 2019e). The Security Bureau of Hong Kong issued the Contingency Plan for Natural Disasters in 2007 and made several revisions later on, in order to strengthen the coordination of government departments to adapt to natural hazards. The Contingency Plan considers evacuation, rescue, shelters, medical supports, food and water supplies, etc. (Yi & Chan, 2017). Moreover, the Hong Kong Jockey Club Disaster Preparedness and Response Institute was established in 2014 based on donations, with the aims to promote community resilience by involving different stakeholders, empower individuals on disaster preparedness awareness, organize training and facilitate knowledge sharing for disaster responders and foster professional education, networking and policy conservations. Chan et al. (2013) suggested to take public participation, equity of flood preparedness and environmental friendly measures into account in sustainable coastal flood mitigation practices. Macao SAR also adopted a variety of flood prevention and protection measures, including establishing the real-time remote controlling systems of stormwater pumping stations, constructing more stormwater pumping stations and culverts in the inner harbour to improve drainage capacities of surface runoff, grout and construct floodwalls along the inner harbour to defend higher storm surges and formulate an evacuation plan for low-lying areas in storm surges during typhoons to enhance the adaptation capacities of the general public. Moreover, to prepare for the greater floods due to the co-occurrence of storm surges and heavy rainstorms, Macao SAR also plans to build movable inner port tide barriers in the Wanzai river channel, with the aim of alleviating flood hazards in the inner harbour for the long-term. The flood prevention standards can reach 200-year return periods after the construction of the barriers.

1.6 Summary

Due to its unique geographical location and rapid urbanization, the PRD megacity region is affected by various types of floods triggered by different mechanisms, including fluvial, pluvial and coastal floods. Therefore, various engineering and non-engineering measures are needed to mitigate these flood problems. Furthermore, the PRD cities have their individual unique geographical setting, socio-economic development and history of adaptation measures. Thus, the adaptation measures, flood management practices and flood prevention standards may vary among the PRD cities. Such differences can help develop adaptation strategies that are suitable for the cities with various geographical settings and socioeconomic conditions on one hand but may bring more challenges for region-scale collaboration for sustainable flood management on the other hand. In general, under the past observed and future projected climate change, the risks of pluvial, fluvial and coastal floods increased due to more extreme rainstorms, higher storm surges and sea level rises. Meanwhile, the PRD cities have adopted and implemented various engineering and non-engineering measures in flood management which have proved to effectively mitigate the flood hazards. However, the current adaptation measures mostly focus on traditional engineering measures. The non-engineering measures can be further improved for sustainable flood management, especially facing potentially greater threats from more extreme floods in the future.

Under global climate change, precipitation extremes in the PRD cities have become more frequent and intense in the past decades and the intensification of precipitation extremes is projected to continue under climate change. The rapid urbanization is another reason for the increasing risks of pluvial floods in the PRD cities through altering urban hydrological processes. Previous studies have also indicated that effective drainage improvement is a way to mitigate urban flood hazards. With effective urban flood management, the water-logging conditions in the PRD cities are improving. The future pluvial floods are very likely to become more severe and frequent due to the intensification of precipitation extremes. The PRD cities need to strengthen their adaptation capacities to deal with the potential of greater threats of pluvial floods in the future.

Due to the intensification of precipitation extremes, the risk of fluvial floods has increased in the past decades. Human activities, such as reservoir regulation in Dongjiang, effectively reduce fluvial flood discharge. Under future climate scenarios, fluvial flood risks in the PRD are projected to increase continuously, although uncertainty remains with regard to the magnitude of the increases in flood risk.

In the past decades, typhoons have become more intense and are accompanied with higher storm surges, leading to greater severity of coastal floods. Furthermore, the sea level rise under global climate change potentially further increases the risks of coastal floods. The increasing trends in coastal flood risks should continue under the future climate scenarios, suggesting that the coastal cities in the PRD should prepare for more severe coastal floods in the future.

The future flood risks in the PRD should increase due to increases in flood hazards under climate change. Furthermore, the anticipated rapid development of cities may increase the population and asset exposures to flood hazards, causing greater flood risks. Therefore, it is necessary to take the climate change factor and future urban growth into account in space planning and flood defence infrastructure design. Moreover, it is insufficient to protect from flood hazards through hard engineering infrastructures alone; non-engineering measures such as early warning and action guidance are also necessary to mitigate flood risks. In summary, more holistic and integrated flood risk management strategies and flexible adaptation options are needed to mitigate the future threats of floods.

References

Adnan, M. S. G., Haque, A., & Hall, J. W. (2019). Have coastal embankments reduced flooding in Bangladesh? *Science of the Total Environment* 682: 405–416.

Ai, H., & Wu, X. (2018). Variation characteristics of rainstorm in Guangzhou. *Guangdong Meteorology* 40(4): 20–33.

Chan, F. K. S., Adekola, O. A., Mitchell, G., & McDonald, A. T. (2013). Appraising sustainable flood risk management in the Pearl River Delta's coastal megacities: A case study of Hong Kong, China. *Journal of Water and Climate Change* 4(4): 390–409.

Chan, F. K. S., Chuah, C. J., Ziegler, A. D., Dąbrowski, M., & Varis, O. (2018). Towards resilient flood risk management for Asian coastal cities: Lessons learned from Hong Kong and Singapore. *Journal of Cleaner Production* 187: 576–589.

Chan, F. K. S., Mitchell, G., & Mcdonald, A. (2012). Flood risk in Asia's urban mega-deltas: Drivers, impacts and response. *Environment and Urbanization Asia* 3(1): 41–61.

Chen, Y., Qin, J., Dong, L., & Zhang, T. (2017). The formation regularity and control measures of urban pluvial floods in Guangzhou City. *China Flood & Drought Management* 24(2): 39–41.

Chen, Y., Zhou, H., Zhang, H., Du, G., & Zhou, J. (2015). Urban flood risk warning under rapid urbanization. *Environmental Research* 139: 3–10.

Church, J. A., Clark, P. U., Cazenave, A., et al. (2013). *Sea Level Change. Climate Change 2013: The Physical Science Basis. Contribution of Working Group I to the Fifth Assessment Report of the Intergovernmental Panel on Climate Change.* Cambridge: Cambridge University Press, pp. 1137–1216.

De Moel, H., Aerts, J. C. J. H., & Koomen, E. (2011). Development of flood exposure in the Netherlands during the 20th and 21st century. *Global Environmental Change* 21 (2): 620–627.

Drainage Services Department (2017). *Sustainability Report 2016–17.* Hong Kong.

Emanuel, K. (2005). Increasing destructiveness of tropical cyclones over the past 30 years. *Nature* 436(7051): 686.

Environment Bureau, Development Bureau, Transport & Housing Bureau, Commerce & Economic Development Bureau, Food & Health Bureau, & Security Bureau (2015). *Hong Kong Climate Change Report 2015.* Hong Kong.

Fischer, T., Gemmer, M., Liu, L., & Su, B. (2012). Change-points in climate extremes in the Zhujiang River Basin, South China, 1961–2007. *Climatic Change* 110(3–4): 783–799.

Gu, X., Zhang, Q., Liu, J., & Zhang, Z. (2014). Characteristics, causes and impact of the changes of the flood frequency in the Pearl River drainage basin from 1951–2010. *Journal of Lake Sciences* 26(5): 661–670.

Hanson, S., Nicholls, R., Ranger, N., et al. (2011). A global ranking of port cities with high exposure to climate extremes. *Climatic Change* 104(1): 89–111.

He, Y. H., Mok, H. Y., & Lai, E. S. (2016). Projection of sea-level change in the vicinity of Hong Kong in the 21st century. *International Journal of Climatology* 36(9): 3237–3244.

Hong Kong Observatory (2014). Climate projections for Hong Kong – Rainfall. Available from www.HKO.gov.hk/en/climate_change/proj_hk_rainfall.htm (Last accessed 24 July 2021).

Hong Kong Observatory (2019a). Super Typhoon Hato (1713) 20 to 24 August 2017. Available from www.weather.gov.hk/informtc/hato17/report.htm (Last accessed 1 May 2020).

Hong Kong Observatory (2019b). Super Typhoon Mangkhut (1822) 7 to 17 September 2018. Available from www.weather.gov.hk/informtc/mangkhut18/report.htm (Last accessed 1 May 2020).

Hong Kong Observatory (2019c). Climate change in Hong Kong: Mean sea level. Available from www.HKO.gov.hk/en/climate_change/obs_hk_sea_level.htm (Last accessed 1 May 2020).

Hong Kong Observatory (2019d). Climate projections for Hong Kong: Mean sea level. Available from www.weather.gov.hk/climate_change/proj_hk_sea_level_e.htm (Last accessed 1 May 2020).

Hong Kong Observatory (2019e). Special announcement on flooding in the northern new territories. Available from www.HKO.gov.hk/wserVICe/warning/flood.htm (Last accessed 1 May 2020).

Huang, H., Chen, X., Zhu, Z., et al. (2018). The changing pattern of urban flooding in Guangzhou, China. *Science of the Total Environment* 622–623: 394–401.
Ikeuchi, H., Hirabayashi, Y., Yamazaki, D., et al. (2017). Compound simulation of fluvial floods and storm surges in a global coupled river-coast flood model: Model development and its application to 2007 Cyclone Sidr in Bangladesh. *Journal of Advances in Modeling Earth Systems* 9(4): 1847–1862.
IPCC (2013). *Climate Change 2013: The Physical Science Basis*. Cambridge: Cambridge University Press.
Johnson, K., Depietri, Y., & Breil, M. (2016). Multi-hazard risk assessment of two Hong Kong districts. *International Journal of Disaster Risk Reduction* 19: 311–323.
Klijn, F., de Bruijn, K. M., Knoop, J., & Kwadijk, J. (2012). Assessment of the Netherlands' flood risk management policy under global change. *AMBIO* 41(2): 180–192.
Le, T. V. H., Nguyen, H. N., Wolanski, E., Tran, T. C., & Haruyama, S. (2007). The combined impact on the flooding in Vietnam's Mekong River delta of local man-made structures, sea level rise, and dams upstream in the river catchment. *Estuarine, Coastal and Shelf Science* 71(1–2): 110–116.
Lee, B. Y., Wong, W. T., & Woo, W. C. (2010). Sea-level rise and storm surge –impacts of climate change on Hong Kong. In *HKIE Civil Division Conference*, 12–14 April 2010, Hong Kong (pp. 12–14).
Lee, T. C., Shun, C. M., & Ma, K. Y. (2016). The great rainstorm of the century in 1889. Observatory's Blog. Available from www.HKO.gov.hk/en/blog/00000208.htm (Last accessed 24 July 2021).
Lenderink, G., Mok, H. Y., Lee, T. C., & van Oldenborgh, G. J. (2011). Scaling and trends of hourly precipitation extremes in two different climate zones – Hong Kong and the Netherlands. *Hydrology and Earth System Sciences* 15(9): 3033–3041.
Li, J., Chen, Y. D., Zhang, L., Zhang, Q., & Chiew, F. H. S. (2016). Future changes in floods and water availability across China: Linkage with changing climate and uncertainties. *Journal of Hydrometeorology* 17(4): 1295–1314.
Li, J., Zhang, Q., Chen, Y. D., & Singh, V. P. (2013). GCMs-based spatiotemporal evolution of climate extremes during the 21st century in China. *Journal of Geophysical Research: Atmospheres* 118(19): 11017–11035.
Li, J., Zhang, L., Shi, X., & Chen, Y. D. (2017). Response of long-term water availability to more extreme climate in the Pearl River Basin, China. *International Journal of Climatology* 37(7): 3223–3237.
Li, J., Zhang, Q., Chen, Y. D., & Singh, V. P. (2015). Future joint probability behaviors of precipitation extremes across China: Spatiotemporal patterns and implications for flood and drought hazards. *Global and Planetary Change*, 124, 107–122.
Li, R. C., Zhou, W., Shun, C. M., & Lee, T. C. (2017). Change in destructiveness of landfalling tropical cyclones over China in recent decades. *Journal of Climate* 30(9): 3367–3379.
Liang, B. (1997). '94.6' Zhujiang liuyu teda baiyuhonglao tezheng fenxi [Analysis on characteristics of '96.4' extraordinary rainstorm and flood in the Pearl River Basin]. *Disaster Reduction in China* 7(4): 21–24.
Liu, F., Yuan, L., Yang, Q., et al. (2014). Hydrological responses to the combined influence of diverse human activities in the Pearl River delta, China. *CATENA* 113: 41–55.
Liu, H., Liu, S., Zhu, J., Yin, Y., & Li, Y. (2014). Chengshi neilao zhengjie tantao ji zhengce jianyi [Discussion on the crux of urban waterlogging and policy suggestions]. *China Flood & Drought Management* 24(2): 39–41.
Liu, L., Fischer, T., Jiang, T., & Luo, Y. (2013). Comparison of uncertainties in projected flood frequency of the Zhujiang River, South China. *Quaternary International* 304: 51–61.

Liu, W., Zhan, J., Zhao, F., et al. (2019). Impacts of urbanization-induced land-use changes on ecosystem services: A case study of the Pearl River Delta Metropolitan Region, China. *Ecological Indicators* 98: 228–238.

Ma, Z., Hu, J., Feng, P., et al. (2017). Assessment of climate technology demands in Chinese Sponge City. *Journal of Geoscience and Environment Protection* 5(12): 102–116.

Murakami, H., Wang, B., & Kitoh, A. (2011). Future change of western North Pacific typhoons: Projections by a 20-km-mesh global atmospheric model. *Journal of Climate*, 24(4), 1154–1169.

Murakami, H., Wang, Y., Yoshimura, H., Mizuta, R., Sugi, M., Shindo, E., et al. (2012). Future changes in tropical cyclone activity projected by the new high-resolution MRI-AGCM. *Journal of Climate*, 25(9), 3237–3260.

Ou, T., Chen, D., Linderholm, H. W., & Jeong, E.-H. (2013). Evaluation of global climate models in simulating extreme precipitation in China. *Tellus A: Dynamic Meteorology and Oceanography* 65(1): 19799.

Qian, W., Fu, J., & Yan, Z. (2007). Decrease of light rain events in summer associated with a warming environment in China during 1961–2005. *Geophysical Research Letters* 34(11): 1–5.

Qu, Y., Jevrejeva, S., Jackson, L. P., & Moore, J. C. (2019). Coastal sea level rise around the China seas. *Global and Planetary Change* 172: 454–463.

Shen, Y., Morsy, M. M., Huxley, C., Tahvildari, N., & Goodall, J. L. (2019). Flood risk assessment and increased resilience for coastal urban watersheds under the combined impact of storm tide and heavy rainfall. *Journal of Hydrology* 579: 124159.

Takagi, H., Xiong, Y., & Furukawa, F. (2018). Track analysis and storm surge investigation of 2017 Typhoon Hato: Were the warning signals issued in Macau and Hong Kong timed appropriately? *Georisk: Assessment and Management of Risk for Engineered Systems and Geohazards* 12(4): 297–307.

Vis, M., Klijn, F., Bruijn, K. M. D., & van Buuren, M. (2003). Resilience strategies for flood risk management in the Netherlands. *International Journal of River Basin Management* 1(1): 33–40.

Walsh, K. J., McBride, J. L., Klotzbach, P. J., et al. (2016). Tropical cyclones and climate change. *Wiley Interdisciplinary Reviews: Climate Change* 7(1): 65–89.

Wang, L., Huang, G., Zhou, W., & Chen, W. (2016). Historical change and future scenarios of sea level rise in Macau and adjacent waters. *Advances in Atmospheric Sciences* 33(4): 462–475.

Wong, M. C., Mok, H. Y., & Lee, T. C. (2011). Observed changes in extreme weather indices in Hong Kong. *International Journal of Climatology* 31(15): 2300–2311.

Wu, C., Huang, G., Yu, H., Chen, Z., & Ma, J. (2014). Impact of climate change on reservoir flood control in the upstream area of the Beijiang River Basin, South China. *Journal of Hydrometeorology* 15(6): 2203–2218.

Wu, C. H., Huang, G. R., & Yu, H. J. (2015). Prediction of extreme floods based on CMIP5 climate models: A case study in the Beijiang River basin, South China. *Hydrology and Earth System Sciences* 19(3): 1385–1399.

Wu, H., Huang, G., Meng, Q., Zhang, M., & Li, L. (2016). Deep tunnel for regulating combined sewer overflow pollution and flood disaster: A case study in Guangzhou City, China. *Water* 8(8): 329.

Wu, J., Zhang, L., Zhao, D., & Tang, J. (2015). Impacts of warming and water vapor content on the decrease in light rain days during the warm season over eastern China. *Climate Dynamics* 45(7): 1841–1857.

Wu, Z.-Y., Lu, G.-H., Liu, Z.-Y., Wang, J.-X., & Heng, X. (2013). Trends of extreme flood events in the Pearl River basin during 1951–2010. *Advances in Climate Change Research* 4(2): 110–116.

Xia, J., Zhang, Y., Xiong, L., et al. (2017). Opportunities and challenges of the Sponge City construction related to urban water issues in China. *Science China: Earth Sciences* 60 (4): 652–658.

Yan, D., Werners, S. E., Ludwig, F., & Huang, H. Q. (2015). Hydrological response to climate change: The Pearl River, China under different RCP scenarios. *Journal of Hydrology: Regional Studies* 4(Part B): 228–245.

Yang, L., Scheffran, J., Qin, H., & You, Q. (2014). Climate-related flood risks and urban responses in the Pearl River Delta, China. *Regional Environmental Change* 15(2): 379–391.

Yi, W., & Chan, A. (2017). Effects of heat stress on construction labor productivity in Hong Kong: A case study of rebar workers. *International Journal of Environmental Research and Public Health* 14(9): 1055.

Yi, Y. (2005). Zhujiang '05.6' kanghongqiangxian lueying [A glimpse of the Pearl River '05.6' flood fighting]. *Pearl River* (4): F0002.

Yin, J., Ye, M., Yin, Z., & Xu, S. (2014). A review of advances in urban flood risk analysis over China. *Stochastic Environmental Research and Risk Assessment* 29(3): 1063–1070.

Yu, Q., Lau, A. K. H., Tsang, K. T., & Fung, J. C. H. (2018). Human damage assessments of coastal flooding for Hong Kong and the Pearl River Delta due to climate change-related sea level rise in the twenty-first century. *Natural Hazards* 92(2): 1011–1038.

Yuan, F., Tung, Y. K., & Ren, L. (2016). Projection of future streamflow changes of the Pearl River basin in China using two delta-change methods. *Hydrology Research* 47 (1): 217–238.

Zhang, A., Xiao, L., Min, C., et al. (2019). Evaluation of latest GPM-era high-resolution satellite precipitation products during the May 2017 Guangdong extreme rainfall event. *Atmospheric Research* 216: 76–85.

Zhang, Q., Gu, X., Singh, V. P., Xiao, M., & Chen, X. (2015). Evaluation of flood frequency and non-stationarity resulting from climate indices and reservoir indices in the East River basin, China. *Journal of Hydrology* 527: 565–575.

Zhang, Q., Li, J., Singh, V. P., & Xu, C.-Y. (2013). Copula-based spatio-temporal patterns of precipitation extremes in China. *International Journal of Climatology* 33(5): 1140–1152.

Zhang, Q., Singh, V. P., Peng, J., Chen, Y. D., & Li, J. (2012). Spatial-temporal changes of precipitation structure across the Pearl River basin, China. *Journal of Hydrology* 440–441: 113–122.

Zhang, Q., Xiao, M., Liu, C.-L., & Singh, V. P. (2014). Reservoir-induced hydrological alterations and environmental flow variation in the East River, the Pearl River basin, China. *Stochastic Environmental Research and Risk Assessment* 28(8): 2119–2131.

Zhang, S. (1997). Catastrophic floods in 1994 and flood control in Guangdong province. *Tropical Geography* 17(1): 30–35.

Zhao, Y., Zou, X., Cao, L., & Xu, X. (2014). Changes in precipitation extremes over the Pearl River Basin, southern China, during 1960–2012. *Quaternary International* 333: 26–39.

Zolina, O., Simmer, C., Gulev, S. K., & Kollet, S. (2010). Changing structure of European precipitation: Longer wet periods leading to more abundant rainfalls. *Geophysical Research Letters* 37(6): L06704.

2

Flooding Risk in the Lancang-Mekong River Basin under Global Change

XIAOBO YUN, JIE WANG, HUAN WU, BINOD BANIYA, HUI LU, SIAO SUN, XIMENG XU, XINGCAI LIU AND QIUHONG TANG

2.1 Introduction

Flooding is one of the most destructive natural disasters that frequently occurs across the world (Figure 2.1). It is a phenomenon of a rapid rise in water level, usually caused by heavy rain, snow melting or storm surge. In a typical flood case, heavy rain and/or snowmelt induce extreme runoff at different locations which subsequently converge in river channels causing high streamflow and flood waves, followed by a decreasing phase until the water in the river channel returns to normal levels.

The occurrence of flood events has seasonal and regional characteristics. In most countries, flood disasters often occur during summer and rainy season. The floodplains prone to severe disasters are widely distributed along river networks or estuary areas. A flood is a double-edged sword process: nearly one billion people in the world live in floodplains and enjoy the benefits while being at risk of flooding (Thomaz et al., 2007). On the one hand, flooding provides abundant freshwater resources for the residents, brings fertile suspended nutrients and erodes an alluvial plain, which is beneficial for agriculture and population development. On the other hand, developing human activities at the proximity of the river also represents the increasing risk from river flooding caused by extreme weather. Floods can submerge large areas of farmland and cities, destroy houses and other infrastructure, and lead to loss of life. The runaway flood tends to cause a lot of human suffering. Studies have shown that more than 110 million people are affected by flood events each year, and the annual average loss caused by the flood is up to 100 billion dollars.[1]

This research is supported by the Strategic Priority Research Program of Chinese Academy of Sciences (XDA20060402), the Second Tibetan Plateau Scientific Expedition and Research (2019QZKK0208), and the International Partnership Program of Chinese Academy of Sciences (131A11KYSB20170113).

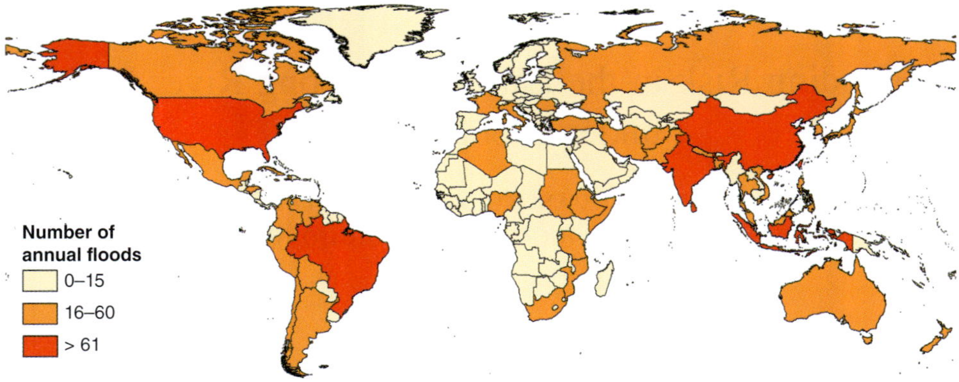

Figure 2.1 Annual flood events distribution during 1985–2016, according to EM-DAT

Since the third millennium BC, humans have migrated or attempted to reduce flood hazards through dykes, canals and elevated structures. However, these actions often came as a result of catastrophic events (Li, 2005). Since the important social development and population growth observed in the recent years, the research community has paid more attention to the risk posed by future floods and their impact on the economy and society. The earth's hydrological cycle is constantly accelerating due to global warming, which may lead to a higher frequency of extreme precipitation leading to an increase in the frequency of weather-related disasters in various parts of the world. The increasing flooding issues caused by continued global warming have threatened human life and property. The world's population and the economic losses due to flood are expected to increase in the following decades (Ward et al., 2013; Alfieri et al., 2017). Under the threat of flooding, people have taken various measures to reduce both the frequency and hazard of floods. Along with post-events remedies, preventive approaches may reduce the impacts of flood. This includes: building physical barriers (such as dams along rivers), increasing water storage areas (such as reservoirs and flood discharge areas) and constructing flood monitoring and warning systems (Ward et al., 2017).

Flood monitoring and warning systems are effective tools to reduce flood impact. Flood forecasting and impact assessment is an important tool to manage a complex phenomenon that is associated with meteorological, hydrological, hydraulics and environmental processes. Forecasting methods can provide real-time information on the location and time of flood events, identify areas at high risk of flood and provide a warning system that helps relevant agencies take effective measures to reduce the adverse impact of floods. A common hydrology method for flood events forecasting consists of the following steps. First, it monitors the atmospheric circulation and calculates the future predicted weather data (such as precipitation, temperature, wind speed, etc.). Second, it inputs these meteorological forcing data into hydrological models while using a function to forecast flood events. Third, it estimates possible hydrological conditions and flood risks in the future. The National Oceanic and Atmospheric Administration (NOAA) provides a set of web interfaces to access real-time observations, short-term weather and water forecasts and extreme weather surveillance and warnings. As for long-term climate change analysis, the Intergovernmental Panel on Climate Change (IPCC) provides future climate scenarios input data Coupled Model Intercomparison Project Phase 5 (CMIP5) to meet the demands for long-term climate change impact assessments. Based on a reliable flood forecasting system, future flood events can be predicted and corresponding practical flood prevention measures can be taken to minimize flood damage. These flood prevention measures include reserving flood storage capacity in reservoirs, issuing flood warnings in affected areas, and opening regional flood discharge channels. Furthermore, a long-term climate change impact assessment system may provide recommendations for regional infrastructure construction to avoid areas that may be negatively affected by extreme floods in the future.

In this present work, we built a flood risk assessment system based on a hydrological forecasting system. Using high quality Digital Elevation Model (DEM) data, we identified flood-prone areas to help implement engineering measures aimed at reducing flood damage. In 2007, the European Union issued a directive on flood risk assessment and management, requiring the establishment of hazard and risk maps of flood events to show potential adverse effects on human health, environment, economic and cultural heritage. Similarly, the United States also spent about 60 million dollars to popularize an Advanced Hydrological Forecasting Service (AHPS), which aims to prevent and detect flood events. With more comprehensive precipitation data, river water level information, weather forecast, climate forecast and relevant additional data, the AHPS can provide a long-term forecast of seasonal flood, through warnings and maps that highlight the areas at risk.

Flood events are one of the main threats to human civilization. In contrast to the passive response commonly adopted in ancient times, modern flood monitoring and warning systems can predict and help mitigate the flood threats and minimize their impact. However, under different climate change, uncertainty in the evolution of hydrological conditions, land use and population growth make the simulation of flood events by hydrological models difficult. This chapter will use the Lancang-Mekong River Basin to illustrate an application of a flood assessment model.

2.2 The Lancang-Mekong River Basin

The Lancang-Mekong River Basin (LMRB) is one of the most important transboundary rivers in Asia. It flows through China, Myanmar, Laos, Thailand, Cambodia and Vietnam, and finally merges into the South China Sea (Figure 2.2). The LMRB is located in the southwest monsoon zone, where the wet season lasts from May to October and the dry season occurs from November to April. The terrain conditions and the climate are not homogenous across the basin. The average annual precipitation is relatively high while the distribution of water resources is heterogenous over time and across space. The upstream Lancang River Basin is dominated by plateau and subtropical monsoon climates. The summer period (June–August) shows a precipitation peak, representing more than 60 per cent of the total annual precipitation. The streamflow is mainly supplied by snowmelt and precipitation. The downstream Mekong River Basin is dominated by

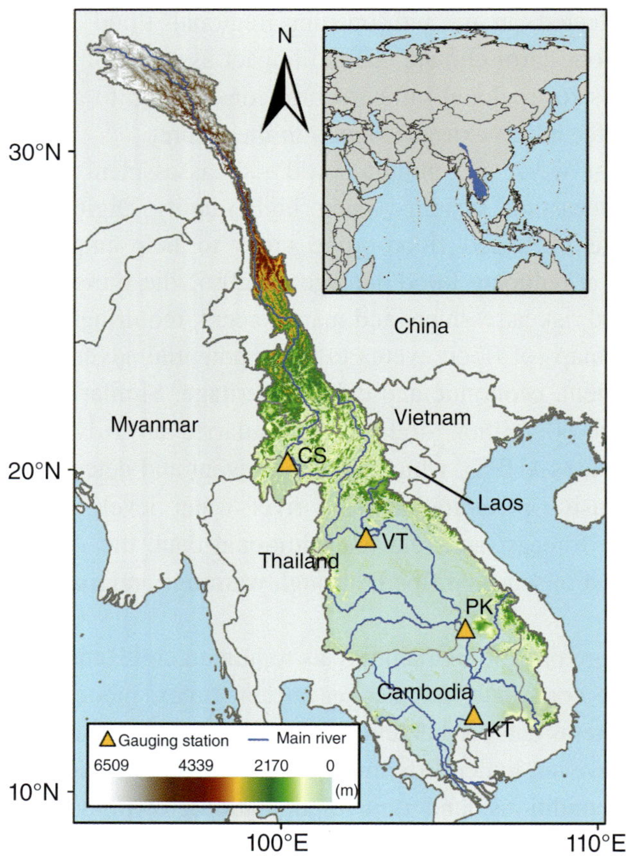

Figure 2.2 The Lancang-Mekong River Basin with the main river system and gauging stations

tropical rainforest and tropical monsoon climate. In the downstream Mekong River Basin, the precipitation peaks in September–October, accounting for more than 75 per cent of annual precipitation. The streamflow is mainly supplied by precipitation and water from upstream (Luo et al., 2019). One usually observes significant changes in the climate and distribution of water resources between upstream and downstream.

The downstream of the LMRB is densely populated, which hosts more than 70 million people. The rural economy, mainly dominated by rainfall-fed agriculture, is the most important and most vulnerable sector. The Vietnam Delta is a world-renowned rice producing region. The rice production accounts for about 20 per cent of the world's total exports, which is also the main source of the delta population's livelihood (Pandey et al., 2010). In the downstream of the LMRB, Tonle Sap lake is the world's largest inland freshwater fishery area, fish produced in this lake contribute the most important source of protein intake for the surrounding residents (Hortle, 2007). Floods are the most frequent and most harmful natural disasters in the LMRB; they destroy farmland and houses, and seriously threaten food production in this basin. Meanwhile, floods can bring a large amount of sediment, which provides rich nutrition for downstream crop cultivation. Furthermore, floods also promote large-scale fish migration. In general, downstream countries are highly reliant on water resources from the LMRB. As such, downstream countries are more vulnerable to their changes. According to the disaster data from the Emergency Events Database (EM-DAT), the LMRB suffered from the most severe flood in history, which was recorded in 2011 and led to the loss of 877 people and caused a loss of 403 million dollars. The basin experienced four huge floods during 2017, which caused the death of 150 people and the loss of 131 million dollars. Changes in water resources are highly related to the life of hundreds of millions of people. Given the economic and geopolitical importance of the LMRB across Asia, assessing the impact of climate change on flood events in the LMRB is of crucial importance and is the topic of this current research work.

2.3 Data and Methods

2.3.1 Data

To assess flood risk, we analyze several relevant datasets. The precipitation, temperature and wind speed data were collected at 0.25° from the Global Meteorological Forcing Dataset (GMDF) (Sheffield et al., 2006, 2012) because models using this data have shown good performance in the LMRB (Tatsumi & Yamashiki, 2015). We consider two representative concentration pathway (RCP) climate change scenarios (RCP4.5 and RCP8.5) and five driven-data models from the Inter-Sectoral Impact Model Intercomparison Project (ISIMIP),[2] which include: GFDL-ESM2M, HadGEM2-ES, IPSL-CM5A-LR, MIROC-ESM-CHEM and NorESM1-M. Forcing data were downscaled from a resolution of 0.5° to a resolution of 0.25° using the nearest neighbour method. We used data from Mohammed et al. (2018) to obtain daily streamflow observations for CS (Chiang Sean), VT (Vientiane), PK (Pakse) and KT (Kratie) in the LMRB. Soil parameter data were obtained from the Harmonized World Soil Database (HWSD) (FAO, 2012). Land cover data were obtained from the Global Land Cover Characterization (GLCC) dataset (Defries & Townshend, 1999). The DEM data were obtained from the Advanced Space borne Thermal Emission and Reflection Radiometer (ASTER).[3] We simulate and forecast flood event changes in the LMRB using these datasets.

2.3.2 SPI-3

The Standardized Precipitation Index (SPI) (Mckee et al., 1993) has been commonly used as an indicator to characterize the relative dryness and wetness in the evaluation of drought and abnormal humidity in a given area. SPI represents the cumulative probability of precipitation under a given time scale (usually, 1, 3, 6 or 12 months), converting the precipitation data from a gamma distribution to a standard normal distribution. The magnitude of the SPI value may represent the relatively dry and wet levels in different periods and has been used to characterize the evolution of long-term water resource changes at different time scales. The commonly used SPI classification (Mckee et al., 1993) is: SPI ≤ -1 for a dry period, and SPI ≥ 1 for an abnormal humidity period. Dry periods are more prone to encounter drought and abnormal humidity periods are more at risk of flooding.

In this study, we used the SPI to assess meteorological drought and abnormal humidity in the LMRB. The three-month scale SPI (SPI-3) showed good performance on analysing the dry and wet characteristics in the monsoon region (Wu et al., 2007). Therefore, the SPI-3 was chosen to measure the water resources changes under dry and wet conditions in the historical period (1985–2016) and future period (2020–2050) of the LMRB. For comparison purposes, the historical period was selected as the baseline period, and the future SPI-3 changes were calculated according to the SPI-3 distribution characteristics in the baseline period. The total precipitation was calculated by summing the precipitation data in each grid, and the corresponding SPI-3 change was also calculated. The performance of SPI-3 was evaluated using the Kolmogorov-Smirnov (Wilks, 1999) method, of which the Pearson-III distribution function had the best fitting effect ($P = 0.91$, the P values of gamma and generalized extreme value distribution were 0.83 and 0.78). Therefore, Pearson-III distribution was used to fit the SPI-3 in this study.

2.3.3 VIC Model

The LMRB has extremely diverse terrain changes, so it was necessary to use a distributed hydrological model to simulate the streamflow under this complex environment. The Variable Infiltration Capacity (VIC) model (Liang et al., 1994; Hamman et al., 2018) is a macro-distributed hydrological model, which can simulate hydro-meteorological processes on the land surface based on water and energy balance theory. The VIC model discretizes the basin into different grids, and each grid follows the principles of energy and water balance to simulate the various processes of the water cycle. We used the VIC model to simulate the streamflow in the LMRB, since it accounts for processes involved in our study area, such as snow and snow melting, soil freezing and thawing processes and vegetation-related evaporation data (Tatsumi & Yamashiki 2015; Dang et al., 2020).

In the process of simulating the streamflow, the performance of the VIC model was validated by the observation value from gauging stations in the LMRB. The performance of the model was evaluated using the Nash-Sutcliffe model Efficiency coefficient (NSE) (Nash & Sutcliffe, 1970) and model bias. The NSE was used to measure the overall performance of the model. The calculation formula is:

$$\mathrm{NSE} = 1 - \frac{\sum_{t=1}^{T}(Q_o^t - Q_m^t)^2}{\sum_{t=1}^{T}(Q_o^t - \overline{Q_o})^2} \tag{2.1}$$

where, Q_o is the observed streamflow, Q_m is the simulated streamflow, Q^t represents the value at time t and $\overline{Q_o}$ is the average of the observed streamflow. NSE can take values from negative infinity to 1. NSE values close to 1 suggest a better simulation effect and higher reliability of the model. The model is unreliable if NSE values are negative. The hydrological model is considered to have a better simulation effect when the NSE coefficient is greater than 0.5.

The model bias was used to measure the matching of the mean value; it represents the difference between the simulation and the observation. The calculation formula is:

$$\mathrm{bias} = \frac{\overline{Q_m} - \overline{Q_o}}{\overline{Q_o}} * 100\% \tag{2.2}$$

in which $\overline{Q_m}$ is the average of the simulated streamflow. Smaller values of the bias indicate better performance of the model simulation.

2.3.4 Definition of the Flood Events

In contrast to the conceptual definition, we analysed the duration and change process of flood events based on relevant characteristics associated with this case study. We used Mean Annual Floods (MAF) (Guo et al., 2014) and Peak Of Threshold (POT) (Hirsch & Archfield, 2015), which have been commonly used to analyse long-term series flood characteristic changes (see Figure 2.3).

In order to measure the change of flood events, we estimate flood size using MAF. MAF selected flood events through several steps. First, the maximum daily streamflow is selected as the peak flood event of each year during the baseline period (e.g. red bars represent three flood peaks during 1991–1993 in Figure 2.3(a)). Second, the average value of the peak flood estimates the magnitude of the flood events during this period. The steps are repeated to compute MAF during the comparison period (average value of blue bars in Figure 2.3(a)).

MAF does not always accurately capture flood events. For example, the peak flow may be small in a dry year, and several large flood peaks will be encountered in a wet year. POT addresses this problem by selecting streamflow values exceeding a threshold that corresponds to two events per year on average during the baseline period (e.g. six red bars in three years, red dotted line represents the threshold in Figure 2.3(b)). This threshold was then used during the comparison period to detect flood events (e.g. blue bars during 1994–1996 in Figure 2.3(b)). We selected only one peak during a 15-day period window to avoid counting the same flood event twice. POT could extract large floods and filter out the drought year. This represents the frequency of flood events during the selected period.

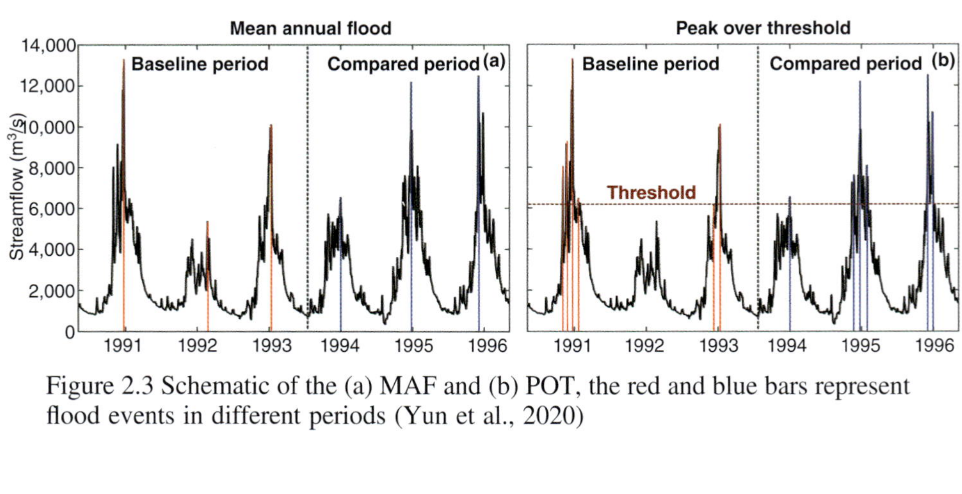

Figure 2.3 Schematic of the (a) MAF and (b) POT, the red and blue bars represent flood events in different periods (Yun et al., 2020)

2.4 Results

2.4.1 Changes of SPI-3

Precipitation is the most important factor that affects flood. Before simulating the model, we analysed the changes in the trend of precipitation in the LMRB. The relative change of annual precipitation during 1985–2016 compared with the period of 1975–1984 in the LMRB is shown in Figure 2.4. In general, the total amount of precipitation did not considerably change over time, however, its distribution varied across space. The precipitations are low upstream, slightly higher in the Tibetan Plateau, where the river originates, and lower in the central basins. We observed a significant change in the precipitation between the midstream and downstream areas. The precipitation in the midstream area and the southeast coastal area increased significantly, while the southwest coastal area showed a decreasing trend. Therefore, the use of the distributed hydrological models could accurately reflect the impact of precipitation changes across the streamflow in the LMRB.

Based on the General Circulation Models (GCMs) data obtained from ISIMIP, we consider several climate change scenario ensembles, which show consistent

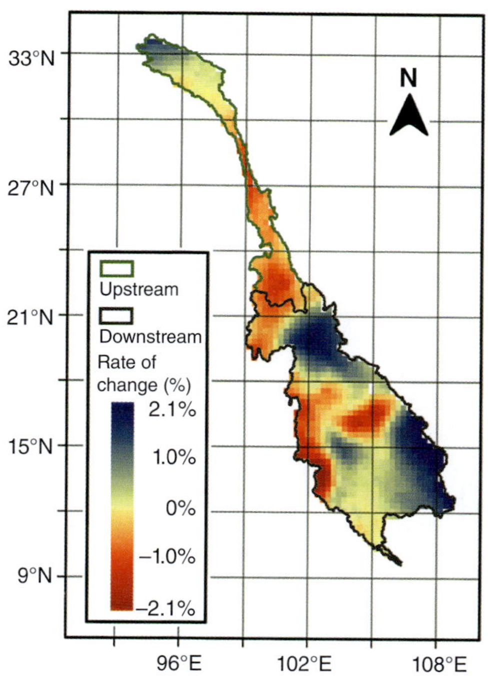

Figure 2.4 Relative change in annual average precipitation during 1985–2016 compared with 1975–1984 over the Lancang-Mekong River Basin. The upstream is the Lancang river basin and the downstream is the Mekong river basin

temperature increases between 1.8°C (RCP4.5) and 2.3°C (RCP8.5) in the LMRB for a future period (2020–2050). RCP4.5 represents the scenario that greenhouse gas emissions and climate would be stabilized by the end of this century; RCP8.5 is the scenario with continued increasing emission of greenhouse gases. Mean precipitation also increases under the majority of ensemble members (i.e. 9 of 10), showing changes between –2 to +6 per cent. Climate change is expected to accelerate the hydrological cycle of the LMRB, which would result in a streamflow increase. Figure 2.5 shows the change of SPI-3 during the historical period and future period in the LMRB. During the historical period, the probability of drought and abnormal humidity is 14.8 and 16.4 per cent, respectively. In the future period, under the scenario of RCP4.5, the probability of average drought and abnormal humidity becomes 15.3 and 17.7 per cent, respectively, while under the scenario of RCP8.5, the probability of average drought and abnormal humidity becomes 14.5 and 18.5 per cent, respectively. The change of probability under the scenario of RCP8.5 is larger than under the scenario of RCP4.5. Compared with the historical period, the average drought probability will decrease and the probability of average abnormal humidity will increase during the future period in the LMRB. This indicates that drought periods will decrease and extreme humid periods will increase, which means that increasing flood risks are expected in the future. Therefore, it is necessary to further analyse the changes of flood characteristics under the impact of climate change in the LMRB.

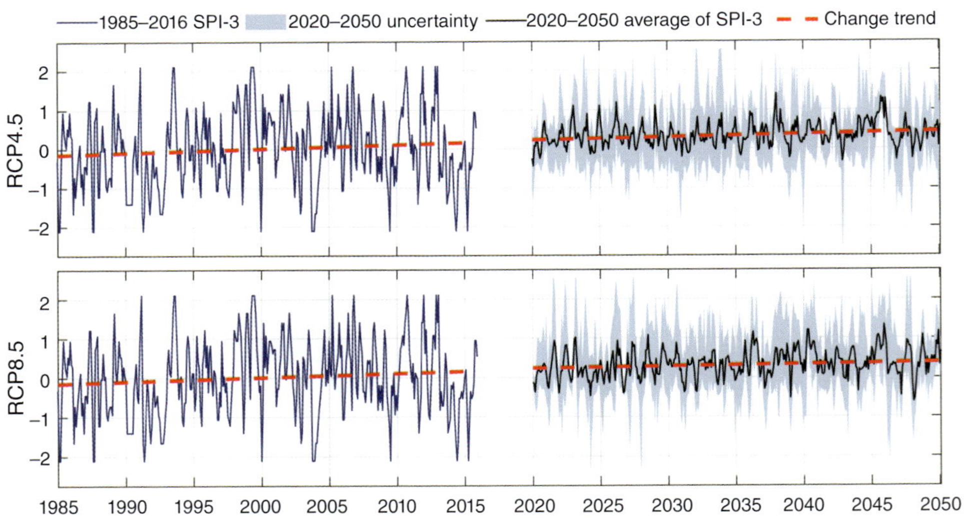

Figure 2.5 Variations of SPI-3 during the historical period (1985–2016) and future period (2020–2050) in the LMRB

2.4.2 VIC Model in the LMRB

Using forcing data, land use data and vegetation distribution data, streamflow has been simulated in the LMRB with the VIC model. Four gauging stations (i.e., CS-KT) were used for calibration, validation and analysis. The simulation results of the VIC model are shown in Figure 2.6, where the black line is the daily-scale observed streamflow data and the red line is the daily-scale simulated streamflow data based on the VIC model. Due to the use of a steeper water retention curve in each grid, the simulated streamflow changed slightly faster from dry season to wet season than the observed streamflow, and the simulated flood peak showed good agreement with the observed value. The daily streamflow data from 1985 to 1993 was used for model calibration, and data from 1994 to 2005 was used for model validation.

Table 2.1 shows the performance of the model in the calibration and the validation period. The NSE coefficient is between 0.72 and 0.76, and the model bias is between –3 and 4 per cent. The hydrological model with NSE > 0.50 could be considered satisfactory, so it was used to simulate the streamflow in the LMRB. The matching of flood characteristic is also very important and Table 2.1 lists the MAF and POT indicators during the history period in the LMRB. The peak value of MAF and the POT threshold shows a good consistency between the simulation and the observation, which suggests that this model can capture the changing characteristics of floods and, hence, be used to analyse the changes in flood events.

2.4.3 Impact of Climate Change on Streamflow

Streamflow has changed in four representative stations in the LMRB under different scenarios from 2020 to 2050 (Figure 2.7). The red line and blue line

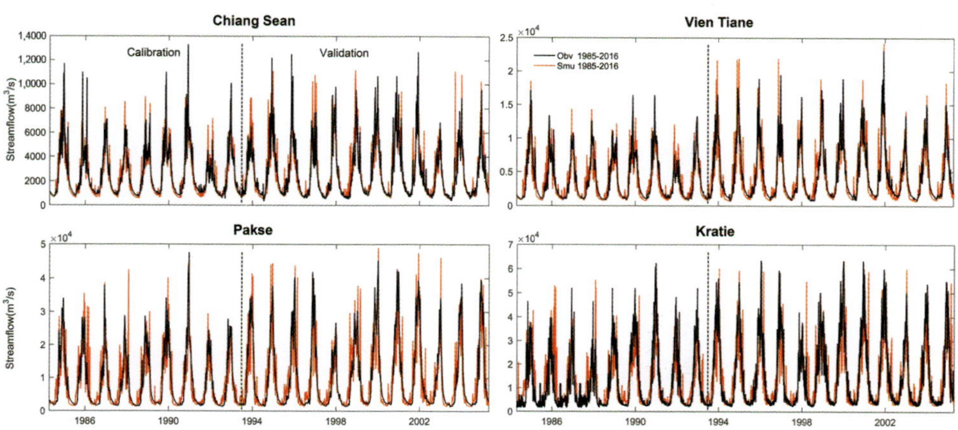

Figure 2.6 The simulation results of the VIC model at four representative stations

Table 2.1. Performance metrics of simulated streamflow at four representative stations in the LMRB for the calibration period (1985–1993) and validation period (1994–2005)

		Calibration Period		Validation Period					
		Performance		Performance		MAF (m^3/s)		Thresholds of POT (m^3/s)	
Station	Name	NSE	RE	NSE	RE	Obv	Smu	Obv	Smu
CS	Chiang Sean	0.76	–0.03	0.7	0.02	10,405	10,219	7,342	7,121
VT	Vien Tiane	0.72	0.02	0.68	0.04	15,193	16,907	11,528	12,737
PK	Pakse	0.74	0.04	0.73	0.03	34,817	33,295	28,138	29,413
KT	Kratie	0.75	0.02	0.71	–0.03	51,921	53,476	43,124	44,195

MAF, Mean Annual Floods; NSE, Nash-Sutcliffe model Efficiency coefficient; Obv, observation; POT, Peak of Threshold; RE, Relative Error; Smu, simulation.

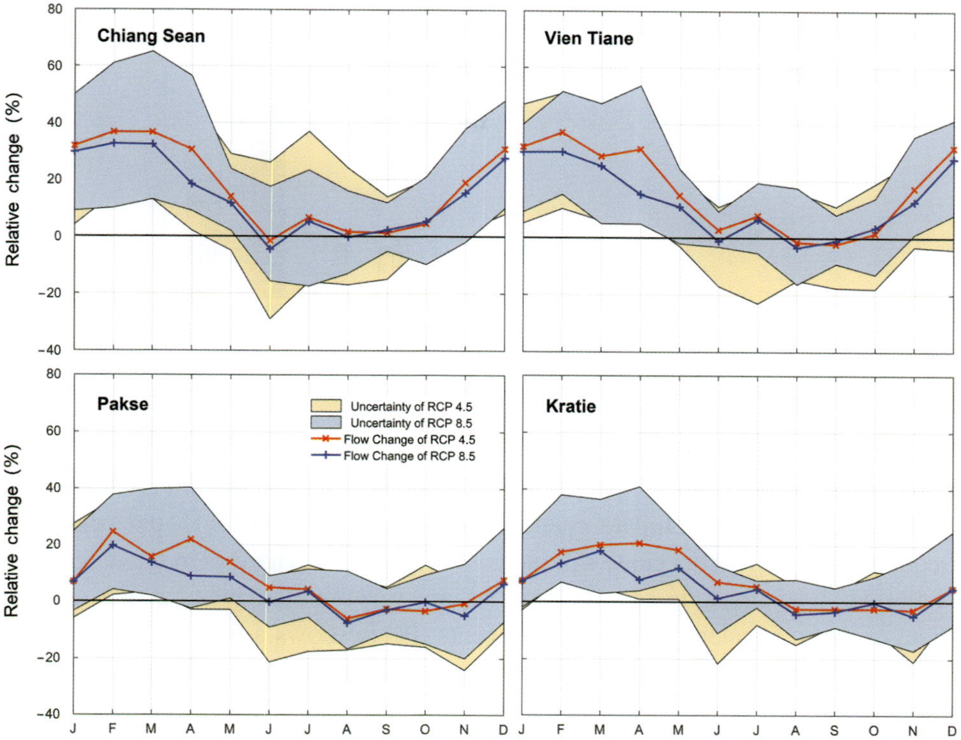

Figure 2.7 Streamflow changes at four representative stations in the LMRB under RCP4.5 and RCP8.5 scenarios from 2020 to 2050

represent the future streamflow changes under the scenarios of RCP4.5 and RCP8.5, respectively, and the yellow and blue areas represent the uncertainty of the five GCMs under the two scenarios, respectively. In the future, with the impact of climate change, streamflow will mainly increase in the dry season (+8 to +39 per cent) but will not change much in the wet season (–7 to ~ +5 per cent).

The uncertainty of the impact of climate change on streamflow decreases from upstream to downstream. At the CS station, climate change has a high fluctuation of –29 to +64 per cent, the uncertainty decreases in subsequent stations and reaches the minimum fluctuation of –21 to +41 per cent at the KT station. This uncertainty is larger in the dry season (December–May) and becomes smaller in the rainy season (June–November). In the future, the streamflow in the LMRB will increase substantially in the dry season.

2.4.4 Impact of Climate Change on Flood Events

Climate change will induce streamflow variability in future periods, which is likely to cause changes in flood event characteristics in the LMRB. Figure 2.8 shows the

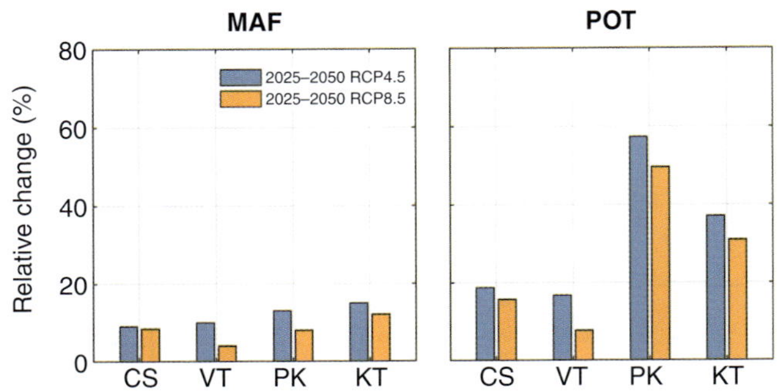

Figure 2.8 Relative changes of flood magnitude (MAF) and frequency (POT) at four representative stations in the future period (2020–2050) compared with the history period (1985–2016)

changes in flood size and frequency at four stations in a future period (2020–2050). With the influence of climate change, the flood size (+3 to +15 per cent) and frequency (+8 to +58 per cent) will increase. The relative changes in the magnitude of the MAF at the four gauging stations are very similar, whereas the relative changes in the frequency of the flood events (POT) shows a huge difference. The relative increase of the flood frequency at the downstream stations (i.e., PK and KT) is much higher than the stations CS and VT. The PK and KT stations are located in the main precipitation core of the LMRB, so the rapidly increasing precipitation in the future will lead to an increase in the frequency of flood events in these areas. In addition, the change in flood frequency is more important than the flood margin at all stations, which indicates that the LMRB will experience more flood events in future periods, mainly in the downstream.

2.5 Discussion

Global climate change and its impact is one of the major challenges that scientists and governments are facing today. Floods are an important hydrological phenomenon that poses a threat to human life and activity across the world. The growing global warming might affect the process of the surface water cycle, which would lead to changes in regional precipitation, evaporation, the acceleration of the hydrological cycle, increasingly frequent extreme weather and disasters. This chapter has reviewed the methods and measures of flood events management and prediction and has demonstrated the utility of predictive methods used in hydrology to simulate history and predict future streamflow and flood events in the LMRB. With the influence of climate change, the probability of the abnormal moisture will increase significantly in the future, whereas the probability of drought will decrease. The magnitude and frequency of flood events will also increase continuously in the future period. The issue of local water resources management in the LMRB will shift from 'lack of water' to 'too much water', which will have a profound impact on the existing water resources security and cooperation in this transboundary river basin.

With the economic development of southeast Asia, the demand for water resources in the LMRB will increase and lead to a larger impact of streamflow changes on regional energy and food security. More than 70 million people live in the downstream of the LMRB, and this number is expected to double by 2050 (more than 145 million) (Pech & Sunada, 2008). After the agricultural reforms in the late 1980s, Vietnam has become one of the world's largest rice exporter countries. Vietnam has been growing 90 per cent of rice for export in the Mekong delta. The rural economy, dominated by rain-fed agriculture, provides 65 per cent of the economic income in the Mekong delta. On the one hand, the decrease of drought probability caused by climate change, the control of seawater intrusion and the treatment of soil salinization will benefit the agriculture and food production in the downstream delta area of the LMRB during the dry season (Smajgl et al., 2015). On the other hand, accelerated melting of glaciers caused by rising temperatures at the source of the LMRB in the future would increase the streamflow supply of rivers in a short-term, and this supply is expected to reach a maximum in the 2050s (Rees & Collins, 2006). The rapid increase in precipitation will lead to accentuated flood peaks during July to September. Meanwhile, the formation of peak flows created by glacier melting water will increase rapidly in a short period of time, leading to an increase of flood risks and affecting grain production in the downstream delta during the rainy season. The rapid increase of flood frequency in the downstream areas will also be detrimental to the breeding of local aquatic species and affect the protein supply in the LMRB (Valbo-Jørgensen et al., 2009).

Under changing climate conditions, there is an urgent need for regional water resources management and networking between the upstream and downstream countries in the LMRB to address the challenges posed by the flood events. Hundreds of large reservoirs built by China, Laos, Vietnam and Cambodia in the LMRB can effectively improve water resources regulation capacity and the uneven distribution of water resources in this basin (Cronin, 2009; Lu et al., 2014). More reservoirs and international multi-reservoir operations will be needed to meet the challenges of substantially increased flood risk in the future. Also, the development of an accurate flood forecasting system will help prevent and manage flood events in the LMRB. Furthermore, the reduction of natural area caused by human development activities in the LMRB and expansion will aggravate the severity of flood events and reduce the effectiveness of flood control measures. The destructive farmland reclamation and massive deforestation in Thailand, Cambodia and Vietnam will increase the probability of flood risk in the basin, while the construction of numerous cities and roads will also affect the permeability of the land (Sohail, 2012; Potapov et al., 2019). Therefore, it is necessary to restore ecosystems to improve the water and soil conservation capacity in this basin. The relationship between humans and nature is of crucial importance on the risk of future flood events in the LMRB. Moreover, implementing the strategy of saving water in industry and agriculture during dry periods and preventing potential flood events during wet periods will contribute to reduce drought and flood disasters in the downstream of this basin.

2.6 Conclusion

This chapter mainly described the floods and the flood analysis method which are commonly used in current research. It evaluated flood event changes impacted by climate change in the Lancang-Mekong River Basin via the VIC model. We divided time period into a history period (1985–2016) and a future period (2020–2050). The major findings of this study are summarized as follows.

1. Under the influence of climate change, the Lancang-Mekong River Basin will experience more exceptional wetness and less drought in the future.
2. In the future, the peak and frequency of flood events in the Lancang-Mekong River Basin will significantly increase, reaching up to +15 per cent and +58 per cent, respectively.

This study provides a reference for assessing the impact of climate change on flood events and for water resources management in the Lancang-Mekong River Basin. This basin needs to take concrete actions (such as building more reservoirs and ecological reforestation) to cope with the increasing flood risk caused by climate change. Further investigations are needed to assess the impact of streamflow and flood events changes on social economy and livelihood activities.

Notes

1 According to EM-DAT, www.emdat.be/.
2 www.isimip.org/protocol/.
3 https://doi.org/10.5067/ASTER/ASTGTM.003.

References

Alfieri, L., Bisselink, B., Dottori, F., et al. (2017). Global projections of river flood risk in a warmer world. *Earth's Future* 5(2): 171–182.
Cronin, R. (2009). Mekong dams, and the perils of peace. *Survival* 51(6): 147–160.
Dang, T. D., Chowdhury, A. M. F. K., & Galelli, S. (2020). On the representation of water reservoir storage and operations in large-scale hydrological models: Implications on

model parameterization and climate change impact assessments. *Hydrology and Earth System Sciences* 24(1): 397–416.
Defries, R. S., and Townshend, J. R. G. (1999). Global land cover characterization from satellite data: From research to operational implementation? *Global Ecology and Biogeography* 8(5): 367–379.
FAO. (2012). ISRIC-World Soil Information, Institute of Soil Science, Chinese Academy of Sciences (ISSCAS), Joint Research Centre of the European Commission (JRC), *Harmonized World Soil Database*, v1.21. Available from www.fao.org/soils-portal/soil-survey/soil-maps-and-databases/harmonized-world-soil-database-v12/en/ (Last accessed March 2020).
Guo, J. L., Hong, Y. L., Leung, L. R., et al. (2014). Links between flood frequency and annual water balance behaviors: A basis for similarity and regionalization. *Water Resources Research* 50(2): 937–953.
Hamman, J. J., Nijssen, B., Bohn, T. J., Gergel, D. R., & Mao, Y. (2018). The Variable Infiltration Capacity model version 5 (VIC-5): Infrastructure improvements for new applications and reproducibility. *Geoscientific Model Development* 11(8): 3481–3496.
Hirsch, R. M., & Archfield, S. A. (2015). Flood trends: Not higher but more often. *Nature Climate Change* 5(3): 198–199.
Hortle, K. G. (2007). MRC Technical Paper No. 16: Consumption and the Yield of Fish and Other Aquatic Animals from the Lower Mekong Basin. *Mekong River Commission.*
Li, R. C. (2005). Flood control history in the Netherlands. *Water Encyclopedia* 2: 524–526.
Liang, X., Lettenmaier, D. P., Wood, E. F., & Burges, S. J. (1994). A simple hydrologically based model of land surface water and energy fluxes for general circulation models. *Journal of Geophysical Research: Atmospheres* 99(D7): 14415–14428.
Lu, X. X., Li, S. Y., Kummu, M., Padawangi, R., & Wang, J. (2014). Observed changes in the water flow at Chiang Saen in the lower Mekong: Impacts of Chinese dams? *Quaternary International* 336(1): 145–157.
Luo, X., Wu, W. Q., He, D. M., Li, Y., & Ji, X. (2019). Hydrological simulation using TRMM and CHIRPS precipitation estimates in the lower Lancang-Mekong river basin. *Chinese Geographical Science* 20(1): 13–25.
Mckee, T. B., Doesken, N. J., & Kleist J. (1993). The relationship of drought frequency and duration to time scales. In *Eighth Conference on Applied Climatology*, January 1993, California.
Mohammed, I. N., Bolten, J. D., Srinivasan, R., et al. (2018). Ground and satellite based observation datasets for the Lower Mekong River Basin. *Data in Brief* 21: 2020–2027.
Nash, J. E., & Sutcliffe J. V. (1970). River flow forecasting through conceptual models part I – A discussion of principles. *Journal of Hydrology* 10(3): 282–290.
Pandey, S., Byerlee, D. R., Dawe, D., et al. (eds.) (2010). *Rice in the Global Economy: Strategic Research and Policy Issues for Food Security*. The Philippines: The International Rice Research Institute (IRRI).
Pech, S., & Sunada, K. (2008). Population growth and natural-resources pressures in the Mekong River Basin. *AMBIO: A Journal of the Human Environment* 37(3): 219–224.
Potapov, P., Tyukavina, A., Turubanova, S., et al. (2019). Annual continuous fields of woody vegetation structure in the Lower Mekong region from 2000–2017 Landsat time-series. *Remote Sensing of Environment* 232: 111278.
Rees, H. G., & Collins, D. N. (2006). Regional differences in response of flow in glacier-fed Himalayan rivers to climatic warming. *Hydrological Processes* 20(10): 2157–2169.

Sheffield, J., Goteti, G., & Wood, E. F. (2006). Development of a 50-year high-resolution global dataset of meteorological forcings for land surface modeling. *Journal of Climate* 19(13): 3088–3111.

Sheffield, J., Wood, E. F., & Roderick, M. L. (2012). Little change in global drought over the past 60 years. *Nature* 491(7424): 435–438.

Smajgl, A., Toan, T. Q., Nhan, D. K., et al. (2015). Responding to rising sea levels in the Mekong Delta. *Nature Climate Change* 5(2): 167–174.

Sohail, A. (2012). Mapping landcover/landuse and coastline change in the Eastern Mekong Delta (Viet Nam) from 1989 to 2002 using remote sensing. Master's Thesis. *Urban Planning & Environment*. Available from www.diva-portal.org/smash/record.jsf?pid=diva2%3A563356&dswid=848 (Last accessed 31 August 2021).

Tatsumi, K., & Yamashiki, Y. (2015). Effect of irrigation water withdrawals on water and energy balance in the Mekong River Basin using an improved VIC land surface model with fewer calibration parameters. *Agricultural Water Management* 159: 92–106.

Thomaz, S. M., Bini, L. M., & Bozelli, R. L. (2007). Floods increase similarity among aquatic habitats in river-floodplain systems. *Hydrobiologia* 579(1): 1–13.

Valbo-Jørgensen, J., Coates, D., & Hortle, K. (2009). Fish diversity in the Mekong River Basin. In I. C. Campbell (ed.), *The Mekong Biophysical Environment of an International River Basin* (pp. 161–196). Cambridge, MA: Academic Press.

Ward, P. J., Jongman, B., Aerts, J. C. J. H., et al. (2017). A global framework for future costs and benefits of river-flood protection in urban areas. *Nature Climate Change* 7 (9): 642–646.

Ward, P. J., Jongman, B., Weiland, F. C. S., et al. (2013). Assessing flood risk at the global scale: Model setup, results, and sensitivity. *Environmental Research Letters* 8(4): 44019.

Wilks, D. S. (1999). Interannual variability and extreme-value characteristics of several stochastic daily precipitation models. *Agricultural and Forest Meteorology* 93(3): 153–169.

Wu, H., Svoboda, M. D., Hayes, M. J., Wilhite, D. A., & Wen, F. (2007). Appropriate application of the standardized precipitation index in arid locations and dry seasons. *International Journal of Climatology* 27(1): 65–79.

Yun, X. B., Tang, Q. H., Wang, J., et al. (2020). Impacts of climate change and reservoir operation on streamflow and flood characteristics in the Lancang-Mekong river basin. *Journal of Hydrology* 590: 125472.

3

Spatial Drought Patterns in East Africa

GEBREMEDHIN GEBREMESKEL HAILE, QIUHONG TANG
AND BINOD BANIYA

3.1 Introduction

Drought is a complex phenomenon with long-lasting impacts on society and natural ecosystems (Sternberg, 2011; Sheffield et al., 2012; Trenberth et al., 2014). Drought leads to large-scale humanitarian, economic and environmental impacts across various regions (Haile et al., 2019; Haile, Tang, Li et al., 2020; Sternberg, 2011). For example, droughts resulted in humanitarian crises leading to more than 450,000 deaths in Ethiopia and Sudan in 1984–1985, over 325,000 deaths in the Sahel in 1974–1975 and around 100,000 deaths in Mozambique in 1979–1981 (Benson & Clay, 1998; Guha-Sapir et al., 2004; Vicente-Serrano et al., 2012). Moreover, the 2012–2013 drought in the US Central Plains, the 1995 drought in Spain and the 1982 drought in Australia cost $US12 billion, $US4.5 billion and $US6 billion, respectively (Guha-Sapir et al., 2004; Mariotti et al., 2013). Growing water demands, deforestation, environmental degradation and the effects of climate change are playing an aggravating role that is extending the damage caused by drought. Hence, there is an urgent need for the implementation of critical management and monitoring of droughts in regions affected by this phenomenon.

Drought is caused by complex natural factors that include changes in climate and anthropogenic processes (AghaKouchak et al., 2015; Schubert et al., 2016; Hao et al., 2018). The naturally occurring high seasonal and interannual variability of the climate system has been the main cause for drought occurrence (Haile et al., 2019; Haile, Tang, Hosseini-Moghari et al., 2020c). Climate variability is mostly linked to land–atmosphere feedback, oceanic oscillations, complex topographies and agro-ecologies with substantial hydrometeorological dynamics (Nicholson, 2000; Funk et al., 2008; Liebmann et al., 2014; Masih et al., 2014; Hua et al.,

Funding for this research is provided by the National Natural Science Foundation of China (41730645, 41790424). Data used for this study was obtained from the SPEI dataset (http://sac.csic.es/spei/database.html). The first author is sponsored by the Chinese Academy of Sciences-The World Academy of Sciences (CAS-TWAS) President's Fellowship Program for his PhD study at the University of Chinese Academy of Sciences.

2016). Similarly, social impacts on the environment, such as urbanization, reservoir construction, deforestation, mining and land-use changes, are emerging drivers associated with increasing drought episodes that have occurred during the Anthropocene (AghaKouchak, 2015b; Van Loon, Gleeson, et al., 2016; Van Loon, Stahl, et al., 2016). Overexploitation of water resources for irrigation, electricity generation, mining and manufacturing may also cause frequent drought (Huang et al., 2018). Moreover, the climate changes induced by global warming are directly linked to human activities (Sheffield et al., 2009, 2012; Schubert et al., 2016). Anthropogenic activities alter underlying hydrological processes which lead to droughts (AghaKouchak, 2015b; Van Loon, Gleeson, et al., 2016; Van Loon, Stahl, et al., 2016). Humans are causing drought, are impacted by it and are working to manage it (Haile, Tang, Li, et al., 2020). In summary, the difficulty to comprehend the complexity of the processes involved in drought leads to large uncertainties associated with the forecasting of the onset, duration or ending of drought events (Touma et al., 2015).

Droughts have been quantified, in time and space, using various types of drought indices, applying different indicators such as rainfall, surface air temperature, soil moisture, evapotranspiration, river flow and groundwater levels (Haile, Tang, Li, et al., 2020). Examples of the commonly used drought indices include the Standardized Precipitation Index (Mckee et al., 1993), the Palmer Drought Severity Index (Palmer, 1965), the Standardized Precipitation-Evapotranspiration Index (SPEI) (Vicente-Serrano et al., 2010) and the Standardized Streamflow Index (Telesca et al., 2012). The selection of drought indices depends on users' interest. However, multivariate drought indices present the advantage of using multiple drought indicators as input variables (Hao & AghaKouchak, 2014; AghaKouchak, 2015a). The use of multivariate drought indices tends to lead to more accurate drought monitoring and early warnings, especially in drought vulnerable regions. Thus, an appropriate selection and use of reliable drought indices at various spatiotemporal levels are required for rigorous drought risk assessment and decision-making.

East Africa is known for its recurrent occurrences of drought over centuries (Kiros, 1991). Dire drought events have occurred in East Africa, with large-scale impacts incorporating substantial regional differences due to varied climate, topography and anthropogenic factors (Funk, 2011; Trenberth et al., 2014; Zhao & Dai, 2015). These droughts have affected society and ecosystem systems, particularly in the last four decades (Muller, 2014; Yang et al., 2014). According to the Emergency Events Database (EM-DAT) (www.emdat.be/database), more than 216.9 million people were affected by droughts, around 572,000 died and about $US1.5 million losses were reported in East Africa over this time period (Guha-Sapir et al., 2004; Grover-Kopec, 2007; Vicente-Serrano et al., 2012; Haile et al., 2019). In East Africa, 100 drought events have been registered since the 1970s (Guha-Sapir et al., 2004; Haile et al., 2019). The 1973–1974, 1984–1985 and 2010–2011 droughts are historically known to have caused huge losses in livelihoods, livestock and the ecosystem, among others (Vicente-Serrano et al., 2012; Solomon et al., 2018; Haile et al., 2019; Haile, Tang, Leng, et al., 2020a).

In East Africa's drought history, ESK have been frequently affected by multiple severe drought events (Funk, 2011; Liebmann et al., 2014; Muller, 2014). As a result, huge humanitarian and economic losses have been reported as a consequence of drought (Haan et al., 2012; Maxwell & Fitzpatrick, 2012; Menkhaus, 2012). For example, in the 2010–2011 drought, around 250,000 deaths were observed in Somalia (Hillbruner & Moloney, 2012; Lyon & Dewitt, 2012; Lyon & Vigaud, 2017). Moreover, drought has become one of the main factors of socio-economic instabilities in ESK (Ayana et al., 2016). This indicates that drought has become a menace for human society in the ESK countries of East Africa. Hence, an in-depth investigation of the evolving drought conditions in ESK is required to suggest possible early warning measures to policymakers. It is suggested that the droughts in the ESK countries need further spatial investigation across multiple timescales.

Despite the evidence collected over time on droughts of East Africa (Muller, 2014; Nicholson, 2014; Omondi et al., 2014; Sheffield et al., 2014; Zhao & Dai, 2015; Nash et al., 2016), the spatial analysis of drought in ESK has been overlooked. Hence, there is an urgent need for investigating droughts that combine various drought parameters over multiple timescales. Furthermore, a systematic detection of severe drought events at fine spatial and temporal scales can help better understand the drought history in East Africa and ESK. To this end, this study attempts to understand the spatial patterns of droughts by systematically analysing various drought parameters under multiple timescales. The main objectives of this paper are: (1) to investigate the spatial drought properties using drought duration, frequency, intensity, severity and area coverage and (2) to understand the spatial coverage of three drought events (during 1973–1974, 1984–1985 and 2010–2011) in East Africa's drought history using the ESK countries as a case study.

3.2 Materials and Methods

3.2.1 Research Area

In East Africa, the countries undergoing the most severe droughts are mainly ESK (Lyon & Dewitt, 2012). ESK are located in the southeastern parts of East Africa, particularly called the Horn of Africa. They are located between the latitudes of 5°S and 15°N, and longitudes of 33°E and 51°E (Figure 3.1). ESK cover around

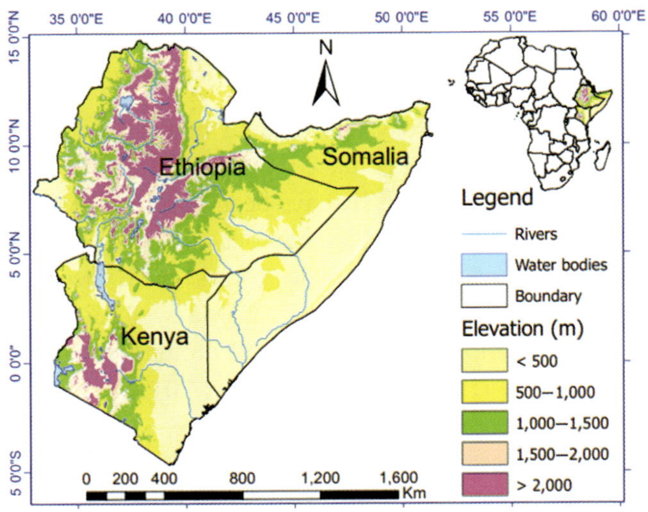

Figure 3.1 Location of ESK (Ethiopia, Somalia and Kenya) countries

2.3 million km^2 with 156.3 million people (United Nations, 2018). The climate and topography of ESK vary from wet highlands to arid lowlands and coastal areas (Dinku et al., 2007, 2011). The mountainous physiography covers the Ethiopian highlands and parts of Kenya, while an arid region covers parts of the Eastern Ethiopian lowlands and Somalia (Dinku et al., 2011). The local population is mainly dependent on agriculture, accompanied by crop production and nomadism.

3.2.2 Drought Index, Dataset and Study Period

In this study, the SPEI (Vicente-Serrano et al., 2010) is used to quantify drought over multiple timescales. The drought index was computed for the period of 1, 3, 6 and 12 months to carry out both short- and medium-term drought analysis. The advantage of SPEI is its capacity to evaluate drought events under global warming (Liu et al., 2016). For this study, the data is taken from the SPEI dataset (http://sac.csic.es/spei/database.html). The data is spatially discretized into a grid composed of $0.5^\circ \times 0.5^\circ$ cells. The SPEI dataset has been widely used for drought analysis at global and regional scales (Schwalm et al., 2017). We selected the SPEI dataset that covers 52 years during 1964–2015 for the drought analysis.

3.2.3 Spatial Drought Analysis Procedures

3.2.3.1 Drought Characterization

We analysed several properties of drought, which include its duration, frequency, intensity and trends (Spinoni et al., 2014; Liu et al., 2016; Z. Wang et al., 2018; Xu

Table 3.1. Drought parameter equations used for drought analysis

Drought Parameter	Equation	Symbol and Units
Drought duration	$D = \frac{\sum_{i=1}^{n} d_i}{n}$	D = drought duration (months) d_i = duration of ith drought event n = total number of drought events
Drought frequency	$F = \frac{n_m}{N_m} \times 100$	F = drought frequency (%), n_m = number of drought months, N_m = total number of drought months (52×12 months)
Drought intensity	$I = \left\vert \frac{1}{n} \sum_{i=1}^{n} SPEI_i \right\vert$	I = drought intensity (–) n = number of drought occurrences in months, $SPEI_i$ = SPEI value below the threshold (–1)

et al., 2018). Table 3.1 shows the equations used to quantify the drought properties. Each drought characterizing parameter (Table 3.1) was quantified by considering –1 as a threshold drought level during 1964–2015. The drought event definition is given by Schwalm et al. (2017), which characterizes drought as a period in which the SPEI value is continuously ≤ -1 for at least three months.

In this study, drought trend is also examined at the pixel level to measure drought changes over time (Eq. 3.1) (Li et al., 2017).

$$Slope = \frac{n \times \sum_{i=1}^{n} (i \times SPEI_i) - \left(\sum_{i=1}^{n} i\right) \times \left(\sum_{i=1}^{n} SPEI_i\right)}{n \times \sum_{i=1}^{n} i^2 - \left(\sum_{i=1}^{n} i\right)^2} \tag{3.1}$$

where $Slope$ = trend change (yr^{-1}), i = series number of the year, n = length of time series and $SPEI_i$ = drought occurring in the ith year.

3.2.3.2 Drought Severity Level Detection

Drought assessment and monitoring is measured using drought levels. In this study, we use the standards recommended for the SPEI drought index (Table 3.2). These drought levels are useful in detecting the severity levels of drought occurrences and help decide on the measures to alleviate it. Severity levels were used to detect and analyse the spatial drought magnitude during the 1973–1974, 1984–1985 and 2010–2011 drought years. Besides, the spatial drought extent (percentage) expresses drought severity and is computed for each investigated time period.

3.3 Results and Discussion

3.3.1 Long-Term Analysis of Precipitation

Figure 3.2 shows the long-term mean annual precipitation. The highlands in western Ethiopia and Kenya show the highest values of the mean annual rainfall. Large areas (52 per cent) in ESK, including Eastern and northeastern Ethiopia, northeastern Kenya and Somalia, receive a lower amount of rainfall (<500 mm) (Figure 3.2) (Fenta et al., 2017). These areas are arid and semiarid regions, which reduce the potential for developmental facilities. The recurrent droughts observed in this region have posed socioeconomical challenges to the local population.

The climate of ESK is divided into three rainy seasons (March–May, July–September and October–November). Kenya, Somalia and southern Ethiopia are affected by a bimodal distribution of the rainfall, with a ‘long rains’ season (i.e.,

Table 3.2. Drought severity levels based on SPEI (F. Wang et al., 2018)

Drought Severity Levels	SPEI Value
Extreme drought	< –2.0
Severe drought	–1.5 to –1.99
Moderate drought	–1 to –1.49
Mild drought	–0.5 to –0.99
No drought	> –0.5

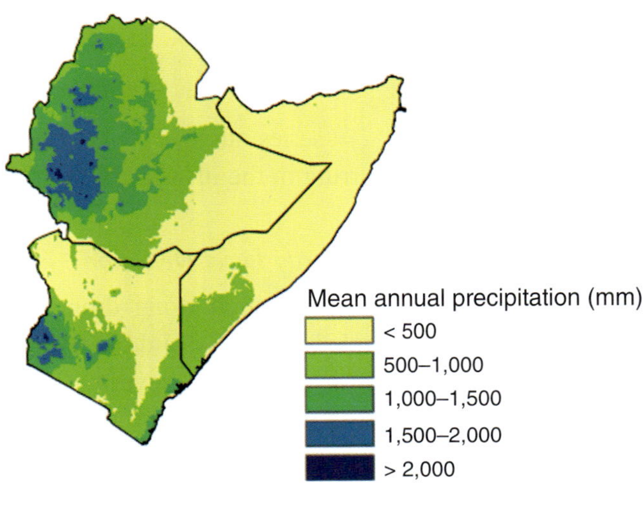

Figure 3.2 Mean annual precipitation of the ESK countries

boreal spring) during March–May and a 'short rains' season (i.e., boreal autumn) during October–November (Tierney et al., 2013, 2015; Liebmann et al., 2014). northern Ethiopia receives most of its rainfall in the boreal summer (July–September) (Camberlin & Okoola, 2003; Dutra et al., 2012, 2013; Camberlin, 2018). The region is highly dependent on rainfall to feed its increasing human population. As a result, drought events are particularly difficult for people living in ESK.

3.3.2 Drought Characterization

3.3.2.1 Spatial Drought Duration

Figure 3.3 shows the long-term spatial variation (1964–2015) in drought duration in the ESK. Higher drought durations were found in eastern Somalia and some parts of Ethiopia and Kenya. However, large areas of Ethiopia and Kenya have lower drought durations. The mean spatial drought duration was 1.49, 2.28, 3.54 and 6.18 months for SPEI-1, SPEI-3, SPEI-6 and SPEI-12, respectively. As the period increases drought duration increases (Figure 3.5). Higher drought durations are also found in longer time periods (SPEI-6 and 12) as compared to the shorter timescales (SPEI-1 and 3). These suggest that annual and interannual drought events could last longer when they occur.

Long consecutive drought months cause large effects on the livelihoods and their surroundings (Chen & Sun, 2015). Longer drought durations may amplify the duration of dry periods in the region. Generally, the long-term drought affects the hydrological cycle and short-term drought may pose a serious threat to agriculture (Rhee & Cho, 2015).

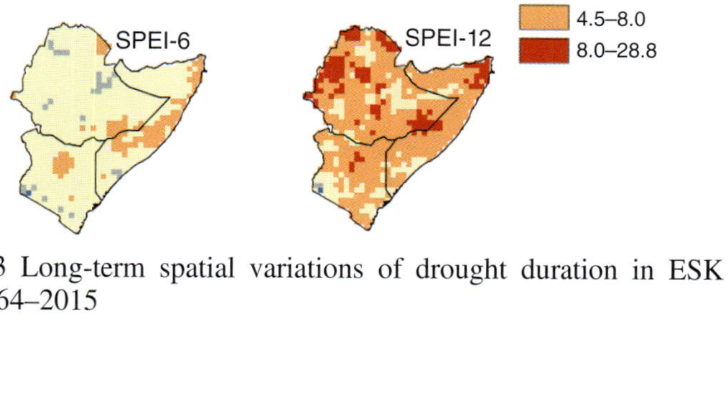

Figure 3.3 Long-term spatial variations of drought duration in ESK countries during 1964–2015

3.3.2.2 Spatial Drought Frequency

Figure 3.4 shows the long-term (1964–2015) spatial variations of drought frequency in the ESK countries. Droughts have occurred more frequently in eastern Somalia, western Ethiopia and central Kenya. However, a large proportion of the area covered by ESK countries do not frequently encounter drought events. The mean spatial drought frequency was 18.20 per cent, 17.19 per cent, 16.33 per cent and 16.15 per cent for SPEI-1, SPEI-3, SPEI-6 and SPEI-12, respectively. The highest frequency in drought was observed in SPEI-1 and 3, while low frequent droughts are found in the longer time periods (SPEI-6 and 12) (Figure 3.5). The variations of SPEI over different time periods suggest that the occurrence of droughts depends on the temporal scales. Higher fluctuations in drought frequency were observed in the shortest time periods (SPEI-1 and 3 months), which is consistent with the findings of Liu et al. (2016).

3.3.2.3 Spatial Drought Intensity

Drought intensity indicates the magnitude of drought and, hence, provides a measure of its impacts on human society and the environment. Figure 3.6 indicates long-term spatial variations (1964–2015) of drought intensity in the ESK countries. We found higher values of drought intensity in eastern and western Somalia, western Ethiopia and eastern and western Kenya. In contrast, large areas composed of central and southern Ethiopia, central Kenya and central Somalia showed lower drought intensity. The mean spatial drought intensity was 0.049, 0.065, 0.089 and 0.126 for SPEI-1, SPEI-3, SPEI-6 and SPEI-12, respectively. The

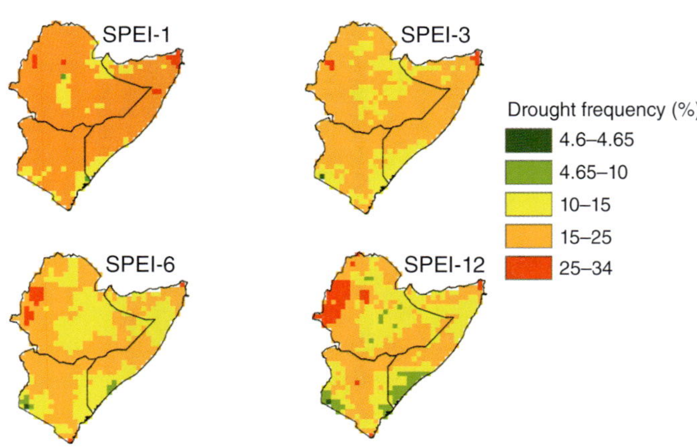

Figure 3.4 Long-term spatial variations of drought frequency ESK countries during 1964–2015

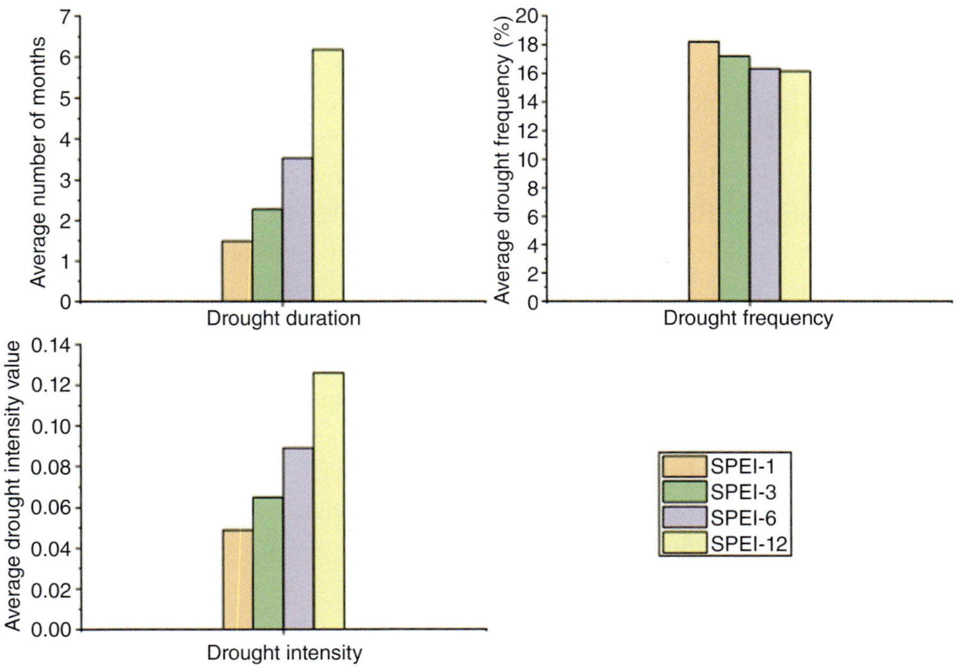

Figure 3.5 Long-term average values of the drought duration, frequency and intensity

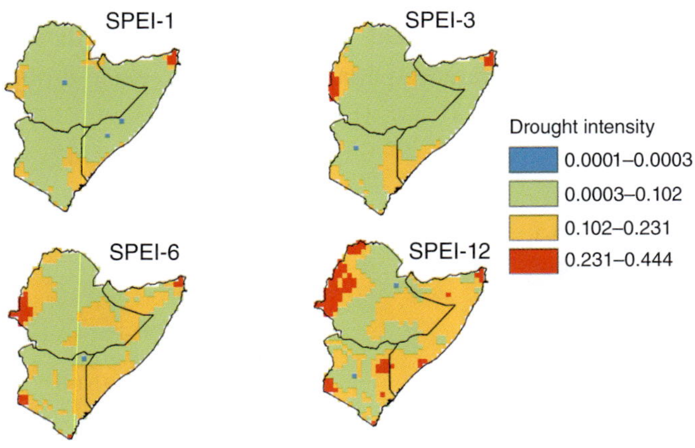

Figure 3.6 Long-term spatial variations of drought intensity in ESK countries during 1964–2015

drought intensity increases as the duration of the investigated period increases (Figure 3.5). A higher magnitude of drought was observed in SPEI-6 and 12, which suggests a potential for hydrological droughts that may have considerable impacts on society and the environment.

3.3.3 Long-Term Spatial Trends

The identification and estimation of drought trends in space and time helps target drought alleviation measures as it may reveal long-term changes over space and time. Figure 3.7 shows a long-term spatial trend during 1964–2015 variations in the ESK countries. Higher spatial drought trends cover large areas of the ESK countries. Unlike SPEI-1, the SPEI-3, SPEI-6 and SPEI-12 showed similar spatial trends of the droughts. Somalia, Kenya and eastern and southwestern Ethiopia revealed higher trends in drought during the last five decades. In contrast, western and northwestern Ethiopia and some parts of south Kenya showed lower drought trends. The mean spatial drought trend was 0.0064, 0.0028, 0.00064 and 0.00095 per year for SPEI-1, SPEI-3, SPEI-6 and SPEI-12, respectively during 1964–2015. This indicates that, as the investigated period increases, drought trend decreases. In summary, long-term drought trends are present in large areas of ESK, which suggests that measures should be taken to alleviate the potential negative consequences of drought in these countries.

3.3.4 Historical Severe Drought Events

In the history of East Africa, droughts have affected the people and their surroundings. The most severe drought events that occurred in East Africa were during 1973–1974, 1984–1985 and 2010–2011 (Haile et al., 2019; Haile, Tang, Leng, et al., 2020). These drought events provoked devastating damage to the ESK countries. We provide a separate spatial analysis for each drought period to highlight their specificities.

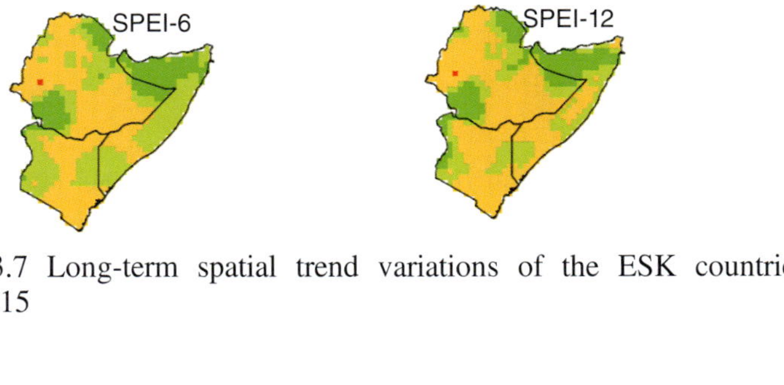

Figure 3.7 Long-term spatial trend variations of the ESK countries during 1964–2015

3.3.4.1 The 1973–1974 Drought

As indicated in Figure 3.8, the drought event during 1973–1974 was explained spatially for the ESK countries based on the drought severity levels given in Table 3.1. Accordingly, extreme and severe droughts were observed in northeastern Somalia and central Ethiopia, while moderate and mild droughts were observed in most parts of Kenya, western Ethiopia and northern Somalia. The total land area of ESK under drought was 43 per cent under SPEI-1, 42 per cent under SPEI-3 and 38 per cent under SPEI-12, during 1973–1974 (Figure 3.9). Large areas of ESK encountered moderate and mild drought severity levels. We observed that drought severity levels increase the duration of the investigated periods. This indicates that the longer the period, the more severe the drought may become.

3.3.4.2 The 1984–1985 Drought

The 1984–1985 drought also had devastating consequences in the ESK countries. Spatially, the drought was concentrated in central and northeastern parts of Ethiopia, central Somalia and north Kenya (Figure 3.10). The drought covered an area of 46 per cent under SPEI-1, 66 per cent under SPEI-3, 80 per cent under SPEI-6 and 75 per cent under SPEI-12 (Figure 3.9). This indicates that the 1984–1985 drought event was the worst as it covered more than half of the total land area of ESK countries over all time periods. It is evident that the area under drought is higher in the interannual timescales than under the short (under SPEI-1)

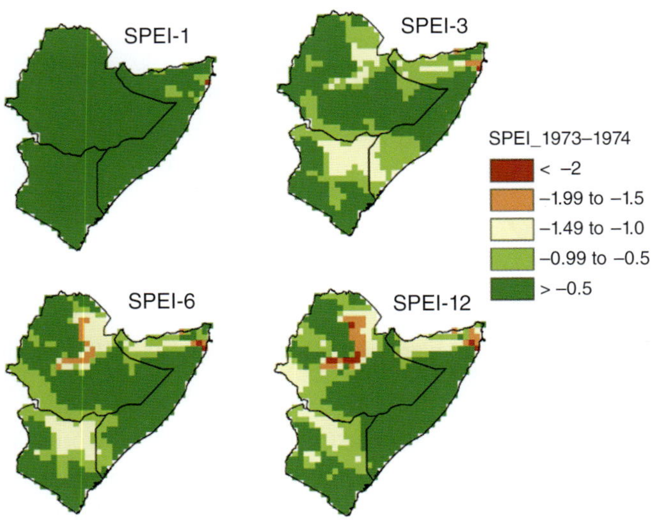

Figure 3.8 Spatial drought severity levels during the 1973–1974 drought year

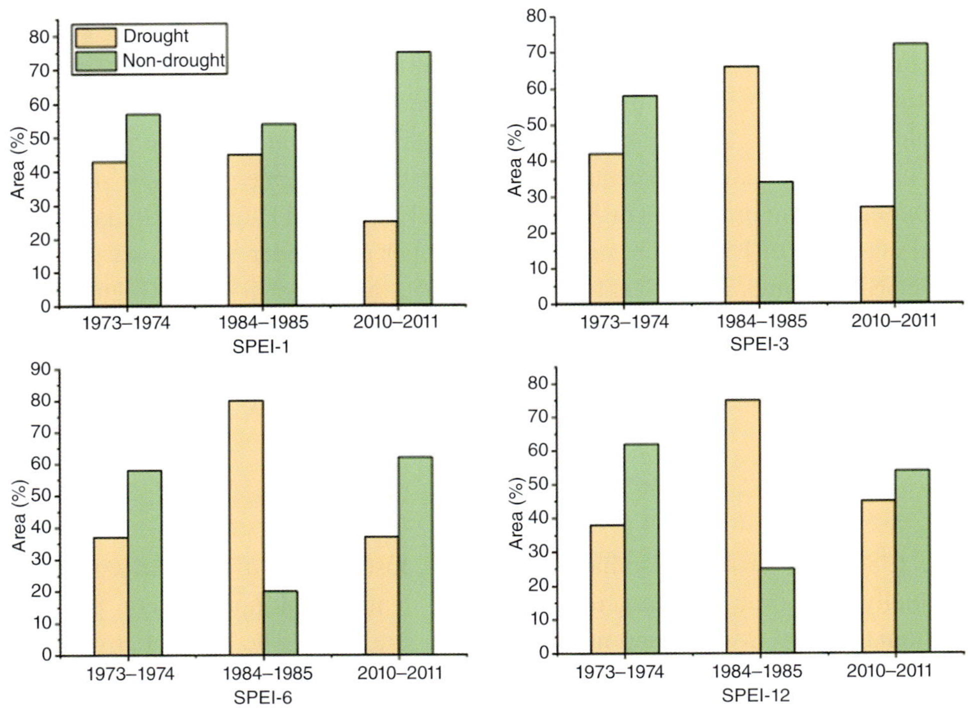

Figure 3.9 Drought and non-drought area coverage of the drought events

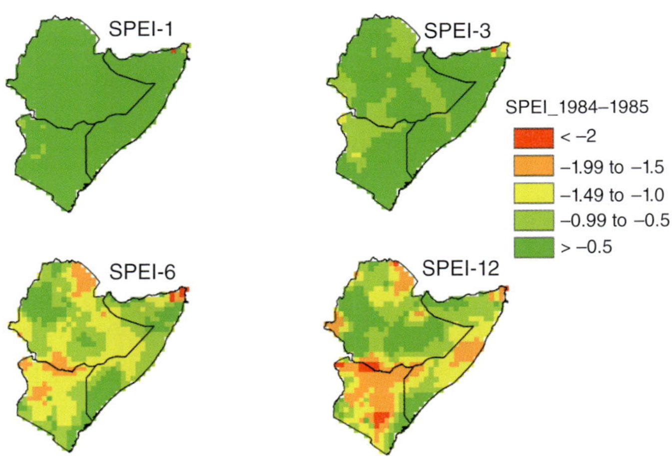

Figure 3.10 Spatial drought severity levels during the 1984–1985 drought year

and long (under SPEI-12) periods. In terms of the drought severity levels, a higher magnitude of extreme and severe drought events tends to occur in longer time periods (10 per cent under SPEI-6 and 23 per cent under SPEI-12). This suggests that the 1984–1985 drought was extremely severe.

3.3.4.3 The 2010–2011 Drought

The 2010–2011 drought event was mainly concentrated in south Ethiopia, western Somalia and north and eastern Kenya (Figure 3.11). The drought covered around 25 per cent under SPEI-1, 28 per cent under SPEI-3, 38 per cent under SPEI-6 and 46 per cent under SPEI-12 (Figure 3.9). As the length of the time period increases, the area under drought increases. This indicates a higher drought magnitude, which represents propagation of the drought in the ESK countries. In addition, the extreme and severe drought events increase as the investigated temporal period increases, which coincides with dire drought events in ESK.

3.3.5 Impacts of the Drought Events

The spatial analysis of dire drought events in the history of East Africa is vital as it helps implement sustainable strategies to mitigate their impacts. Historically, the 1973–1974 drought incident took the lives of 250,000 people and 30 per cent of the domestic animals in Ethiopia (Kebbede & Jacob, 1988, Degefu & Bewket 2015). The 1984–1985 drought event affected about 7–10 million people, of which 250,000–750,000 people died in Ethiopia alone (Bayissa et al., 2015; Degefu & Bewket, 2015). This drought also forced 5.8 million people to become completely dependent on food aid in Ethiopia (Gebrehiwot et al., 2011). Ethiopia faced a severe drought between 2010 and 2011 but its impacts were mitigated by the government and the citizens (Bayissa et al., 2015; Degefu & Bewket, 2015). The 2010–2011 drought event led to a devastating famine in

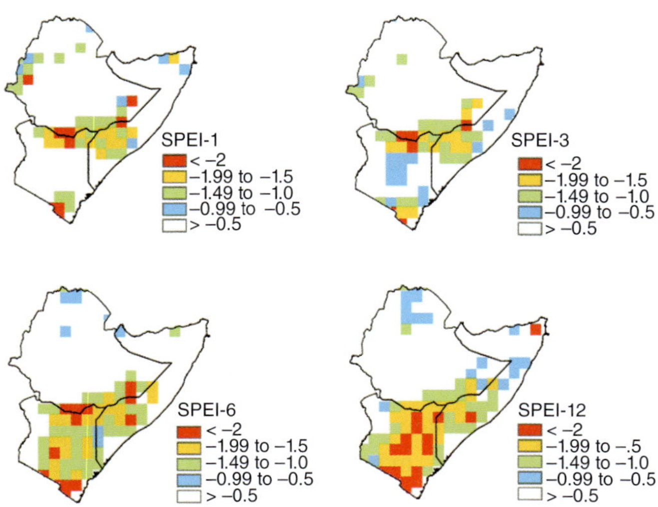

Figure 3.11 Spatial drought severity levels during the 2010–2011 drought year

Somalia and affected around 3.1 million people, of which 258,000 died, half a million children were malnourished and 1.46 million Somalis were displaced (Funk, 2011; Hillbruner & Moloney, 2012; Maxwell & Fitzpatrick, 2012). This drought also severely affected Kenya, with substantial effects (Tadesse et al., 2008; Uhe et al., 2018).

Therefore, a comprehensive analysis of historical droughts over East Africa helps understand the underlying mechanisms behind the processes that generate drought. Results from the analysis of historical data that account for different time scales can be used to help policymakers find solutions to monitor and mitigate drought events and reduce the consequences of future events.

3.4 Conclusion

This chapter analysed the spatial extent of the drought properties, which include drought duration, frequency, intensity and trends for several time periods associated with a particular time scale. As the investigated period increases, drought duration and intensity increase. In contrast, drought frequency decreases in larger periods of time. These results imply that the frequency of hydrological droughts is lower than meteorological and agricultural droughts. Meanwhile, higher spatial drought trends are concentrated in large areas of the ESK countries, with mean spatial drought trend values of 0.0064, 0.0028, 0.00064 and –0.00095 per year for SPEI-1, SPEI-3, SPEI-6 and SPEI-12, respectively. Furthermore, we evaluated the severity of the droughts that occurred in ESK countries during 1973–1974, 1984–1985 and 2010–2011 using drought severity levels. The large spatial extent of the drought years indicates a severe risk for society and the environment. The total land area of the ESK countries under drought was 43 per cent under SPEI-1, 42 per cent under SPEI-3 and 38 per cent under SPEI-12 during 1973–1974. In 1984–1985 the drought covered an area of 46 per cent under SPEI-1, 66 per cent under SPEI-3, 80 per cent under SPEI-6 and 75 per cent under SPEI-12. Similarly, the drought covered 25 per cent under SPEI-1, 28 per cent under SPEI-3, 38 per cent under SPEI-6 and 46 per cent under SPEI-12 during the 2010–2011 drought year. Generally, northeastern and southern Ethiopia, eastern Somalia and north-eastern Kenya were affected by dire drought impacts during the drought years. The results of this study could have an important implication for the ESK countries to manage and alleviate drought risks locally and regionally in areas vulnerable to drought. A better management of the risk using evidence-based research work may reduce the impact of droughts on humans and ecosystems, which are exacerbated by global warming.

References

AghaKouchak, A. (2015a). A multivariate approach for persistence-based drought prediction: Application to the 2010–2011 East Africa drought. *Journal of Hydrology* 526 (5): 127–135.

AghaKouchak, A. (2015b). Recognize anthropogenic drought. *Nature* 524(7566): 409–411.

AghaKouchak, A., Farahmand, A., Melton, F. S., et al. (2015). Remote sensing of drought: Progress, challenges and opportunities. *Reviews of Geophysics* 53(3): 1–29.

Ayana, E. K., Ceccato, P., Fisher, J. R. B., & DeFries, R. (2016). Examining the relationship between environmental factors and conflict in pastoralist areas of East Africa. *Science of the Total Environment* 557–558(7): 601–611.

Bayissa, Y. A., Moges, S. A., Xuan, Y., et al. (2015). Spatio-temporal assessment of meteorological drought under the influence of varying record length: The case of Upper Blue Nile Basin. *Ethiopia. Hydrological Sciences Journal* 60(11): 1927–1942.

Benson, C., & Clay, E. (1998). Drought and sub-Saharan African economies. *Findings: Africa Region* 118(5): 831–852.

Camberlin, P. (2018). *Climate of Eastern Africa* (Vol. 1). Oxford: Oxford University Press.

Camberlin, P., & Okoola, R. E. (2003). The onset and cessation of the 'long rains' in eastern Africa and their interannual variability. *Theoretical and Applied Climatology* 54(1–2): 43–54.

Chen, H., & Sun, J. (2015). Changes in drought characteristics over China using the standardized precipitation evapotranspiration index. *Journal of Climate* 28(13): 5430–5447.

Degefu, M. A., & Bewket, W. (2015). Trends and spatial patterns of drought incidence in the Omo-Ghibe River Basin, Ethiopia. *Geografiska Annaler, Series A: Physical Geography* 97(2): 395–414.

Dinku, T., Ceccato, P., & Connor, S. J. (2011). Challenges of satellite rainfall estimation over mountainous and arid parts of east Africa. *International Journal of Remote Sensing* 32(21): 5965–5979.

Dinku, T., Ceccato, P., Grover-Kopec, E., et al. (2007). Validation of satellite rainfall products over East Africa's complex topography. *International Journal of Remote Sensing* 28(7): 1503–1526.

Dutra, E., Di Giuseppe, F., Wetterhall, F., & Pappenberger, F. (2012). Seasonal forecasts of drought indices in African basins. *Hydrology and Earth System Sciences Discussions* 9(9): 11093–11129.

Dutra, E., Magnusson, L., Wetterhall, F., et al. (2013). The 2010–2011 drought in the Horn of Africa in ECMWF reanalysis and seasonal forecast products. *International Journal of Climatology* 33(7): 1720–1729.

Fenta, A. A., Yasuda, H., Shimizu, K., et al. (2017). Spatial distribution and temporal trends of rainfall and erosivity in the Eastern Africa region. *Hydrological Processes* 31(25): 4555–4567.

Funk, C. (2011). We thought trouble was coming. *Nature* 476(7358): 7.

Funk, C., Dettinger, M. D., Michaelsen, J. C., et al. (2008). Warming of the Indian Ocean threatens eastern and southern African food security but could be mitigated by agricultural development. *Proceedings of the National Academy of Sciences* 105 (32): 11081–11086.

Gebrehiwot, T., van der Veen, A., & Maathuis, B. (2011). Spatial and temporal assessment of drought in the Northern highlands of Ethiopia. *International Journal of Applied Earth Observation and Geoinformation* 13(3): 309–321.

Grover-Kopec, E. (2007). *Documenting Drought-Related Disasters* (pp. 328–344). Washington, DC: World Bank.

Guha-Sapir, D., Hargitt, D., & Hoyois, P. (2004). *Thirty Years of Natural Disasters 1974–2003: The Numbers*. Leuven, Belgium: Presses universitaires de Louvain.

Haan, N., Devereux, S., & Maxwell, D. (2012). Global implications of Somalia 2011 for famine prevention, mitigation and response. *Global Food Security* 1(1): 74–79.

Haile, G. G., Tang, Q., Hosseini-Moghari, S., et al. (2020). Projected impacts of climate change on drought patterns over East Africa. *Earth's Future* 8(7): 1–23.

Haile, G. G., Tang, Q., Leng, G., et al. (2020). Long-term spatiotemporal variation of drought patterns over the Greater Horn of Africa. *Science of the Total Environment* 704: 135299.

Haile, G. G., Tang, Q., Li, W., Liu, X., & Zhang, X. (2020). Drought: Progress in broadening its understanding. *WIREs Water* 7(2): e1407.

Haile, G. G., Tang, Q., Sun, S., et al. (2019). Droughts in East Africa: Causes, impacts and resilience. *Earth-Science Reviews*, 193(2018): 146–161.

Hao, Z., & AghaKouchak, A. (2014). A nonparametric multivariate multi-index drought monitoring framework. *Journal of Hydrometeorology* 15(1): 89–101.

Hao, Z., Singh, V. P., & Xia, Y. (2018). Seasonal drought prediction: Advances, challenges, and future prospects. *Reviews of Geophysics* 56(1): 108–141.

Hillbruner, C., & Moloney, G. (2012). When early warning is not enough – Lessons learned from the 2011 Somalia famine. *Global Food Security* 1(1): 20–28.

Hua, W., Zhou, L., Chen, H., et al. (2016). Possible causes of the Central Equatorial African long-term drought. *Environmental Research Letters* 11(12): 124002.

Huang, Z., Hejazi, M., Li, X., et al. (2018). Reconstruction of global gridded monthly sectoral water withdrawals for 1971–2010 and analysis of their spatiotemporal patterns. *Hydrology and Earth System Sciences* 22(4): 2117–2133.

Kebbede, G., & Jacob, M. J. (1988). Drought, famine and the political economy of environmental degradation in Ethiopia. *JSTOR* 73(1): 65–70.

Kiros, F. G. (1991). Economic consequences of drought, crop failure and famine in Ethiopia, 1973–1986. *Ambio (Sweden)* 20(5): 183–185.

Li, Z., Chen, Y., Fang, G., & Li, Y. (2017). Multivariate assessment and attribution of droughts in Central Asia. *Scientific Reports* 7(1): 1–12.

Liebmann, B., Hoerling, M. P., Funk, C., et al. (2014). Understanding recent eastern Horn of Africa rainfall variability and change. *Journal of Climate* 27(23): 8630–8645.

Liu, Z., Li, C., Zhou, P., & Chen, X. (2016). A probabilistic assessment of the likelihood of vegetation drought under varying climate conditions across China. *Scientific Reports* 6(May): 1–10.

Lyon, B., & Dewitt, D. G. (2012). A recent and abrupt decline in the East African long rains. *Geophysical Research Letters* 39(2): 1–5.

Lyon, B., & Vigaud, N. (2017). Unraveling East Africa's climate paradox. In S.-Y. S. Wang, J.-H. Yoon, C. C. Funk, & R. R. Gillies (eds.), *Climate Extremes: Patterns and Mechanisms, Geophysical Monography 226* (pp. 265–287). Hoboken, NJ: John Wiley & Sons.

Mariotti, A., Schubert, S., Mo, K., et al. (2013). An Interpretation of the Origins of the 2012 Central Great Plains Drought. Assessment Report. Available from https://psl.noaa.gov/csi/factsheets/pdf/noaa-gp-drought-assessment-report.pdf (Last accessed 31 August 2021).

Masih, I., Maskey, S., Mussá, F. E. F., & Trambauer, P. (2014). A review of droughts on the African continent: A geospatial and long-term perspective. *Hydrology and Earth System Sciences* 18(9): 3635–3649.

Maxwell, D., & Fitzpatrick, M. (2012). The 2011 Somalia famine: Context, causes, and complications. *Global Food Security* 1(1): 5–12.

Mckee, T. B., Doesken, N. J., & Kleist, J. (1993). The relationship of drought frequency and duration to time scales. In *Eighth Conference on Applied Climatology*, 17–22 January 1993, Anaheim, CA. Available from www.droughtmanagement.info/literature/AMS_Relationship_Drought_Frequency_Duration_Time_Scales_1993.pdf (Last accessed 31 August 2021).

Menkhaus, K. (2012). No access: Critical bottlenecks in the 2011 Somali famine. *Global Food Security* 1(1): 29–35.

Muller, J. C. Y. (2014). Adapting to climate change and addressing drought – Learning from the Red Cross Red Crescent experiences in the Horn of Africa. *Weather and Climate Extremes* 3: 31–36.

Nash, D. J., De Cort, G., Chase, B. M., et al. (2016). African hydroclimatic variability during the last 2000 years. *Quaternary Science Reviews* 154: 1–22.

Nicholson, S. E. (2000). Land surface processes and land use change land. *Reviews of Geophysics* 38(1): 117–139.

Nicholson, S. E. (2014). A detailed look at the recent drought situation in the Greater Horn of Africa. *Journal of Arid Environments* 103: 71–79.

Omondi, P. A., Awange, J. L., Forootan, E., et al. (2014). Changes in temperature and precipitation extremes over the Greater Horn of Africa region from 1961 to 2010. *International Journal of Climatology* 34(4): 1262–1277.

Palmer, W. C. (1965). Meteorological Drought. Research Paper No. 45, US Department of Commerce, Washington D.C., 58 pp. Available from www.ncdc.noaa.gov/temp-and-precip/drought/docs/palmer.pdf (Last accessed 31 August 2021)

Rhee, J., & Cho, J. (2015). Future changes in drought characteristics: Regional analysis for South Korea under CMIP5 projections. *Journal of Hydrometeorology* 17(1): 437–451.

Schubert, S. D., Stewart, R. E., Wang, H., et al. (2016). Global meteorological drought: A synthesis of current understanding with a focus on SST drivers of precipitation deficits. *Journal of Climate* 29(11): 3989–4019.

Schwalm, C. R., Anderegg, W. R. L., Michalak, A. M., et al. (2017). Global patterns of drought recovery. *Nature* 548(7666): 202–205.

Sheffield, J., Andreadis, K. M., Wood, E. F., & Lettenmaier, D. P. (2009). Global and continental drought in the second half of the twentieth century: Severity-area-duration analysis and temporal variability of large-scale events. *Journal of Climate* 22(8): 1962–1981.

Sheffield, J., Wood, E. F., Chaney, N., et al. (2014). A drought monitoring and forecasting system for sub-Sahara African water resources and food security. *Bulletin of the American Meteorological Society* 95(6): 861–882.

Sheffield, J., Wood, E. F., & Roderick, M. L. (2012). Little change in global drought over the past 60 years. *Nature* 491(7424): 435–438.

Solomon, N., Birhane, E., Gordon, C., et al. (2018). Environmental impacts and causes of conflict in the Horn of Africa: A review. *Earth-Science Reviews* 177: 284–290.

Spinoni, J., Naumann, G., Carrao, H., Barbosa, P., & Vogt, J. (2014). World drought frequency, duration, and severity for 1951–2010. *International Journal of Climatology* 34(8): 2792–2804.

Sternberg, T. (2011). Regional drought has a global impact. *Nature* 472(7342): 169.

Tadesse, T., Senay, G., Wardlow, B. D., Knutson, C. L., & Haile, M. (2008). The need for integration of drought monitoring tools for proactive food security management in sub-Saharan Africa. *Natural Resources Forum* 32: 265–279.

Telesca, L., Lovallo, M., Lopez-Moreno, I., & Vicente-Serrano, S. (2012). Investigation of scaling properties in monthly streamflow and Standardized Streamflow Index (SSI) time series in the Ebro basin (Spain). *Physica A* 391(4): 1662–1678.

Tierney, J. E., Smerdon, J. E., Anchukaitis, K. J., & Seager, R. (2013). Multidecadal variability in East African hydroclimate controlled by the Indian Ocean. *Nature* 493 (7432): 389–392.

Tierney, J. E., Ummenhofer, C. C., & DeMenocal, P. B. (2015). Supplementary materials for: Past and future rainfall in the Horn of Africa. *Science Advances* 1(9): e1500682.

Touma, D., Ashfaq, M., Nayak, M. A., Kao, S. C., & Diffenbaugh, N. S. (2015). A multi-model and multi-index evaluation of drought characteristics in the 21st century. *Journal of Hydrology* 526: 196–207.

Trenberth, K. E., Dai, A., Van Der Schrier, G., et al. (2014). Global warming and changes in drought. *Nature Climate Change* 4(1): 17–22.

Uhe, P., Philip, S., Kew, S., et al. (2018). Attributing drivers of the 2016 Kenyan drought. *International Journal of Climatology* 38: e554–e568.

United Nations (2018). World Economic Situation and Prospects 2018. Available from www.un.org/development/desa/dpad/wp-content/uploads/sites/45/publication/WESP2018_Full_Web-1.pdf (Last accessed June 2019).

Van Loon, A. F., Gleeson, T., Clark, J., et al. (2016). Drought in the anthropocene. *Nature Geoscience* 9(2): 89–91.

Van Loon, A. F., Stahl, K., Di Baldassarre, G., et al. (2016). Drought in a human-modified world: Reframing drought definitions, understanding, and analysis approaches. *Hydrology and Earth System Sciences* 20(9): 3631–3650.

Vicente-Serrano, S. M., Beguería, S., Gimeno, L., et al. (2012). Challenges for drought mitigation in Africa: The potential use of geospatial data and drought information systems. *Applied Geography* 34: 471–486.

Vicente-Serrano, S. M., Beguería, S., & López-Moreno, J. I. (2010). A multiscalar drought index sensitive to global warming: The standardized precipitation evapotranspiration index. *Journal of Climate* 23(7): 1696–1718.

Wang, F., Wang, Z., Yang, H., & Zhao, Y. (2018). Study of the temporal and spatial patterns of drought in the Yellow River basin based on SPEI. *Science China Earth Sciences* 61(8): 1098–1111.

Wang, Z., Li, J., Lai, C., et al. (2018). Increasing drought has been observed by SPEI_pm in Southwest China during 1962–2012. *Theoretical and Applied Climatology* 133 (1–2): 23–38.

Xu, L., Chen, N., & Zhang, X. (2018). Global drought trends under 1.5 and 2°C warming. *International Journal of Climatology* 39(4): 2375–2385.

Yang, W., Seager, R., Cane, M. A., & Lyon, B. (2014). The East African long rains in observations and models. *Journal of Climate* 27(19): 7185–7202.

Zhao, T., & Dai, A. (2015). The magnitude and causes of global drought changes in the twenty-first century under a low–moderate emissions scenario. *Journal of Climate* 28 (11): 4490–4512.

4

Assessment of Global Water Erosion Vulnerability under Climate Change

MUQI XIONG AND GUOYONG LENG

4.1 Introduction

Water-driven soil erosion is the most widespread form of soil degradation worldwide (Maetens, Poesen, et al., 2012; García-Ruiz et al., 2017) and it has been a major threat to soil quality and productivity, sustainable agriculture, crop production (Lobell & Gourdji, 2012; Challinor et al., 2014; Wuepper et al., 2019) and carbon stocks across the globe since the beginning of agriculture (Li & Fang, 2016). Water erosion also brings about off-site damage, which can cause significant economic loss (Li & Fang, 2016).

Global climate change as a hotspot has been noticed worldwide in the twenty-first century (Lu et al., 2019). The IPCC Fourth Assessment Report pointed out that global temperature has increased by 0.6–0.8°C over the past century and there will be more global extreme precipitation events in upcoming decades (IPCC, 2007). Climate change affects precipitation, temperature and land cover, which is likely to strongly influence the processes of water erosion (Litschert et al., 2014). The impacts of climate change on water erosion have been subject to considerable attention, as illustrated by a wide-range of case studies (Li & Fang, 2016). Changes in rainfall amount (Bangash et al., 2013), rainfall intensity (Zhang et al., 2012; Panagos et al., 2017) and precipitation patterns (Maeda et al., 2010) inevitably and directly affect water erosion (Li & Fang, 2016). Li and Fang (2016) showed that water erosion can be affected indirectly by rising temperatures. Furthermore, changes in land use and management practices that adapt to the recent consequences of climate change can indirectly affect water erosion (Li & Fang, 2016). However, these indirect impacts are more complex. The impacts of climate change on water erosion are critical for planning water erosion risk mitigation and soil resource management.

This work was supported by Chinese Academy of Sciences (No. YJRCPT2019–001; Y9V90222YZ).

Modelling is a common and useful way to assess the response of water erosion to climate change (Li et al., 2010; Li & Fang, 2016). With the development of climate and erosion models, we have observed an increasing number of studies that have focused on climate change impacts on water erosion worldwide since 2000 (Li & Fang, 2016). These studies have assessed the impacts of climate change on water erosion based on various water erosion models, including empirical models, conceptual models and physically based models such as EROSION 3D (Routschek et al., 2014), WEPP (Zhang & Nearing, 2005), STREAM (Paroissien et al., 2015; Simonneaux et al., 2015), SWAT (Mukundan et al., 2013; Nunes et al., 2013; Serpa et al., 2015; Parajuli et al., 2016) and USLE (Yang et al., 2003; Zhang, 2012; Liu et al., 2019) at various temporal and spatial scales.

The universal soil loss equation (USLE) (Wischmeier & Smith, 1978) and its revised version (RUSLE) (Renard et al., 1997) have been widely applied to assess water erosion around the world, owing to its simple structure and empirical basis (Naipal et al., 2015). Especially, it is endowed with the capability of large-scale applicability (Rao et al., 2014) and has been used on a global scale (Pham et al., 2001; Yang et al., 2003; Ito, 2007; Van Oost et al., 2007; Doetterl et al., 2012; Naipal et al., 2015; Scherer & Pfister, 2015; Borrelli et al., 2017; Liu et al., 2019). In this study, a Geographic Information Systems (GIS)-based RUSLE is used to simulate average annual water erosion rate globally. The value of the water erosion rate indicates the level of water erosion vulnerability (Panagos et al., 2015; Liu et al., 2019).

Climate change impacts on water erosion have been observed across the world and many studies suggest that rainfall is the most direct influencing factor (Li & Fang, 2016). Here, we focus on the change of rainfall to analyse the spatial pattern of global water erosion vulnerability change over the past 30 years based on USLE model to provide a global view of impacts of climate change on water erosion. The results might help adaptation planning land managers design and implement policies aimed at mitigating the effects of climate change. Furthermore, our analysis could be used to identify and answer the future needs of communities impacted by this global threat.

4.2 Materials and Methods

4.2.1 RUSLE Model

The evaluation of the long-term annual water erosion was carried out using the RUSLE model in a GIS environment. RUSLE is an empirical model which belongs to the detachment-limited models. The soil loss (Mg ha^{-1} yr^{-1}) resulting from the sheet and rill erosion process is given by the following equation (Renard et al., 1997):

$$A = R \cdot K \cdot L \cdot S \cdot C \cdot P \tag{4.1}$$

where A [Mg ha^{-1} yr^{-1}] is the annual soil loss rate (SLa), R [MJ mm h^{-1} ha^{-1} yr^{-1}] is the rainfall erosivity factor, K [Mg h MJ^{-1} mm^{-1}] is the soil erodibility factor, L [dimensionless] is the slope length factor, S [dimensionless] is the slope steepness factor, C [dimensionless] is the land cover and management factor and P [dimensionless] is the support practices factor (Renard et al., 1997). The support practices have high uncertainty at large scales and identifying the yearly tendency is difficult. Therefore, we set the value of the support practices factor as 1.

4.2.1.1 Rainfall Erosivity Factor (R)

Rainfall erosivity is determined by rainfall kinetic energy and rainfall intensity. The rainfall erosivity factor (R) is computed originally from rainfall amount and intensity (Wischmeier & Smith, 1978; Renard et al., 1997). However, this kind of precipitation record is difficult to obtain on a large scale (Naipal et al., 2015). We applied some alternative methods that use monthly precipitation or annual precipitation (Renard & Freimund, 1994; Liu et al., 2019) to calculate the R factor. First, we calculated the R factor according to the method of Renard and Freimund (1994) (Eq. 4.2).

The change in precipitation is expected to impact water erosion directly, in terms of rainfall amount and intensity (Li & Fang, 2016). The method of Renard and Freimund (1994) does not consider rainfall intensity, therefore we used the method of Naipal et al. (2015), which adjusted for erosivity using precipitation, elevation and the simple precipitation intensity index (SDII) (Figure 4.1(a)) in four climate zones (CZs) according to the updated world KöppenGeiger climate classification (Peel et al., 2007) (Figure 4.1(b)).

$$R = \begin{cases} 0.0483 \times P^{1.61} & P \leq 850\,\text{mm} \\ 587.8 - 1.219 \times P + 0.004105 \times P^2 & P > 850\,\text{mm} \end{cases} \tag{4.2}$$

In Eq. (4.2), R is the rainfall erosivity and P is the annual precipitation in millimetres.

4.2.1.2 Soil Erodibility Factor (K)

The soil erodibility K-factor is generally measured based on soil properties such as texture, organic matter and the structure of the topsoil profile (Shabani et al., 2014). Soil erodibility describes the susceptibility of soil particles to detachment and movement by water and reflects the sensitivity of soils to erosion (Zhang et al., 2008; Rao et al., 2014). In this study, the Erosion-Productivity Impact Calculator (EPIC) model was used to estimate soil erodibility. The equation is as follows (Sharpley & Williams, 1990):

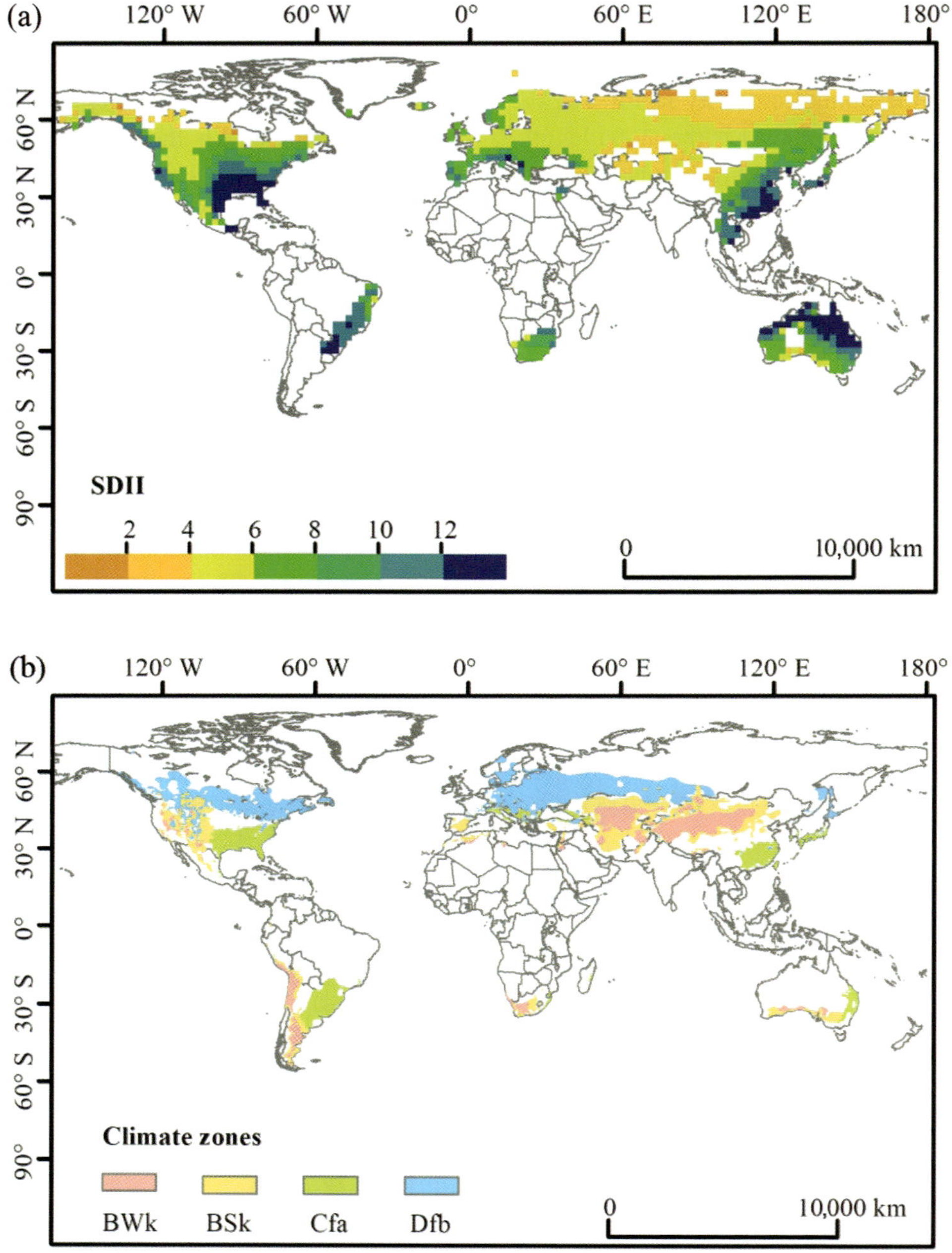

Figure 4.1 (a) Global averaged SDII from 1992 to 2015; (b) Climate zones of the Köppen–Geiger climate classification system of which R values are modified according to the method of Naipal et al. (2015)

$$K = 0.137 \times \left\{0.2 + 0.3\exp\left[-0.0256 \times \text{SAN}\left(1 - \frac{\text{SIL}}{100}\right)\right]\right\}\left(\frac{\text{SIL}}{\text{CLA} + \text{SIL}}\right)^{0.3} \times \left(1 - \frac{0.25 \times C}{C + \exp(3.72 - 2.95 \times C)}\right)\left(1 - \frac{0.7 \times \text{SN1}}{\text{SN1} + \exp(-5.51 + 22.9 * \text{SN1})}\right) \quad (4.3)$$

In Eq. (4.3), K is the soil erodibility, SAN is the percentage of sand content, SIL is the percentage of silt content, CLA is the percentage of clay content and C is the percentage of organic carbon content. In addition, SN1 = 1 – SAN/100. In this study, we used the topsoil properties of the ISRIC SoilGrids (Hengl et al., 2017) database at 250 m spatial resolution.

4.2.1.3 The Slope Length and Steepness Factor (LS)

The LS-factor represents the influence of the terrain topography on the sediment transport capacity of the overland flow (Wischmeier & Smith, 1978; Renard et al., 1997). The slope length factor (L) and slope steepness factor (S) are given by

$$L = \left(\frac{\lambda}{22.13}\right)^{m} \quad (4.4)$$

$$m = \beta(1 + \beta) \text{ and } \beta = (\sin\theta/0.0896)/[3 \times (\sin\theta)^{0.8} + 0.56] \quad (4.5)$$

$$S = 65.41 * \sin\theta^2 + 4.56 * \sin\theta + 0.065 \quad (4.6)$$

where λ is the slope length (m), m is the slope length index and θ is the slope gradient (°). We computed the morphometric parameters from the Global Multi-resolution Terrain Elevation Data 2010 (Danielson & Gesch, 2011).

4.2.1.4 The Land Cover and Management Factor (C)

The C-factor measures the combined effect of all interrelated cover and management variables. For this global assessment, we followed the suggestions from previous panEuropean and global studies (Panagos et al., 2015; Borrelli et al., 2017). Accordingly, two different approaches were undertaken to estimate C-factor values for agricultural and non-agricultural land. The agricultural and non-agricultural lands were classified from the global land cover maps of the European Space Agency (ESA) Climate Change Initiative Land Cover (CCI-LC). The ESA CCI-LC maps were released as the first time series of annual global land cover maps at 300 metres spanning a 24-year period, from 1992 to 2015 (Liu et al., 2018). For agricultural lands, we estimate the C-factor using statistical data describing the crop harvested area of each country. We used the Global Inventory Modelling and Mapping Studies (GIMMS) NDVI3g data for non-agricultural-lands.

4.2.2 Data Analyses

4.2.2.1 Spatial Analyses

We employed the *Spatial Analyst* tool from ArcGIS to identify and analyse the spatial distribution of water erosion across different countries and climate zones and within different ranges of altitude. The mean SLa values in corresponding units were consequently calculated.

4.2.2.2 Ordinary Least Squares Linear Regression

The linear regression method was introduced for trend analysis of water erosion vulnerability over the entire temporal domain (Li et al., 2017; Pan et al., 2018):

$$Y_i = \alpha + \beta T_i + \varepsilon \tag{4.7}$$

where Y_i is the Sla or R values, T_i is the year, ε is the residual, α is the intercept and β is the slope. A *t*-test was used to examine the significance level of the trend. In this study, a positive slope indicates an increase and a negative slope suggests a decrease; and we set $p < 0.05$ as the significance threshold.

4.3 Results

4.3.1 Model Validation

To generalize the model on a global scale, we validated the USLE model with SLa values estimated from observed data sampled from five major land uses across four climate zones (Xiong et al., 2019). We used the study by Xiong et al. (2019), which provides a comprehensive database representing 1,531 field runoff plots in 27 countries, to estimate the mean SLa values associated with different land use types (including bare land, cropland, grassland, shrubland and forest land) and four climate zones (including tropical, temperate, cold and arid). Based on the simulation by RUSLE during 1992–2015, we gathered the mean SLa values of bare lands, cropland, grasslands, shrublands and forest lands in tropical, temperate, cold and arid CZs, respectively. The simulations (USLE) of SLa values agreed well with the corresponding estimates ($R^2 = 0.877$, $n = 15$, $p < 0.001$; Figure 4.2). This suggests that RUSLE and its parameters used in this study are suitable to assess general water erosion conditions. Indirectly, the estimation from the model also matches the tendency of water erosion vulnerability.

4.3.2 Spatial Characteristics of Global Water Erosion Vulnerability

According to the method of Naipal et al. (2015), we modified the R factor in four climate zones (Figure 4.1(b)). The total area with increasing water erosion

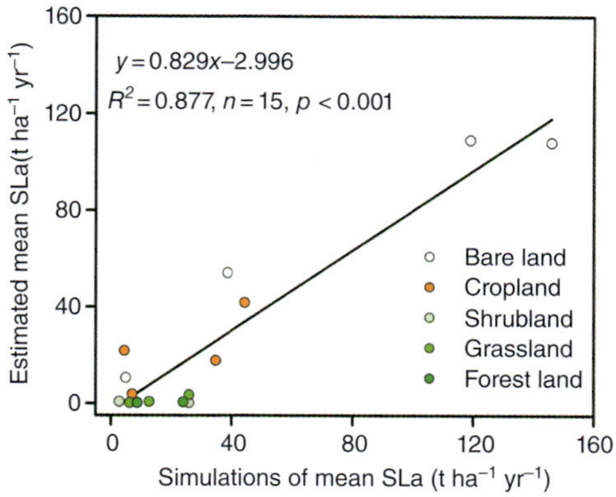

Figure 4.2 Validation of the USLE model. The simulations (USLE) and observations (Xiong et al., 2019) of mean SLa correspond to five land uses of four climate zones

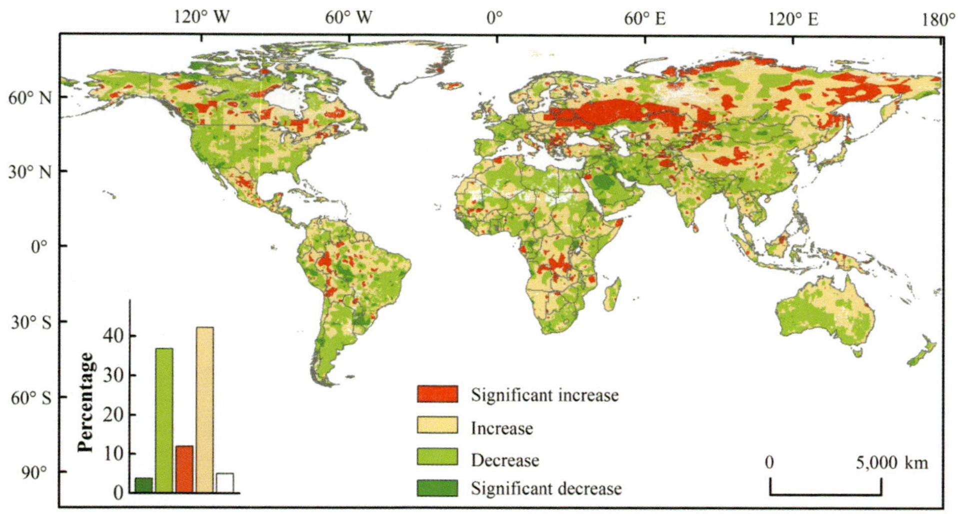

Figure 4.3 The change tendency of vulnerability of global water erosion from 1992 to 2015

vulnerability was larger than the total area with decreasing vulnerability (54.23 per cent versus 40.78 per cent, Figure 4.3). However, this contrast was larger in areas with statistically significant trends (11.96 per cent versus 3.84 per cent for *t*-test, Figure 4.3). The SLa of the areas with significantly increasing trends (11.96 per cent, $p < 0.05$) increased at a rate of $0.27\,\mathrm{t\,ha^{-1}\,yr^{-2}}$, and the areas were mainly distributed in cold (8.14 per cent) and arid (2.83 per cent) climate zone. The SLa of the areas with significantly decreasing trends (3.84 per cent, $p < 0.05$) decreased at a rate of $-0.85\,\mathrm{t\,ha^{-1}\,yr^{-2}}$, and the areas were mainly distributed in arid (1.33 per cent) climate zones.

4.3.3 Global Water Erosion Change Associated with Various Land Uses

The percentage of the area with statistically significant trends ($p < 0.05$) was 15.80 per cent (11.96 per cent increasing versus 3.84 per cent decreasing). We divided all areas with statistically significant trends into five different CZs, some of which, in polar CZ, were associated with a low change rate, according to the updated world KöppenGeiger climate classification (Peel et al., 2007). Therefore, we analysed the global water erosion change associated with various land uses in four CZs: tropical, arid, temperate and cold. A large proportion of the analysed areas were associated with the cold CZ (48.38 per cent), followed by arid (17.24 per cent) and tropical (10.03 per cent). The analysed areas were mainly composed of forest lands (37.62 per cent) and croplands (20.09 per cent), followed by bare lands (12.09 per cent), grasslands (8.57 per cent) and shrublands (5.23 per cent).

Figure 4.4 shows that the area with significantly increasing water erosion vulnerability were mainly forest lands (31.93 per cent), followed by croplands (15.75 per cent), grasslands (6.47 per cent) and bare lands (5.87 per cent). However, the mean SLa of forest lands only increased at a rate of 0.013 t ha^{-1} yr^{-2}. and the mean SLa of bare lands, croplands and grasslands increased at a rate of 0.954 t ha^{-1} yr^{-2}, 0.405 t ha^{-1} yr^{-2} and 0.228 t ha^{-1} yr^{-2}, respectively. While most croplands, bare lands and grasslands were in cold and arid CZs, only a few of these

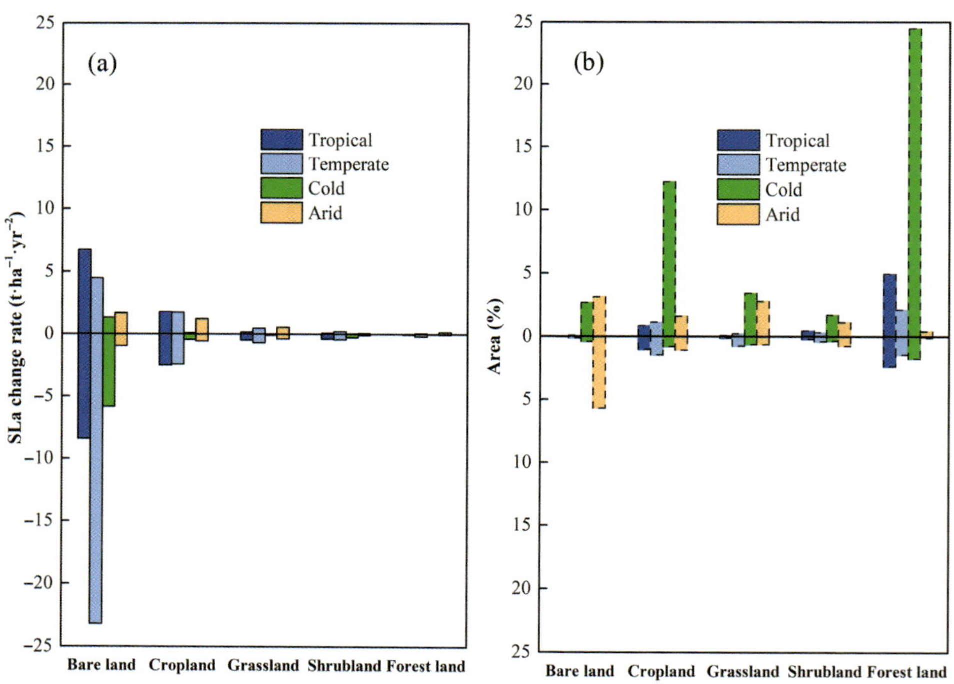

Figure 4.4 The change tendency of global water erosion ($p < 0.05$) associated with various land use and climate types. (a) The changing rate; (b) the corresponding area

were encountered in tropical and temperate CZs. The percentage of the bare lands, croplands and grasslands in cold CZs was 2.67 per cent, 12.2 per cent and 3.44 per cent, respectively, with an increasing SLa rate of 1.313 t ha^{-1} yr^{-2}, 0.083 t ha^{-1} yr^{-2} and 0.032 t ha^{-1} yr^{-2}, respectively. And the percentage of the bare lands, croplands and grasslands in arid CZs was 3.13 per cent, 1.6 per cent and 2.77 per cent, respectively. Each land use had a higher mean SLa rate in arid CZs than cold CZs. Areas showing significantly decreasing water erosion vulnerability include mainly bare lands (6.22 per cent), followed by forest lands (5.67 per cent) and croplands (4.33 per cent). The mean SLa of bare lands and croplands decreased at a rate of −1.547 t ha^{-1} yr^{-2} and −1.621 t ha^{-1} yr^{-2}, respectively.

For areas with a significant water erosion vulnerability, forest lands and grasslands were larger than croplands and bare lands. However, the change rate (both increasing and decreasing trends) of croplands and bare lands was higher than those observed in forest areas. Furthermore, most areas were in cold and arid CZs. These results may indicate that water erosion of bare lands and croplands in cold and arid CZs are more sensitive to climate change.

4.3.4 Effects of SDII on Water Erosion

We adjusted erosivity using the simple precipitation intensity index SDII in four climate zones (CZs). The lands in which the R factor was modified using the method of Naipal et al. (2015) correspond to about 5.60 per cent of the global area (Figure 4.1(b)). For the modified lands, the mean SLa increased at an average rate of 37.60×10^{-3} t ha^{-1} yr^{-2} using the method of Naipal et al. (2015) (Figure 4.5a). The total area with increasing SLa trends was larger than the total area with decreasing trends (61.20 per cent versus 38.80 per cent, Figure 4.2). This difference was more pronounced in areas with statistically significant trends (23.12 per cent versus 3.26 per cent for *t*-test, Figure 4.2). However, the mean SLa of the same lands increased at an average rate of 3.41×10^{-3} t ha^{-1} yr^{-2} with the R factor calculated using the method of Renard and Freimund (1994) (Figure 4.5(b)). The total area with increasing Sla trends was similar to the total area with decreasing trends (50.30 per cent versus 49.70 per cent, Figure 4.3). This difference was also small in areas with statistically significant trends (4.06 per cent versus 2.23 per cent for *t*-test, Figure 4.3). The results show that water erosion is not only dependent on the amount of rainfall; rainfall intensity can significantly impact water erosion.

4.4 Discussion

4.4.1 Impacts of Rainfall Change on Water Erosion

Based on the previous studies, climate change impacts on water erosion have been observed in various locations across the world (Li & Fang, 2016). The most direct

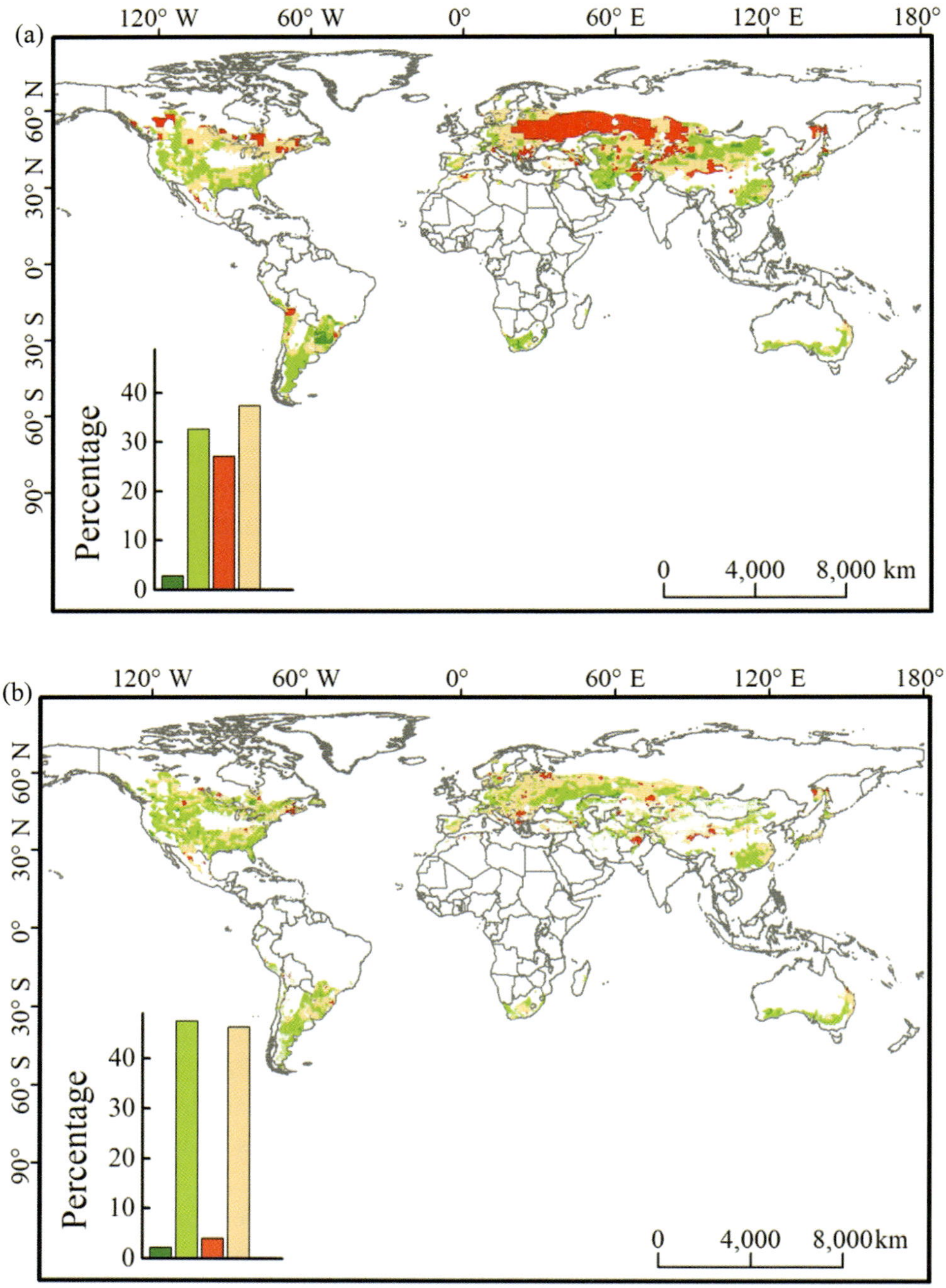

Figure 4.5 The change tendency of vulnerability of global water erosion from 1992–2015. (a) With R factor using the method of Naipal et al. (2015); (b) with R factor using the method of Renard and Freimund (1994)

impact of climate change on water erosion is the increasing precipitation, rainfall intensity and the frequency of extreme events (Nearing et al., 2004; Achite & Ouillon, 2016; Panagos et al., 2017). Lu et al. (2013) presented a quantitative estimate of changes in sediment loads (1990–2007) in response to climate change in eight large Chinese rivers, and found that every 1 per cent change in precipitation has led to a 2 per cent change in sediment loads and a 1.3 per cent change in water discharge (Lu et al., 2013). Previous studies showed that water erosion has increased with an increasing rainfall amount in the UK (Longfield & Macklin, 1999), Greece (Maas & Macklin, 2002) and China (Dai et al., 2008) over the past decades. Furthermore, García-Ruiz et al. (2015) suggested that erosion rates tend to increase with increasing mean annual precipitation across the world. However, González-Hidalgo et al. (2010) analysed the contribution of the largest events to suspended sediment transport across the United States and found that most precipitation indicators do not generate water erosion. In particular, the mean precipitation value is not a good predictor of water erosion (González-Hidalgo et al., 2010). Some studies indicated that the impacts of rainfall intensity are more important than rainfall amount (Shi & Wang, 2015; Tang et al., 2015; Wang et al., 2015; Yao et al., 2015) because the high erosivity of raindrops in extreme events can cause rain splash erosion (Mohamadi & Kavian, 2015). Furthermore, high-intensity rainfall may also lead to gully erosion (Poesen et al., 2003).

However, some studies showed that a positive relationship between rainfall and water erosion cannot always be identified (Li & Fang, 2016). Ruiz-Sinoga and Daiz (2010) reported an increase in erosion associated with a decrease in rainfall amount in Mediterranean study areas. Achite and Ouillon (2016) investigated the changes of temperature, precipitation, river runoff and sediment transport in the Mediterranean basin from 1970 to 2010 and found that water erosion was not significantly correlated with precipitation.

This present study found that the areas with significantly increasing and decreasing water erosion vulnerability were mainly associated with cold and arid CZs. The results suggest a difference in the levels of erosion rate among CZs. A 100 per cent increase in erosion rate in lands in cold CZs corresponds to a small increase in the same lands located in tropical and temperate CZs. Amplitude changes of the mean SLa were rather small in forest lands, grasslands and shrublands, which may be attributed to their low baseline erosion rate. Despite the highly variable response of water erosion rates to a changing rainfall associated with various land uses in different regions, our study demonstrates that a close link exists between rainfall change and water erosion. Furthermore, it shows that water erosion of bare lands and croplands in cold and arid CZs are more sensitive to climate change.

4.4.2 Impacts of Other Factors Associated with Climate Change

As for soil erosion, there is great heterogeneity in the trends across the world, as the processes linking climate and erosion are dependent on many variables (Serpa et al., 2015). Except for rainfall amount and intensity, rising temperatures also influence water erosion (Li & Fang, 2016). Lu et al. (2013) found that lower precipitation coupled with rising temperatures has significantly reduced the sediment loads delivered into the sea from the large rivers located in the semi-arid and arid climatic zones of China. However, Li et al. (2011) pointed out that, when the temperature exceeds the optimum temperature for plants, plant growth may decrease, which leads to a reduction of the vegetation coverage and, therefore, an aggravation of the losses of water and soil in China. Furthermore, water erosion can be affected indirectly by atmospheric CO_2 concentration (Xu, 2003), soil water content, evapotranspiration and plant cover (García-Ruiz et al., 2013). The increase in the atmospheric CO_2 concentrations may contribute to the growth of plants (García-Ruiz et al., 2013), which can increase canopy interception and reduce runoff and soil erosion (Guo et al., 2019). However, the effects of higher CO_2 concentration are not necessarily positive on water erosion (Li & Fang, 2016; Guo et al., 2019). Based on the above results, we cannot confirm whether soil erosion will increase with increased temperatures or CO_2 concentrations. One of the possible reasons is that the temperature or CO_2 concentrations only indirectly affect soil erosion by changing the vegetation cover in various ways.

Moreover, the freeze–thaw cycles may increase the potential for water erosion and results in higher sediment supply (Maetens, Vanmaercke, et al., 2012). Recent studies have estimated that spring snowmelt and frequent melting periods during winter are important erosion sources in cold climate regions (Guo et al., 2019). Severe soil erosions were found under the freeze–thaw cycle conditions of Russia (Komissarov & Gabbasova, 2014), Canada (Kieta et al., 2018) and northern China (Wu et al., 2018). In addition, a more indirect impact is the adoption of new land use and crop management practices that adapt to recent climate regimes (Li & Fang, 2016). These indirect impacts are complex (Mullan et al., 2012) and can affect land use and crop management in various ways (Zhang & Nearing, 2005).

4.4.3 Limitations of this Study

This work presents several limitations that should be addressed in future work. First, this study is constrained by the limitations of the predictive ability of the selected models, as already discussed for RUSLE. The cascade application of these models could further lead to a propagation of errors, culminating in the results obtained with RUSLE. These limitations could be minimized by improving model calibration and validation. The acquisition of additional water erosion data (in particular on field scale) would improve the reliability of the results presented in this research work. Second, we did not include gully erosion, despite it often being considered as one form of water erosion, along with rain splash erosion (García-Ruiz et al., 2015). Gully erosion is an important form of water erosion in many places around the world, including the United States, Australia, Europe and China (Li & Fang, 2016). Third, we only focus on the changes of rainfall on water erosion. Water erosion can be affected indirectly by temperature, CO_2 concentration and land use change. However, the combined direct and indirect impacts of changes on water erosion are very complex because most factors can cause positive and negative impacts in different contexts (Li & Fang, 2016). Despite the advances from the literature, it remains difficult to determine the extent to which some factors offset each other and predict the response of water erosion to climate change. Further studies would be required to bring insight into this complex phenomenon.

4.5 Conclusions

This study provided a systematic analysis of the global water erosion vulnerability change over the past 30 years based on RUSLE and ordinary least squares linear regression methods. The main conclusions of this work are as follows: (1) the total area with increasing water erosion vulnerability was larger than the total area with decreasing vulnerability; (2) water erosion was not only dependent on the amount of rainfall; rainfall intensity can significantly impact of water erosion; (3) there was great heterogeneity in the trends across the world. In areas showing a significant water erosion vulnerability, the change rate (both increasing and decreasing trends) of croplands and bare lands was higher than that of forest lands, shrublands and grasslands and (4) most areas with statistically significant trends were in cold and arid CZs, which means that the water erosion of bare lands and croplands in cold and arid CZs was more sensitive to climate change. The results provide a global view of the impacts of rainfall change on water erosion. Future studies that would account for the interactions of both direct and indirect impacts of climate change on water erosion might bring additional insight into this complex phenomenon.

References

Achite, M., & Ouillon, S. (2016). Recent changes in climate, hydrology and sediment load in the Wadi Abd, Algeria (1970–2010). *Hydrology and Earth System Sciences* 20(4): 1355–1372.

Bangash, R. F., Passuello, A., Sanchez-Canales, M., et al. (2013). Ecosystem services in Mediterranean river basin: Climate change impact on water provisioning and erosion control. *Science of the Total Environment* 458–460: 246–255.

Borrelli, P., Robinson, D. A., Fleischer, L. R., et al. (2017). An assessment of the global impact of 21st century land use change on soil erosion. *Nature Communications* 8: 2013.

Challinor, A. J., Watson, J., Lobell, D. B., et al. (2014). A meta-analysis of crop yield under climate change and adaptation. *Nature Climate Change* 4: 287–291.

Dai, S. B., Lu, X. X., Yang, S. L., & Cai, A. M. (2008). A preliminary estimate of human and natural contributions to the decline in sediment flux from the Yangtze River to the East China Sea. *Quaternary International* 186(1): 43–54.

Danielson, J. J., & Gesch, D. B. (2011). Global multi-resolution terrain elevation data 2010 (GMTED2010). *US Geological Survey Open-File Report 2011–1073*, 26 p.

Doetterl, S., Van Oost, K., & Six, J. (2012). Towards constraining the magnitude of global agricultural sediment and soil organic carbon fluxes. *Earth Surface Processes and Landforms* 37(6): 642–655.

García-Ruiz, J. M., Begueria, S., Lana-Renault, N., Nadal-Romero, E., & Cerda, A. (2017). Ongoing and emerging questions in water erosion studies. *Land Degradation & Development* 28(1): 5–21.

García-Ruiz, J. M., Beguería, S., Nadal-Romero, E., et al. (2015). A meta-analysis of soil erosion rates across the world. *Geomorphology* 239: 160–173.

García-Ruiz, J. M., Nadal-Romero, E., Lana-Renault, N., & Beguería, S. (2013). Erosion in Mediterranean landscapes: Changes and future challenges. *Geomorphology* 198: 20–36.

González-Hidalgo, J. C., Batalla, R. J., Cerdà, A., & de Luis, M. (2010). Contribution of the largest events to suspended sediment transport across the USA. *Land Degradation & Development* 21(2): 83–91.

Guo, Y., Peng, C., Zhu, Q., et al. (2019). Modelling the impacts of climate and land use changes on soil water erosion: Model applications, limitations and future challenges. *Journal of Environmental Management* 250: 109403.

Hengl, T., Mendes de Jesus, J., Heuvelink, G. B. M., et al. (2017). SoilGrids250m: Global gridded soil information based on machine learning. *PLoS One* 12(2): e0169748.

IPCC (2007). Climate change: Impacts, adaptation and vulnerability. Contribution of Working Group II to the Fourth Assessment Report of IPCC, Cambridge, UK.

Ito, A. (2007). Simulated impacts of climate and land-cover change on soil erosion and implication for the carbon cycle, 1901 to 2100. *Geophysical Research Letters* 34(9): 1–5.

Kieta, K. A., Owens, P. N., Lobb, D. A., Vanrobaeys, J. A., & Flaten, D. N. (2018). Phosphorus dynamics in vegetated buffer strips in cold climates: A review. *Environmental Reviews* 26(3): 255–272.

Komissarov, M. A., & Gabbasova, I. M. (2014). Snowmelt-induced soil erosion on gentle slopes in the southern Cis-Ural region. *Eurasian Soil Science* 47(6): 598–607.

Li, Y., Chen, B. M., Wang, Z. G., & Peng, S. L. (2011). Effects of temperature change on water discharge, and sediment and nutrient loading in the lower Pearl River basin based on SWAT modelling. *Hydrological Sciences Journal* 56(1): 68–83.

Li, Z., & Fang, H. (2016). Impacts of climate change on water erosion: A review. *Earth-Science Reviews* 163: 94–117.

Li, Z., Liu, W., Zhang, X., & Zheng, F. (2010). Assessing the site-specific impacts of climate change on hydrology, soil erosion and crop yields in the Loess Plateau of China. *Climatic Change* 105(1–2): 223–242.

Li, Z., Sun, R., Zhang, J., & Zhang, C. (2017). Temporal-spatial analysis of vegetation coverage dynamics in Beijing-Tianjin-Hebei metropolitan regions. *Acta Ecologica Sinica* 37(22): 7418–7426.

Litschert, S. E., Theobald, D. M., & Brown, T. C. (2014). Effects of climate change and wildfire on soil loss in the Southern Rockies Ecoregion. *Catena* 118: 206–219.

Liu, X., Yu, L., Si, Y., et al. (2018). Identifying patterns and hotspots of global land cover transitions using the ESA CCI Land Cover dataset. *Remote Sensing Letters* 9(10): 972–981.

Liu, Y., Fu, B., Liu, Y., Zhao, W., & Wang, S. (2019). Vulnerability assessment of the global water erosion tendency: Vegetation greening can partly offset increasing rainfall stress. *Land Degradation & Development* 30(9): 1061–1069.

Lobell, D. B., & Gourdji, S. M. (2012). The influence of climate change on global crop productivity. *Plant Physiology* 160(4): 1686–1697.

Longfield, S. A., & Macklin, M. G. (1999). The influence of recent environmental change on flooding and sediment fluxes in the Yorkshire Ouse basin. *Hydrological Processes* 13(7): 1051–1066.

Lu, S., Bai, X., Li, W., & Wang, N. (2019). Impacts of climate change on water resources and grain production. *Technological Forecasting and Social Change* 143: 76–84.

Lu, X. X., Ran, L. S., Liu, S., et al. (2013). Sediment loads response to climate change: A preliminary study of eight large Chinese rivers. *International Journal of Sediment Research* 28(1): 1–14.

Maas, G. S., & Macklin, M. G. (2002). The impact of recent climate change on flooding and sediment supply within a Mediterranean mountain catchment, southwestern Crete, Greece. *Earth Surface Processes and Landforms* 27(10): 1087–1105.

Maeda, E. E., Pellikka, P. K. E., Siljander, M., & Clark, B. J. F. (2010). Potential impacts of agricultural expansion and climate change on soil erosion in the Eastern Arc Mountains of Kenya. *Geomorphology* 123(3–4): 279–289.

Maetens, W., Poesen, J., & Vanmaercke, M. (2012). How effective are soil conservation techniques in reducing plot runoff and soil loss in Europe and the Mediterranean? *Earth-Science Reviews* 115(1–2): 21–36.

Maetens, W., Vanmaercke, M., Poesen, J., et al. (2012). Effects of land use on annual runoff and soil loss in Europe and the Mediterranean. *Progress in Physical Geography* 36(5): 599–653.

Mohamadi, M. A., & Kavian, A. (2015). Effects of rainfall patterns on runoff and soil erosion in field plots. *International Soil and Water Conservation Research* 3(4): 273–281.

Mukundan, R., Pradhanang, S. M., Schneiderman, E. M., et al. (2013). Suspended sediment source areas and future climate impact on soil erosion and sediment yield in a New York City water supply watershed, USA. *Geomorphology* 183: 110–119.

Mullan, D., Favis-Mortlock, D., & Fealy, R. (2012). Addressing key limitations associated with modelling soil erosion under the impacts of future climate change. *Agricultural and Forest Meteorology* 156: 18–30.

Naipal, V., Reick, C., Pongratz, J., & Van Oost, K. (2015). Improving the global applicability of the RUSLE model – Adjustment of the topographical and rainfall erosivity factors. *Geoscientific Model Development* 8(9): 2893–2913.

Nearing, M. A., Pruski, F. F., & O'Neal, M. R. (2004). Expected climate change impacts on soil erosion rates: A review. *Journal of Soil and Water Conservation* 59(4): 43–50.

Nunes, J. P., Seixas, J., & Keizer, J. J. (2013). Modeling the response of within-storm runoff and erosion dynamics to climate change in two Mediterranean watersheds: A multi-model, multi-scale approach to scenario design and analysis. *Catena* 102: 27–39.

Pan, N., Feng, X., Fu, B., et al. (2018). Increasing global vegetation browning hidden in overall vegetation greening: Insights from time-varying trends. *Remote Sensing of Environment* 214: 59–72.

Panagos, P., Borrelli, P., Meusburger, K., et al. (2015). Estimating the soil erosion cover-management factor at the European scale. *Land Use Policy* 48: 38–50.

Panagos, P., Borrelli, P., Meusburger, K., et al. (2017). Global rainfall erosivity assessment based on high-temporal resolution rainfall records. *Scientific Reports* 7: 4175.

Parajuli, P. B., Jayakody, P., Sassenrath, G. F., & Ouyang, Y. (2016). Assessing the impacts of climate change and tillage practices on stream flow, crop and sediment yields from the Mississippi River Basin. *Agricultural Water Management* 168: 112–124.

Paroissien, J. B., Darboux, F., Couturier, A., et al. (2015). A method for modeling the effects of climate and land use changes on erosion and sustainability of soil in a Mediterranean watershed (Languedoc, France). *Journal of Environmental Management* 150: 57–68.

Peel, M. C., Finlayson, B. L., & McMahon, T. A. (2007). Updated world map of the Köppen-Geiger climate classification. *Hydrology and Earth System Sciences* 11: 1633–1644.

Pham, T. N., Yang, D., Kanae, S., Oki, T., & Musiake, K. (2001). Application of RUSLE model on global soil erosion estimate. *Annual Journal of Hydraulic Engineering* 45: 811–816.

Poesen, J., Nachtergaele, J., Verstraeten, G., & Valentin, C. (2003). Gully erosion and environmental change: Importance and research needs. *Catena* 50(2–4): 91–133.

Rao, E., Ouyang, Z., Yu, X., & Xiao, Y. (2014). Spatial patterns and impacts of soil conservation service in China. *Geomorphology* 207: 64–70.

Renard, K. G., Foster, G. R., McCool, W. D. K., & Yoder, D. C. (1997). Predicting soil erosion by water: A guide to conservation planning with the Revised Universal Soil Loss equation. In *Agricultural Handbook* (p. 703). Washington, DC: US Department of Agriculture.

Renard, K. G., & Freimund, J. R. (1994). Using monthly precipitation data to estimate the R-factor in the revised USLE. *Journal of Hydrology* 157(1–4): 287–306.

Routschek, A., Schmidt, J., & Kreienkamp, F. (2014). Impact of climate change on soil erosion – A high-resolution projection on catchment scale until 2100 in Saxony/Germany. *Catena* 121: 99–109.

Ruiz-Sinoga, J. D., & Daiz, A. R. (2010). Soil degradation factors along a Mediterranean pluviometric gradient in Southern Spain. *Geomorphology* 118(3–4): 359–368.

Scherer, L., & Pfister, S. (2015). Modelling spatially explicit impacts from phosphorus emissions in agriculture. *International Journal of Life Cycle Assessment* 20(6): 785–795.

Serpa, D., Nunes, J. P., Santos, J., et al. (2015). Impacts of climate and land use changes on the hydrological and erosion processes of two contrasting Mediterranean catchments. *Science of the Total Environment* 538: 64–77.

Shabani, F., Kumar, L., & Esmaeili, A. (2014). Improvement to the prediction of the USLE K factor. *Geomorphology* 204: 229–234.

Sharpley, A. N., & Williams, C. F. (1990). EPIC: Erosion/Productivity Impact Calculator. U.S. Department of Agriculture Technical Bulletin No. 1768. 235 pp.

Shi, H., & Wang, G. (2015). Impacts of climate change and hydraulic structures on runoff and sediment discharge in the middle Yellow River. *Hydrological Processes* 29(14): 3236–3246.

Simonneaux, V., Cheggour, A., Deschamps, C., et al. (2015). Land use and climate change effects on soil erosion in a semi-arid mountainous watershed (High Atlas, Morocco). *Journal of Arid Environments* 122: 64–75.

Tang, J. L., Cheng, X. Q., Zhu, B., et al. (2015). Rainfall and tillage impacts on soil erosion of sloping cropland with subtropical monsoon climate: A case study in hilly purple soil area. *Journal of Mountain Science* 12(1): 134–144.

Van Oost, K., Quine, T. A., Govers, G., et al. (2007). The impact of agricultural soil erosion on the global carbon cycle. *Science* 318: 626–629.

Wang, J., Wang, K., Zhang, M., & Zhang, C. (2015). Impacts of climate change and human activities on vegetation cover in hilly southern China. *Ecological Engineering* 81: 451–461.

Wischmeier, W. H., & Smith, D. D. (1978). Predicting rainfall erosion losses. A guide to conservation planning. In *Agricultural Handbook* (p. 537). Washington, DC: US Department of Agriculture.

Wu, Y., Ouyang, W., Hao, Z., et al. (2018). Assessment of soil erosion characteristics in response to temperature and precipitation in a freeze-thaw watershed. *Geoderma* 328: 56–65.

Wuepper, D., Borrelli, P., & Finger, R. (2019). Countries and the global rate of soil erosion. *Nature Sustainability* 3(1): 51–55.

Xiong, M. Q., Sun, R. H., & Chen, L. D. (2019). A global comparison of soil erosion associated with land use and climate type. *Geoderma* 343: 31–39.

Xu, J. X. (2003). Sediment flux to the sea as influenced by changing human activities and precipitation: Example of the Yellow River, China. *Environmental Management* 31 (3): 328–341.

Yang, D. W., Kanae, S., Oki, T., Koike, T., & Musiake, K. (2003). Global potential soil erosion with reference to land use and climate changes. *Hydrological Processes* 17 (14): 2913–2928.

Yao, H., Shi, C., Shao, W., Bai, J., & Yang, H. (2015). Impacts of climate change and human activities on runoff and sediment load of the Xiliugou Basin in the Upper Yellow River. *Advances in Meteorology* 2015: 1–12.

Zhang, K. L., Shu, A. P., Xu, X. L., Yang, Q. K., & Yu, B. (2008). Soil erodibility and its estimation for agricultural soils in China. *Journal of Arid Environments* 72(6): 1002–1011.

Zhang, X. C., & Nearing, M. A. (2005). Impact of climate change on soil erosion, runoff, and wheat productivity in central Oklahoma. *Catena* 61(2–3): 185–195.

Zhang, X. C. J. (2012). Cropping and tillage systems effects on soil erosion under climate change in Oklahoma. *Soil Science Society of America Journal* 76(5): 1789–1797.

Zhang, Y., Hernandez, M., Anson, E., et al. (2012). Modeling climate change effects on runoff and soil erosion in southeastern Arizona rangelands and implications for mitigation with conservation practices. *Journal of Soil and Water Conservation* 67 (5): 390–405.

5

Water Erosion and Its Controlling Factors in the Anthropocene

XIMENG XU AND QIUHONG TANG

5.1 Introduction

Soil erosion is one of the most important global environment hazards which has influenced the survival and development of human society and restricted social sustainable development worldwide. Soil erosion not only has on-site impacts on local natural resources (loss of valuable soil and water resources), environment (nutrient loss, ecological imbalance) and social economy, but has also off-site influences (reservoir sedimentation and non-point chemical pollution) on related adjacent regions. According to the Food and Agriculture Organization of the United Nations (FAO), over 33 per cent of the Earth's soils are already degraded and 90 per cent could become degraded by 2050. Soil erosion can lead to an up to 50 per cent loss in crop yields, while it can take up to 1,000 years to produce 2–3 centimetres of soil (www.un.org/en/observances/world-soil-day). The theme of World Soil Day 2019 was 'Stop Soil Erosion, Save Our Future', which aims to raise awareness of the importance of sustaining healthy ecosystems and human wellbeing. Without the implementation of sustainable measures to mitigate soil erosion, soil fertility will continue to be adversely affected at an alarming rate, threatening global food supplies and food safety.

Soil erosion can be divided by the erosion agent, such as water erosion, wind erosion, tillage erosion, gravitational erosion, freeze–thaw erosion and complex erosion. Water erosion is one of the most important erosion agents. According to the previous estimation, 75 billon tons of soil are removed from agriculture land by water and wind every year, which causes $400 billion economic lost in agricultural production (Pimentel et al., 1995; Montanarella, 2015). More recently, Borrelli et al. (2017) conducted a global soil erosion

This research is supported by the International Partnership Program of the Chinese Academy of Sciences (131A11KYSB20180034) and National Natural Science Foundation of China (41790424), the Strategic Priority Research Program of Chinese Academy of Sciences (XDA20060402).

estimation by using the Revised Universal Soil Loss Equation (RUSLE) model applied to data acquired from different sources and concluded that 35.9 billion tons of soil were eroded by water in 2012 worldwide.

The USLE/RUSLE models have been extensively used during the last 40 years in 109 countries (Alewell et al., 2019), having originated from the United States Department of Agriculture (USDA) based on thousands of controlled studies on field plots and small watersheds since 1930. The empirical equation underlying the USLE/RUSLE models is:

$$A = RKLSCP \tag{5.1}$$

where A ($Mg\ ha^{-1}\ yr^{-1}$) is the annual average soil erosion, R ($MJ\ mm\ h^{-1}\ ha^{-1}\ yr^{-1}$) is the rainfall-runoff erosivity factor, K ($Mg\ h\ MJ^{-1}\ mm^{-1}$) is the soil erodibility factor, L (dimensionless) is the slope length factor, S (dimensionless) is the slope steepness factor, C (dimensionless) is the land cover and management factor and P (dimensionless) is the soil conservation or prevention practices factor.

Recent research conducted by the Joint Research Centre of European Commission estimated a 33.7 million tons reduction in global agri-food production due to lost soil and an annual loss of $8 billion provoked by water erosion (Sartori et al., 2019). Thus, water erosion is a serious problem threatening the agriculture sustainability and food security globally. Some measures like no-till agriculture are beneficial for sustainable agriculture since they lead to reduced erosion rates to levels similar to those associated with soil production rates (Montgomery, 2007). Soil erosion rates are strongly related to social economies and the less developed economies experience the higher soil erosion rates (Sartori et al., 2019). Water erosion is predicted to increase greatly in Sub-Saharan Africa, South America and southeast Asia (Borrelli et al., 2017), which makes the control of soil erosion harder and requires actions targeting these relatively less developed countries. Wuepper et al. (2020) quantitively described the differences of soil erosion rates and control practice effectiveness among countries worldwide by utilizing a spatial regression discontinuity design and high-resolution global data. They concluded that soil erosion has a surprisingly large country effect and the average rate of soil erosion among countries may vary by about $1.4\,t\,ha^{-1}\,yr^{-1}$, which can be explained by differences in the agricultural characteristics of the countries. Many countries have sufficient resources and capabilities to mitigate soil erosion problems, which calls for the international cooperation on soil erosion control practices.

Under global change, rapid changes of the climate characteristics and earth surface landscape may lead to accelerated water erosion processes and erosion of higher magnitude. Mitigating the effect of future climate change on water erosion requires a better understanding and comprehensive review of the water erosion processes at stake and the water erosion control practices.

5.2 Water Erosion Processes and their Relations to Climate Change

Water erosion processes are driven by a series of complex sub-processes on the hillslope and in the river channel. Figure 5.1 shows the typical processes, including raindrop splash, inter-rill and rill erosion, ephemeral gully erosion and permanent gully erosion on the hillslope scale, which are discussed in the following sections.

These erosion processes can be vertically distributed along the slope in various regions like the Loess plateau of China. The formation of a vertical zoning system and different zones are associated with different erosion types. This distinction helps design and implement optimal erosion control practices.

5.2.1 Raindrop Splash Erosion

Raindrop splash refers to the process of raindrop fall on the soil surface which causes a movement of the soil particles. It is composed of dispersion, detachment and transition processes of soil particles. In natural conditions, soil particles can be displaced as high as 1.5 metres vertically by raindrop slash (Ryżak et al., 2015) and can be carried more than 5 metres horizontally by wind (Erpul et al., 2009). The raindrop splash process often happens before or at the beginning of the surface runoff generating process, which is the start of water erosion processes. Raindrop splash force can change the topsoil structure, increase surface runoff turbulence, and enhance the dispersion and transportation ability of runoff. Splashed soil particles can block soil pores and infiltration and increase surface runoff and erosion rates.

Raindrop impact was first suggested as a dominant agent of water erosion by Ellison (1944) using a simulated experimental investigation. Since then, many researchers have conducted abundant simulated and natural on-site experiments to

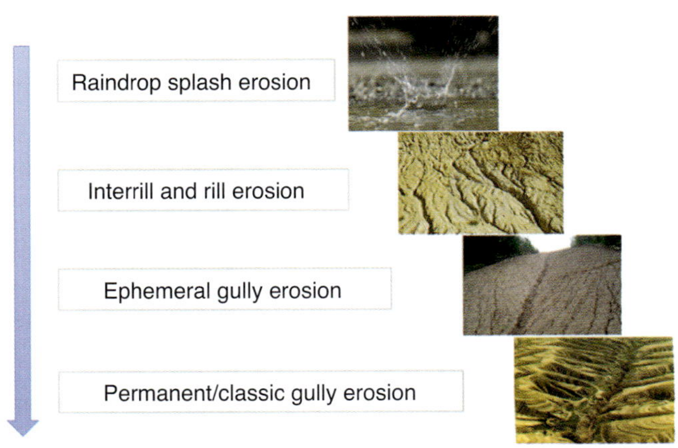

Figure 5.1 Typical water erosion processes on the hillslope scale

investigate and understand the physical process and mechanisms of raindrop splash erosion. Research work has mainly attempted to better understand raindrop characteristics and factors of splash erosion, measure and model splash erosion and mitigate the effects of splash erosion (Fernández-Raga et al., 2017).

Raindrop kinetic energy greatly determines the splash erosion process and is closely related to the physical properties of raindrop (size, shape, terminal velocity, momentum). Thus, raindrop kinetic energy and raindrop characteristics have long been recognized as important parameters to quantify the splash erosion and many empirical formulas were developed to quantify theses parameters and their relationship. Raindrops with diameters smaller than 0.25 mm are called small raindrops usually shaped as a ball. When raindrops gradually increase in size, the shape tends to be a spindle. When raindrop diameter is larger than 5.5 mm, the shape of the raindrop becomes instable and the raindrop may break or merge during the falling process. Every rainfall contains raindrops with different diameters. The median volume diameter (D_{50}) is usually used to quantify the raindrop distribution in one rainfall event and it is found to be related with rainfall intensity (Laws & Parsons, 1943)

$$D_{50} = aI^b \tag{5.2}$$

where I is the rainfall intensity (mmh^{-1}), a and b are the constants varied with rainfall pattern and geological regions.

Raindrop terminal velocity is determined by the raindrop diameter (d, mm) and shape. Considering the raindrop shape change, raindrop terminal velocity $(V_m, \text{m s}^{-1})$ can be expressed as:

$$V_m = 0.495 \times 10^{\left(\sqrt{28.32+6.524lg0.1d-(lg0.1d)^2}-3.665\right)} \quad (d < 1.9\,\text{mm})$$
$$V_m = (17.2 - 0.844d)\sqrt{0.1d} \quad (d \geq 1.9\,\text{mm}) \tag{5.3}$$

Then, raindrop kinetic energy can be easily calculated based on the raindrop terminal velocity by accumulating all raindrops in one rainfall event. Because of the difficulty of accurately measuring raindrop diameter and the fuzzy calculation process, Wischmeier and Smith (1960) developed an equation describing the relationship between rainfall kinetic energy (E, J) and rainfall intensity $(I, \text{mm h}^{-1})$ based on the investigation data by Laws and Parsons (1943):

$$E = 1.213 + 0.89\,\text{lg}I \tag{5.4}$$

While developing the Universal Soil Loss Equation, Wischmeier (1959) comprehensively analysed observation data from 8,000 natural plots and abundant simulated rainfall experiment data. He finally selected a combined parameter, that is, the product of total raindrop kinetic energy (E) and maximum 30-minutes rainfall intensity (I_{30}), to quantify the soil erosion rate response to rainfall, called the rainfall erosivity factor ($R = EI_{30}$) in Eq. (5.1).

It is hard to acquire high precision rainfall graph data on a large scale like the whole river basin, a country scale or even a global scale, let alone compute the correct rainfall kinetic energy data. Thus, many empirical relationships between rain intensity and kinetic energy with coarser resolution rainfall data were used to quantify the rainfall erosivity around the world. However, these empirical methods are not very reliable, since they depend on rain gauges located near cities and at low altitude (Nyssen et al., 2005). They exclude measures of rainfall erosivity in mountainous environments and in sparsely populated regions and therefore fail to generalize results to larger areas (Poesen, 2018).

Under climate change, the number of extreme rainfall intensity events and excess raindrop kinetic energy are expected to increase and to impact wider areas and therefore increase water erosion. Nearing et al. (2004) reported that the rate of soil erosion and runoff would increase by 1.7-times that of the expected increase of rainfall amount. While an estimation for Europe showed that rainfall intensity will increase on average by 18 per cent in 2050 compared with 2010, the expected changes are expected to exhibit a large spatial heterogeneity (Panagos et al., 2017). Zhang and Nearing (2005) also made a prediction about the relationship between climate change and soil erosion in central Oklahoma, suggesting an increase of the frequency of large rainfall events. They predicted an increasing future erosion rate despite an expected decrease in annual precipitation. This apparent paradox can be explained by a higher concentration of extreme rain events expected in specific months. A higher frequency of high-intensity rain events and larger variability of daily rainfall amount is expected during fall when the soils are more exposed to erosion risk. Future work on rainfall erosivity that accounts for the complex role of extreme rainfall events on erosion should lead to more accurate water erosion predictions.

5.2.2 Inter-rill and Rill Erosion Processes

Along with raindrop impact, rainfall may also detach soil indirectly by means of runoff in rills and inter-rill areas. At the beginning stage of the runoff generation process, the fraction of the rainfall which does not infiltrate (soak into) the soil flows downslope because of gravity. At this initial stage, the flow depth is very shallow and may be dispersed by the topographic relief without a fixed flow path. The overland flow has lower velocity, kinetic energy and power to detach soil particles. If rain continues, the flow depth will eventually overtop the microtopographic depressions. The overland flow has higher velocity and power to detach soil particles at this stage. Then, the soil surface will offer lower and preferential flow paths in lower areas. Gradually, small rills will form on the slope, which illustrate the end of surface erosion and start the linear erosion features toward gully erosion. When rill is formed on the slope, the soil erosion can then be divided into sheet flow erosion in inter-rill areas and concentrated flow erosion in rills.

Horton et al. (1934) pioneered research on the laminar sheet flow and introduced the equations used for river flow to the sheet flow on the hillslope surface. Flow depth, velocity and shear stress were all calculated on the slope surface accordingly. It is well recognized that the soil detachment rate by runoff is highly related to the difference between flow shear stress and critical shear stress of soil particles (Wischmeier, 1962; Meyer & Monke, 1965; Foster et al., 1977).

For shallow sheet flow in inter-rill areas, open channel flow theory is usually applied although it does not suffice to comprehend the complexity observed in real conditions. Sediment detachment and transport ability of shallow sheet flow are both influenced by the runoff and raindrop impacts, which are not accounted for in the open channel flow theory.

For rill flow and rill erosion estimation, rill initiation is determined by the flow hydraulic parameters and soil conditions. Rills can only occur when the hydraulic parameters exceed the critical values and these hydraulic parameters are composed of the flow rates, Froude number, flow velocity and flow shear stress (Meyer et al., 1975; Rauws & Covers, 1988; Sun et al., 2013). Among these parameters, critical shear stress is a good indicator to quantitatively describe the threshold conditions of the soil to trigger rill initiation. Thus, the new generation and most widely applied process-based soil erosion model, named the Water Erosion Prediction Project (WEPP), includes the critical shear stress component. In this framework, the estimation of the rill detachment capacity and shear stress is used to calculate the sediment transportation capacity in the rill. The rill erosion rate (D_f, kg s^{-1} m^{-2}) can be calculated as:

$$
\begin{aligned}
D_f &= D_c(1 - G/T_c) \\
D_c &= K_r(\tau_f - \tau_c) \\
T_c &= k_t \tau_f^{3/2}
\end{aligned}
\tag{5.5}
$$

where T_c is the sediment transport capacity in the rill (kg s^{-1} m^{-2}), K_r is the rill soil erodibility parameter (s m^{-1}), τ_f is the flow shear stress acting on the soil (Pa), τ_c is the critical shear stress of the soil (Pa) and k_t is a transport coefficient.

Net soil erosion rates can be obtained by the steady state sediment continuity equation, which is also considered as the basic governing equation of the erosion process (Foster & Meyer, 1972), used in WEPP by Nearing et al. (1989):

$$\mathrm{d}G/\mathrm{d}x = D_f + D_i \tag{5.6}$$

where G(kg s^{-1} m^{-1}) is sediment load, x (m) is distance downslope, D_i is the inter-rill erosion rate (kg s^{-1} m^{-2}) and D_f is the rill erosion rate (kg s^{-1} m^{-2}).

Accurate calculation of the inter-rill erosion and rill erosion rates allows for quantifying hillslope soil erosion. However, many influencing factors on these processes remained not well understood. These include rainfall, soil conditions, topographic characteristics and soil management practices. Under an accelerated global warming and increasing extreme weather events scenario, rill erosion and inter-rill erosion processes need to be better quantified to protect soil resources from degradation.

5.2.3 Ephemeral Gully Erosion

As the typical rill channels increase in size and become large gullies, the processes of migration of headcut and gravitational collapse of channel walls become more important. In most situations, erosion by concentrated flow in a gully is the main agent of erosion by water. Based on the criteria of whether a gully can be temporarily removed by normal tillage practice, gullies can be divided into ephemeral gullies and permanent/classical gullies.

The Soil Science Society of America (2008) defined ephemeral gullies as small channels eroded by concentrated flow that can be easily filled by normal tillage only to be reformed again in the same location, which agrees well with the definition of mega-rill erosion and shallow gully erosion phenomena observed around the world (Zhu, 1956; Foster, 1986). Unlike rill erosion, that can be permanently removed by tillage, an ephemeral gully cannot be removed completely and will leave a mark on the land surface, which initiates an irreversible linear erosion. Ephemeral gully erosion affects various areas across the world. Typical observation and estimation show that ephemeral gully erosion contributes to 30–100 per cent of total soil erosion in actively eroding areas in the United States (Bennett et al., 2000). It contributes to more than half of the total erosion observed in the loess belt and Mediterranean region of Western Europe (Casalí et al., 1999; Capra & La Spada, 2015) and more than 80 per cent of the total erosion observed in areas composed of highly erodible soils in Australia (Hancock & Evans, 2006). Its contribution to total erosion ranges from 30 per cent to 70 per cent or sometimes to more than 90 per cent on the Loess Plateau (Cheng et al., 2007) as well as in the northeastern black soil region in China (Y. G. Zhang et al., 2007).

Because of its large contributions to soil erosion, a large number of researches have focused on the influencing factors associated with ephemeral gullies, however little attention has been devoted to quantitative studies of this phenomenon. Hydrologic processes like rainfall (Capra et al., 2009; Han et al., 2017), overland flow and subsurface flow (Wilson, 2011; Xu et al., 2019) are the main driving forces governing the formation and growth of ephemeral gullies. The ephemeral gully has long played an important role in connecting hillslope soil erosion and more permanent land morphology like gullies, canyons and river systems. Topography, soil type and land use/management are also important factors exhibiting high spatial heterogeneity, but a systematic quantitative assessment of their impacts has not been conducted worldwide (Xu et al., 2017). Future climate change might considerably change the hydrologic processes and land use by altering rainfall and temperature patterns spatially and temporarily and, therefore, increase the risk of an ephemeral gully becoming a permanent one, which deserve more research attention.

5.2.4 Permanent Gully Erosion

The development of a permanent gully is composed of two stages. The first stage is defined as the 'active stage', with quick incise and growth in gully length and width, while occupying only about 5 percent of the gully life. The second stage is considered as the 'stable stage', which occupies most of the gully life period. Sediment detachment and deposition at the second stage are both weak and gully morphology is kept constant. A permanent gully, especially one that is still in its active stage, is one of the most important sediment sources around the world (Poesen et al., 2003; Wu et al., 2018). Soil loss due to active stage gullies is supposed to be half the sediment yield of gully-erosion-dominated watersheds in China (X. H. Zhang et al., 2007) and its contribution to erosion represents more than 30 per cent on average in some European loess regions (Casalí et al., 2009). Gullies destroy cultivated land, limit mechanical tillage and incise roads as well as buildings, which are harmful to social life and economic development. Extensive gully distribution can destruct ecosystem functions and sustainability (Stavi et al., 2010). The development of modern technologies has increased quantitative research on gullies, which has contributed to better understanding of the role and interactions of the factors associated with the development of gullies and modelling in varied eco-environments (Poesen et al., 2003; Valentin et al., 2005; Castillo & Gómez, 2016). These research paths can be categorized as follows: (1) gully development process, (2) technologies for gully accurate measuring and mapping, (3) influencing factors of gully erosion and (4) gully erosion prediction model (Zheng et al., 2016). For the gully development process research, it is generally recognized that gully development is dominated by three main processes (i.e., headcut retreat/migration/advance, sidewall/bank expansion and vertical bed incision) (Martınez-Casasnovas et al., 2004; Simon & Rinaldi, 2006; Allen et al., 2018; Qin et al., 2019). Subsurface hydrological processes are also found to be closely related to the gully formation and development (Bocco, 1991; Nichols et al., 2016).

For the technologies of quantifying the gully morphology and position, a wide range of new techniques (e.g., 3D laser scanning, high precision GPS, photogrammetry) have been developed and applied over the last decades. High resolution DEMs are generated using these methods in a wide spatial range, which greatly helps in understanding the gully erosion process, influencing factors and erosion prediction modelling. The main factors that influence gully erosion have been widely investigated and sampled across the world. This includes topographical (drainage area, slope length, gradient and shape), environmental (rainfall, overland runoff, groundwater, vegetation) and artificial factors. Due to the stochastic nature of its development, gully erosion predictions have been challenging. A few empirical equations can be applied to estimate gully erosion based on the morphological characteristics of the gullies. Promising modelling attempts include a dynamic gully model (DIMGUL) and static gully model (STAGUL) developed by Sidorchuk (1999). These models need to be tested in a wider range of environmental conditions. Research on gully erosion has made important progress in recent decades. However, the mechanisms behind the process of gully erosion remain poorly understood and need further investigation. The rate of gully erosion varies according to the development stage under various environmental backgrounds. Predictive models of gully erosion that account for these variations should offer a more accurate picture of gully erosion and hence provide decision-making tools to enhance control practice settlements.

The spatial distribution of the risk associated with gully development is greatly influenced by rapid environmental changes and atmospheric extreme events included by climate change. Recently, machine learning and artificial intelligence are promising approaches (Tien Bui et al., 2019) that could help predict the rapid increase of gully erosion susceptibility and help protect valuable land resources from the degradation caused by gullies.

5.2.5 From Hillslope to River System

During the abovementioned erosion processes in a hillslope system, sediment can be detached by the force of raindrop and concentrated flow, then transported to the river system, and finally carried through rivers to the coastline and delta regions. At the catchment scale, sediment yield in a river system is not only determined by the foregoing stated water erosion process on the hillslope, but also by the mass movement and river channel erosion (Vanmaercke et al., 2012).

Mass movements like landslide and mudflow are discrete events triggered by many factors such as extreme rainfall, land use change and seismic activity. Landslide connected to the drainage flow path can greatly contribute to the sediment yield at the outlet of watershed. Landsides that happen at an incised river channel can provide abundant sediment sources for the river flow. The sediment yield downstream is determined by the sediment transport capacity of the river flow (Ouimet et al., 2007). In the catchment affected by mass movement, sediment yield is not significantly related to the watershed area (de Vente et al., 2006). The influencing factors of mass movement are mainly rainfall and land use change, whose variabilities are highly uncertain under future climate change. An increase of extreme climate events might pose great challenges to the mass movement process and the total sediment yield of the entire catchment area.

River channel erosion, similar to bank collapse and incision, can also contribute to the sediment yield (Walling & Collins, 2005). Especially in urbanizing watersheds, stream channel erosion has long been suspected to be the major contributor to sediment yield. For example, in the San Diego Creek of southern California, stream channel erosion furnishes about two thirds of the total sediment yield, which highlights the importance of channel stabilization (Trimble, 1997). The sediment detached by the river is dependent on the stream power and energy of the river flow, which is positively related to the catchment area (Birkinshaw & Bathurst, 2006). The future impact of climate change on the river channel erosion and riverbank stability is still not well understood. Most research works have been focused on the sediment yield and transport in the system under a changing climate (Li & Fang, 2017; Shrestha & Wang, 2018). Thus, further assessments of river channel erosion variations under climate risk are needed.

5.2.6 Other Agents Coupled with the Water Erosion Process

Complex erosion generated by multiple agents has been observed in many regions worldwide. The magnitude and duration of erosion is increased compared to that of a single agent erosion and therefore leads to more serious problems. Common agents include wind, gravitational and freeze–thaw, which interact with each other under different environmental contexts. Complex erosion can be divided into two types based on the chronology of the agent interaction. The first type is observed when two or more agents occur and interact simultaneously with each other. For example, gravity and water agents can simultaneously contribute to the sediment yield in landslide and debris flow events (Guerra et al., 2017; Hürlimann et al., 2019). The second type is observed when agents occurred alternatively. In this case, one agent detaches and transports soil particles providing the erosion materials required for other agents to initiate their process. This second type of

Figure 5.2 Soil surface conditions after freeze–thaw and wind impact

complex erosion can be found in the Mollisol region of northeast China, which exhibits a soil erosion process influenced by water, wind and freeze–thaw. Freeze–thaw processes during winter and wind in spring can change the soil structure characteristics and increase surface soil erodibility. Figure 5.2 shows the soil surface condition after freeze–thaw and wind impact during the early rainy season. Abundant loose soil materials on the surface can be detached and transported by overland flow that usually occurs in summer, which greatly increases the amount of soil erosion (Xu et al., 2010; Zhao et al., 2012).

Complex soil erosion, especially water erosion, is a global environmental problem. Arid and semi-arid areas are often exposed to complex erosion driven by water and wind, which affects 23.74 million km^2, representing about 17.5 per cent of land surface worldwide (Bullard & McTainsh, 2003). Typical studies on the Loess Plateau and Yellow River of China have shown that the most intensive erosion magnitude did not happen in the regions where water erosion was severest. The severest erosion occurs at the belt zone, which is dominated by water–wind complex erosion. The rainfall amount in this region is only 400 millimetres per year, but alternative occurrences of water and wind erosion are found in this region. The Pisha stone area exhibits a typical water–wind–freeze–thaw complex erosion, in which the erosion modulus can reach up to 30,000–40,000 t km^{-2} yr^{-1}. Although the Pisha stone area only occupies 2 per cent of the total area of the Yellow River basin, coarse sand from this region can occupy 25 per cent of the sediment deposition in the lower reach of Yellow River, which threatens human and property security downstream (Wang et al., 2007). Intense complex erosion in these special areas increases the volatility and vulnerability of the environment, making these regions the spatial centres of the high erosion modulus and sediment concentration. Many questions about the interaction mechanisms of different agents in complex erosion remain unanswered (Song et al., 2006). Furthermore, the literature lacks quantitative research work that assesses the contribution of each agent (Zhang et al., 2019). Future studies on complex erosion processes are therefore of crucial importance, especially in the context of rapid global change (Poesen, 2018).

Soil erosion is closely related to the climate conditions that vary among regions. It has been shown that complex erosion varies under different climate regimes. Future climate change is expected to have a considerable impact on the risk and magnitude of complex erosion. Variations in the temperature may change the freeze–thaw circulations in high latitude and altitude areas. As snow packs are sensitive to temperature increases, an increase in temperatures may lead to faster snow melting, higher snowmelt runoff and an increase in the magnitude of soil erosion (Grossi et al., 2007; Shen et al., 2015).

5.3 Water Erosion Control Practices

In the widely applied USLE-type models, the cover management (C) factor and support practice (P) factor are model parameters quantifying the impacts of land cover, soil management and soil conservation practices on the soil loss (Renard et al., 1997; Alewell et al., 2019). These models are useful tools to help control water erosion. However, they have been calibrated using US data and the determination of C and P factors for applications outside the United States would require a global assessment in various environments (Maetens et al., 2012; Buendia et al., 2016; Xiong et al., 2018). An alternative empirical approach suggests a soil loss equation called the Chinese Soil Loss Equation (CSLE), which has been used for national water resource surveys in China. In this model, C and P factors are replaced by the biological-control (B), engineering-control (E) and tillage (T) factors. Biological-control practices include forest or grass plantation for reducing runoff and soil loss. Engineering-control practices refer to changes of topography to reduce runoff and soil loss by engineering construction like terraces or check dams. Tillage practices are measures taken using farmland equipment (Liu et al., 2002).

5.3.1 Biological Techniques

The most natural and effective way to prevent erosion is planting vegetation. The canopy of frost and crop can prevent raindrops falling on bare land and reduce the rainfall energy acting on the soil surface, hence reducing the soil loss rates. Rainfall energy reduction can greatly contribute to the reduction of soil erosion because of the reduction of flow hydraulic parameters that occur after the elimination of raindrop impact (Kinnell, 2005; An et al., 2012). Under the canopy, the litter layer reduces runoff, delays the beginning of runoff and significantly reduces soil loss (Li et al., 2014). The root system of plants can grip soil and contribute to stabilizing the inclined terrain, thus effectively preventing the excess movement of soil throughout the ground (Gyssels & Poesen, 2003; Reubens et al., 2007).

Biological techniques include various management measures like grass cover/strip, afforestation, crop cover/strip, hedge barriers and mulching. The most common and widespread biological techniques are vegetative strips, cover (mostly grass) and agroforestry systems. Perennial grasses, shrubs or trees are often used to lay on the slope based on various spatial distribution patterns to optimize the reduction of soil erosion (Hurni et al., 2008).

C. Chen et al. (2019) showed a striking greening pattern in China and India using recent satellite data (2000–2017). Their results showed that China alone contributed to 25 per cent of the global net increase in leaf area, which represents only 6.6 per cent of the vegetation cover worldwide. The greening trend in China has been mainly driven by the grain-for-green programme (GGP) implemented by the Chinese government in 1999 to control the severe problem of soil erosion in China. An evaluation of the effectiveness of this programme showed that both runoff and soil erosion significantly decreased because of the increase of the forest land cover. Compared with the situation in 1998–2002, runoff and soil erosion were reduced by 18 per cent and 45 per cent, respectively, which illustrates the important role of forests in controlling soil erosion and producing ecological benefits (Deng et al., 2012).

Mulching practices using vegetation residues is also an important biological technique for reducing soil water erosion (Prosdocimi et al., 2016). Mulching can break up the raindrop energy, prevent splash erosion, slow down the flow and decrease flow energy as well as erosion rate. Straw mulch improves soil physicochemical properties, such as bulk density, porosity and aggregate stability (Jordán et al., 2010), thus decreases soil erodibility and soil erosion. Mulching practice has been spatially distributed as a strip following contour lines, acting like a vegetated buffer strip (Heede, 1990; Xu et al., 2017).

5.3.2 Engineering Methods

Engineering-control practices refer to changes of the topography to reduce runoff and soil loss by engineering constructions such as terraces and check dams, to name but two (Liu et al., 2002).

Terracing has long been considered as an effective method for soil and water conservation in many mountainous regions across the world (Wei et al., 2016). First, terraces can decrease water erosion by intercepting the overland flow and weakening rainfall-runoff erosivity (Chen et al., 2012). Second, terraces can increase the soil water content and nutrients to support plant growth, which contribute to the reduction of soil erosion. Third, special structures like ridges and embankments in the terrace system can also contribute to soil conservation (Wei et al., 2016). The effectiveness of terracing is determined by many factors, such as climate, soil properties, land use and socioeconomic status. However, their effects have not been well quantified yet. Due to their considerable effects on soil erosion, terraces must be considered in predictive models on soil erosion. In USLE-type soil erosion predictive models, the effects of terraces are simulated by changes in the P factor. Moreover, a process-based algorithm was also developed and incorporated into the Soil and Water Assessment Tool model (SWAT), which obtained satisfactory simulation results (Shao et al., 2013). Chen et al. (2017) evaluated the water erosion control effects of terraces in China and confirmed the significant and positive effects of terraces in erosion control, in which bench terraces and those associated with tree crops obtained the largest erosion reduction rates.

Check dam is another widespread engineering method to stabilize ephemeral and permanent gullies or streams, which can reduce gully incision and upstream sediment running into the downstream river channel (Castillo et al., 2007; Wang et al., 2011; Abedini et al., 2012). Figure 5.3 shows a typical stone check dam in a

Figure 5.3 A stone check dam for controlling the gully erosion in a permanent channel and an earth check dam for creating fertilized farmlands

permanent stream to control gully erosion in northeast China, an earth check dam and a dam cropland with maize on the Loess Plateau where gullies are widely spread. Check dams have been used to prevent flooding, reduce sediment yield and create fertilized farmland on the Loess Plateau (Xu et al., 2004; Y. Chen et al., 2019). More than 110,000 check dams have been constructed, which intercept 21 billion m^3 of sediments, creating 320,000 hectares of dam croplands that benefit the environment and economy (Wang et al., 2011; Jin et al., 2012). However, the dam construction poses various problems: (1) many dams have been fully silted or present a high risk of breaking and therefore require further protective measures; (2) the choice of spatial distribution of dams in watersheds with dam system remains unclear and (3) future climate change and afforestation might significantly decrease soil erosion which may lead to dams losing their primary function of intercepting sediments and creating farmland. These problems may need further discussion. Moreover, the effectiveness of check dams and their impacts on the channel morphology have been examined and reported in various environments, such as in a semiarid Mediterranean catchment located in Murcia, Spain (Castillo et al., 2007), in a tropical catchment called the Ulu Kinta Basin (Abedini et al., 2012), in Tigray, located in northern Ethiopia (Nyssen et al., 2004), and in the Italian Alpine mountain river channel (Lenzi & Comiti, 2003).

5.3.3 Tillage Practices

Tillage practices encompass all measures using farmland equipment. The difference between engineering and tillage is that the latter does not change the topography and is only applied on farmland (Liu et al., 2002). According to the Conservation Technology Information Center, the tillage system can be classified into three categories associated with the values of residue percentage after planting: (1) conventional tillage, (2) reduced tillage and (3) conservation tillage (CTIC, 2002). In the last few decades, intensive agricultural practices have been applied worldwide, which have caused severe soil erosion on bare farmland in some previous problem-free areas like northwest Europe and southeast Asia. Thus, the corresponding conservation tillage practice is more relevant for future erosion control in these intensive agriculture regions (Derpsch, 2003).

Conservation tillage methods are currently viewed as effective measures to protect soil from erosion worldwide (Holland, 2004; Blevins et al., 2018; Seitz et al., 2019). Conservation tillage refers to any tillage system that maintains at least 30 per cent of cover on the soil surface, such as reduced tillage, no tillage, strip tillage, ridge tillage and mulch tillage. It can reduce agronomy costs and increase the efficiency of machine entry (Van den Putte et al., 2010), while reduced tillage also requires minimizing the frequency of tillage practices for crop growth. Both the reduced tillage and no tillage can maintain a relatively high cover throughout the whole year and protect the soil surface as well as its structure. These measures provide large benefits by preventing soil erosion and improving soil quality, especially when crop rotation and permanent soil cover are applied simultaneously (Goebes et al., 2014; Pittelkow et al., 2015). Zhang and Nearing (2005) evaluated the potential impacts of climate change on soil erosion as well as the role of tillage practice in central Oklahoma. They concluded that the conservation tillage system would be effective in combating soil erosion under projected climate scenarios. However, conservation tillage may induce soil compaction, whose environmental effects need further understanding, especially under future climate change conditions (Peigné et al., 2018).

Ridge tillage is also a common agronomic practice widely used across the world, which prepares a seedbed that is elevated above the natural land surface (Lal, 1990). Ridge direction can affect the soil erosion processes in these kinds of farmland greatly. Contour ridging has long been recognized for effectively intercepting the runoff and reducing erosion (Stevens et al., 2009). However, contour ridges may breach and foster ephemeral gully erosion under extreme rainfall conditions (Hatfield et al., 1998; Xu et al., 2018). The stability of a contour ridge can be affected by ridge distance, ridge height, row grade, slope gradient, soil properties of the ridge and rainfall characteristics (Liu et al., 2014), whose effects in various soil, climatic and environment conditions have not been quantified yet.

5.4 Water Erosion under Future Climate Risk

It is a general concern that both rainfall pattern and temperature are affected by global change, which is driven by an increase in greenhouse gas concentration worldwide due to anthropogenic activities (Trenberth et al., 2007). Temperatures are expected to continue rising in the future, while there is no consensus with regard to the future trends in precipitation (Segura et al., 2014). These projected variations of rainfall and temperature will be very likely to affect water erosion directly and indirectly.

5.4.1 Direct Impacts of Climate Change: Change in Rainfall Erosivity

The most direct impact of climate change on the soil erosion is the increasing precipitation, rainfall intensity and frequency of extreme events (Nearing et al., 2004). Water erosion is mostly induced by rainfall events with a short duration and high intensity, which is mainly observed in semiarid regions (Han et al., 2017). Changes in the rainfall characteristics will be the most important climate risk factor affecting water erosion rates (Nearing, 2001; Pruski & Nearing, 2002), because intense rainfall is likely to be more frequent in the coming decades according to the predictions made by the IPCC (2007). Under climate change, rising temperatures may also reduce snowfall in favour of an increase in rainfall, which is likely to contribute to increased rainfall erosivity.

Rain storms have become more intense than before (Nearing et al., 2005). In both North and South America and northern Europe, the annual rainfall amount among years has shown little variation over time. However, the frequency of intense rainfall has increased, which has led to higher water erosion rates (SWCS, 2003). The model results show that rainfall erosivity is projected to increase significantly under different climate change scenarios. The Soil and Water Conservation Society estimated that the frequency of intense rainfall events would increase by 20–60 per cent during the twenty-first century. Remember that rainfall erosivity is the product of rainfall energy and the maximum rainfall intensity over a timeframe. A higher number of intense rainfalls may lead to more serious erosion, even if the total rainfall amount does not vary. By 2050, rainfall erosivity in Europe is expected to increase on average by 18 per cent compared to 2010, while these changes are expected to exhibit a high spatial heterogeneity (Panagos et al., 2017). Rainfall erosivity in the central part of India is expected to increase by 21–202 per cent in 2080 with regard to past data (Mondal et al., 2016). In northeast China, rainfall erosivity is expected to increase at a higher rate than that of precipitation (Zhang et al., 2010). Mitigating the increasing water erosion risk under future climate change requires the development of management strategies adapted to face future climate risks. Further quantitative research on the impact of climate change on rainfall erosivity should bring light into the complex processes associated with the changes in rainfall depth, rain intensity, number of rainy days and, in some regions, the ratio of rain-to-snow (Poesen, 2018).

5.4.2 Indirect Climate Change Impact: Change in Land Use and Land Cover

Climate change also indirectly affects water erosion, by changing the land use and land cover, including vegetation cover variations and cropping system management. Land use and land cover changes are strongly associated with rising temperatures, increasing concentrations of carbon dioxide and rising moisture by precipitation (Li & Fang, 2016). Most current studies consider the direct effects of climate on rainfall erosivity without considering its indirect effects on erosion via changes on the land surface (Maeda et al., 2010; Litschert et al., 2014; Ramos & Martínez-Casasnovas, 2015).

Land cover changes induced by climate change are related to rainfall and temperature variations, which may have larger impacts on soil erosion under a future changing climate (Yang et al., 2003; Nearing et al., 2005). The water erosion response to different land use can vary greatly among regions under future climate change (Nunes et al., 2013; Xiong et al., 2019). Nunes et al. (2013) compared the soil erosion response to land use and climate change of two Mediterranean watersheds in central Portugal by using multiple models on various scales. They found that erosion decreased in forest and vineyard lands while it increased on farmland. A study comparing two contrasting Mediterranean catchments, with a humid and a dry environment, showed evidence that permanent vegetation cover lowered the erosion in the humid site but increased the erosion in the dry site (Serpa et al., 2015). Another study in a semi-arid mountainous watershed of Morocco showed that land use change has a larger impact on erosion than the future precipitation changes. These findings have important consequences for future land use management strategies (Simonneaux et al., 2015). Liu et al. (2019) assessed the vulnerability of global water erosion tendency during 1982–2015 and concluded that vegetation can offset the water erosion stress (i.e., rainfall erosivity stress) caused by climate change. Improving the vegetation growth condition can globally reduce the erosion threat in regions highly sensitive to water erosion.

Crop system change is also related to land use which needs adjustment in both crop rotation and varieties according to the future changes in temperature and precipitation. Temperature increases will change the freeze and thaw processes in the temperate zone, which will influence the soil erosion from farmlands. Crop systems can affect the erosion rates in farmland through the crop canopy cover and residues left after harvest. The percentage of crop canopy coverage determines the soil loss rate, which is highly related to the crop types, crop growth stage and plant density (O'Neal et al., 2005). Crop residue can intercept rainfall, decrease surface sealing, affect the equilibrium of runoff rate and significantly reduce the inter-rill and rill erosion. It is therefore an effective measure that requires a better utilization and management in a future facing the risk of erosion on farmland (Alberts & Neibling, 1994). O'Neal et al. (2005) predicted the erosion rate variations influenced by crop management changes under future climate change scenarios in the Midwest USA. The results showed that the combined effect of soybeans would have higher positive yield effects than maize. Another study in the same region indicated that a continuous replacement of soybeans with corn (corn–soybean rotation) may exacerbate the erosion rate (Southworth et al., 2000). Paroissien et al. (2015) investigated the impact of global warming on the growth of inter-row grass used to prevent soil erosion and concluded that erosion may increase. However, the effects of the harvesting date on the erosion rates remain poorly understood. Thus, future research is still needed to better understand the complex mechanisms associated with the impacts of crop management impacts on water erosion under future climate risk (Li & Fang, 2016).

Land use changes induced by human activities are a primary cause of soil erosion variations. Borrelli et al. (2017) highlighted the effects of global land use changes on the soil erosion. The results showed that human activities have induced cropland expansion and forestland reduction which could be the main cause of accelerated soil erosion. Human activities like infrastructure construction projects and soil and water conservation practices can also strongly influence water erosion worldwide. The construction processes of infrastructure, including highways, train lines, coal mining and buildings, create land degradation on construction sites. A wide range of studies have been conducted to evaluate the impacts of these intensive human construction activities on soil erosion (Evans et al., 1997; Dong et al., 2012; Meng et al., 2012; Bakr et al., 2015; Navarrohevia et al., 2016). Research has mainly focused on the cut slope and spoil soils which need urgent protection during and after the construction processes. Human activities are not always accelerating soil erosion; soil and water conservation practices can also prevent soil loss to some extent. Thus, the future impact of human activities on water erosion risk will be determined by human society. In many countries, soil conservation laws have been implemented to reduce water erosion, while some countries facing quick urbanization and extensive development are still neglecting this issue. Thus, future cooperation among countries on water erosion controls are needed.

5.5 Future Research Needs

Water erosion processes and mechanisms need to be better understood to build more accurate process-based predictive models that can provide support to implement optimal erosion control practices and sustainable development in various ecosystems. The research should put emphasis on: (1) natural and anthropogenic controls on water erosion processes and their interactions; (2) the connections between hillslope water erosion processes and the river channel sediment transportation; (3) better utilization or development of process-based erosion predictive models on a watershed scale for regional erosion control, since capturing variations of climatic and environmental conditions at large spatial scales are not suitable to implement scientific erosion control practices and (4) water erosion prediction at large scale with high resolution.

Global change (both natural and anthropogenic factors) has affected the water cycle and may increase water-related hazard risk (Tang, 2020). However, few studies have investigated the impact of global change on water erosion at large scales. Large-scale studies require the availability of global data and cooperation among relevant countries. The General Circulation Models (GCMs) appear the most direct and efficient way to project climate characteristics to predict future water erosion under different climate scenarios, which can be combined with dynamical downscaling approaches to investigate phenomena on large scales. The most recent model, CMIP6, released by the Working Group on Coupled Modelling (WGCM) or future updates should provide more relevant data to predict water erosion under climate risk on large scales.

Socioeconomical evaluations of water erosion related to climate change are key in modern society. The rapid expansion of cities and increase in population and GDP can lead to serious soil erosion problems, which can be observed in many parts across the world. Finding an equilibrium between economic development and soil protection is challenging and requires an accurate quantification of economic values of soil and water resources. For example, the identification of the economic benefits from soil and water conservation practices at different scales can help build resilient socioecological systems.

References

Abedini, M., Said, M. A. M., & Ahmad, F. (2012). Effectiveness of check dam to control soil erosion in a tropical catchment (The Ulu Kinta Basin). *Catena* 97: 63–70.

Alberts, E. E., & Neibling, W. H. (1994). Influence of crop residues on water erosion. In P. W. Unger (ed.), *Managing Agricultural Residues* (pp. 19–39). Boca Raton, FL: Lewis Publishers.

Alewell, C., Borelli, P., Meusburger, K., & Panagos, P. (2019). Using the USLE: Chances, challenges and limitations of soil erosion modelling. *International Soil and Water Conservation Research* 7(3): 203–225.

Allen, P. M., Arnold, J. G., Auguste, L., White, J., & Dunbar, J. (2018). Application of a simple headcut advance model for gullies. *Earth Surface Processes and Landforms* 43(1): 202–217.

An, J., Zheng, F., Lu, J., & Li, G. (2012). Investigating the role of raindrop impact on hydrodynamic mechanism of soil erosion under simulated rainfall conditions. *Soil Science* 177(8): 517–526.

Bakr, N., Elbana, T. A., Arceneaux, A., et al. (2015). Runoff and water quality from highway hillsides: Influence compost/mulch. *Soil & Tillage Research* 150: 158–170.

Bennett, S. J., Casalí, J., Robinson, K. M., & Kadavy, K. C. (2000). Characteristics of actively eroding ephemeral gullies in an experimental channel. *Transactions of the ASAE* 43(3): 641.

Birkinshaw, S. J., & Bathurst, J. C. (2006). Model study of the relationship between sediment yield and river basin area. *Earth Surface Processes and Landforms* 31(6): 750–761.

Blevins, R. L., Lal, R., Doran, J. W., Langdale, G. W., & Frye, W. W. (2018). Conservation tillage for erosion control and soil quality. In F. J. Pierce, & W. W. Frye (eds.), *Advances in Soil and Water Conservation* (pp. 51–68). Boca Raton, FL: Routledge.

Bocco, G. (1991). Gully erosion: Processes and models. *Progress in Physical Geography* 15(4): 392–406.

Borrelli, P., Robinson, D. A., Fleischer, L. R., et al. (2017). An assessment of the global impact of 21st century land use change on soil erosion. *Nature Communications* 8(1): 2013.

Buendia, C., Bussi, G., Tuset, J., et al. (2016). Effects of afforestation on runoff and sediment load in an upland Mediterranean catchment. *Science of the Total Environment* 540: 144–157.

Bullard, J. E., & McTainsh, G. H. (2003). Aeolian-fluvial interactions in dryland environments: Examples, concepts and Australia case study. *Progress in Physical Geography* 27(4): 471–501.

Capra, A., & La Spada, C. (2015). Medium-term evolution of some ephemeral gullies in Sicily (Italy). *Soil and Tillage Research* 154: 34–43.

Capra, A., Porto, P., & Scicolone, B. (2009). Relationships between rainfall characteristics and ephemeral gully erosion in a cultivated catchment in Sicily (Italy). *Soil and Tillage Research* 105(1): 77–87.

Casalí, J., Giménez, R., & Bennett, S. (2009). Gully erosion processes: Monitoring and modelling. *Earth Surface Processes and Landforms* 34(14): 1839–1840.

Casalí, J., López, J. J., & Giráldez, J. V. (1999). Ephemeral gully erosion in southern Navarra (Spain). *Catena* 36(1–2): 65–84.

Castillo, C., & Gómez, J. A. (2016). A century of gully erosion research: Urgency, complexity and study approaches. *Earth-Science Reviews* 160: 300–319.

Castillo, V. M., Mosch, W. M., García, C. C., et al. (2007). Effectiveness and geomorphological impacts of check dams for soil erosion control in a semiarid Mediterranean catchment: El Cárcavo (Murcia, Spain). *Catena* 70(3): 416–427.

Chen, C., Park, T., Wang, X., et al. (2019). China and India lead in greening of the world through land-use management. *Nature Sustainability* 2(2): 122–129.

Chen, D., Wei, W., & Chen, L. (2017). Effects of terracing practices on water erosion control in China: A meta-analysis. *Earth-Science Reviews* 173: 109–121.

Chen, S. K., Liu, C. W., & Chen, Y. R. (2012). Assessing soil erosion in a terraced paddy field using experimental measurements and universal soil loss equation. *Catena* 95: 131–141.

Chen, Y., Wu, J., Wang, H., et al. (2019). Evaluating the soil quality of newly created farmland in the hilly and gully region on the Loess Plateau, China. *Journal of Geographical Sciences* 29(5): 791–802.

Cheng, H., Zou, X., Wu, Y., et al. (2007). Morphology parameters of ephemeral gully in characteristics hillslopes on the Loess Plateau of China. *Soil & Tillage Research* 94 (1): 4–14.

CTIC (Conservation Technology Information Center) (2002). *National Crop Residue Management Survey*. West Lafayette, IN: CTIC.

Deng, L., Shangguan, Z. P., & Li, R. (2012). Effects of the grain-for-green program on soil erosion in China. *International Journal of Sediment Research* 27(1): 120–127.

Derpsch, R. (2003). Conservation tillage, no-tillage and related technologies. In G. T. Luis, B. José, M. V. Armando, & H. C. Antonio (eds.), *Conservation Agriculture* (pp. 181–190). Dordrecht: Springer.

Dong, J., Zhang, K., & Guo, Z. (2012). Runoff and soil erosion from highway construction spoil deposits: A rainfall simulation study. *Transportation Research Part D-transport and Environment* 17(1): 8–14.

Ellison, W. D. (1944). Studies of raindrop erosion. *Agricultural Engineering* 25(4): 131–136.

Erpul, G., Gabriels, D., Cornelis, W. M., Samray, H., & Guzelordu, T. (2009). Average sand particle trajectory examined by the Raindrop Detachment and Wind-driven Transport (RD-WDT) process. *Earth Surface Processes and Landforms* 34(9): 1270–1278.

Evans, K. G., Loch, R. J., Aspinall, T. O., & Bell, L. C. (1997). Laboratory rainfall simulator studies of selected open-cut coal mine overburden spoils from Central Queensland. *Soil Research* 35(1): 15–30.

Fernández-Raga, M., Palencia, C., Keesstra, S., et al. (2017). Splash erosion: A review with unanswered questions. *Earth-Science Reviews* 171: 463–477.

Foster, G. R. (1986). Understanding ephemeral gully erosion. In National Research Council, Board on Agriculture (ed.), *Soil Conservation: Assessing the National Research Inventory* (pp. 90–118). Washington, DC: National Academy Press.

Foster, G. R., & Meyer, L. D. (1972). A closed-form soil erosion equation for upland areas. In Shen, H. W. (ed.), *Sedimentation: Symposium to Honor Professor H.A. Einstein* (pp. 12.1–12.19). Fort Collins, CO: Einstein.

Foster, G. R., Meyer, L. D., & Onstad, C. A. (1977). An erosion equation derived from basic erosion principles. *Transactions of the ASAE* 20(4): 678–682.

Goebes, P., Seitz, S., Geißler, C., et al. (2014). Momentum or kinetic energy – How do substrate properties influence the calculation of rainfall erosivity? *Journal of Hydrology* 517: 310–316.

Grossi, C. M., Brimblecombe, P., & Harris, I. (2007). Predicting long term freeze–thaw risks on Europe built heritage and archaeological sites in a changing climate. *Science of the Total Environment* 377(2–3): 273–281.

Guerra, A. J. T., Fullen, M. A., Jorge, M. D. C. O., Bezerra, J. F. R., & Shokr, M. S. (2017). Slope processes, mass movement and soil erosion: A review. *Pedosphere* 27(1): 27–41.

Gyssels, G., & Poesen, J. (2003). The importance of plant root characteristics in controlling concentrated flow erosion rates. *Earth Surface Processes and Landforms: The Journal of the British Geomorphological Research Group* 28(4): 371–384.

Han, Y., Zheng, F. L., & Xu, X. M. (2017). Effects of rainfall regime and its character indices on soil loss at loessial hillslope with ephemeral gully. *Journal of Mountain Science* 14(3): 527–538.

Hancock, G. R., & Evans, K. G. (2006). Gully position, characteristics and geomorphic thresholds in an undisturbed catchment in northern Australia. *Hydrological Processes* 20(14): 2935–2951.

Hatfield, J. L., Allmaras, R. R., Rehm, G. W., & Lowery, B. (1998). Ridge tillage for corn and soybean production: Environmental quality impacts. *Soil and Tillage Research* 48(3): 145–154.

Heede, B. H. (1990). *Vegetation Strips Control Erosion in Watersheds* (Vol. 499). Fort Collins, CO: USDA Forest Service, Rocky Mountain Forest and Range Experiment Station.

Holland, J. M. (2004). The environmental consequences of adopting conservation tillage in Europe: Reviewing the evidence. *Agriculture, Ecosystems & Environment* 103(1): 1–25.

Horton, R. E., Leach, H. R., & Van Vliet, R. (1934). Laminar sheet-flow. *Eos, Transactions American Geophysical Union* 15(2): 393–404.

Hürlimann, M., Coviello, V., & Bel, C. (2019). Debris-flow monitoring and warning: Review and examples. *Earth-Science Reviews* 199(1): 102981.

Hurni, H., Herweg, K., Portner, B., & Liniger, H. (2008). Soil erosion and conservation in global agriculture. In A. K. Braimoh, & P. L. G. Vlek (eds.), *Land Use and Soil Resources* (pp. 41–71). Dordrecht: Springer.

Intergovernmental Panel on Climate Change (IPCC) (2007). *Climate Change: Impacts, Adaptation and Vulnerability. Contribution of Working Group II to the Forth Assessment Report of IPCC*. Cambridge: IPCC.

Jin, Z., Cui, B., Song, Y., et al. (2012). How many check dams do we need to build on the Loess Plateau? *Environmental Science & Technology* 46(16): 8527–8528.

Jordán, A., Zavala, L. M., & Gil, J. (2010). Effects of mulching on soil physical properties and runoff under semi-arid conditions in southern Spain. *Catena* 81(1): 77–85.

Kinnell, P. I. A. (2005). Raindrop-impact-induced erosion processes and prediction: A review. *Hydrological Processes: An International Journal* 19(14): 2815–2844.

Lal, R. (1990). Ridge-tillage. *Soil & Tillage Research* 18(2–3): 107–111.

Laws, J. O., & Parsons, D. A. (1943). The relation of raindrop-size to intensity. *Eos, Transactions American Geophysical Union* 24(2): 452–460.

Lenzi, M. A., & Comiti, F. (2003). Local scouring and morphological adjustments in steep channels with check-dam sequences. *Geomorphology* 55(1–4): 97–109.

Li, X., Niu, J., & Xie, B. (2014). The effect of leaf litter cover on surface runoff and soil erosion in Northern China. *PloS One* 9(9): e107789.

Li, Z., & Fang, H. (2016). Impacts of climate change on water erosion: A review. *Earth-Science Reviews* 163: 94–117.

Li, Z., & Fang, H. (2017). Modeling the impact of climate change on watershed discharge and sediment yield in the black soil region, northeastern China. *Geomorphology* 293: 255–271.

Litschert, S. E., Theobald, D. M., & Brown, T. C. (2014). Effects of climate change and wildfire on soil loss in the Southern Rockies Ecoregion. *Catena* 118: 206–219.

Liu, B., Zhang, K., & Xie, Y. (2002). An empirical soil loss equation. In *12th International Soil Conservation Organization Conference*, May 2002 (pp. 21–25). Beijing, China: Tsinghua University Press.

Liu, Q. J., Shi, Z. H., Yu, X. X., & Zhang, H. Y. (2014). Influence of microtopography, ridge geometry and rainfall intensity on soil erosion induced by contouring failure. *Soil and Tillage Research* 136: 1–8.

Liu, Y., Fu, B., Liu, Y., Zhao, W., & Wang, S. (2019). Vulnerability assessment of the global water erosion tendency: Vegetation greening can partly offset increasing rainfall stress. *Land Degradation & Development* 30(9): 1061–1069.

Maeda, E. E., Pellikka, P. K., Siljander, M., & Clark, B. J. (2010). Potential impacts of agricultural expansion and climate change on soil erosion in the Eastern Arc Mountains of Kenya. *Geomorphology* 123(3–4): 279–289.

Maetens, W., Poesen, J., & Vanmaercke, M. (2012). How effective are soil conservation techniques in reducing plot runoff and soil loss in Europe and the Mediterranean? *Earth-Science Reviews* 115(1–2): 21–36.

Martınez-Casasnovas, J. A., Ramos, M. C., & Poesen, J. (2004). Assessment of sidewall erosion in large gullies using multi-temporal DEMs and logistic regression analysis. *Geomorphology* 58(1–4): 305–321.

Meng, L., Feng, Q., Wu, K., & Meng, Q. (2012). Quantitative evaluation of soil erosion of land subsided by coal mining using RUSLE. *International Journal of Mining Science and Technology* 22(1): 7–11.

Meyer, L. D., & Monke, E. J. (1965). Mechanics of soil erosion by rainfall and overland flow. *Transactions of the ASAE* 8(4): 572–577.

Meyer, L. D., Foster, G. R., & Romkens, M. J. M. (1975). Source of soil eroded by water from upland slopes. In ARS-S-40 (ed.), *Present and Prospective Technology for Predicting Sediment Yields and Sources* (pp. 177–189). Oxford, MS: US Department of Agriculture, National Sedimentation Laboratory.

Mondal, A., Khare, D., & Kundu, S. (2016). Change in rainfall erosivity in the past and future due to climate change in the central part of India. *International Soil and Water Conservation Research* 4(3): 186–194.

Montanarella, L. (2015). Agricultural policy: Govern our soils. *Nature* 528: 32–33.
Montgomery, D. R. (2007). Soil erosion and agricultural sustainability. *Proceedings of the National Academy of Sciences* 104(33): 13268–13272.
Navarrohevia, J., Limafarias, T. R., De Araujo, J. C., Osoriopelaez, C., & Pando, V. (2016). Soil erosion in steep road cut slopes in Palencia (Spain). *Land Degradation & Development* 27(2): 190–199.
Nearing, M. A. (2001). Potential changes in rainfall erosivity in the US with climate change during the 21st century. *Journal of Soil and Water Conservation* 56(3): 229–232.
Nearing, M. A., Foster, G. R., Lane, L. J., & Finkner, S. C. (1989). A process-based soil erosion model for USDA-Water Erosion Prediction Project technology. *Transactions of the ASAE* 32(5): 1587–1593.
Nearing, M. A., Jetten, V., Baffaut, C., et al. (2005). Modeling response of soil erosion and runoff to changes in precipitation and cover. *Catena* 61(2–3): 131–154.
Nearing, M. A., Pruski, F. F., & O'Neal, M. R. (2004). Expected climate change impacts on soil erosion rates: A review. *Journal of Soil and Water Conservation* 59(1): 43–50.
Nichols, M. H., Nearing, M., Hernandez, M., & Polyakov, V. O. (2016). Monitoring channel head erosion processes in response to an artificially induced abrupt base level change using time-lapse photography. *Geomorphology* 265: 107–116.
Nunes, J. P., Seixas, J., & Keizer, J. J. (2013). Modeling the response of within-storm runoff and erosion dynamics to climate change in two Mediterranean watersheds: A multi-model, multi-scale approach to scenario design and analysis. *Catena* 102: 27–39.
Nyssen, J., Vandenreyken, H., Poesen, J., et al. (2005). Rainfall erosivity and variability in the Northern Ethiopian highlands. *Journal of Hydrology* 311(1–4): 172–187.
Nyssen, J., Veyret-Picot, M., Poesen, J., et al. (2004). The effectiveness of loose rock check dams for gully control in Tigray, northern Ethiopia. *Soil Use and Management* 20(1): 55–64.
O'Neal, M. R., Nearing, M. A., Vining, R. C., Southworth, J., & Pfeifer, R. A. (2005). Climate change impacts on soil erosion in midwest United States with changes in crop management. *Catena* 61(2–3): 165–184.
Ouimet, W. B., Whipple, K. X., Royden, L. H., Sun, Z., & Chen, Z. (2007). The influence of large landslides on river incision in a transient landscape: Eastern margin of the Tibetan Plateau (Sichuan, China). *Geological Society of America Bulletin* 119 (11–12): 1462–1476.
Panagos, P., Ballabio, C., Meusburger, K., et al. (2017). Towards estimates of future rainfall erosivity in Europe based on REDES and WorldClim datasets. *Journal of Hydrology* 548: 251–262.
Paroissien, J. B., Darboux, F., Couturier, A., et al. (2015). A method for modeling the effects of climate and land use changes on erosion and sustainability of soil in a Mediterranean watershed (Languedoc, France). *Journal of Environmental Management* 150: 57–68.
Peigné, J., Vian, J. F., Payet, V., & Saby, N. P. (2018). Soil fertility after 10 years of conservation tillage in organic farming. *Soil and Tillage Research* 175: 194–204.
Pimentel, D., Harvey, C., Resosudarmo, P., et al. (1995). Environmental and economic costs of soil erosion and conservation. *Science* 267(5201): 1117–1123.
Pittelkow, C. M., Linquist, B. A., Lundy, M. E., et al. (2015). When does no-till yield more? A global meta-analysis. *Field Crops Research* 183: 156–168.
Poesen, J. (2018). Soil erosion in the Anthropocene: Research needs. *Earth Surface Processes and Landforms* 43(1): 64–84.

Poesen, J., Nachtergaele, J., Verstraeten, G., & Valentin, C. (2003). Gully erosion and environmental change: Importance and research needs. *Catena* 50(2–4): 91–133.
Prosdocimi, M., Tarolli, P., & Cerdà, A. (2016). Mulching practices for reducing soil water erosion: A review. *Earth-Science Reviews* 161: 191–203.
Pruski, F. F., & Nearing, M. A. (2002). Climate-induced changes in erosion during the 21st century for eight US locations. *Water Resources Research* 38(12): 34-1–34-11.
Qin, C., Zheng, F., Wilson, G. V., Zhang, X. J., & Xu, X. (2019). Apportioning contributions of individual rill erosion processes and their interactions on loessial hillslopes. *Catena* 181: 104099.
Ramos, M. C., & Martínez-Casasnovas, J. A. (2015). Climate change influence on runoff and soil losses in a rainfed basin with Mediterranean climate. *Natural Hazards* 78(2): 1065–1089.
Rauws, G., & Covers, G. (1988). Hydraulic and soil mechanical aspects of rill generation on agricultural soils. *Journal of Soil Science* 39(1): 111–124.
Renard, K. G., Foster, G. R., Weesies, G. A., McCool, D. K., & Yoder, D. C. (1997). *Predicting Soil Erosion by Water: A Guide to Conservation Planning with the Revised Universal Soil Loss Equation (RUSLE).* Agricultural Handbook 703. Washington, DC: United States Government Printing.
Reubens, B., Poesen, J., Danjon, F., Geudens, G., & Muys, B. (2007). The role of fine and coarse roots in shallow slope stability and soil erosion control with a focus on root system architecture: A review. *Trees* 21(4): 385–402.
Ryżak, M., Bieganowski, A., & Polakowski, C. (2015). Effect of soil moisture content on the splash phenomenon reproducibility. *PloS One* 10(3): 1–15.
Sartori, M., Philippidis, G., Ferrari, E., et al. (2019). A linkage between the biophysical and the economic: Assessing the global market impacts of soil erosion. *Land Use Policy* 86: 299–312.
Segura, C., Sun, G., McNulty, S., & Zhang, Y. (2014). Potential impacts of climate change on soil erosion vulnerability across the conterminous United States. *Journal of Soil and Water Conservation* 69(2): 171–181.
Seitz, S., Goebes, P., Puerta, V. L., et al. (2019). Conservation tillage and organic farming reduce soil erosion. *Agronomy for Sustainable Development* 39(1): 4.
Serpa, D., Nunes, J. P., Santos, J., et al. (2015). Impacts of climate and land use changes on the hydrological and erosion processes of two contrasting Mediterranean catchments. *Science of the Total Environment* 538: 64–77.
Shao, H., Baffaut, C., Nelson, N. O., et al. (2013). Development and application of algorithms for simulating terraces within SWAT. *Transactions of the ASABE* 56(5): 1715–1730.
Shen, W., Zou, C., Liu, D., et al. (2015). Climate-forced ecological changes over the Tibetan Plateau. *Cold Regions Science and Technology* 114: 27–35.
Shrestha, N. K., & Wang, J. (2018). Predicting sediment yield and transport dynamics of a cold climate region watershed in changing climate. *Science of the Total Environment* 625: 1030–1045.
Sidorchuk, A. (1999). Dynamic and static models of gully erosion. *Catena* 37(3–4): 401–414.
Simon, A., & Rinaldi, M. (2006). Disturbance, stream incision, and channel evolution: The roles of excess transport capacity and boundary materials in controlling channel response. *Geomorphology* 79(3–4): 361–383.
Simonneaux, V., Cheggour, A., Deschamps, C., et al. (2015). Land use and climate change effects on soil erosion in a semi-arid mountainous watershed (High Atlas, Morocco). *Journal of Arid Environments* 122: 64–75.

Soil and Water Conservation Society (SWCS) (2003). *Conservation Implications of Climate Change: Soil Erosion and Runoff from Cropland*. Ankeny, IA: Soil and Water Conservation Society.

Soil Science Society of America (2008). *Glossary of Soil Science Terms*. Madison, WI: ASA-CSSA-SSSA.

Song, Y., Yan, P., & Liu, L. (2006). A review of the research on complex erosion by wind and water. *Journal of Geographical Sciences* 16(2): 231–241.

Southworth, J., Randolph, J. C., Habeck, M., et al. (2000). Consequences of future climate change and changing climate variability on maize yields in the midwestern United States. *Agriculture, Ecosystems & Environment* 82(1–3): 139–158.

Stavi, I., Perevolotsky, A., & Avni, Y. (2010). Effects of gully formation and headcut retreat on primary production in an arid rangeland: Natural desertification in action. *Journal of Arid Environments* 74(2): 221–228.

Stevens, C. J., Quinton, J. N., Bailey, A. P., et al. (2009). The effects of minimal tillage, contour cultivation and in-field vegetative barriers on soil erosion and phosphorus loss. *Soil and Tillage Research* 106(1): 145–151.

Sun, L., Fang, H., Qi, D., Li, J., & Cai, Q. (2013). A review on rill erosion process and its influencing factors. *Chinese Geographical Science* 23(4): 389–402.

Tang, Q. (2020). Global change hydrology: Terrestrial water cycle and global change. *Science China Earth Sciences* 63(3): 459–462.

Tien Bui, D., Shirzadi, A., Shahabi, H., et al. (2019). A novel ensemble artificial intelligence approach for gully erosion mapping in a semi-arid watershed (Iran). *Sensors* 19 (11): 2444.

Trenberth, K. E., Jones, P. D., Ambenje, P., et al. (2007). Observations: Surface and atmospheric climate change. In S. Solomon, D. Qin, M. Manning, et al. (eds.), *Climate Change 2007: The Physical Science Basis* (pp. 235–336). Cambridge: Cambridge University Press.

Trimble, S. W. (1997). Contribution of stream channel erosion to sediment yield from an urbanizing watershed. *Science* 278(5342): 1442–1444.

Valentin, C., Poesen, J., & Li, Y. (2005). Gully erosion: Impacts, factors and control. *Catena* 63(2–3): 132–153.

Van den Putte, A., Govers, G., Diels, J., Gillijns, K., & Demuzere, M. (2010). Assessing the effect of soil tillage on crop growth: A meta-regression analysis on European crop yields under conservation agriculture. *European Journal of Agronomy* 33(3): 231–241.

Vanmaercke, M., Maetens, W., Poesen, J., et al. (2012). A comparison of measured catchment sediment yields with measured and predicted hillslope erosion rates in Europe. *Journal of Soils and Sediments* 12(4): 586–602.

de Vente, J., Poesen, J., Bazzoffi, P., Rompaey, A. V., & Verstraeten, G. (2006). Predicting catchment sediment yield in Mediterranean environments: The importance of sediment sources and connectivity in Italian drainage basins. *Earth Surface Processes and Landforms* 31(8): 1017–1034.

Walling, D. E., & Collins, A. L. (2005). Suspended sediment sources in British rivers. *Sediment Budgets 1* IAHS Publication 291.

Wang, Y., Fu, B., Chen, L., Lü, Y., & Gao, Y. (2011). Check dam in the Loess Plateau of China: Engineering for environmental services and food security. *Environmental Science & Technology* 45(24): 10298–10299.

Wang, Y., Wu, Y., Kou, Q., et al. (2007). Definition of arsenic rock zone borderline and its classification. *Science of Soil and Water Conservation* 5(1): 4–8.

Wei, W., Chen, D., Wang, L., et al. (2016). Global synthesis of the classifications, distributions, benefits and issues of terracing. *Earth-Science Reviews* 159: 388–403.

Wilson, G. (2011). Understanding soil-pipe flow and its role in ephemeral gully erosion. *Hydrological Processes* 25(15): 2354–2364.

Wischmeier, W. H. (1959). A rainfall erosion index for a universal soil-loss equation. *Soil Science Society of America Journal* 23(3): 246–249.

Wischmeier, W. H. (1962). Rainfall erosion potential. *Agricultural Engineering* 43(4): 212–225.

Wischmeier, W. H., & Smith, D. D. (1960). A universal soil-loss equation to guide conservation farm planning. *Transactions of the 7th International Congress on Soil Sciences*, 1: 418–425.

Wu, H., Xu, X., Zheng, F., Qin, C., & He, X. (2018). Gully morphological characteristics in the loess hilly–gully region based on 3D laser scanning technique. *Earth Surface Processes and Landforms* 43(8): 1701–1710.

Wuepper, D., Borrelli, P., & Finger, R. (2020). Countries and the global rate of soil erosion. *Nature Sustainability* 3: 51–55.

Xiong, M., Sun, R., & Chen, L. (2018). Effects of soil conservation techniques on water erosion control: A global analysis. *Science of the Total Environment* 645: 753–760.

Xiong, M., Sun, R., & Chen, L. (2019). A global comparison of soil erosion associated with land use and climate type. *Geoderma* 343: 31–39.

Xu, X., Zhang, H., & Zhang, O. (2004). Development of check-dam systems in gullies on the Loess Plateau, China. *Environmental Science & Policy* 7(2): 79–86.

Xu, X., Zheng, F., Qin, C., Wu, H., & Wilson, G. V. (2017). Impact of cornstalk buffer strip on hillslope soil erosion and its hydrodynamic understanding. *Catena* 149: 417–425.

Xu, X., Zheng, F., Wilson, G. V., et al. (2018). Comparison of runoff and soil loss in different tillage systems in the Mollisol region of Northeast China. *Soil and Tillage Research* 177: 1–11.

Xu, X., Zheng, F., Wilson, G. V., et al. (2019). Quantification of upslope and lateral inflow impacts on runoff discharge and soil loss in ephemeral gully systems under laboratory conditions. *Journal of Hydrology* 579: 124174.

Xu, X. Z., Xu, Y., Chen, S. C., Xu, S. G., & Zhang, H. W. (2010). Soil loss and conservation in the black soil region of Northeast China: A retrospective study. *Environmental Science & Policy* 13(8): 793–800.

Yang, D., Kanae, S., Oki, T., Koike, T., & Musiake, K. (2003). Global potential soil erosion with reference to land use and climate changes. *Hydrological Processes* 17 (14): 2913–2928.

Zhang, P., Yao, W., Liu, G., & Xiao, P. (2019). Research progress and prospects of complex soil erosion. *Transactions of the Chinese Society of Agricultural Engineering* 35(24): 154–161.

Zhang, X. C., & Nearing, M. A. (2005). Impact of climate change on soil erosion, runoff, and wheat productivity in central Oklahoma. *Catena* 61(2–3): 185–195.

Zhang, X. H., Zheng, F. L., & Li, J. (2007). Current situation and existing problems of gully erosion research. *Research of Soil Water Conservation* 14(4): 31–32.

Zhang, Y. G., Nearing, M. A., Zhang, X. C., Xie, Y., & Wei, H. (2010). Projected rainfall erosivity changes under climate change from multimodel and multiscenario projections in Northeast China. *Journal of Hydrology* 384(1–2): 97–106.

Zhang, Y. G., Wu, Y. Q., Liu, B. Y., Zheng, Q. H., & Yin, J. Y. (2007). Characteristics and factors controlling the development of ephemeral gullies in cultivated catchments of black soil region, Northeast China. *Soil & Tillage Research* 96(1–2): 28–41.

Zhao, Y., Wang, E., Cruse, R. M., & Chen, X. (2012). Characterization of seasonal freeze–thaw and potential impacts on soil erosion in northeast China. *Canadian Journal of Soil Science* 92(3): 567–571.

Zheng, F., Xu, X., & Qin, C. (2016). A review of gully erosion process research. *Transactions of the Chinese Society for Agricultural Machinery* 47(8): 48–59.

Zhu, X. M. (1956). Classification on the soil erosion in the loess region. *Acta Pedologica Sinica* 4(2): 99–115.

6

Climate Change Impacts on Saltwater Intrusion into Coastal Aquifers

HAN XIAO, HAIMING LI, YIN TANG, QIUHONG TANG AND MARWAN KHEIMI

6.1 Introduction

Coastal aquifers are bounded by or connected with saline waterbodies such as sea/ocean and/or saltwater lagoons and fresh groundwater beneath the land surface is usually threatened by the encroachment of saline groundwater beneath the sea floor due to the direct hydraulic link between them. Fresh and saline groundwater are usually kept in a dynamic equilibrium, although the freshwater/saltwater interface/boundary does not always remain in the same place but moving landward or seaward dynamically, mainly depending on hydro-climatic conditions and hydrogeological conditions (Bear, 1979). In general, the interface/boundary can shift seaward when freshwater supply increases (e.g., increasing fresh groundwater recharge due to increased infiltrated rainwater in the wet season), while it can shift landward conversely when freshwater supply decreases (e.g., decreasing fresh groundwater recharge due to decreased infiltrated rainwater in the dry season). A landward shift of the interface/boundary can cause a mixing of saline and fresh groundwater in the location that previously contained fresh groundwater. This process is recognized as an effective indicator of landward encroachment of saline groundwater and is referred to as saltwater intrusion into coastal aquifers (Freeze & Cherry, 1979).

Densely populated coastal areas around the world often encounter acute problems of water supply shortage attributed to huge freshwater demand and severe surface water contamination (which greatly limit its utilization for water supply), and groundwater usually serves as the primary source of freshwater supply for human consumption (Bear et al., 1999). However, degradation of water quality of coastal aquifers due to saltwater intrusion poses an important threat to

This chapter was partly supported by the (1) Youth Program of the Natural Science Foundation of Tianjin (19JCQNJC08400); (2) Start-up Fund of Tianjin University of Science and Technology; and (3) Young Talent Fund of College of Marine and Environmental Sciences of TUST. The opinions and findings expressed in this chapter are those of the author(s) and not necessarily those of these agencies.

freshwater supply for agricultural, industrial and domestic utilization in many coastal areas globally and large amounts of coastal wells have been abandoned because of the increased salt content. It is indicated that a 2–3 per cent mixing with seawater renders freshwater inadequate for human consumption, while a 5 per cent mixing results in abstraction wells being abandoned and no longer being used for freshwater supply (Sherif & Singh, 1999). In recent decades, the phenomenon of saltwater intrusion has been seriously aggravated, resulting from exacerbated human activities such as groundwater over-exploitation due to the ever-increasing water demand because of population growth and economic and social development. Groundwater over-exploitation can significantly lower water tables of coastal aquifers and break the existing equilibrium between fresh and saline groundwater, causing hydrostatic pressure to become unbalanced, which compels saline groundwater moving landward to form a new equilibrium. Besides, groundwater overexploitation can also change local-scale groundwater flow patterns, acting as a driving force of saline groundwater 'up-coning' toward abstraction wells. The former process of landward encroachment of saline groundwater is referred to as horizontal (lateral) saltwater intrusion and the latter process is referred to as vertical (upward) saltwater intrusion.

Intensified human activities such as groundwater over-exploitation are commonly recognized as the principle factor causing saltwater intrusion into coastal aquifers. However, phenomena induced by climate change such as sea-level rise, more frequent and intensified tropical cyclones and hurricanes, as well as changing precipitation patterns and regimes have recently become dominant drivers of saltwater intrusion. Thermal expansion of the ocean as well as glacial melting (primarily due to global warming) coupled with land subsidence (primarily due to excessive groundwater pumping and urban construction) have significantly contributed to relative sea-level rises in numerous coastal areas worldwide. A sea-level rise can increase the saltwater hydraulic head causing a disequilibrium of the existing hydrostatic pressure, resulting in lateral saltwater intrusion into coastal aquifers due to advection/dispersion. An increase in both the frequency and intensity of extreme weather events such as tropical storms and hurricanes, in combination with sea-level rise, can bring more frequent and intense storm surges to coastal areas causing saltwater overtopping in low-lying areas, resulting in downward saltwater intrusion into coastal aquifers due to infiltration/dispersion (Van Biersel et al., 2007). In comparison to sea-level rise, the duration of saltwater inundation in coastal lowlands is short-lived since the event of storm surge is temporary. The consequences of saltwater inundation within a relatively short period of time are two-fold: (1) an increase in root zone salinity due to infiltration from the land surface, which may threaten the survival of vegetation species; and (2) a high risk of physical damage or permanent abandonment of coastal water-supply wells due to water quality degradation. Changes in precipitation regimes have the potential to aggravate the extent of saltwater intrusion due to alterations of freshwater infiltration/recharge. For example, the total amount of precipitation over a time period may remain relatively unchanged, however its temporal variability within the temporal framework may change, with less but more intense rainfall events, for example. In this scenario, a higher proportion of rainwater loss in the form of runoff is inevitable. This may reduce the amount of freshwater infiltration/recharge, resulting in lowering of the groundwater level and saltwater intrusion. In the future, the combined effects of intensified human activities and global climate change are likely to amplify the extent of saltwater intrusion into coastal aquifers which would inevitably lead to more severe consequences for the environment.

In the context of increasing freshwater demand, sea-level rise, increased frequency and intensity of extreme weather events, coastal erosion, warmer temperatures and changing precipitation patterns and regimes, the fast-growing risks of further saltwater intrusion into coastal aquifers menace multiple coastal areas in Australia, Canada, China, Egypt, Germany, India, Israel, Italy, the Netherlands, Senegal, the United States, etc. Numerous studies have explored the mechanism, pathway and scope of saltwater intrusion into coastal aquifers and attempted to predict the future extent of groundwater salinization in vulnerable coastal areas worldwide (Barlow & Reichard, 2010; Bocanegra et al., 2010; Custodio, 2010; Steyl & Dennis, 2010; Werner, 2010; White & Falkland, 2010; Shi & Jiao, 2014).

6.2 Theory

Two different theories provide a complementary explanation of the issue of saltwater intrusion into coastal aquifers, that is, the sharp-interface theory and the transition zone theory. The two theories differ from each other in the way they model the underlying phenomenon, and their divergence is expressed through different equations (Bear et al., 1999). Both are however relevant depending on the context in which they are applied.

The sharp-interface theory is based on an important assumption that freshwater and saltwater are immiscible fluids and a sharp interface/boundary exists between fresh and saline groundwater. Based on this assumption, Drabbe and Badon-Ghyben (1888–1889) and Herzberg (1901) investigated the shape and position of the sharp interface under hydrostatic equilibrium conditions and demonstrated that the slope of the interface/boundary between fresh and saline groundwater is 40-times greater than the slope of the water table. For example, a 1 metre water table above the mean sea level (MSL) can ensure 40 metres of freshwater below the MSL at any distance from the shoreline, and a lowering of 0.1 metre on the water table can cause a rising of 4 metres of the interface/boundary between fresh and saline groundwater. This theorical limitation implies that, in practice, the sharp-interface theory cannot accurately simulate the saltwater intrusion process, especially if coastal aquifers are closely connected with the sea/ocean hydraulically.

The transition zone theory is developed based on the actual situation that freshwater and saltwater are miscible fluids and can mix, forming a relatively broad variable-density freshwater/saltwater transition/dispersion zone between fresh and saline groundwater. Within the transition/dispersion zone, groundwater densities and solute concentrations vary gradually from those of freshwater at the land side to those of saltwater at the sea/ocean side, and the groundwater flow under hydraulic gradient and solute transport under concentration gradient are coupled. The dimensions and migration of the transition zone rely on hydroclimatic and hydrogeological conditions (Anderson & Woessner, 1991). A landward migration of the transition zone is an effective indicator of the occurrence of saltwater intrusion.

6.3 Method

The sharp-interface theory is applied and the analytical method is utilized to solve simplified and idealized saltwater intrusion problems such as estimation of the depth of the interface/boundary between fresh and saline groundwater. The transition zone theory along with numerical method are widely utilized to solve real-world complex saltwater intrusion problems such as how far inland the saltwater wedge can encroach and what is the amount of saline groundwater entering the fresh groundwater system. We suggest sophisticated numerical models based on finite-difference or finite-element methods to account for the variable-density conditions observed in our case study.

6.3.1 Finite-Difference Method

The finite-difference method is a traditional and widely used numerical method for solving (partial) differential equations governing the coupled groundwater flow and solute transport processes under variable-density conditions. A system of nodal points is superimposed over a discretized gridded domain, composed of (irregular) rectangular or square cells, with defined initial and boundary conditions. Solutions of the (partial) differential equations yield numerical values at the specific nodal points, including groundwater flow rates, flow directions, hydraulic heads and solute concentrations. The primary advantages of finite-difference methods are their simplicity and accuracy of high-order approximations. However, their applications are limited. Phenomena acknowledged as numerical dispersion can cause an over-estimation of the solute transport resulting from solving hydrodynamic dispersion equations. Therefore, minimizing numerical oscillations and obtaining numerical stability are crucial objectives of the simulation process.

Common software using finite-difference methods to simulate groundwater flow and solute transport under variable-density conditions for modelling the saltwater intrusion process include: FAST-C (Holzbecher, 1998), HST3D (Kipp, 1986), MOCDENS3D (Oude Essink, 1998), SEAWAT (Guo & Langevin, 2002), SWIFT (Ward, 1991), etc.

6.3.2 Finite-Element Method

Finite-element methods provide more flexibility to represent natural boundaries, since the study domain can be discretized into an irregular polygonal mesh (Zienkiewicz, 1977). Triangular elements are extensively employed, and groundwater flow and solute transport can be approximated by a series of small flat or curved triangular surfaces (if triangular elements are applied) and each element has its own physical and hydraulic parameters. Sometimes, quadrilateral elements are used rather than triangular elements. Finite-element methods are a powerful numerical tool based on a mathematically elegant technique. They offer high flexibility that allows the user to refine the grid size if necessary, however they may exhibit numerical instability. For example, when solving saltwater intrusion problems, finite-element methods can be used to design a dense grid near the boundary of sea/ocean where groundwater flow is rotational and salt concentration gradient is relatively high. This greatly increases the accuracy of simulation results, the implementation of the stability criteria is not straightforward as for the finite-difference method since it requires calculating hydraulic parameters for each grid-cell, which remains computationally intense.

Common software using finite-element methods to simulate groundwater flow and solute transport under variable-density conditions for the modelling saltwater intrusion process include: FEFLOW (Diersch, 1996), METROPOL (Sauter et al., 1993), SUTRA (Voss, 1984), SWICHA (Huyakorn et al., 1987), etc.

6.4 Case Study

In our case study described here, we model the processes of saltwater intrusion into coastal aquifers in east-central Florida (USA), which are mainly driven by sea-level rise and storm surge (Xiao et al., 2016, 2019). The SEAWAT computer code is utilized to simulate the dimension and migration of saltwater/freshwater transition zone, which allows us to assess the extent of saltwater intrusion.

SEAWAT has been widely acknowledged as an efficient and effective numerical tool for modelling coupled groundwater flow and solute transport under variable-density conditions. The objective of this case study is to quantify the impacts of sea-level rise as well as storm surge on the extent of saltwater intrusion into the surficial aquifer of the low-lying coastal alluvial plains and barrier islands in coastal east-central Florida. This study explores the impacts of climate change and aims at determining how far inland the toe of the subterranean saltwater wedge can encroach.

6.4.1 Study Area and Model Domain

The study area covers Cape Canaveral Barrier Island Complex, an area located in the low-lying coastal alluvial plains and barrier islands in coastal east-central Florida (USA). This region is composed of multiple barrier islands, saltwater/freshwater lagoons and the Atlantic Ocean coastline, covering an area of approximately 1,000 km^2 bounded by the Atlantic Ocean to the east, Mosquito Lagoon to the northeast and north, Indian River Lagoon to the west and Banana River to the southeast and south (Figure 6.1). The climate is humid subtropical

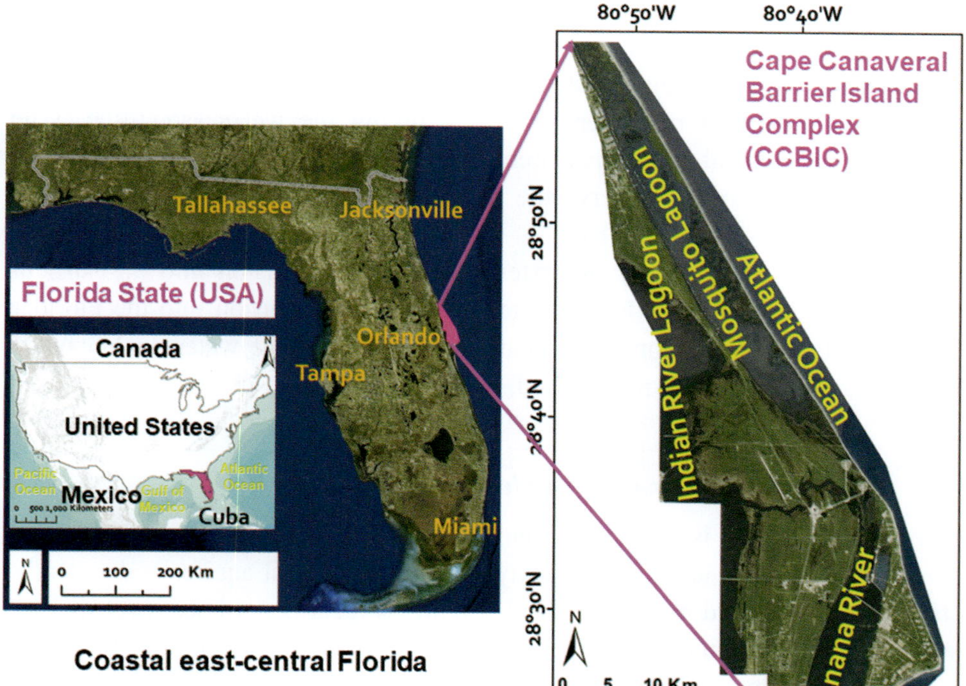

Figure 6.1 Location of the study area and model domain

with hot/humid summers (mean temperature varying from 22°C to 33°C) and mild/dry winters (mean temperature varying from 10°C to 22°C), with a mean annual rainfall of 1,366 mm (annual rainfall varying from 848 mm to 2,075 mm). The hydrostratigraphic units are composed of (from top to bottom) the surficial aquifer, the intermediate confining unit, the Floridan aquifer and the lower confining unit. The surficial aquifer occurs in the saturated part of the moderate-to-low-permeability Holocene and Pleistocene sediments of fine-to-medium sand, coquina, silt, shell and marl. It is mainly composed of fine-to-medium sand which has the water table as its upper boundary and the top of the intermediate confining unit as its lower boundary. The primary inflow is infiltrated rainwater and the primary outflow is evapotranspiration and discharge to surface waterbodies, that is, fresh and/or salty marshes/wetlands, coastal lagoons and the Atlantic Ocean. The water level, which rises to its highest level late in the wet season (September/October) and drops to its lowest level late in the dry season (April/May), is mainly controlled by the temporal variation of rainfall and the seasonal thermal expansion and contraction of the Atlantic Ocean. The saltwater/freshwater transition zone migrates either landward (saltwater intrusion) or seaward (saltwater recession) according to the lowering or rising water table, and its thickness is mainly dependent on the characteristics of the hydrogeologic settings. Although pumping from shallow groundwater abstraction wells for citrus irrigation is conducted from time to time, the water table usually remains at its natural state and is not disturbed from a relatively low pumping rate and large amounts of infiltrated rainwater, especially during the wet season. The water quality of the surficial aquifer is extremely important to maintain the biodiversity of the ecosystem and to preserve the life of threatened and endangered species of wildlife.

In this study, we quantify the extent of saltwater intrusion into the surficial aquifer without accounting for the extent of saltwater intrusion into the Floridan aquifer. This model assumption is made mainly because the water quality of the Floridan aquifer is highly mineralized, which greatly limits its consumptive use.

6.4.2 Model Development

A reference model is developed and calibrated with field-measured data to simulate the spatial variation of salinity within the surficial aquifer under steady-state hydrological and hydrogeological conditions regardless of sea-level rise and storm surge. Based on the calibrated steady-state reference model, three predictive models are built by varying the boundary conditions to simulate the impacts of sea-level rise on saltwater intrusion into the surficial aquifer. A diagnostic model uses various boundary conditions to simulate the impacts of storm surge on saltwater intrusion into the surficial aquifer. The temporal variation of total dissolved solids (TDS) concentrations within the surficial aquifer are quantified to assess the processes of saltwater intrusion and recession (infiltrated saltwater being diluted and flushed by infiltrated rainwater). We use SEAWAT Version 4 (SEAWAT_V4) code (Langevin et al., 2008) for the simulation. Groundwater Vistas, a well-known graphical user interface developed by Environmental Simulations Inc., is used to create the model input files.

6.4.2.1 Spatial Discretization

The topographic variation is relatively small since the region is mainly composed of broad flat lowlands, and the land surface elevation varies from –0.2 to 10 metres with a regional average of about 1.2 metres (NAVD 88) (LiDAR data from NASA) (Figure 6.2(a)). The determination of horizontal and vertical discretization that can effectively represent the topographic variation of land surface and the dimension and movement of saltwater/freshwater transition zone is of great importance. In this context, the criterion is to improve simulation accuracy while maintaining reasonable computer runtimes. A uniform grid spacing of 100 metres in both the x- and y-directions (646 rows and 373 columns) is assigned horizontally and the model domain is vertically divided into five layers. The top elevation of Layer 1 is set to the land-surface/sea-floor elevation and the bottom elevation of Layer 1 is set to –2 metres (NAVD 88). From Layers 2–5, the layers are flat and have a

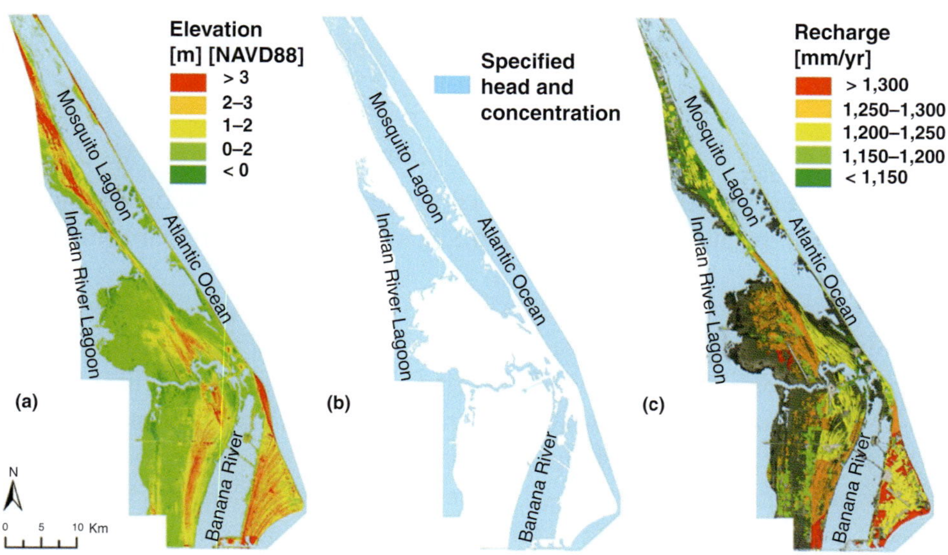

Figure 6.2 (a) Spatial variation of land surface elevation (LiDAR data); (b) Specified head and concentration boundary assigned to Layer 1; (c) Recharge boundary assigned to Layer 1

uniform thickness of 2 metres. The bottom elevation of Layer 5 is set to –10 metres (NAVD 88). Note that the criteria of spatial discretization designed for the reference model, the predictions model and the diagnostic model are identical.

6.4.2.2 Temporal Discretization

The reference model is a steady-state model which simulates the annual-averaged steady-state TDS concentrations in the surficial aquifer based on the assumptions that the groundwater systems are in equilibrium with the annual-averaged steady-state hydro-climatic conditions. In this model, temporal discretization is introduced in terms of transport time steps for a better simulation of the solute transport. The length of the transport time step starts with 0.01 day and increases by a time step multiplier of 1.2, with a maximum transport time step of 100 days. The process does not terminate until a steady state is reached. The predictive models are designed to be 'transient' modelling of the temporal variation of surficial aquifer TDS concentrations under sea-level rise impacts, and the diagnostic model is designed to be 'transient' modelling of the temporal variation of surficial aquifer TDS concentrations under storm surge impacts.

6.4.2.3 Parameters

Hydrogeological parameters describing the characteristics of each hydrostratigraphic unit including horizontal/vertical hydraulic conductivity, porosity, longitudinal/transverse/vertical dispersivity, and diffusion coefficient are obtained from the literature and case reports (see Table 6.1). The hydrogeological parameters assigned for the reference model, the predictions model and the diagnostic model are identical.

6.4.2.4 Boundary Conditions

To minimize the boundary effects, the model domain extends offshore to simulate the surface water and groundwater interactions between the surficial aquifer and

Table 6.1. Hydrogeological parameters

Parameters	Value	Unit
Horizontal Hydraulic Conductivity (K_h)	15	m d^{-1}
Anisotropy (K_h/K_z)	10	–
Porosity (n)	0.2	–
Longitudinal Dispersivity (a_L)	6	m
Transverse Dispersivity (a_T)	0.01	m
Vertical Dispersivity (a_V)	0.025	m
Diffusion Coefficient (D^*)	0.00128	m^2 d^{-1}

coastal lagoons as well as the Atlantic Ocean. The specified head and concentration boundary, the time-variant specified head and concentration boundary, the no-flow boundary, the general-head boundary, the well boundary, the recharge boundary and the evapotranspiration boundary are used for model implementation. The specified head and concentration boundary is assigned to the coastal lagoons and the Atlantic Ocean. Water levels of the coastal lagoons and the Atlantic Ocean are obtained from observations. TDS concentrations of the coastal lagoons and the Atlantic Ocean are assumed to be equal to the TDS concentration of seawater ($35\,kg\,m^{-3}$). The horizontal view of the specified head and concentration boundary assigned to Layer 1 is shown in Figure 6.2(b). The no-flow boundary is assigned to the lateral boundaries in Layers 1, 2, 3, 4 and 5 where groundwater flux through the boundaries is zero, and is also assigned to the bottom of Layer 5 since groundwater flux between the surficial aquifer and the underlying Floridan aquifer through the intermediate confining unit is negligible. The general-head boundary is assigned to the lateral boundaries in Layers 1, 2, 3, 4 and 5 where groundwater flux through the boundaries is non-zero. The well boundary is assigned to the groundwater pumping wells. The recharge boundary and the evapotranspiration boundary are assigned to the top of Layer 1 (Figure 6.2(c)). The boundary conditions assigned for the reference model, the predictions model and the diagnostic model are identical. Due to the ever-changing water levels of coastal lagoons and the Atlantic Ocean because of sea-level rises and storm surges, the head values of the specified head and concentration boundary vary according to sea-level rise and storm surge scenarios.

6.4.2.5 Initial Conditions

For the reference model, accurate starting heads and concentrations (starting heads and concentrations are consistent with the designated aquifer properties and boundary conditions) are not required since it is a steady-state condition. Reasonable estimations of the starting heads and concentrations based on the observed data are used as the initial conditions to avoid numerical instability during simulation. For the prediction models and the diagnostic model, accurate starting heads and concentrations are required since they are transient and the simulated groundwater hydraulic heads and TDS concentrations from the reference model are used as the initial conditions before starting the modelling process.

6.4.2.6 Calibration

The output of the reference model contains hydraulic heads and TDS concentrations. The simulated hydraulic heads are calibrated with field-measured groundwater levels obtained from 10 active observation wells, whereas the simulated TDS concentrations are not calibrated because groundwater salinity measurements are lacking. The daily variation of water table elevations monitored by the observation wells are averaged to compute the annual-averaged steady-state hydraulic heads to serve as calibration targets shown in Figure 6.3(a). During the calibration process, the horizontal and vertical hydraulic conductivities are adjusted using a trial-and-error method aiming at minimizing the differences between the simulated hydraulic heads and the monitored water table elevations. The calibration process does not terminate until the simulated heads match the observed water table elevations to a satisfactory degree.

After calibration, the Nash-Sutcliffe model efficiency (NSE) coefficient reaches 0.96 (Figure 6.3(b)), demonstrating a successful agreement between the simulated and the observed groundwater hydraulic heads. To further improve model performance, the concepts of 'simulated' and 'true' marshes/wetlands are introduced herein. The 'simulated' marshes/wetlands refer to the land areas that have a simulated water table depth less than 0.2 metres (including the land areas that have simulated water tables higher than the land surface). The 'true' marshes/wetlands refer to the marshes/wetlands classified by the map of land use and land cover determined by field observation and surveying. The results suggest a good model performance, with 64.6 per cent consistency between the 'simulated' and 'true' marshes/wetlands (Figure 6.3(c)).

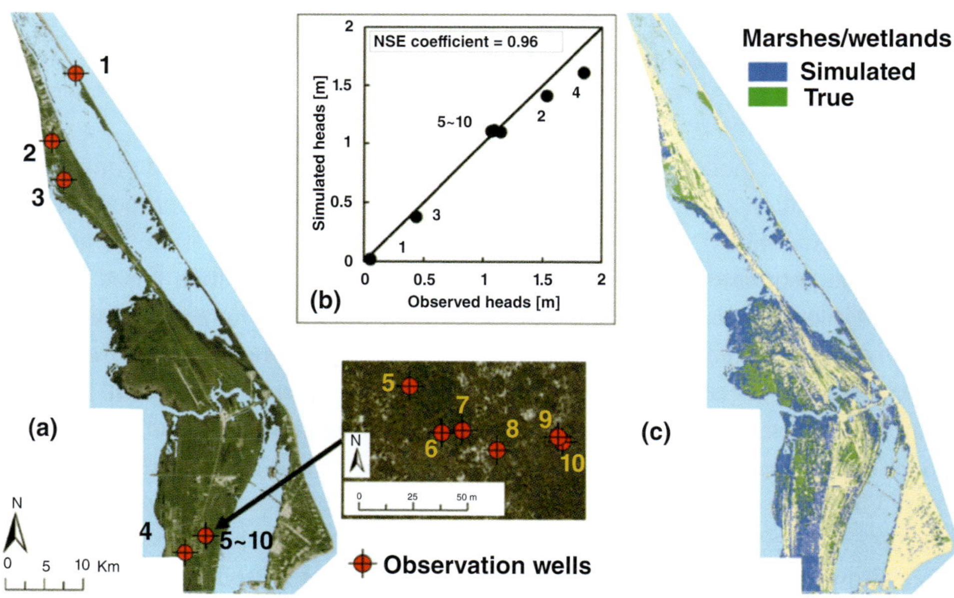

Figure 6.3 (a) Locations of the in-situ observation wells; (b) Scatter diagram showing the goodness of fit between the observed and simulated hydraulic heads; (c) Comparison between the 'simulated' and 'true' marshes/wetlands

Note that marshes/wetlands are dynamic natural systems of high complexity composed of swamps, marshes and bogs dependent not only on water table depth but also topography, vegetation cover and soil type. Thus, defining 'simulated' marshes/wetlands solely on water table depth might oversimplify the processes at stake. However, the comparison between the 'simulated' and 'true' marshes/wetlands remains valid since the in-situ observation wells are limited. Calibration could be improved with future additional data on field-measured groundwater hydraulic heads and TDS concentrations.

6.4.3 Results and Discussion

The TDS concentration of groundwater within the saltwater/freshwater transition zone appears higher than freshwater (TDS concentration no greater than 1 kg m^{-3}) but lower than seawater (TDS concentration equivalent to 35 kg m^{-3}). Based on the TDS concentration, the National Groundwater Association defines the transition zone as the slightly saline zone (TDS concentration varying from 1 to 3 kg m^{-3}), moderately saline zone (TDS concentration varying from 3 to 10 kg m^{-3}), and highly saline zone (TDS concentration varying from 10 to 35 kg m^{-3}). The contours of 1, 3, 10 and 35 kg m^{-3} TDS concentrations are usually used to determine the location and migration of the transition zone for determination of the extent of saltwater intrusion.

Based on the output from the reference model, the spatial distribution of TDS concentrations of the surficial aquifer under steady-state hydro-climatic and hydrogeological conditions are shown in Figure 6.4. From the plan view, the transition zone is mainly located at the western part where the saltwater wedge toe trespasses 3–4 kilometres inland. From the cross-sectional view along the N-S transect, the widths of the slightly saline zone and the moderately saline zone are relatively thin, while the width of the highly saline zone is thick. Note that the areas of the highly, moderately and slightly saline zones might expand and shrink dynamically due to the seasonal landward/seaward migration of the transition zone because of seasonal variation of sea level and rainfall.

6.4.3.1 Sea-Level Rise Impacts on Saltwater Intrusion

Under three sea-level rise scenarios of 13.2, 31.0 and 58.5 centimetres (Rosenzweig et al., 2014) (downscaled projections for the year of 2050 from the year of 2010 based on low, intermediate and high ice melt projections), the projected increase of percentages of saline zones are calculated (see Table 6.2). The projected spatial distribution of highly, moderately and slightly saline zones are visualized (see Figure 6.5(a–c)) based on the output from the prediction models. Under that scenarios, Figure 6.5 shows that the landward migration of

Table 6.2. Projected increase of percentages of saline zones

Sea-Level Rise Scenarios	Percentage (%)	Growth Rate (%)
0	13.6	–
13.2 (cm)	21.8	8.2
31.0 (cm)	34.0	20.4
58.5 (cm)	47.9	34.3

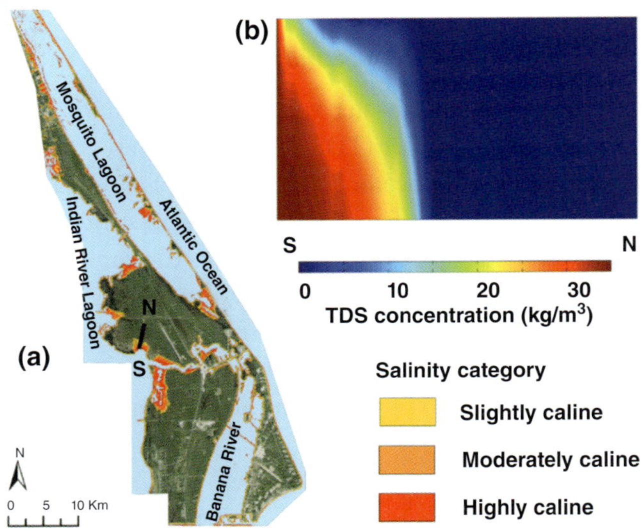

Figure 6.4 Surficial aquifer TDS concentration under steady-state hydro-climatic and hydrogeological conditions: (a) Plan view; (b) South-to-north cross-sectional view

saline water would mostly occur in the western part, and the saltwater wedge toe could trespass 8–10 kilometres inland in some areas.

The predicted results indicate that the effects of sea-level rise can greatly exacerbate the extent of saltwater intrusion into the surficial aquifer in coastal east-central Florida (USA). This area is particularly vulnerable to sea-level rise because of low land surface elevation with flat topography. In the western part, land cover is mainly composed of fresh marsh, intermediate marsh (less saline than brackish), brackish marsh and saline marsh. Landward migration of seawater due to sea-level rise into the traditionally fresh marsh can cause degradation of the ecologic system and altering of the distribution and productivity of vegetation communities. Salt tolerance of plant communities is dependent on vegetation type, duration of exposure to saline water, rate of salinity increase, mineral content of soil and degree of submergence. Some species can tolerate a wide range of salinity and can

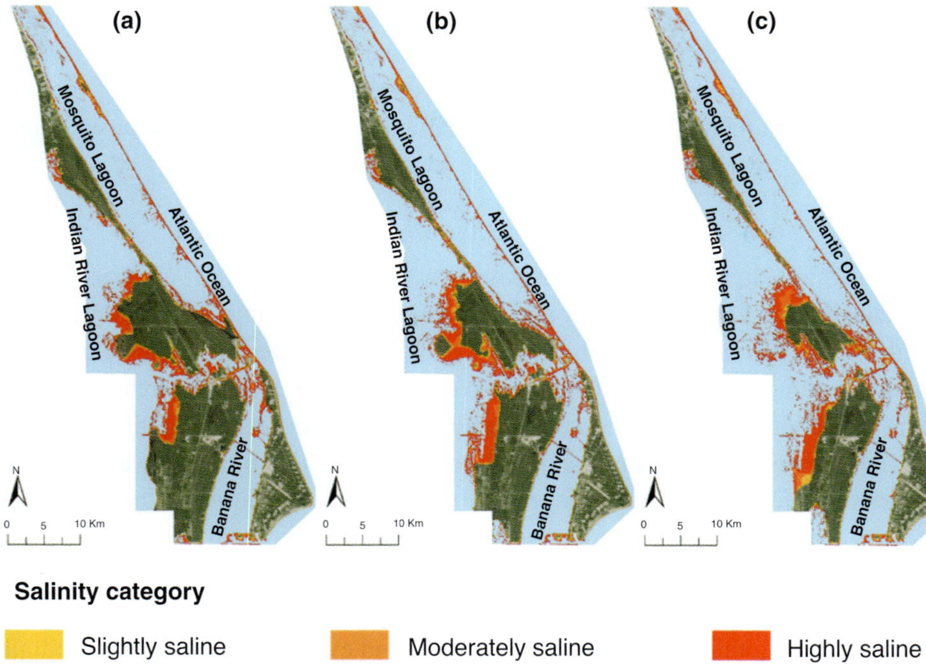

Figure 6.5 Spatial distribution of highly, moderately and slightly saline zones: (a) 13.2 cm sea-level rise; (b) 31.0 cm sea-level rise; (c) 58.5 cm sea-level rise

recover quickly once the salinity declines. However, other species die off quickly and cannot recover from high levels of salinity. Potential consequences of exposure to salinity include, but are not limited to, shift of fresh or less saline marsh to brackish or saline marsh, vegetation species dieback and limited recovery, shift of vegetation species composition from less salinity-tolerant species to more salinity-tolerant species and reduction in biomass production. Saltwater intrusion not only affects marshes/wetlands, but also affects agriculture as well. Citrus is the main agricultural product in this area and a reduction in citrus production due to increased groundwater salinity might cause socioeconomic issues. The impacts of sea-level rise are not significant in the eastern part due to high land surface elevation. Besides, highly permeable sand can facilitate rainwater infiltration and generate a high freshwater pressure head resulting in a high rate of submarine groundwater discharge, which builds up an effective freshwater hydraulic barrier to prevent the landward migration of saline groundwater so as to alleviate the negative effects of saltwater intrusion to some extent.

In order to prevent saltwater intrusion, it is of great importance to minimize the effect of sea-level rise. In order to 'balance' the rising sea level, it is necessary to increase the inland fresh groundwater pressure head by artificial recharge.

Recharge wells can be constructed close to the coastline, along with detention ponds designed for temporarily storing the excess rainwater during the wet season while draining to the coastal recharge wells for artificial recharge during the dry season. Artificial recharge is even more important in the dry season because of less infiltrated rainwater for fresh groundwater supplement.

6.4.3.2 Storm Surge Impacts on Saltwater Intrusion

Based on the output from the diagnostic model, the temporal variation of the simulated surficial aquifer TDS concentrations after the occurrence of storm surge and the temporal variation of the locations of the saltwater/freshwater transition zone are highlighted in Figure 6.6. An expansion of the highlighted areas indicates inland migration of the transition zone, while a contraction of the highlighted areas indicates seaward migration of the transition zone. Note that the storm surge in this case study refers to the one associated with Hurricane Jeanne (the deadliest Category 3 hurricane with exceptionally large size and strength in the 2004 Atlantic hurricane season and the thirteenth costliest hurricane in US history), which swept over the Caribbean area including Puerto Rico, the Dominican Republic, Haiti and the Bahamas beginning in mid-September 2004 and eventually provoked a landfall near Stuart, Florida (USA) in late-September 2004 that caused five direct deaths and \$6.8 billion in property damage in the mainland US.

From Figure 6.6, large amounts of coastal low-lying areas are overtopped by seawater temporally primarily due to low altitude and flat topography. Overtopping seawater can infiltrate into the unsaturated zone rapidly, developing salt plumes moving downwards and reaching the surficial aquifer, eventually resulting in vertical saltwater intrusion. However, due to sufficient rainfall and high aquifer permeability, dilution/flushing of the infiltrated saltwater is also rapid because a high rate of fresh groundwater recharge from large amounts of infiltrated rainwater can generate a downgradient groundwater flow contributing to transporting saltwater to the surrounding waterbodies, i.e., the coastal lagoons and the Atlantic Ocean. Therefore, the severe extent of saltwater intrusion is alleviated rapidly and the dilution/flushing rate is: (1) relatively fast in the first stage (from the years of 2004 to 2011), as shown in Figure 6.6(a–h); and (2) relatively slow in the second stage (from the years of 2012 to 2024), as shown in Figure 6.6(i–n). The models suggest that large amounts of saltwater could be removed from the surficial aquifer from 2024. The process of saltwater recession based on dilution/flushing by infiltrated rainwater is usually called ‘natural remediation’.

Based on the output from the diagnostic model, the temporal variation of percentage of saline zones are calculated (see Table 6.3) and plotted (see Figure 6.7). No increases in the percentage of saline zones are expected before the occurrence of a storm surge (25 September 2004). Right after the occurrence of a

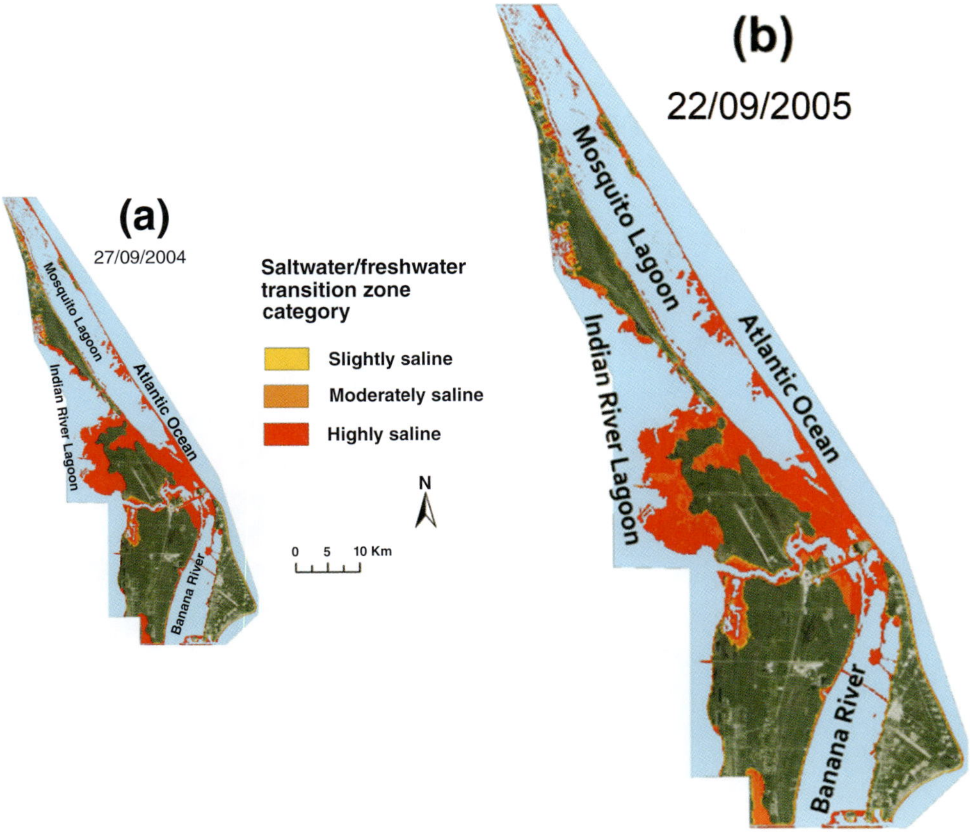

Figure 6.6 Temporal variation of the simulated surficial aquifer TDS concentrations after the occurrence of storm surge (plan view): (a) 27 September 2004; (b) 22 September 2005; (c) 17 September 2006; (d) 12 September 2007; (e) 6 September 2008; (f) 1 September 2009; (g) 1 September 2010; (h) 1 September 2011; (i) 31 August 2012; (j) 31 August 2013; (k) 3 August 2014; (l) 30 August 2016; (m) 29 August 2020; (n) 29 August 2024

storm surge, the percentage of saline zones exhibit a significant increase of 18.79 per cent. Then, the percentage increase rises steadily until a peak (19.42 per cent) is reached (25 May 2005), and declines following a first-order exponential decay curve. From the years 2005 to 2011, the decline rate of the percentage increase curve was relatively fast (annual decline rate equivalent to 2.70 per cent), indicating that large amounts of infiltrated saltwater were diluted and flushed to the surrounding water bodies. From the years of 2012 to 2024, the decline rate of the percentage increase curve is relatively slow (annual decline rate approximately 0.16 per cent), indicating that the dilution/flushing process continues at a much slower rate and the surficial aquifer TDS concentrations return to normal gradually.

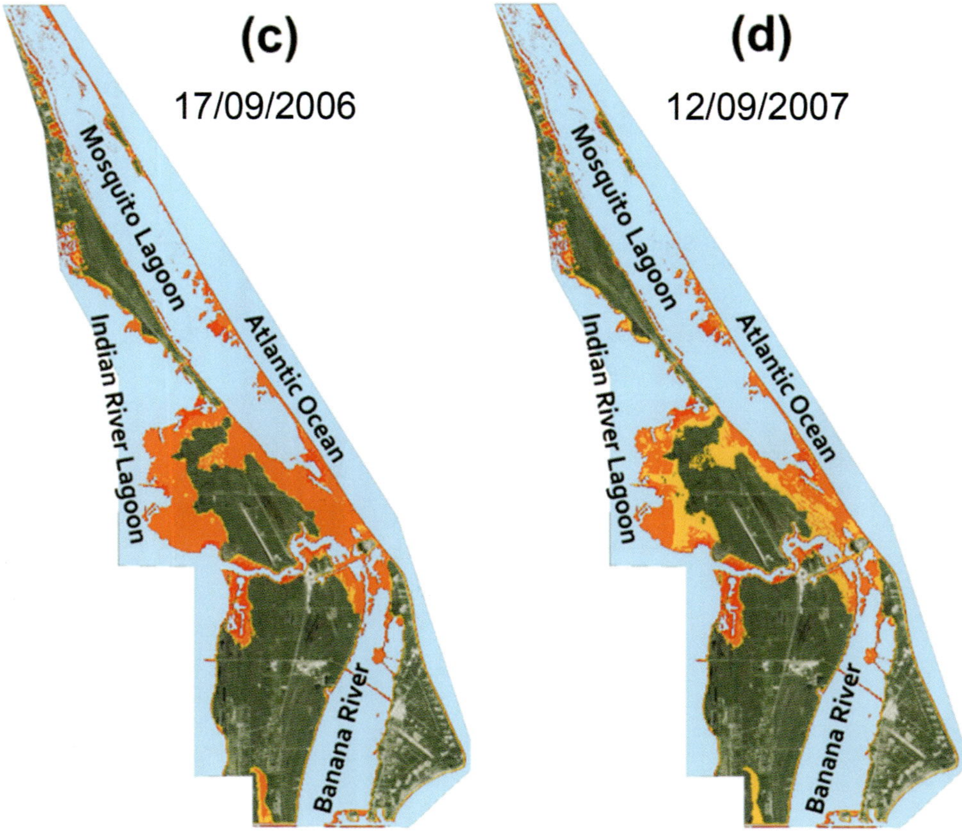

Figure 6.6 (*cont.*)

It is estimated that it takes approximately eight years for the infiltrated rainwater to dilute most of the infiltrated saltwater to the surrounding waterbodies, that is, the coastal lagoons and the Atlantic Ocean. However, it might still take several years for a total recovery of the surficial aquifer water quality through natural remediation processes.

The effects of a storm surge from tropical cyclones on the extent of saltwater intrusion into the surficial aquifer is a major concern for the local ecosystem in the study area. Storm surge sweeping over the study area can cause saltwater overtopping on coastal lowlands in that waves can reach and pass over the crest of the coastal defence structure (dunes) if wave run-up and storm tide levels are high enough. After a storm surge has receded, stagnant seawater that is trapped in the topographic depressions in the coastal low-lying areas can act as effective point-sources of contamination because ponding saltwater can infiltrate into the porous media rapidly. In this scenario, open wells in the coastal lowlands can expedite the

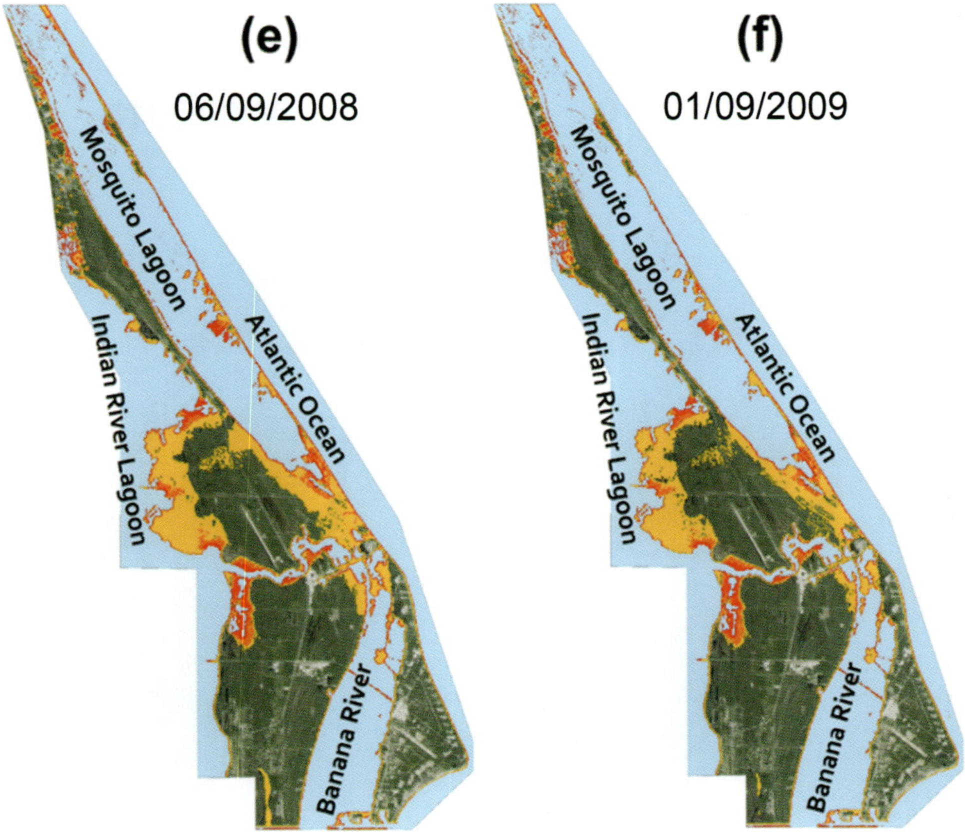

Figure 6.6 (*cont.*)

rapid vertical transmission of saltwater, resulting in saltwater intrusion into the surficial aquifer.

Following the occurrence of a storm surge, the infiltrated saltwater forms a layer of denser fluid above the lighter freshwater layer, and the formed saltwater plume is 'forced' downgradient (seaward) by infiltrated rainwater to the surrounding water bodies, that is, the coastal lagoons and the Atlantic Ocean, resulting in a reduction of surficial aquifer TDS concentrations. Usually, the natural remediation process takes several months to several years, dependent on hydro-climatic and hydrogeological conditions. It can be fast if the aquifer permeability is high and large amounts of infiltrated rainwater are available, or slow if the aquifer permeability is low and infiltrated rainwater is limited. From the simulation results, it is estimated that the natural remediation process might take approximately eight years. It is very likely that the natural remediation process might be faster and the recovery time might be shorter if storm surge from tropical cyclones and

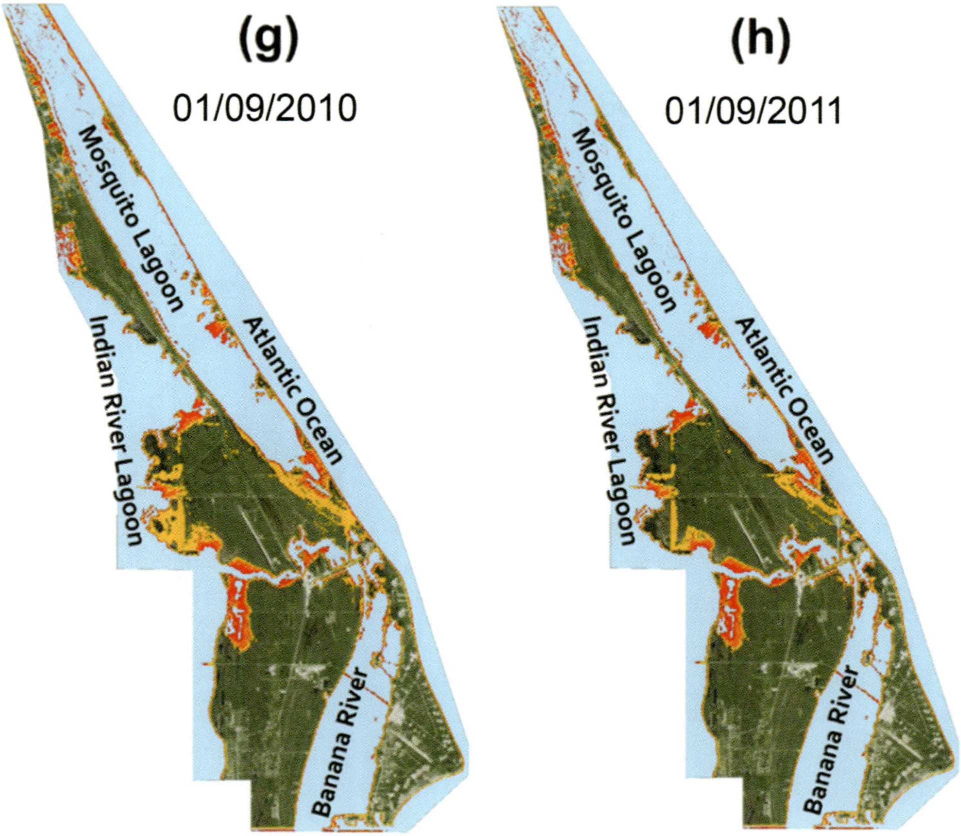

Figure 6.6 (*cont.*)

hurricanes struck the study area earlier in the wet season, say May or June, since a high rate of freshwater infiltration caused by large rainfall/storms later in the wet season could contribute to saltwater dilution/flushing.

Increasing salinity can create problems for plant species by altering their metabolic pathways and rates of activities and affect an area at various scales depending on the magnitude of the changes in salinity level. Some plant species can tolerate a wide range of salinity and can recover soon once salinity declines, while some plant species cannot tolerate increasing salinity and cannot recover even if salinity declines afterwards. Increasing salinity can reduce citrus production, which is the main agricultural product in the study area. In addition, increasing salinity could affect the water supply from abstraction wells for agricultural and domestic use, since up-coning of deeper saline groundwater can occur and coastal wells are particularly at high risk of being submerged by overtopping saltwater.

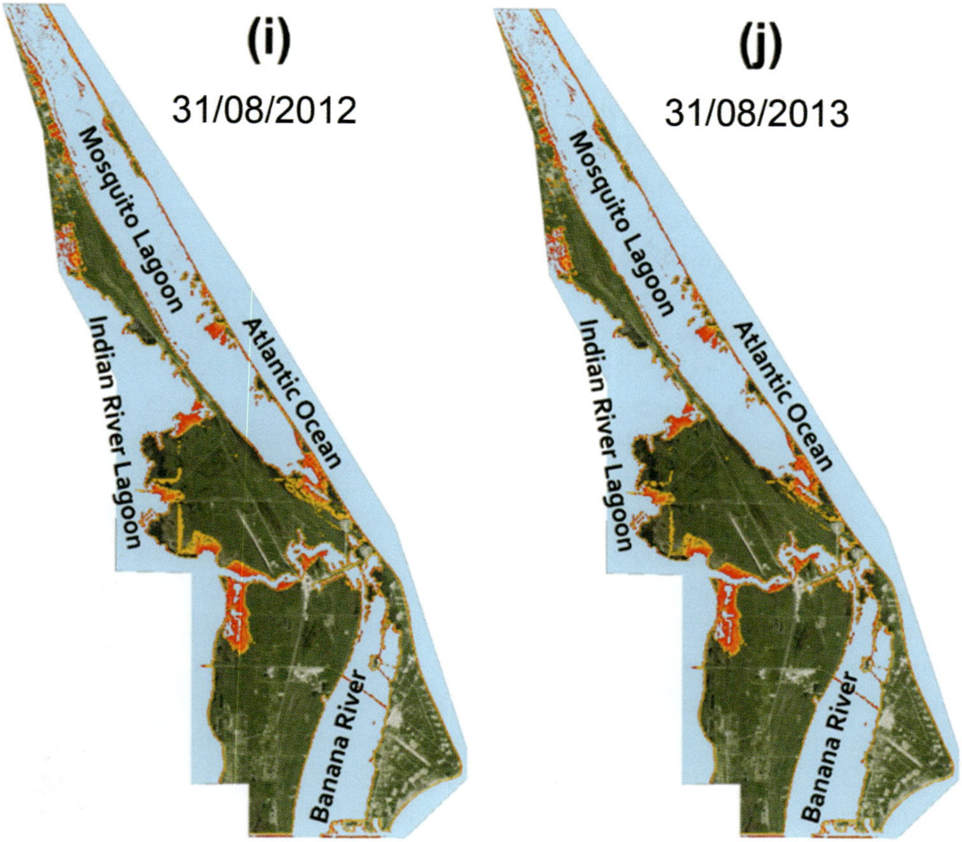

Figure 6.6 (*cont.*)

Complementary to natural remediation, the extent of saltwater intrusion could be further mitigated by artificial remediation. This includes the implementation of freshwater injection barriers through deep-well injection, infiltration of freshwater, reclaimed water near the shoreline, as well as increasing artificial recharge in upland areas to enlarge submarined groundwater discharge. In this area, engineering solutions can effectively generate a hydraulic barrier for impeding further inland migration of saltwater and provide a downgradient freshwater discharge for saltwater dilution and flushing. Furthermore, it is extremely important and necessary to prevent saltwater overtopping in coastal low-lying areas. Construction and maintenance of artificial coastal defence structures such as seawalls can stabilize beach and shoreline and protect coastal lowlands from saltwater overtopping. However, the construction of breakwaters/seawalls is usually costly and might not be necessary. In this study area, where the harbour, dock and channel inlet/outlet are not artificial, the main role of breakwaters/

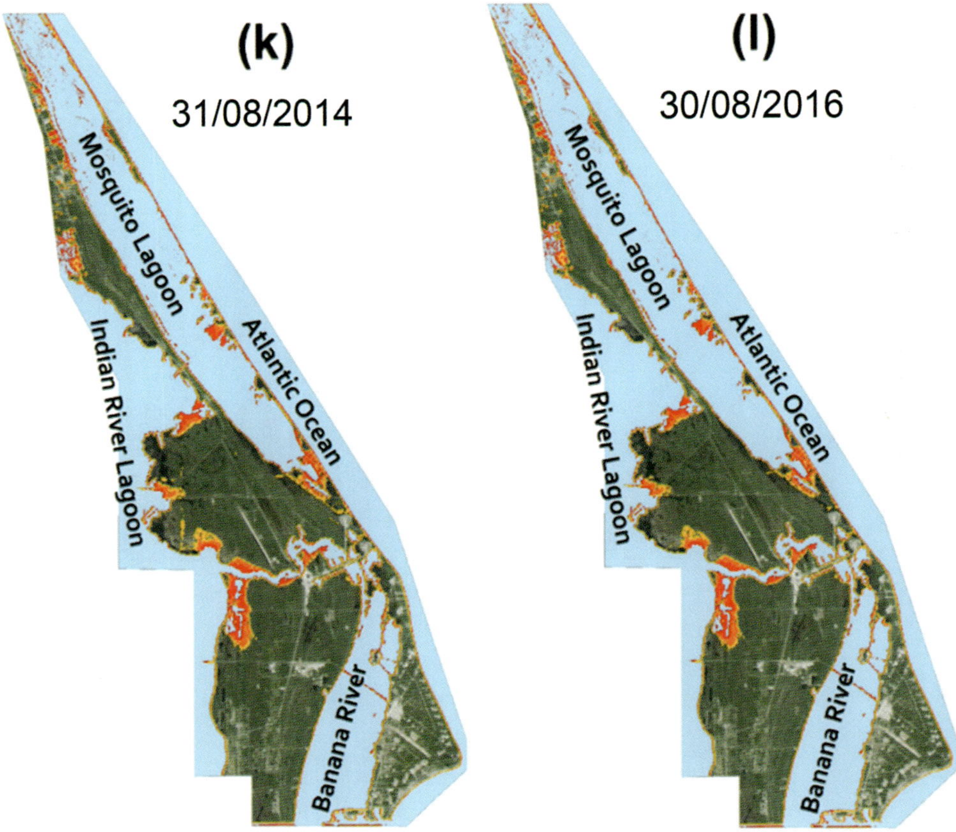

Figure 6.6 (*cont.*)

seawalls would be to absorb the shock of waves to protect the coastline from high energy waves. Thus, an economical but efficient measure could be to heighten and strengthen the existing sand dunes and build up 'new' dunes where sand dunes are destroyed or missing near the shoreline. It is expected that the extent of saltwater intrusion could be greatly compensated if saltwater overtopping in coastal lowland does not occur.

6.4.3.3 Uncertainty Analysis

The simulation results provide an initial exploration of the impacts of sea-level rise and storm surge on the extent of saltwater intrusion into the surficial aquifer. However, it should be noted that the simulation results might be biased in that uncertainties can be associated with the reference, prediction and diagnostic models.

First, limited sampling data for calibration leads to large uncertainty resulting in a less robust model calibration process. The simulated hydraulic heads output from

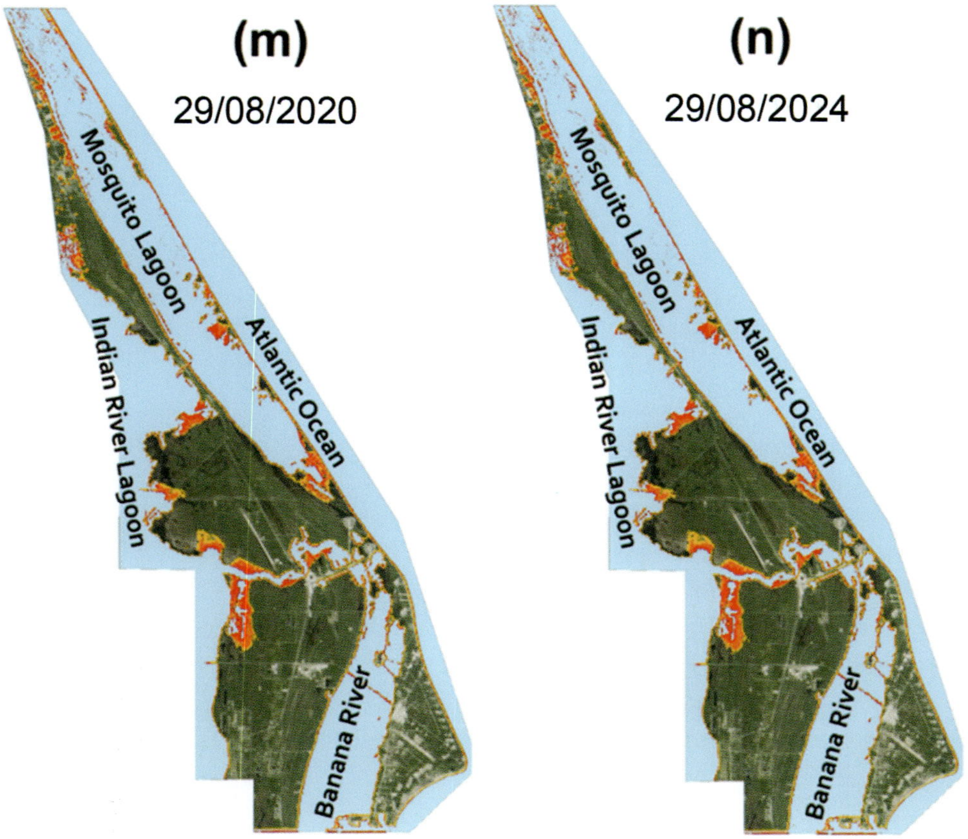

Figure 6.6 (*cont.*)

the reference model are calibrated with field-measured groundwater levels while the simulated TDS concentrations are not calibrated due to the lack of salinity measurements. Moreover, both the simulated hydraulic heads and TDS concentrations output from the diagnostic model are not calibrated due to the lack of groundwater level and salinity measurements. It is expected that more observed groundwater level and salinity data would be available for future model calibration. However, a future groundwater monitoring system implemented in this study area might not be beneficial since the data required to calibrate the diagnostic model will not be able to capture the crucial information resulting from a storm surge that occurred several years ago.

Second, there is uncertainty associated with the spatial variation of hydrogeologic parameters (horizontal/vertical hydraulic conductivity, porosity, longitudinal/transverse/vertical dispersivity and diffusion coefficient). Due to the lack of geophysical surveys and borehole tests, the spatial variation of

Table 6.3. Temporal variation of percentages of saline zones

Time	Percentage (%)	Growth Rate (%)
August 2004	13.6	–
September 2004	32.3	18.7
January 2005	32.8	19.2
May 2005	33.0	19.4
September 2005	33.0	19.4
January 2006	32.8	19.2
May 2006	32.6	19.0
September 2006	32.2	18.6
January 2007	31.7	18.1
May 2007	31.1	17.5
September 2007	30.6	17.0
January 2008	29.9	16.3
May 2008	29.1	15.5
September 2008	28.1	14.5
January 2009	27.1	13.5
May 2009	26.1	12.5
September 2009	25.1	11.5
September 2010	19.4	5.8
September 2011	16.8	3.2
September 2012	15.8	2.2
September 2013	15.1	1.5
September 2014	14.6	1.0
September 2016	14.3	0.7
September 2018	14.1	0.5
September 2020	14.0	0.4
September 2022	14.0	0.4
September 2024	13.9	0.3

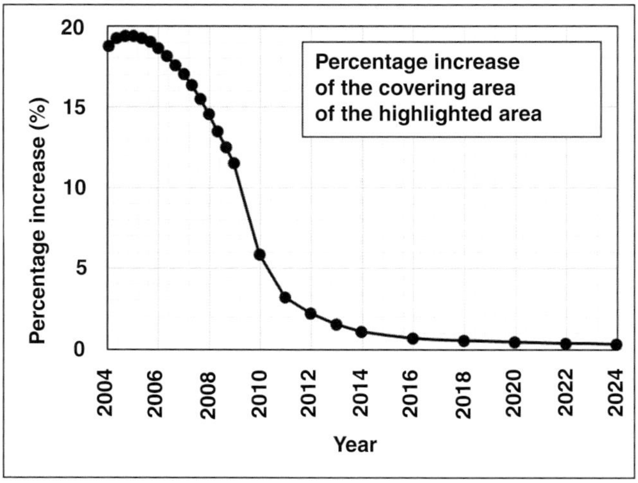

Figure 6.7 Percentage increase of the covering area of the highlighted areas

hydrogeologic parameters are unknown and the surficial aquifer is deemed as homogenous with uniform values initially assigned to all model grid cells. Ignoring the impacts of heterogeneity of the surficial aquifer simplifies the model implementation and is considered appropriate and reasonable for the implementation of a regional-scale model although it reduces ineluctably the accuracy of the simulated salinity at a local scale.

Third, we acknowledge that the presented models remain a simplification of the complex processes involved in saltwater intrusion associated with variable-density conditions and changing climate. They underestimate the uncertainty of various parameters as a direct consequence of their assumptions, which include that: (1) surface runoff is negligible and the overtopping seawater infiltrates downwards vertically and reaches the water table eventually; (2) the residual overtopping seawater is drained out after the 3.5-day effective storm surge duration so that the saltwater ponding depth is zero afterwards; (3) the inland encroachment of the coastline after the storm surge is negligible; (4) continuous sea-level rise and additional storm surges associated with other tropical cyclones under the complex changing climate are ignored; and (5) the effects of astronomic tides are negligible. Among the aforementioned assumptions, (4) has a larger range of uncertainty and is expected to have a larger effect on the simulation results than (1), (2), (3) and (5). Due to the El Niño conditions in a greenhouse-enhanced world, an increase in both the frequency and intensity of extreme weather events such as tropical storms and hurricanes, in combination with continuous sea-level rise, are expected to bring more frequent and intense storm surges to coastal areas. Their magnitude depends on the dynamics of storms, coastal topography and the position of the coastline relative to the storm track. Consequently, if stronger and more persistent El Niño conditions are projected for the Caribbean area over the next few years, the surficial aquifer of the study area will be under greater threat from storm surge and sea-level rise, as well as subsequent further saltwater intrusion of its limited freshwater resources.

Fourth, the differences in the water levels and TDS concentrations between the coastal lagoons and the Atlantic Ocean are not considered. The mean water levels of the coastal lagoons are usually slightly higher than the Atlantic Ocean (10 centimetres approximately). The TDS concentrations of the coastal lagoons and the Atlantic Ocean vary spatially and are affected by climatic and anthropogenic factors. In some regions, the TDS concentrations of the coastal lagoons are higher than 35 kg m^{-3}, particularly during the dry season, because of large amounts of water loss due to evaporation and drainage. The TDS concentrations of the nearshore Atlantic Ocean might be lower than 35 kg m^{-3} because of submarine groundwater discharge. However, due to a lack of relevant data, water levels of the coastal lagoons are assumed to be exactly the same as the Atlantic Ocean, and TDS concentrations of the coastal lagoons and the Atlantic Ocean are assumed to be equal to 35 kg m^{-3}.

Fifth, upward migration of saline groundwater from the underlying highly mineralized Floridan aquifer is not considered. An upward groundwater seepage from the Floridan aquifer is not excluded. If true, saltwater intrusion is likely to be under-estimated by our predictive models.

Sixth, the 'new' coastline determined for the predicted models under various sea-level rise scenarios contains high levels of uncertainty, resulting in the predicted values of saltwater intrusion being over- or under-estimated-probably.

In short, uncertainties mainly resulted from deficiencies in sampling data for model calibration, limitations in spatial variation of hydrogeological parameters and assumptions made for simplification of model implementation. However, it is necessary to simplify the model development procedure to achieve the objectives of this case study. It is well known that the salinity level of the surficial aquifer is extremely important to the local ecosystem, rich in biodiversity, and saltwater intrusion can greatly threaten the survival of endangered species. Considering the importance of the surficial aquifer salinity level, the extent of saltwater intrusion should be investigated periodically, as the location and migration of the saltwater/freshwater transition zone may change over time. Therefore, there is an urgent need for systems to continuously monitor changes in groundwater levels and TDS concentrations of the surficial aquifer, as well as conduct geophysical surveys and borehole tests to determine soil characteristics.

6.5 Mitigation Processes

Saltwater intrusion mitigation measures take time to become effective. Replacing an exceptionally large amount of saline groundwater by fresh groundwater needs time. Controlling or reducing saltwater intrusion can be greatly improved by monitoring coastal aquifer water quality through the identification of the spatial and temporal distribution of saline groundwater. Conventional approaches usually implement a network of observation wells at various depths around water supply wells to analyse and sample piezometer heads and solute concentrations. Novel approaches such as geo-electrical sounding techniques, which detect specific resistivity values of subsurface layers (highly affected by porewater salinity), have already been applied in various coastal areas worldwide. Common mitigation processes include, but are not limited to, a reduction of groundwater withdrawal rate and/or landward relocation of coastal pumping wells, implementation of fresh groundwater hydraulic barriers through direct recharge or deep-well infiltration of rainwater or treated reclaimed water near the shoreline, extraction of saline/brackish groundwater, land reclamation, increasing upland artificial freshwater recharge, construction of physical barriers such as sheet piles and clay trenches, etc.

References

Anderson, M. P., & Woessner, W. W. (1991). *Applied Groundwater Modeling: Simulation of Flow and Advective Transport*. San Diego, CA: Academic Press.

Barlow, P. M., & Reichard, E. G. (2010). Saltwater intrusion in coastal regions of North America. *Hydrogeology Journal* 18: 247–260.

Bear, J. (1979). *Hydraulics of Groundwater*. New York: McGraw-Hill.

Bear, J., Cheng, A., Sorek, S., et al. (1999). *Seawater Intrusion in Coastal Aquifers: Concepts, Methods and Practices (Theory and Applications of Transport in Porous Media)*. Netherlands: Kluwer Academic Publishers.

Bocanegra, E., Da Silva, G. C., Custodio, E., et al. (2010). State of knowledge of coastal aquifer management in South America. *Hydrogeology Journal* 18: 261–267.

Custodio, E. (2010). Coastal aquifers of Europe: An overview. *Hydrogeology Journal* 18: 269–280.

Diersch, H. J. G. (1996). Interactive, graphics-based finite-element simulation system FEFLOW for modeling groundwater flow, contaminant mass and heat transport processes, *FEFLOW User's Manual Version 4.5*. Institute for Water Resources Planning and System Research, Ltd.

Drabbe, J., & Badon Ghyben, W. (1888–1889). Notes on the probable results of the proposed well drilling near Amsterdam: The Hague. *Koninklijk Instituut Instagram Tijdschrift* 8–22.

Freeze, R. A., & Cherry, J. A. (1979). *Groundwater*. Englewood Cliffs, NJ: Prentice Hall.

Guo, W., & Langevin, C. D. (2002). User's guide to SEAWAT: A computer program for simulation of three-dimensional variable-density ground-water flow. *Techniques of Water-Resources Investigations* 06-A7: 77.

Herzberg, A. (1901). The water supply on parts of the North Sea coast: Munich. *Journal of Gasbeleucht and Wasserversorg* 44: 815–819, 842–844.

Holzbecher, E. (1998). *Modeling Density-Driven Flow in Porous Media: Principles, Numerics, Software*. Berlin: Springer-Verlag.

Huyakorn, P. S., Anderson, P. F., Mercer, J. W., et al. (1987). Saltwater intrusion in aquifers: Development and testing of a three-dimensional finite-element model. *Water Resources Research* 23(2): 293–312.

Kipp, K. L. (1986). HST3D – A computer code for simulation of heat and solute transport in three-dimensional groundwater flow systems. *International groundwater modeling center, U.S. Geological Survey Water Resources Investigations Report 86-4095*.

Lin, J., Snodsmith, B., Zheng, C., et al. (2006). A modeling study of seawater intrusion in Alabama Gulf coast, USA. *Environmental Geology* 57: 119–130.

Oude Essink, G. H. P. (1998). Simulating density dependent groundwater flow: The adapted MOC3D. In *Proceedings of the 15th Saltwater Intrusion Meeting*, Ghent, Belgium.

Rosenzweig, C., Horton, R. M., Bader, D. A., et al. (2014). Enhancing climate resilience at NASA centers: A collaboration between science and stewardship. *Bulletin of the American Meteorological Society* 95(9): 1351–1363.

Sauter, F. J., Leijnse, A., & Beusen, A. H. W. (1993). *METROPOL's User Guide*. Netherlands: National Institute of Public Health and Environmental Protection.

Sherif, M. M., & Singh, V. P. (1999). Effect of climate change on sea water intrusion in coastal aquifers. *Hydrological Processes* 13(8): 1277–1287.
Shi, L., & Jiao, J. J. (2014). Seawater intrusion and coastal aquifer management in China: A review. *Environmental Earth Sciences* 72: 2811–2819.
Steyl, G., & Dennis, I. (2010). Review of coastal-area aquifers in Africa. *Hydrogeology Journal* 18: 217–225.
Van Biersel, T. P., Carlson, D. A., & Milner, L. R. (2007). Impact of hurricane storm surges on the groundwater resources. *Environmental Geology* 53: 813–826.
Voss, C. I. (1984). A finite element simulation model for saturated-unsaturated, fluid-density-dependent groundwater flow with energy transport or chemically-reactive single species solute transport. *Water Resources Investigation Report 84-4369, U.S. Geological Survey.*
Ward, D. S. (1991). Date input for SWIFT/386 (version 2.50). *Geo-trans Technical Report*, Sterling, VA.
Werner, A. D. (2010). A review of seawater intrusion and its management in Australia. *Hydrogeology Journal* 18: 281–285.
White, I., & Falkland, T. (2010). Management of freshwater lenses on small Pacific islands. *Hydrogeology Journal* 18: 227–246.
Xiao, H., Wang, D., Hagen, S. C., et al. (2016). Assessing the impacts of sea-level rise and precipitation change on the surficial aquifer in the low-lying coastal alluvial plains and barrier islands, east-central Florida (USA). *Hydrogeology Journal* 24(7): 1791–1806.
Xiao, H., Wang, D., Medeiros, S. C., et al. (2019). Exploration of the effects of storm surge on the extent of saltwater intrusion into the surficial aquifer in coastal east-central Florida (USA). *Science of the Total Environment* 648: 1002–1017.
Zienkiewicz, O. C. (1977). *The Finite Element Method* (3rd ed.). London: McGraw-Hill.

Part II

Climate Risk to Human and Natural Systems

7

Observed Urban Effects on Temperature and Precipitation in Southeast China

SIAO SUN, GUANGDONG LI AND QIUHONG TANG

7.1 Introduction

Over half of the world's population are now residing in urban areas and, by 2050, 66 per cent of the world's population is projected to be urban (UN, 2014). Urbanization has been recognized as an important contributor to local and regional climate change. The impact of urbanization on the thermal environment has been extensively studied, with the majority of studies reporting an urban warming trend in many regions across the world (e.g., Jones et al., 1989; Zhou et al., 2004; Ren et al., 2008; Kataoka et al., 2009; Ren & Zhou, 2014; Santamouris, 2015; Sun et al., 2016). The change of land surfaces (which modifies surface albedo, roughness and heat capacity) and the generation of anthropogenic waste heat in urban areas (Rizwan et al., 2008; Huang et al., 2009; Li et al., 2013) can lead to a temperature excess in urban areas, termed an urban heat island (UHI) effect. The UHI is of great concern as it may cause many side effects on the environment and ecosystems, such as augmenting global warming, intensifying pollutant concentration, impacting local wind, cloud, fog and precipitation patterns and elevating ground-level ozone (e.g., Rosenfeld et al., 1998; Sarrat et al., 2006; Mirzaei & Haghighat, 2010), and will likely lead to further economic and social implications.

While UHI is perhaps the best documented example of an urban effect on climate, precipitation change due to urbanization has been less studied in the literature. Evidence shows that the range of impacts from urbanization on precipitation may vary over a large spectrum, from negligible to increasing or decreasing effects (e.g., Rosenfeld, 2000; Burian & Shepherd, 2005; Kaufmann et al., 2007; Grimm et al., 2008). In particular, increased precipitation in urban areas is termed an urban rainfall island (URI) effect, which has been mostly associated with the intense convective rainfall in warmer seasons rather than

Funding for this research is provided by the National Natural Science Foundation of China (41790424, 41730645).

lighter, stratiform rainfall (Changnon, 1979; Burian & Shepherd, 2005; Mote et al., 2007; Kishtawal et al., 2009; Shepherd et al., 2010). The identification and quantification of URI are important because a modified precipitation pattern may affect weather prediction, flood forecasting, urban drainage design and water management at the urban catchment scale (Burian & Shepherd, 2005).

The influence that urbanization exerts on temperature or precipitation is often regarded as a bias in observational data. The detection of the urban bias based on observational data is mostly made via the comparison of climatic parameters (either temperature or precipitation) between rural and urban observations, which typically involves a station pairing technique (i.e., assigning stations to pairs of rural and urban sites), based on surrogates such as land use/land cover (Gallo et al., 1996), population data (Jones et al., 1990; Huang et al., 2009), satellite night lights (Gallo et al., 1999; Hansen et al., 2001) or a combination of them (Ren & Zhou, 2014). For instance, temperature changes due to urban effect were often assessed as the difference in observed temperature in pairs of urban and rural sites (Jones et al., 1990; Wang et al., 1990; Gallo et al., 1996; Peterson, 2003; Ren et al., 2008; Ren & Zhou, 2014) or urban and rural grids interpolated from observational data (Jones et al., 1989). Previous URI effect studies relied on comparative analysis of precipitation variables in urban and non-urban locations (or upwind and downwind urban control regions) where different impacts from urbanization were observed (e.g., Huff & Changnon, 1972; Changnon et al., 1991; Burian & Shepherd, 2005; Mote et al., 2007). The results of this approach, however, depend heavily on the classification method used to distinguish rural from urban sites. Another prevalent approach for the quantification of the urban effect on temperature change uses reanalysis data, which is compiled based on a large number of observations from various sources (Kistler et al., 2001; Kanamitsu et al., 2002). The temperature difference between the trends in observations and reanalysis data is considered to represent the impact of land-use changes including urbanization on temperature changes, as the reanalysis temperature data estimated mainly from atmospheric parameters are not sensitive to changes in land surfaces (Kalnay & Cai, 2003; Zhou et al., 2004; Jones et al., 2008). This approach, however, is not suitable for investigating urban impacts on precipitation, because precipitation reanalysis data are determined by the model using data assimilation techniques and should be compared with model-independent estimates before being used (Kanamitsu et al., 2002). Recently, a new approach of using continuous empirical relationships between urbanized indicator and temperature has been developed to investigate the urban effect on temperature (Schatz & Kucharik, 2015; Wang et al., 2017), with the advantage of not requiring the classification of observational sites into urban and rural categories. However, the application of this approach is currently limited to studies that focus on the effects on temperatures on a city scale. It is also worth mentioning that, while the UHI effect was analysed across a wide range of spatial scales including global, regional and city scales, the URI effect was rarely examined beyond a city scale (Kishtawal et al., 2009).

The main literature studies have focused on urban effects on one climatic factor (either temperature or precipitation). Yet, a comprehensive study on urban effects in both temperature and precipitation on a regional/national scale in China is still missing. China has experienced fast economic growth and urban expansion during the last few decades, in particular from 1978 when the reform process started. The number of cities (including province-level, prefecture-level and county-level cities according to official statistics) soared from 190 in 1978 to 663 in 2010; and the proportion of urban population increased from 18 per cent to 50 per cent (National Bureau of Statistics of China, 2011). Changes in temperature and precipitation are two important environmental effects caused by rapid urbanization. As mentioned earlier in this section, the methods used to categorize observations into urban and rural classes may affect the estimation of the urban bias. This is particularly problematic since most meteorological stations in China are located in or near cities (Zhou et al., 2004). So far, the URI studies have been restricted to small areas, such as a single city or a metropolitan area, but have overlooked larger scales.

This study aims to examine UHI and URI effects on a large regional scale by using the relatively new approach that characterizes urbanization degrees with continuous indicators in the vicinity of observational sites in southeast China. The urban effects in temperature and precipitation are assessed using statistical analysis and linear regression techniques. The results are essential for understanding local climate changes induced by urbanization and for further counteracting their subsequent environmental, social and economic effects.

7.2 Data

We used temperature and precipitation data from the Chinese meteorological network which comprises a total of 671 meteorological stations in China. Data are available with a daily temporal resolution in 1961–2010, including daily maximum, mean and minimum temperatures and daily precipitation. The study area includes southeast China, which exhibits a relatively high urbanization rate, and the meteorological records of the region do not suffer much inhomogeneity (Li et al., 2004; Zhou et al., 2004). The climate of this region is mainly subtropical and marginal tropical humid. The study area counts 223 meteorological stations (after excluding those in islands and those with substantial data gaps).

The variables T_{min}, T_{max} and T_{mean} are annual averages of the daily maximum, mean and minimum temperature data, respectively. The DTR is defined as the difference between T_{max} and T_{min}. To explore potential seasonal variability in the data, we also compute the seasonal values of these four variables for spring (March–May), summer (June–August), autumn (September–November) and winter (December–February).

For precipitation variables, we considered annual precipitation P_{annual}, seasonal precipitation in each season (P_{spring}, P_{summer}, P_{autumn} and P_{winter}) and annual maximum precipitation in one and three days, noted as P_{daily} and $P_{3\text{-}day}$, respectively. P_{daily} and $P_{3\text{-}day}$ are used to study extreme precipitation properties.

We used data on land-use and the population of China provided on a 1 × 1 kilometre spatial grid to estimate urbanization information at grid-cell level in the study area. The land-use map of China in 2010, which was developed based on remote sensing images of Landsat TM and Landsat ETM scenes (www.resdc.cn),[1] provides the percentage value of urban areas for each pixel. The population map of China in 2010 was generated based on census data and land-use information (www.geodoi.ac.cn/WebEn/doi.aspx?Id=131).[2]

7.3 Methodology

While most Chinese meteorological stations are located in and near cities (Zhou et al., 2004), the vicinities of observational sites still present variations in urbanization levels in terms of built-up area sizes, populations, human activities, etc. This study uses two continuous variables to measure the degree of urbanization degrees within a circular buffer area of each meteorological station. The variables include: (1) the urban area fraction, based on the land-use map, and (2) population density, based on the population density map. To identify the optimal influential range of the urban effect, we employed and compared buffer areas of different sizes using different radii: 1, 2, 3, 5, 10 and 15 kilometres. We acknowledge site-specific uncertainty of urbanization, which depends on land-use and population density data in 2010.

For each site, we compute the trend of temperature and precipitation in 1961–2010 and the average of temperature and precipitation in a recent period (i.e., 2001–2010). The Mann-Kendall method (Mann, 1945; Kendall, 1955) is applied to identify trends in the time series using a 5 per cent significance level. Significant trends are quantified through a linear regression model, while non-significant trends are assigned the value 0. We consider the average of climatic variables to represent the mean level during the short period (2001–2010) for two reasons: the correlation between the mean level in a long period and an urbanization indicator may not be independent of the correlation between an inherent trend and the urbanization indicator; the urbanization indicators are quantified based on data in 2010, which characterize better urbanization degrees in periods close to 2010.

To examine a potential association between climatic variables and urbanization indicators, we compute two types of correlation coefficients (CCs), including the Pearson CC (r) and the Spearman's rank CC(ρ), and estimate their level of significance. r measures the linear dependency of two variables, while the non-parametric ρ assesses how well a relation between two variables can be described using a monotonic function. If r or ρ between a climatic factor and an urbanization indicator is significant (at the 5 per cent significance level in this study), the difference in the climatic factor at different sites can possibly be linked to an urban effect. It is worth noting that significant CCs between climatic and urbanization parameters are not equivalent to causal relations between two variables. They only provide empirical evidence of urban bias in temperature or precipitation, like other approaches that assess UHI and URI effects based on observational temperature and precipitation data.

If the CC analysis indicates that variation in a climatic variable can be linked to an urbanization indicator, urban bias in that climatic factor is approximately quantified using the linear regression result. The linear regression can be written:

$$x_i = \beta u_i + \alpha \tag{7.1}$$

where x_i represents the climatic variable at site i (either trend or mean value of temperature or precipitation), u_i is the urbanization level at site i, and β α are the linear coefficient and intercept, respectively, estimated in the linear regression model framework. The average urban bias over all sites associated with temperature or precipitation is estimated as the product of the linear coefficient β and the mean urbanization degree across all sites.

$$B_{\text{urban}} = \beta \frac{\sum_{i=1}^{n} u_i}{n} \tag{7.2}$$

where B_{urban} indicates the average change of a corresponding climatic variable across all observational sites due to urban effect, when compared to a rural site with urban area fraction or population density equal to 0, and n is the total number of sites.

7.4 Trends and Mean Values of Temperature and Precipitation

According to the Mann-Kendall test results, the annual T_{min}, T_{max} and T_{mean} in most meteorological stations (over 70 per cent) present a rising trend significant at the 5 per cent significance level (Figure 7.1). The average increasing rates in T_{min}, T_{max} and T_{mean} in different sites are 0.24, 0.15 and 0.18°C/decade, respectively

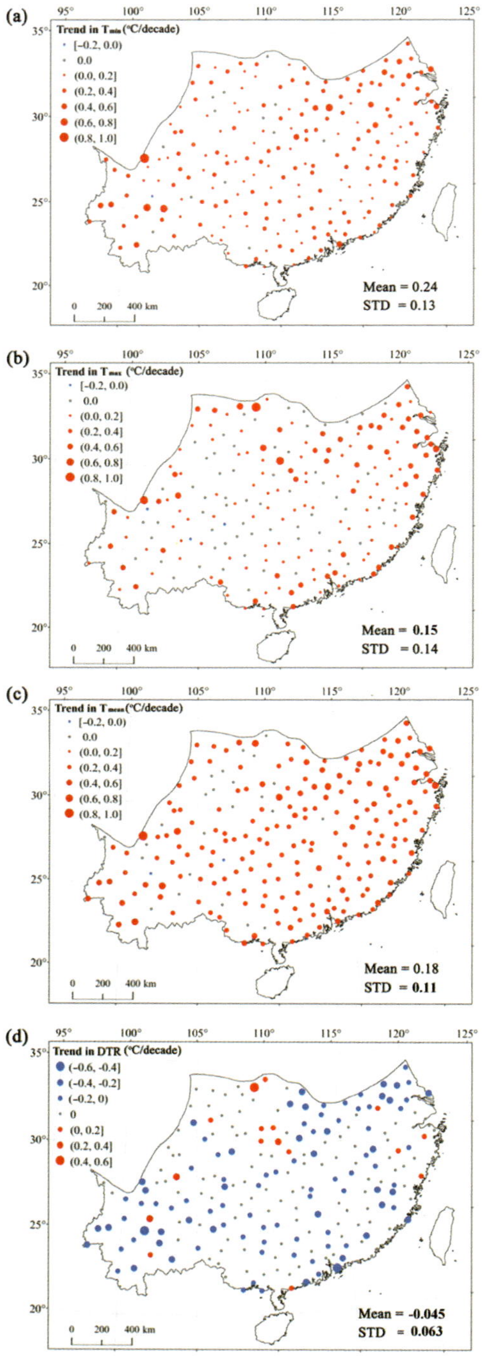

Figure 7.1 Trends in temperature variables at different meteorological sites (a total of 223 sites) in southeast China from 1961 to 2010: (a) minimum daily temperature (T_{min}): data in 210, 1 and 12 stations present, respectively, increasing,

(Figure 7.1). T_{min} has risen faster than T_{max}, in line with previous findings (Easterling et al., 1997; Zhou et al., 2004), which results in an overall downward trend in DTR of $-0.045°C/decade$.

We also inferred seasonal trends associated with temperature changes (see Table 7.1). In all seasons, T_{min} shows a faster warming trend compared to T_{max} and T_{mean}. A more pronounced warming trend is observed in winter, which is in line with previous results (e.g., Zhou et al., 2004). Summer shows the least pronounced warming trend. The average trends in T_{min} are estimated at between 0.38 and 0.16°C/decade in winter and summer, respectively.

Mean values of the T_{min}, T_{max}, T_{mean} and DTR in 2001–2010 at individual sites characterize local temperature levels (Figure 7.2). The temperature levels gradually decrease from south to north in T_{min}, T_{max} and T_{mean} (Figure 7.2(a–c)). The mean values of DTR tend to rise with increasing distances to the southeast coastline (Figure 7.2(d)). The mean values of temperature variables exhibit seasonal difference. The mean levels of T_{min} are 22.7 and 5.3°C in summer and winter, respectively (see Table 7.1).

The Mann-Kendall test results show that, in most stations, the precipitation data do not exhibit any significant trend in any of the investigated periods. More specifically, there is no significant trend in 208, 203, 200, 171, 202, 208 and 211 stations (out of a total number of 223 stations) associated with the precipitation variables P_{annual}, P_{spring}, P_{summer}, P_{autumn}, P_{winter}, P_{daily} and $P_{3\text{-}day}$, respectively.

Figure 7.3 shows the average precipitation values (2001–2010) of P_{annual}, P_{daily} and $P_{3\text{-}day}$ for each observational site. We observe a spatial pattern, suggesting a decline in precipitation that increases with distance to the southeast coastline. The $P_{3\text{-}day}$ is around 50 per cent higher than P_{daily} and these two variables present a high correlation ($r = 0.96$). Table 7.2 shows that high seasonal variability is present, with the wettest season in summer (with an average precipitation level of 603 mm in 2001–2010 over the study region), followed by spring (372 mm), autumn (222 mm) and winter (128 mm).

7.5 Urban Effect on Temperature

We examine CCs including r and ρ between different temperature variables (including the trends and mean values) and urbanization indicators (i.e., the urban

---

Figure 7.1 (*cont.*) decreasing and no trends at the 5 per cent significance level; (b) maximum daily temperature (T_{max}): data in 164, 3 and 56 stations present, respectively, increasing, decreasing and no trends; (c) mean daily temperature (T_{mean}): data in 200, 2 and 21 stations present, respectively, increasing, decreasing and no trends; (d) daily temperature range (DTR): data in 14, 122 and 87 stations present, respectively, increasing, decreasing and no trends.

Table 7.1. Summarized statistics of trends and mean levels of temperature variables at different meteorological sites

		Year		Spring		Summer		Autumn		Winter	
		Mean	Standard deviation	Mean	Standard deviation	Mean	Standard deviation	Mean	Standard deviation	Mean	Standard deviation
Trends in temperature (°C/decade)	T_{min}	0.24	0.13	0.20	0.18	0.16	0.11	0.19	0.15	0.38	0.17
	T_{max}	0.15	0.14	0.17	0.22	0.054	0.14	0.16	0.15	0.12	0.17
	T_{mean}	0.18	0.11	0.17	0.17	0.078	0.11	0.17	0.13	0.28	0.14
	DTR	−0.045	0.063	−0.014	0.20	−0.097	0.14	−0.016	0.12	−0.14	0.19
Mean values of temperature (°C)	T_{min}	14.4	3.2	14.1	3.3	22.6	3.0	15.3	3.3	5.3	3.9
	T_{max}	22.9	2.9	23.2	2.8	30.8	2.9	23.8	3.4	13.5	4.5
	T_{mean}	17.9	3.0	18.0	2.9	25.9	3.0	18.7	3.3	8.7	3.9
	DTR	8.5	1.5	9.2	2.0	8.1	1.2	8.5	1.5	8.2	2.4

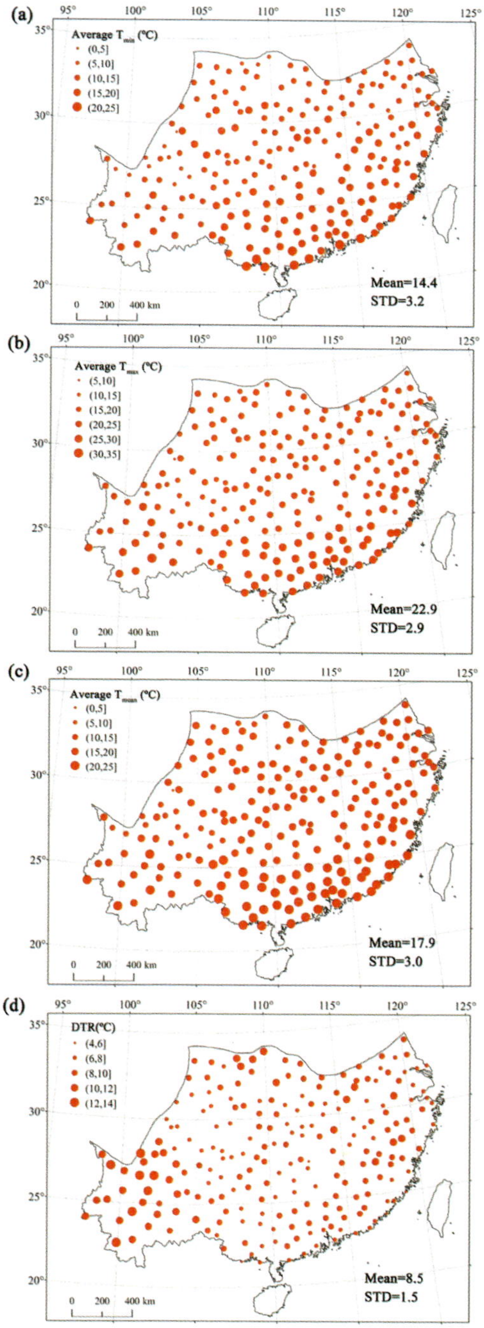

Figure 7.2 Mean values of temperature variables at different meteorological sites (a total number of 223 sites) in southeast China in 2001–2010 at individual sites: (a) minimum daily temperature (T_{min}); (b) maximum daily temperature (T_{max}); (c) mean daily temperature (T_{mean}); (d) daily temperature range (DTR)

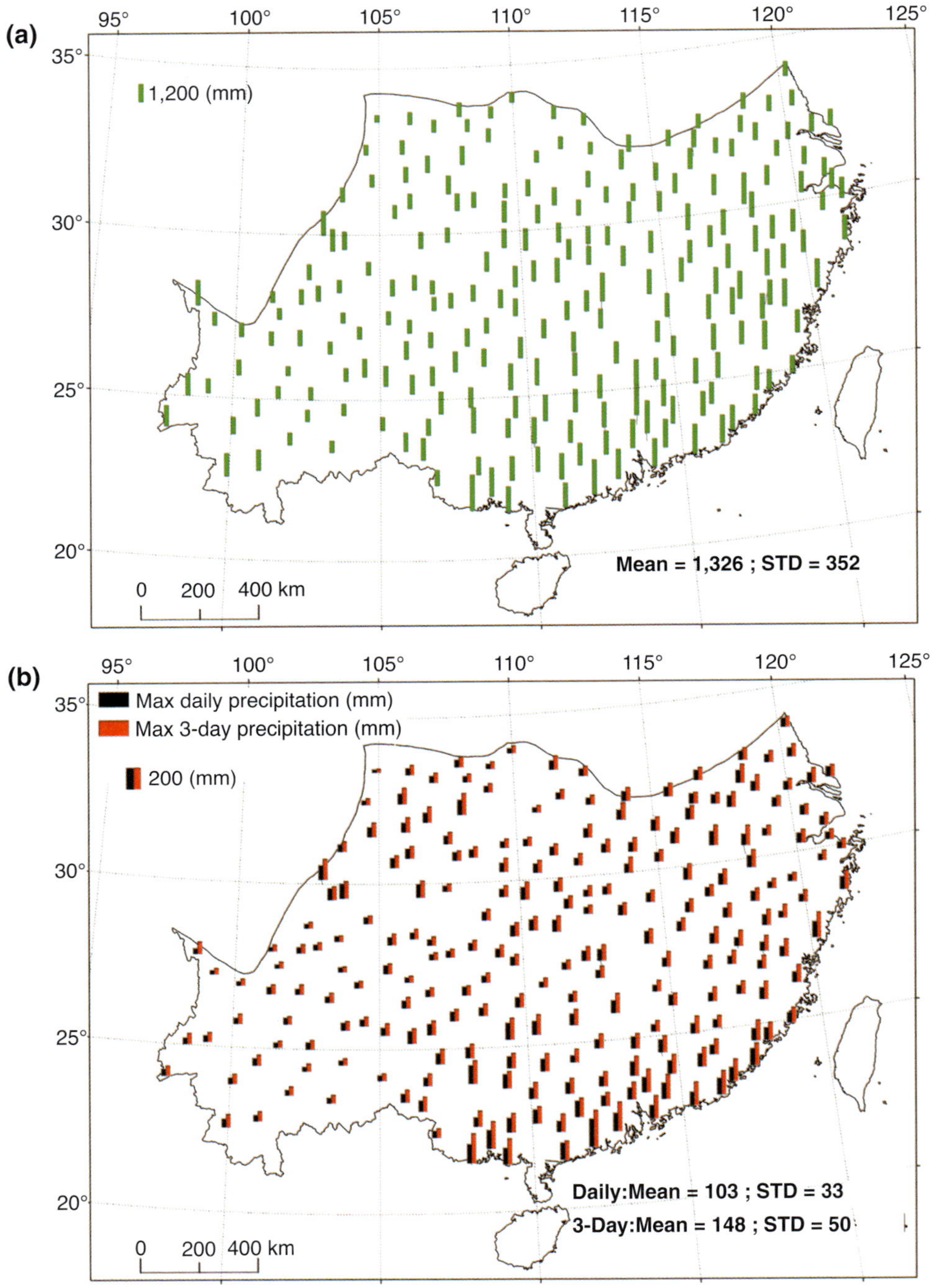

Figure 7.3 Mean values of precipitation variables at different meteorological sites (a total of 223 sites) in southeast China in 2001–2010: (a) annual precipitation (P_{annual}); (b) maximum daily precipitation (P_{daily}) and maximum three-day precipitation ($P_{\text{3-day}}$).

Table 7.2. Summarized statistics of mean levels of precipitatoin variables at different meteorological sites (unit: mm)

	Mean	Standard deviation
P_{annual}	1,326	352
P_{spring}	372	164
P_{summer}	603	185
P_{autumn}	222	58
P_{winter}	128	74
P_{daily}	103	33
$P_{\mathrm{3\text{-}day}}$	148	50

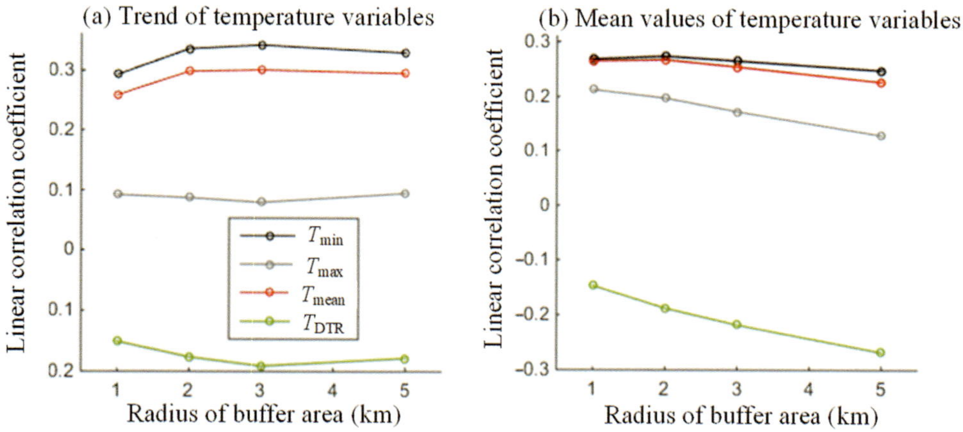

Figure 7.4 Linear correlation coefficients between annual temperature variables and urbanization fractions in buffer areas of different sizes: (a) trend of temperature variables in 1961–2010; (b) mean values of temperature variables in 2001–2010. Linear correlation coefficients for buffer areas of radius 10 and 15 kilometres, which are lower than those for a 5-kilometre buffer area, are not shown

area fraction and population density in buffer areas with different sizes) computed for each individual site. In many cases, statistically significant CCs suggest that variations in both trends and mean values of temperature variables can be partly linked to urban effects. The temperature variables tend to be more correlated with urban area fraction than population density, implying that the urban effect on temperature is likely to be more associated with the size of built-up areas than the population density in the study area. Figure 7.4 shows the linear CCs as a function of the size of buffer areas where the urban effect is analysed. For the sake of clarity, results of buffer sizes of 10 and 15 kilometres are not shown, as the corresponding linear CCs are lower than those with buffer sizes between 1 and 5

Table 7.3. Correlation coefficients between trends in temperature and urban area fractions in 2-kilometre buffer areas of meteorological sites and urban bias assessment of the trend in temperature variables

	Variable	Linear corrlation coeffcient r	Rank correlation coefficient ρ	Linear coefficient with urban area fraction	Urban bias (°C/decade)
Trends in annual temperature	T_{min}	0.34^{***}	0.35^{***}	0.17	0.061
	T_{max}	–	0.15^{**}	–	–
	T_{mean}	0.30^{***}	0.35^{***}	0.13	0.046
	DTR	-0.18^{***}	-0.16^{***}	−0.042	−0.015
Trends in temperature in spring	T_{min}	0.35^{***}	0.38^{***}	0.25	0.088
	T_{max}	0.19^{***}	0.19^{***}	0.16	0.057
	T_{mean}	0.32^{***}	0.29^{***}	0.21	0.074
	DTR	–	-0.13^{**}	–	–
Trends in temperature in summer	T_{min}	0.25^{***}	0.24^{***}	0.10	0.036
	T_{max}	0.11^{*}	0.15^{**}	–	–
	T_{mean}	0.17^{***}	0.19^{***}	0.075	0.027
	DTR	-0.16^{**}	-0.14^{**}	−0.082	−0.029
Trends in temperature in autumn	T_{min}	0.32^{***}	0.31^{***}	0.18	0.065
	T_{max}	–	–	–	–
	T_{mean}	0.27^{***}	0.33^{***}	0.13	0.047
	DTR	-0.26^{***}	-0.22^{***}	−0.025	−0.041
Trends in temperature in winter	T_{min}	0.31^{***}	0.32^{***}	0.20	0.070
	T_{max}	–	–	–	–
	T_{mean}	0.24^{***}	0.28^{***}	0.13	0.045
	DTR	-0.33^{***}	-0.32^{***}	−0.24	−0.084

Note: *, ** and *** indicate 10, 5 and 1 per cent significance levels, respectively; – indicates an insignificant correlation coefficient at the 10 per cent significance level.

kilometres. For most temperature variables (except the mean value of T_{DTR}), the highest correlation coefficient is observed with the urban area fraction in buffer areas with radii between one and three kilometres. Therefore, urban area fractions with two kilometre-buffer areas are used to further analyse urban effects on temperature hereinafter.

The CCs between the trends in temperature variables and urban area fractions in two-kilometre buffer areas are listed in Table 7.3. Positive CCs indicate positive urban biases in upward trends of annual T_{min} and T_{mean}. The urban effect on the trend of annual T_{max} appears insignificant. Negative urban bias is shown in the trend of DTR. The values of r and ρ are generally close, revealing that a linear function and monotonic function describe equally well the relation between the temperature variables and urban area fractions. Based on linear regression results, approximate estimates of urban effects in trends in annual T_{min}, T_{mean} and DTR are on average 0.061, 0.046 and −0.015°C/decade, respectively (Table 7.1), which account for 25, 26 and 33 per cent of the total trends detected in these three variables (i.e., 0.24, 0.18 and −0.045°C/decade, respectively, see Figure 7.1). T_{min} is more sensitive to urban effect than T_{max}, consistent with previous results in China (e.g., Zhang et al., 2005; Ren & Zhou, 2014). This can be explained by the fact that, over a day, the UHI effect typically peaks at night, when stored daytime heat prevents the urban environment from cooling as fast as rural areas (Oke, 1982; Schatz & Kucharik, 2015).

The urban effect on the trend of temperature shows a seasonal pattern (see Table 7.3 for the results in spring, summer, autumn and winter), because of complex interactions among surface properties and surface radiation and energy budgets (Schatz & Kucharik, 2014). The urban bias associated with the trend in temperature is more pronounced in winter than summer, which suggests that the bias has a larger effect on increasing temperatures during cold periods. The urban biases in T_{min} in winter and summer are on average 0.070 and 0.036°C/decade, respectively, accounting for 18.4 and 22.5 per cent of the overall warming trends. The urban bias in T_{min} peaks in spring, with an average positive effect of 0.088°C/decade, accounting for 44.0 per cent of the overall warming trend.

The CCs between the mean values of temperature variables in 2001–2010 and urban area fractions are listed in Table 7.4. Variations in the mean values of T_{min}, T_{max} and T_{mean} are positively linked to urbanization degrees, suggesting that locations with a higher degree of urbanization tend to have higher temperature levels, which confirms the UHI effect. Likewise, the mean level of DTR tends to decline at sites with increasing urbanization levels. In the study area, the estimated average effect of the urban bias over 2001–2010 on the mean annual temperatures T_{min}, T_{max}, T_{mean} and DTR are 1.18, 0.79, 1.08 and −0.39°C, respectively (Table 7.4). These results imply that T_{min}, T_{max} and T_{mean} are on average 1.18, 0.79 and 1.08°C higher in the meteorological sites in comparison to rural areas (with urban area fraction of 0 per cent), and DTR is 0.39°C lower. The largest effect of urban bias on temperatures is observed on the mean annual T_{min}, which is consistent with the result from the temperature trend analysis.

The UHI effect is stronger in summer compared to winter with regard to the mean temperature levels (Table 7.4), which is in line with many previous studies (e.g., Wang et al., 1990; Schatz & Kucharik, 2014). The estimated average urban bias on T_{min} in 2001–2010 is on average 1.55 and 0.82°C in summer and winter, respectively. The values of urban bias on T_{min} in 2001–2010 in spring and autumn are within the range of those in winter and summer.

Table 7.4. Correlation coefficients between mean levels of temperature and urban area fractions in two-kilometre buffer areas that surround meteorological sites and urban bias assessment of the mean level of temperature variables

	Variable	Linear corrlation coeffcient r	Rank correlation coefficient ρ	Linear coefficient with urban area fraction	Urban bias (°C)
Mean value of annual temperature	T_{min}	0.26^{***}	0.27^{***}	3.31	1.18
	T_{max}	0.20^{***}	0.19^{***}	2.22	0.79
	T_{mean}	0.27^{***}	0.26^{***}	3.02	1.08
	DTR	-0.19^{***}	-0.17^{**}	−1.09	−0.39
Mean value of temperature in spring	T_{min}	0.23^{***}	0.23^{***}	2.98	1.06
	T_{max}	0.15^{**}	0.14^{**}	1.64	0.58
	T_{mean}	0.22^{***}	0.24^{***}	2.50	0.89
	DTR	-0.17^{***}	0.13^{*}	−1.34	−0.48
Mean value of temperature in summer	T_{min}	0.38^{***}	0.40^{***}	4.34	1.55
	T_{max}	0.28^{***}	0.27^{***}	3.15	1.13
	T_{mean}	0.35^{***}	0.39^{***}	4.02	1.44
	DTR	-0.26^{***}	-0.31^{***}	−1.19	−0.43
Mean value of temperature in autumn	T_{min}	0.28^{***}	0.30^{***}	3.77	1.35
	T_{max}	0.24^{***}	0.24^{***}	3.09	1.11
	T_{mean}	0.30^{***}	0.28^{***}	3.76	1.35
	DTR	-0.13^{*}	-0.11^{*}	–	–
Mean values of temperature in winter	T_{min}	0.15^{**}	0.14^{**}	2.29	0.82
	T_{max}	0.13^{*}	0.12^{*}	–	–
	T_{mean}	0.12^{*}	0.11^{*}	–	–
	DTR	-0.13^{*}	–	–	–

Note: *, ** and *** indicate 10, 5 and 1 per cent significance levels, respectively; – indicates an insignificant correlation coefficient at the 10 per cent significance level.

7.6 Urban Effect on Precipitation

We did not find any significant correlation (both ρ and r) between precipitation trends and urbanization indicators in most stations. The mean levels of a few precipitation variables are significantly correlated to urbanization indicators but not to their trends. This is not surprising as a trend reflects changes in precipitation over time, while the mean precipitation characterizes the local climate as a function integrating impacts of all influential factors (including the urban effect).

Similar to the temperature analysis, precipitation levels are generally more correlated with the urban area fraction than population density. Annual precipitation and most seasonal precipitation (except P_{winter}) are not significantly correlated to urban area fractions, while maximum precipitation variables characterizing extreme events are correlated to urban area fractions. The highest correlation coefficients occur with urban area fractions in the two-kilometre buffer

Table 7.5. Correlation coefficients between precipitation variables (mean values of precipitation in 2001–2010) and urban area fractions and urban bias assessment of precipitation

Variable	Linear correlation coefficient r	Buffer area size	Rank correlation coefficient ρ	Linear coefficient with urban area fraction	Urban bias (mm)
P_{winter}	0.19***	2 km	0.19***	52.5	18.8
P_{daily}	0.19***	5 km	0.21***	34.9	6.1
$P_{3\text{-}day}$	0.16***	5 km	0.16***	43.8	7.7

Note: *** indicates 1 per cent significance level.

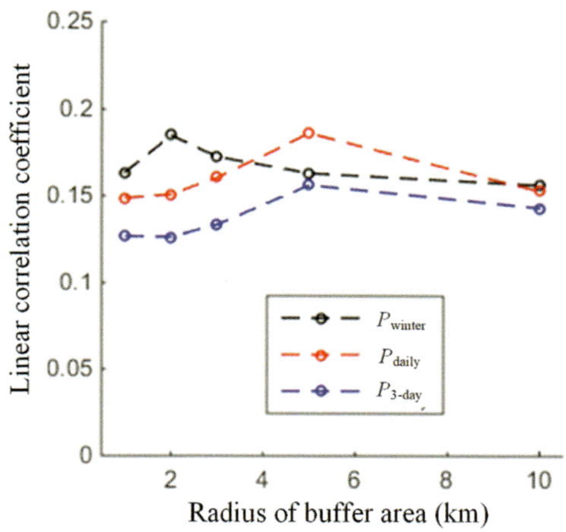

Figure 7.5 Linear correlation coefficients between precipitation variables (including P_{Winter}, P_{daily} and $P_{3\text{-}day}$) and urban area fractions in buffer areas of different sizes. Linear correlation coefficients for the 15-kilometre buffer area, which are lower than those for the 10-kilometre buffer area, are not shown

area for P_{winter} and in the five-kilometre buffer area for P_{daily} and $P_{3\text{-}day}$, when different buffer area sizes are tested (see Figure 7.5). Urban effects on these precipitation variables are assessed based on the urban area fractions in buffer areas with the highest correlation coefficients.

Table 7.5 lists the CCs between the mean values of precipitation variables and urban area fractions in buffer areas with the CCs reaching the maximum. ρ is close to r. Based on linear regression results applied on average precipitation in the period 2001–2010, the urban effect leads to an average increase of 18.8 millimetres of precipitation in winter, contributing to 14.7 per cent of the total mean winter precipitation (128 millimetres). There is no evidence of an urban bias on precipitation in other seasons and annual precipitation.

The mean values of P_{daily} and $P_{3\text{-}day}$ are positively correlated to the urban area fraction, suggesting that densely urbanized areas are more likely to be exposed to larger extreme rainfall, which is consistent to evidence of an upward trend of intensified heavy rainstorms in urban environment elsewhere (e.g., Changnon, 1979; Jauregui & Romales, 1996; Changnon & Westcott, 2002; Li et al., 2011). The estimates of urban bias in P_{daily} and $P_{3\text{-}day}$ are on average 6.1 and 7.7 millimetres, respectively, contributing to 5.9 and 5.2 per cent precipitation, respectively.

7.7 Discussion

This study assessed UHI and URI effects based on observational data. In contrast to physically based process models (e.g., Georgescu et al., 2011; Benson-Lira et al., 2015; Nie et al., 2017), which may contribute to improve our understanding of physical based processes, this observational study shows empirical evidence of urban effects on climatic factors and illustrates the state of our evolving urban environment. Both approaches are important alternatives to provide knowledge on how and to what degree the urban climate can be reshaped by urbanization and may also provide guidance to improve urban management practices.

In this study focused on southeast China, our results show positive correlations between rising trends and mean levels of temperature and urbanization degree, which brings evidence of a positive effect of UHI on temperature. The estimated effects of urbanization on temperature lie within the interval of plausible values (0.01–0.25°C/decade, see Table 7.6) obtained from previous studies in China. The variability of the effect is affected by changes in the study region, temporal periods and methodologies. Assessment results should not be interpreted independent of the method used. In this study, urban bias represents the average bias of temperature (or precipitation) in all meteorological sites (mostly located in or near cities with the average urban area fraction in two kilometre buffer areas of 36 per cent) in comparison to a rural site (with the urban area fraction of zero). In contrast, if one dichotomizes the variable urbanity (urban and rural), the urban bias would refer to the difference in two categorized sites instead of expressing a continuum of urbanization. Our assessments show that over 25 per cent of the rising trends in annual temperature variables can be attributed to the UHI. In addition to rising temperature trend analysis, we also provide assessments of urban effect on the mean level of temperature in 2001–2010. Night temperature in urban environments can be on average 1.18°C higher than in rural areas. The literature suggests that UHI is mainly driven by surface albedo, roughness and heat capacity modifications due to land surface change and anthropogenic heat releases in urban areas (Rizwan et al., 2008; Huang et al., 2009; Li et al., 2013).

Table 7.6. Estimates on urban effects on rising temperature trends in China based on meteorological data from selected literature studies

Study	Region	Period	Method	Variable	Urban bias (°C/decade)
Wang et al. (1990)	China	1954–1983	Classifying rural and urban sites	T_{mean}	0.08
Zhou et al. (2004)	Southeast China	1979–1998	Comparison to reanalysis data	T_{max} T_{min} DTR	–0.016, 0.116 and –0.132
Li et al. (2004)	China	1951–2001	Classifying rural and urban sites	T_{mean}	<0.012
Zhang et al. (2005)	Eastern China	1960–1999	Comparison to reanalysis data	T_{max} T_{min} T_{mean}	0.03, 0.20 and 0.12
Ren et al. (2008)	North China	1961–2000	Classifying rural and urban sites	T_{mean}	0.11
Jones et al. (2008)	China	1951–2004	Comparison to sea surface temperature	T_{mean}	0.1
Yang et al. (2009)	China	1960–1999	Comparison to reanalysis data	T_{mean}	0.05–0.25 depending on land uses
Wang and Ge (2012)	China	1980–2009	Classifying rural and urban sites	T_{mean}	0.099–0.26 depending on urbanization levels
Ren and Zhou (2014)	China	1961–2008	Classifying rural and urban sites	T_{max} T_{min} T_{mean}	0.023, 0.070 and 0.047
Sun et al. (2016)	China	1961–2013	Classifying rural and urban sites, and climate modelling	T_{mean}	0.094

This study investigates the URI effect on a large regional scale in China, in contrast to most literature studies, which have focused on single cities or metropolitan scales. The UHI effect is a combined consequence of thermodynamics changes due to the built environment replacing natural landscape and generating heat in the urban environment (Oke, 1982; Rizwan et al., 2008; Huang et al., 2009; Li et al., 2013). Possible causal factors leading to urban precipitation pattern change include modifications in local thermodynamics associated with the UHI and elevated aerosols in urban areas (Burian & Shepherd, 2005). The mechanisms behind the observed urban effect on precipitation are complex. In some cases, precipitation may be enhanced because of strengthened convection with additional latent heat released in urban areas when cloud droplets are lifted to the freezing levels; in other cases, high aerosol concentrations may lead to an increase of smaller cloud droplets, further resulting in suppressed precipitation (Tao et al., 2012). These two opposite effects of urbanization superpose each other. Enhanced or suppressed precipitation occurs in urban areas when one effect dominates the other. This explains varying results in previous URI studies, reporting urban precipitation increase (e.g., Shepherd, 2006; Mote et al., 2007; Li et al., 2011) and precipitation deficit (e.g., Rosenfeld, 2000; Kaufmann et al., 2007). The role of urbanization in shaping precipitation change depends on regions, environmental conditions and precipitation characteristics (Tao et al., 2007). This study examines a wide range of precipitation variables. While no significant urban bias is found in annual and most seasonal precipitation variables, it appears positively associated with extreme precipitation. This results may be explained by a higher propensity of urban areas to initiate an increasing convective precipitation process, which is usually associated with strong precipitations (Changnon, 1979; Jauregui & Romales, 1996; Changnon & Westcott, 2002; Li et al., 2011).

This study also provides quantification of urban influence on precipitation at a large regional scale in southeast China. Similar to the UHI effect, the quantified values should be interpreted based on the method used. The estimated urban biases in extreme precipitation P_{daily} and $P_{\text{3-day}}$ are 6.1 and 7.7 millimetres, respectively, contributing to around 5 per cent of the amount of corresponding extreme precipitation. Elsewhere, rainfall increases by 5–28 per cent above background values were reported in and downwind of cities based on city-scale studies (Huff & Changnon, 1972; Changnon, 1979). The large range of plausible values of the URI effects are explained by variations of their drivers, which include: the location, time period, precipitation characteristics and method used.

7.8 Conclusions

This study analysed the urban effect on both temperature and precipitation in southeast China based on long-term observational data in 1961–2010, by examining the correlation between climatic variables and continuous urbanization levels. The urban area fraction is generally a better indicator of the urbanization level (compared to population density). Therefore, we assessed the urban effects on temperature and precipitation using with urban area fractions in buffer areas of the meteorological sites between a range of 1–15 kilometres. The radius of the buffer area is determined so that it maximizes the effect of the correlation coefficient between a climatic factor and the buffer area. This study showed evidence of both UHI and URI effects in southeast China. The estimated average urban biases on the trends of T_{min}, T_{mean} and DTR in 1961–2010 are 0.061, 0.046

and $-0.015°C/decade$, respectively. The estimated average urban biases on the mean levels of T_{min}, T_{max}, T_{mean} and DTR in 2001–2010 are 1.18, 0.79, 1.08 and $-0.39°C$, respectively. We did not find any significant urban–rural difference with regard to the annual precipitation, while winter precipitation was estimated to increase by 18.8 millimetres in urban areas in comparison to rural areas in 2001–2010. The average urban biases in P_{daily} and $P_{3\text{-}day}$ are on average 6.1 and 7.7 mm, indicating that more extreme rainfall occurs in more densely urbanized areas. Furthermore, our results showed that both UHI and URI effects vary over the seasons.

Notes

1 Website last accessed 16 June 2016.
2 Website last accessed 16 June 2016.

References

Benson-Lira, V., Georgescu, M., Kaplan, S., & Vivoni, E. R. (2015). Loss of a lake system in a megacity: The impact of urban expansion on seasonal meteorology in Mexico City. *Journal of Geophysical Research: Atmospheres* 121(7): 3079–3099.

Burian, S. J., & Shepherd, J. M. (2005). Effect of urbanization on the diurnal rainfall pattern in Houston. *Hydrological Processes* 19(5): 1089–1103.

Changnon, S. A. (1979). Rainfall changes in summer caused by St. Louis. *Science* 205 (4404): 402–404.

Changnon, S. A., Shealy, R. T., & Scott, R. W. (1991). Precipitation changes in fall, winter, and spring caused by St. Louis. *Journal of Applied Meteorology* 30(1): 126–134.

Changnon, S. A., & Westcott, N. E. (2002). Heavy rainstorms in Chicago: Increasing frequency, altered impacts and future implications. *Journal of the American Water Resources Association* 38(5): 1467–1475.

Easterling, D. R., Horton, B., Jones, P. D., et al. (1997). Maximum and minimum temperature trends for the globe. *Science* 277(5324): 364–367.

Gallo, K. P., Easterling, D. R., & Peterson, T. C. (1996). The influence of land use/land cover on climatological values of the diurnal temperature range. *Journal of Climate* 9 (11): 2941–2944.

Gallo, K. P., Owen, T. W., Easterling, D. R., & Jamason, P. F. (1999). Temperature trends of the US historical climatology network based on satellite-designated land use/land cover. *Journal of Climate* 12(5): 1344–1348.

Georgescu, M., Moustaoui, M., Mahalov, A., & Dudhia, J. (2011). An alternative explanation of the semiarid urban area "oasis effect". *Journal of Geophysical Research: Atmospheres* 116(D24): 113–125.

Grimm, N. B., Faeth, S. H., Golubiewski, N. E., et al. (2008). Global change and the ecology of cities. *Science* 319(5864): 756–760.

Hansen, J., Ruedy, R., Sato, M., et al. (2001). A closer look at United States and global surface temperature change. *Journal of Geophysical Research: Atmospheres* 106 (D20): 23947–23963.

Huang, S., Taniguchi, M., Yamano, M., & Wang, C.-H. (2009). Detecting urbanization effects on surface and subsurface thermal environment – A case study of Osaka. *Science of the Total Environment* 407(9): 3142–3152.

Huff, F. A., & Changnon, S. A. (1972). Climatological assessment of urban effects on precipitation at St. Louis. *Journal of Applied Meteorology* 11(5): 823–842.
Jauregui, E., & Romales, E. (1996). Urban effects on convective precipitation in Mexico city. *Atmospheric Environment* 30(20): 3383–3389.
Jones, P., Kelly, P., Goodess, C., & Karl, T. (1989). The effect of urban warming on the Northern Hemisphere temperature average. *Journal of Climate* 2(3): 285–290.
Jones, P. D., Groisman, P. Y., Coughlan, M., et al. (1990). Assessment of urbanization effects in time series of surface air temperature over land. *Nature* 347: 169–172.
Jones, P. D., Lister, D. H., & Li, Q. (2008). Urbanization effects in large-scale temperature records, with an emphasis on China. *Journal of Geophysical Research Atmospheres* 113(16): D16122.
Kalnay, E., & Cai, M. (2003). Impact of urbanization and land-use change on climate. *Nature* 423: 528–531.
Kanamitsu, M., Ebisuzaki, W., Woollen, J., et al. (2002). NCEP-DOE AMIP-II reanalysis (R-2). *Bulletin of the American Meteorological Society* 83(11): 1631–1643.
Kataoka, K., Matsumoto, F., Ichinose, T., & Taniguchi, M. (2009). Urban warming trends in several large Asian cities over the last 100 years. *Science of the Total Environment* 407(9): 3112–3119.
Kaufmann, R. K., Seto, K. C., Schneider, A., et al. (2007). Climate response to rapid urban growth: Evidence of a human-induced precipitation deficit. *Journal of Climate* 20 (10): 2299–2306.
Kendall, M. G. (1955). *Rank Correlation Methods*. London: Griffin.
Kishtawal, C. M., Niyogi, D., Tewari, M., Pielke Sr, R. A., & Shepherd, J. M. (2009). Urbanization signature in the observed heavy rainfall climatology over India. *International Journal of Climatology* 30(13): 1908–1916.
Kistler, R., Kalnay, E., Collins, W., et al. (2001). The NCEP_NCAR 50-year reanalysis: Monthly means CD-ROM and documentation. *Bulletin of the American Meteorological Society* 82(2): 247–268.
Li, Q., Liu, X., & Zhang, H. (2004). Detecting and adjusting temporal inhomogeneity in Chinese mean surface air temperature data. *Advances in Atmospheric Sciences* 21(2): 260–268.
Li, W., Chen, S., Chen, G., et al. (2011). Urbanization signatures in strong versus weak precipitation over the Pearl River Delta metropolitan regions of China. *Environmental Research Letters* 6(3): 034020.
Li, Y., Zhu, L., Zhao, X., Li, S., & Yan, Y. (2013). Urbanization impact on temperature change in China with emphasis on land cover change and human activity. *Journal of Climate* 26(22): 8765–8780.
Mann, H. B. (1945). Nonparametric tests against trend. *Econometrica* 13(3): 245–259.
Mirzaei, P. A., & Haghighat, F. (2010). Approaches to study Urban Heat Island – Abilities and limitations. *Building and Environment* 45(10): 2192–2201.
Mote, T. L., Lacke, M. C., & Shepherd, J. M. (2007). Radar signatures of the urban effect on precipitation distribution: A case study for Atlanta, Georgia. *Geophysical Research Letters* 34(20): L20710.
National Bureau of Statistics of China (2011). *China Statistical Yearbook*. Beijing: China Statistics Press.
Nie, W., Zaitchik, B. F., Ni, G., & Sun T. (2017). Impacts of anthropogenic heat on summertime rainfall in Beijing. *Journal of Hydrometeorology* 18(3): 693–712.
Oke, T. R. (1982). The energetic basis of the urban heat island. *Quarterly Journal of the Royal Meteorological Society* 108(455): 1–24.
Peterson, T. C. (2003). Assessment of urban versus rural in situ surface temperatures in the contiguous United States: No difference found. *Journal of Climate* 16(18): 2941–2959.

Ren, G., & Zhou, Y. (2014). Urbanization effect on trends of extreme temperature indices of national stations over mainland China, 1961–2008. *Journal of Climate* 27(6): 2340–2360.

Ren, G., Zhou, Y., Chu, Z., et al. (2008). Urbanization effects on observed surface air temperature trends in North China. *Journal of Climate* 21(6): 1333–1348.

Rizwan, A. M., Dennis, Y. C., & Liu, C. (2008). A review on the generation, determination and mitigation of Urban Heat Island. *Journal of Environmental Sciences* 20(1): 120–128.

Rosenfeld, A. H., Akbari, H., & Romm, J. J. (1998). Cool communities: Strategies for heat island mitigation and smog reduction. *Energy and Building* 28(1): 51–62.

Rosenfeld, D. (2000). Suppression of rain and snow by urban and industrial air pollution. *Science* 287(5459): 1793–1796.

Santamouris, M. (2015). Analyzing the heat island magnitude and characteristics in one hundred Asian and Australian cities and regions. *Science of the Total Environment* 512–513: 582–298.

Sarrat, C., Lemonsu, A., Masson, V., & Guedalia, D. (2006). Impact of urban heat island on regional atmospheric pollution. *Atmospheric Environment* 40(10): 1743–1758.

Schatz, J., & Kucharik, C. J. (2014). Seasonality of the urban heat island effect in Madison, Wisconsin. *Journal of Applied Meteorology and Climatology* 53(10): 2371–2386.

Schatz, J., & Kucharik, C. J. (2015). Urban climate effects on extreme temperatures in Madison, Wisconsin, USA. *Environmental Research Letters* 10(9): 094024.

Shepherd, J. (2006). Evidence of urban-induced precipitation variability in arid climate regimes. *Journal of Arid Environments* 67: 607–628.

Shepherd, J., Stallins, J., Jin, M., & Mote, T. (2010). Urbanization: Impacts on clouds, precipitation, and lightning. In J. Aitkenhead-Peterson, & A. Volder (eds.), *Urban Ecosystem Ecology* (pp. 1–27). Madison, WI: American Society of Agronomy, Crop Science Society of America, Soil Science Society of America.

Sun, Y., Zhang, X., Ren, G., Zwiers, F. W., & Hu, T. (2016). Contribution of urbanization to warming in China. *Nature Climate Change* 6: 706–710.

Tao, W.-K., Chen, J. P., Li, Z., Wang, C., & Zhang, C. (2012). Impact of aerosols on convective clouds and precipitation. *Reviews of Geophysics* 50(2): RG2001.

Tao, W.-K., Li, X., Khain, A., et al. (2007). Role of atmospheric aerosol concentration on deep convective precipitation: Cloud-resolving model simulations. *Journal of Geophysical Research* 112(D24): D24S18.

United Nations (UN) (2014). *World Urbanization Prospects: The 2014 Revision, Highlights*. New York: UN.

Wang, F., & Ge, Q. (2012). Estimation of urbanization bias in observed surface temperature change in China from 1980 to 2009 using satellite land-use data. *Chinese Science Bulletin* 57(14): 1708–1715.

Wang, J. A., Hutrya, L. R., Li, D., & Friedl, M. A. (2017). Gradients of atmospheric temperature and humidity controlled by local urban land use intensity in Boston. *Journal of Applied Meteorology and Climatology* 56(4): 817–831.

Wang, W.-C., Zeng, Z., & Karl, T. R. (1990). Urban heat islands in China. *Geophysical Research Letters* 17(12): 2377–2380.

Yang, X., Zhang, Y., Liu, L., et al. (2009). Sensitivity of surface air temperature change to land use/cover types in China. Science in China Series D. *Earth Sciences* 52(8): 1207.

Zhang, J., Dong, W., Wu, L., et al. (2005). Impact of land use changes on surface warming in China. *Advances in Atmospheric Sciences* 22(3): 343–348.

Zhou, L., Dickinson, R. E., Tian, Y., et al. (2004). Evidence for a significant urbanization effect on climate in China. *Proceedings of the National Academy of Sciences (USA)* 101(26): 9540–9544.

8

Vegetation Dynamics, Land Use and Ecological Risk in Response to NDVI and Climate Change in Nepal

BINOD BANIYA, QIUHONG TANG, GYAN CHHIPI-SHRESTHA, HOM BAHADUR BANIYA AND GEBREMEDHIN GEBREMESKEL HAILE

8.1 Introduction

Nepal, located between 26° 22′ and 30° 27′ N latitude and 80° 04′ and 88° 12′ E longitude in the central part of the Himalayan region, is characterized by unique topography, climatic variations and vegetation distribution. The total area of the country is 147,181 square kilometres with 800 kilometres parallel length from west to east and 200 kilometres width from north to south. The altitude varies from 70 metres in the south to 8,848 metres above sea level at Mount Everest in the north. Based on this altitudinal variation, Nepal has been divided into six bio climatic zones ranging from Tropical to Nival (Dobremez, 1976; LRMP, 1986) under three broad ecological zones (i.e., Terai, Hill and Mountain). Since 2015, the Constitution Assembly of Nepal adopted the federal structure of governance with 7 provinces, 77 districts and 753 local bodies, where Terai, Hill and Mountain occupied 21, 36 and 20 districts, respectively (MoFALD, 2017). The main land use types are forest (39.1 per cent), cultivated land (29.83 per cent), shrublands (3.40 per cent), grasslands (7.90 per cent), snow and glacier (8.20 per cent), barren land (10.65 per cent), lake and river (0.6 per cent) and 0.32 per cent built up area (Uddin et al., 2015). The abrupt changes of the climate (i.e., increased temperature and rainfall variability due to high topographic variation of Nepal) affected in vegetation changes in Nepal. The forest, biodiversity and natural ecosystems are main thematic sectors which are more susceptible to climate change (NAPA, 2010). Large volumes of top soils from high lands are eroded and washed down annually and a large intensity of rainwater flows down instead of being retained as moisture (Sigdyal, 1999). This soil erosion has been increased throughout the last

This research was funded by the National Natural Science Foundation of China (41425002) and the Second Tibetan Plateau Scientific Expedition and Research (2019QZKK0208). Data used for this study was provided by the NOAA GIMMS and MODIS team, ICIMOD and DHM, Nepal. The first author is sponsored by the Chinese Academy of Sciences-The World Academy of Sciences (CAS-TWAS) President's PhD Fellowship program at the University of Chinese Academy of Sciences.

decade. In 2010, the annual soil erosion was identified at 110.55 million tons at the rate of 7.49 ton ha^{-1} yr^{-1} (Uddin et al., 2018).

Climate change risk Atlas 2010 ranks Nepal as one of the world's most vulnerable countries, indicating extreme vulnerabilities of forest, bio-diversity, agriculture and water resources to climate change (NRRC, 2013). On the one hand, heavy rainfall events are causing massive landslides, on the another hand frequent and intense droughts have been causing severe damage to the natural ecosystem and agricultural systems (Dahal et al., 2015). Increasing human activities, urbanization (i.e., the total urban population is 38.26 per cent with a 6 per cent annual growth rate) (CBS, 2011), cropland expansion, deforestation, increased grazing, transhumance activities and ecological restoration also accelerate the widespread changes in land use types and the vegetation system in different ecological regions (DoFRS, 2017). The average temperature rise from 1972 to 1994 was 1.5°C at the rate of 0.06°C yr^{-1} (Shrestha et al., 1999), 0.04°C yr^{-1} from 1975 to 2007 (Sharma, 2009) and 0.05°C during 2000–2015 (Paudel et al., 2019). The temperature has significantly increased across the country while pre-monsoon precipitation has significantly decreased in the high-Himalaya where very wet days and extremely wet days decreased significantly (DHM, 2017). The summer monsoon dominates the total precipitation whereas the winter monsoon contributes to only 3 per cent of total annual precipitation (Sigdel & Ikeda, 2012). Nepal is one of the highly vulnerable countries facing rapid alterations of its ecosystem and changes in land surface processes resulting from climate change (NRRC, 2013). Heavy rainfall events, natural disasters and severe droughts exert problems on the structure and function of the ecosystem (Karki et al., 2017). A high population growth (CBS, 2011), cropland expansion, deforestation and land uses changes have unevenly interrupted the vegetation system in different geographical regions (DoFRS, 2017).

There is an increasing interest in understanding vegetation dynamics in Nepal (Chhetri & Cairns, 2016; Gaire et al., 2017a, 2017b; Kharal et al., 2017; Shrestha et al., 2017; Thapa et al., 2017; Tiwari et al., 2017b; Sigdel et al., 2018). So far, scholars have mainly focused on studies performed in relatively small areas which make the generalization of their results to the entire Himalayan region challenging. In this context, we used satellite-based NDVI products as a proxy for vegetation dynamics in Nepal. The satellite-derived NDVI which is the normalized ratio of red and near-infrared (NIR) reflectance (Tucker, 1979). It has been widely used and eliminated noise (Zhou et al., 2001; Piao et al., 2003; Yuan et al., 2015). This index has the largest temporal coverage and is considered as the most suitable data for the analysis of vegetation dynamics (Zhao et al., 2011). The NDVI has multidimensional implications for land cover classification (He et al., 2017), ecological productivity (Pettorelli et al., 2005), drought monitoring (Kogan &

Sullivan, 1993; Liang et al., 2017), land degradation (Yengoh et al., 2015) and forest carbon dynamics (Myneni et al., 2001; Dong et al., 2003; Piao et al., 2005). Land use and land cover (LULC) vary according to natural and anthropogenic activities (Foley et al., 2005; Paudel et al., 2016; Li et al., 2019). Land use and land cover indicate the interaction between humans and the environment (Chapagain et al., 2018). In the last decades, important changes in land use and land cover have been noticed at global, regional and local scales (Klein Goldewijk et al., 2011). In Nepal, human footprint pressure (HFP), which indicates the human pressure on natural resources, has been found to have increased between 1993 and 2009 (Venter et al., 2016). These observed rapid changes in physical environment have increased the ecological risk in Nepal. Ecological risk can be linked with changes in NDVI and can be measured in terms of LULC, HFP and climatic variations. Ecological risk can be represented by the probability of occurrence of undesirable events and expected ecological damage which includes mainly stressor, exposure and receptor (EPA, 1998). A key point of the ecological risk is the change associated with the characteristic of ecological components that may be affected by exposure to stressors. Thus, exposure can be considered as an indicator of ecological risk which can be measured using NDVI or vegetation dynamics. The aim of this study is to identify spatial and temporal vegetation dynamic from different satellite-derived NDVI products (e.g., NOAA NDVI3g.v1 and MODIS NDVI), land use land cover change and possible environmental risk based on vegetation, land cover change, HFP and climate variables (i.e., temperature and precipitation) variations in Nepal. The higher values of NDVI represent a good ecological status and low NDVI indicates ecological risk. NDVI changes, variances of NDVI, land use land cover changes, HFP and climate variables (i.e., temperature and precipitation) are considered as the main indicators of environmental changes in Nepal.

8.2 Research Approach and Methods

In this study, vegetation dynamics based on NDVI, LULC, HFP and climate variables (i.e., temperature and precipitation) were analysed at the national scales. These environmental changes determine the ecological risk in Nepal. For this purpose, the MODIS NDVI available from 2000 to 2017 and NOAA, NDVI from 1982 to 2015, national land use land cover data and climatic observation data for 1982–2015 were used. The vegetation changes are derived in annual time scale (1982–2015; 2000–2017), land use and land cover changes based on 1990, 2000 and 2010 and climate variation during 1982–2015 for the entire country were used for the analysis. Human footprint pressure on environment was analysed using global terrestrial human footprint pressure for 1993 and 2009 because this year falls under our study period. The values of global human footprint pressure ranges from 0 to 46, in which high values correspond with the high level of human pressure and vice versa

8.2.1 Data

The data sets used are NOAA, AVHRR, NDVI3g.v1 and MODIS NDVI. The NDVI data set used in this study is the latest version of data sets, namely NDVI3g.v1 downloaded from https://ecocast.arc.nasa.gov/data/pub/gimms/3g.v1 (Tucker et al., 2005). The GIMMS NDVI product has an 8 × 8 kilometres spatial and 15 days temporal resolution. The data set covers the period spanning from 1982 to 2015. Monthly data was generated from bi-monthly data using maximum value composite (MVC) method (Holben, 1986). The MODIS NDVI (MOD13Q1) Terra product was downloaded from https://modis.gsfc.nasa.gov (Didan, 2015). The data has 16 days temporal resolution and 250 metres spatial resolution spanning from 2000 to 2017. The four horizontal and vertical tiles (i.e., h24 v06, h25 v06, h24 v05 and h25 v05), Land cover data sets of Nepal prepared by the International Centre for Integrated Mountain Development (ICIMOD) for the years 1990 (ICIMOD, 2014a), 2000 (ICIMOD, 2014b) and 2010 using Landsat TM data of 2010 (Uddin et al., 2015) were adopted. Similarly, the monthly precipitation and temperature data obtained from meteorological stations for the period of 1982–2015 were collected from the Department of Hydrology and Meteorology (DHM), Nepal. Monthly precipitation data from 272 meteorological stations and maximum and minimum air surface temperature data from 115 meteorological stations were used. Similarly, the global terrestrial HFP data sets developed by Venter et al. (2016) for the years 1993 and 2009 were used.

8.2.2 Methods

The linear NDVI trends (Piao et al., 2011; Fensholt & Proud, 2012; Liu et al., 2015; Baniya et al., 2018), Mann Kendell test (Mann 1945; Kendall 1975) and Sen's slope (Sen, 1968) were used and identified time series NDVI trends for both NOAA, NDVI during 1982–2015, MODIS NDVI during 2000–2017 and climate change. The presence of statistically significant trends was evaluated using both p values and t test. The null hypothesis was tested at 95 per cent confidence level. Null hypothesis (H_o) (no trend) is rejected if the p value is less than a predefined significance level of 0.05. A statistically significant positive NDVI refers to greening and a statistically significant negative NDVI refers to browning trends. The ecological risk was categorized based on standard deviation of each pixel from spatially averaged annual mean NDVI during 1982–2015.

8.3 Results

8.3.1 Spatial and Temporal Vegetation Dynamics

The annual NDVI has significantly increased at the rate of 0.0008 yr^{-1} during 1982–2015 and 0.0018 yr^{-1} during 2000–2017 (Figure 8.1(c) and (d)). The spatially average mean annual NDVI during 1982–2015 was 0.50 and during 2000–2017 it was 0.44 in Nepal. The NDVI are higher in the lower hills and lower in the plains and the mountain. The maximum NDVI of 0.8 was mainly observed in the hills and Siwalik regions (i.e., above the Terai). The annual mean NDVI in the northern part was very low, with a negative NDVI in some pixels, and the NDVI value was higher (above 0.4) in the middle zone of Nepal (Figure 8.1). Mostly, the NDVI values range from 0.4 to 0.5 in lowlands present in southern regions.

The average NDVI values in western Nepal (from the western boundary to 83°E) and central Nepal (from 83° to 86° 3′ E) were 0.481 and 0.485, respectively, whereas it was 0.551 in eastern Nepal (from 86° 3′ E to the eastern boundary) which was higher than those in western and central Nepal. The spatial distribution of average annual NDVI is almost similar in both NOAA and MODIS NDVI data sets. An average annual NDVI of less than 0.1 is considered as a non-vegetated

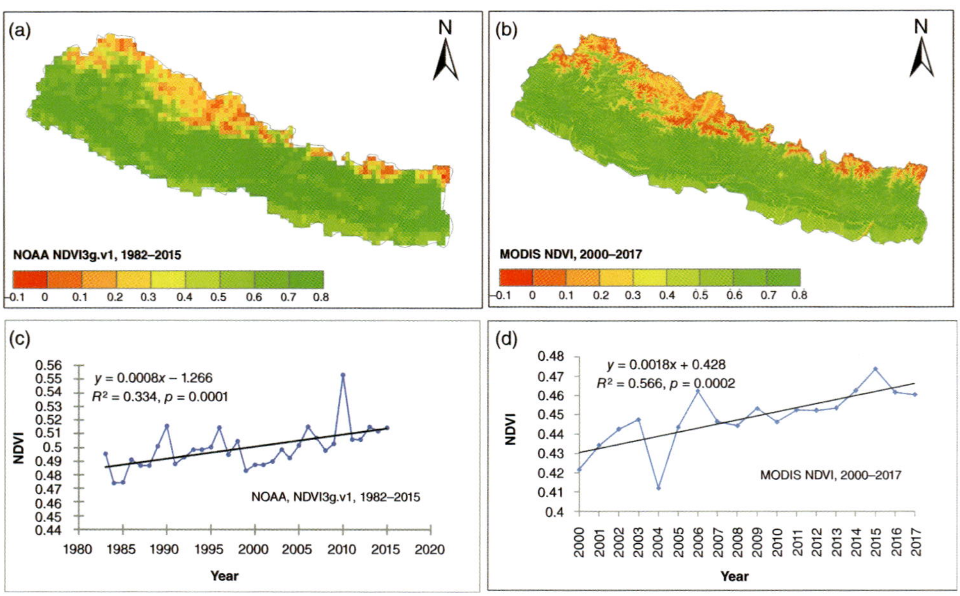

Figure 8.1 Spatial variation of average annual NDVI and NDVI trends in Nepal. (a) NOAA NDVI3g.v1; (b) MODIS NDVI; (c) NOAA NDVI3g.v1 trends during 1982–2015; and (d) MODIS NDVI trends during 2000–2017 in Nepal

area and these areas are found to be in the mountains, mainly in the Trans-Himalayan region in both datasets. Both MODIS and NOAA NDVI identify an increase in the density of the vegetation in mid hills during 2000–2017 and 1982–2015, respectively.

8.3.1.1 Spatio-Temporal Trends and Net NDVI Changes

The NDVI shows an increasing trend in 57.35 per cent ($p < 0.05$) of the study area and 16.10 per cent (4.68 per cent significant, $p < 0.05$) of the area presented a decreasing trend during 1982–2015 (Figure 8.2(a) and (c)). During 2000–2017, the MODIS NDVI increased in 84.24 per cent of areas, of which 27.88 per cent had significantly increased. Only, 1.18 per cent of the area was found to be browning (Figure 8.2(b) and (d)). Thus, NOAA NDVI shows 57.35 per cent greening and 4.68 per cent browning during 1982–2015 and MODIS NDVI shows 27.88 per cent greening and 1.18 per cent browning during 2000–2017 in Nepal (Figure 8.2).

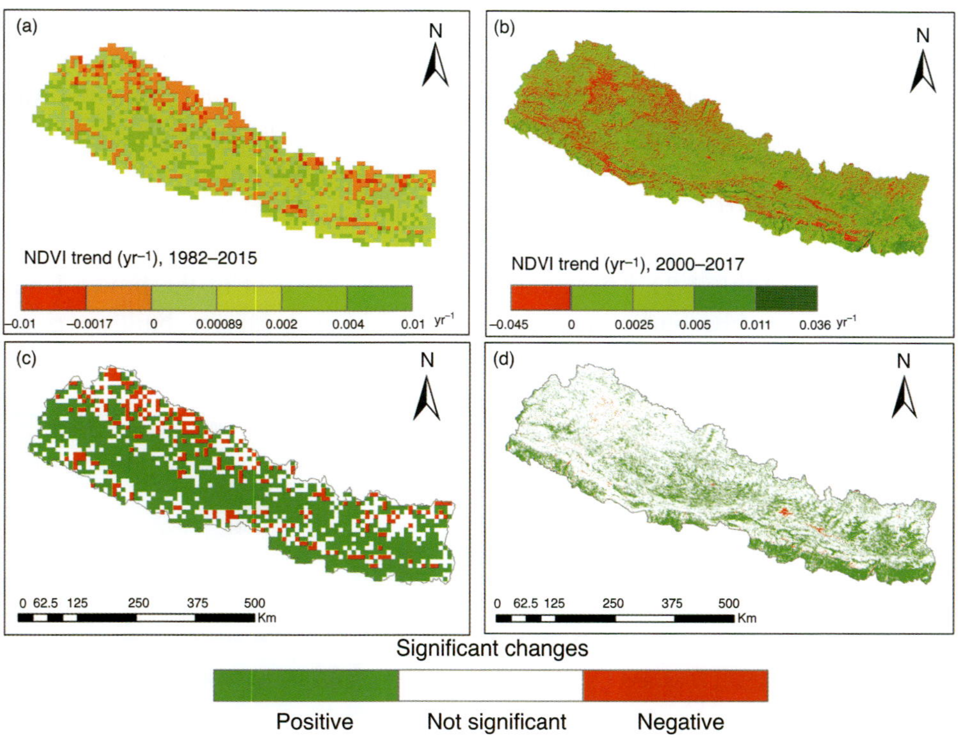

Figure 8.2 Annual and significant NDVI trends in different periods. (a) NOAA NDVI trends; (b) MODIS NDVI trends; (c) NOAA NDVI significant trends; and (d) MODIS NDVI significant trends during 1982–2015 and 2000–2017

The greening was higher than the browning in the last three decades. The annual NDVI trends were negative mainly in the northwest and some parts of southern regions, with a maximum negative NDVI changes of 0.01 yr^{-1} (Figure 8.2(a)) and –0.045 yr^{-1} (Figure 8.2(b)). The annual significant trend was mostly found in the middle hills and southeastern parts of Nepal. Both data sets showed a similar pattern with significant NDVI trends in southeastern Terai. The MODIS NDVI during 2000–2017 showed that the eastern parts are more greening and central regions near the capital city have browning trends. The large fraction of the significant trends showed high vegetation dynamic or ecosystem changes during the last three decades in Nepal.

8.3.2 Land Use Land Cover Change and HFP

In Nepal, changes in land use and land cover have been considerable. The agricultural land has increased from 151.2×10^2 km^2 to 438.8×10^2 km^2 over the past 100 years (Paudel et al., 2018). The overall pattern of the land use and land cover in Nepal was observed by a satellite-based study in 2010 (Uddin et al., 2015) which reported the forest area as the largest land cover of the country in 2010, followed by agricultural land (Figure 8.3(c)). Over the history of Nepal, the most noticeable changes in land cover were found in the built-up areas, forest regions and agricultural lands. Green lands such as forest, croplands, grassland and shrublands had increased in 2000 but slightly decreased in 2010. After 2010, satellite derived vegetation index showed an increase in green lands. Green lands showed an increase in 2010 as compared to 1990 by 2.17 per cent (Table 8.1). This indicates that 2.17 per cent of the brown lands had decreased and become green. The brown lands, such as bare, built-up, river,

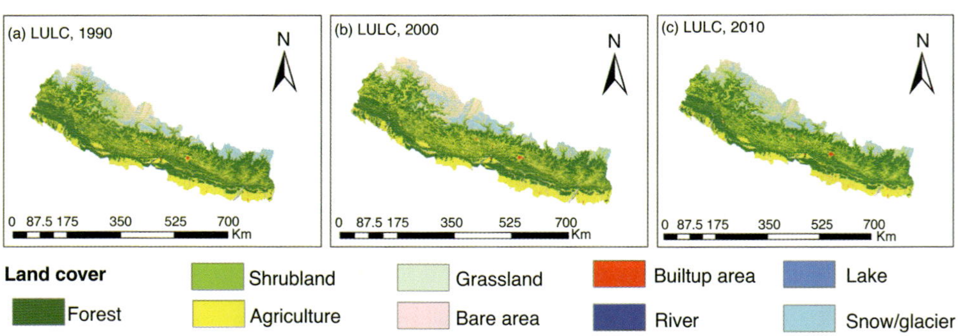

Figure 8.3 Land use and land cover (LULC) pattern of Nepal in (a) 1990 (ICIMOD, 2014a), (b) 2000 (ICIMOD, 2014b) and (c) 2010 (Uddin et al., 2015)

Table 8.1. Land use land cover changes in Nepal

Land Use Types	LULC, 1990	LULC, 2000	LULC, 2010
Green lands (vegetated)	118,053.88 km^2 (80.21%)	122,675.36 km^2 (83.35%)	121,247.70 km^2 (82.38%)
Brown lands (non-vegetated)	29,127.11 km^2 (19.79%)	24,505.63 km^2 (16.65%)	25,933.29 km^2 (17.62%)

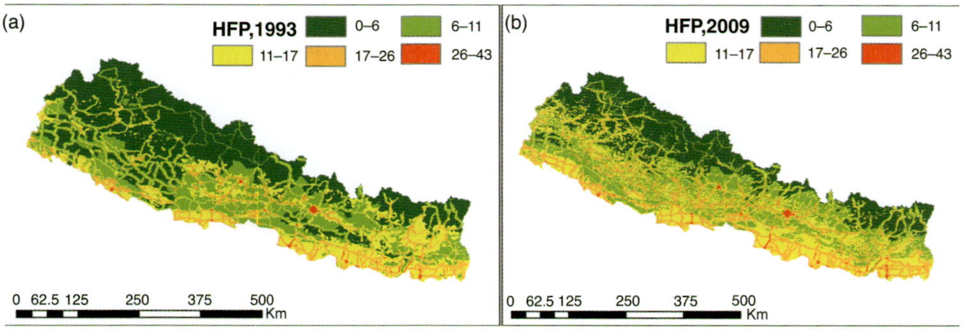

Figure 8.4 One kilometre gridded human footprint pressure in natural system in Nepal; (a) 1993 and (b) 2009

lake and snow/glacier lands covered 19.79 per cent of the territory in 1990 and reduced to 17.62 per cent in 2010.

Spatially, forests are dominant mainly in the mid hills, while croplands are mainly located in the lowlands and hills. Shrublands and grassland are present mostly in mountain areas (Figure 8.3). The increased areas covered by green lands and decreased brown lands illustrate a greening process that has taken place in Nepal in the recent decades. However, a decrease in the number of rivers, lakes, glacier and snow coverage has had negative impacts on the eco-hydrological systems and livelihood in Nepal. While dense vegetation dominates mid hills, which have become greener, the vegetation has decreased mainly in the lower mountains and Terai. Specifically, built-up areas have exhibited a rapid increase through the conversion of agricultural lands, especially in the peripheral area of large cities in Nepal (Thapa & Murayama, 2012; Rimal, 2013). Furthermore, glaciers have decreased in Nepal (Bajracharya et al., 2014). The historical trend of forest cover in the last three decades in Nepal has increased with a peak observed in 2018 (MoFE, 2019). Over the past one and half decade, the human footprint pressure has significantly increased in Nepal. The maximum footprints have reached 43, mainly in the large cities, such as Kathmandu, Pokhara and southern plain region. Footprints as low as 0–6 were found in the western mountain regions. On temporal pattern, the human footprint pressure was higher in 2009 compared to in 1993 (Figure 8.4).

Spatially, the footprint pressure was found to be increased toward western and central regions, indicating exerting pressures on the natural system. The increased human footprint pressure also corresponded to the population growth rate. The total population in Nepal was 26,494,504, with an annual growth rate of 1.35 per cent, while the average and maximum population density were 260 and 15,355 people km^{-2}, respectively (CBS, 2011). The spatial expansion of the human footprint pressure from southwest to east during 2009 compared with 1993 warned of ecological risk in Nepal. Higher HFP in the south and central regions indicates higher ecological risk. Similarly, the rapid increase of HFP in the mid hills also indicates high ecological risk that would affect both upstream and downstream regions.

8.3.3 Ecological Risk based on NDVI and Climate

Mainly, there are 16 ecoregions under three broad ecological zones (i.e., Terai, Hill and Mountain) in Nepal. Nepal lies at the border of the tropics. The bio-climatic zones in Nepal are tropical (<1,000 metres), subtropical (1,000–2,000 metres), temperate (2,000–3,000 metres), sub-alpine (3,000–4,000 metres), alpine (4,000–5,000 metres) and nival (>5,000 metres above sea level). The mean NDVI, NDVI Relative Change Ratio (RCR), Vegetation Condition Index (VCI) derived from NDVI, Standard deviation (SD) of annual NDVI and climate variables (i.e., temperature and precipitation) were computed for each ecoregion as a parameter of the environmental risk in Nepal (Table 8.2). Over the last 34 years, in Western and mid-Western Mountain the mean annual NDVIs were only 0.12 and 0.27, respectively. These values of surface greenness index reflect critical condition for a good ecosystem.

Within this time period, the NDVI RCR was negative in the Western Mountains which is the indication of environmental degradation. The VCI (percentage) showed that the Western Mountains experienced extreme drought (i.e., VCI < 30 per cent) and Mid-Western Mountains experienced normal drought (30–50 per cent VCI). In contrast, the NDVI and VCI values indicate stability in the other Nepalese regions. The highest value of precipitation was recorded in Eastern and Far-Western Terai, Eastern, Western and Far-western Hill. The temperature in the Terai ecoregions was higher than hills and hills ecoregions exhibit higher temperatures than in the mountains. In the Western Mountains, the average annual temperature was negative (–0.54°C) during the last 34 years and the average temperature was very low (3.11°C) in the Mid-Western Mountains. The vegetation growth was dormant given the low and sometimes negative temperature mainly

Table 8.2. Variation of environmental risk factors in different eco-regions in Nepal

Eco-regions	Mean NDVI Values	NDVI RCR (%)	VCI (%)	SD of NDVI (10^{-4})	Temp (°C)	Precipitation (mm)
Eastern Mountain	0.44	7.09	62.48	495.22	5.57	1,291.05
Eastern Hills	0.64	5.48	84.49	348.99	18.21	1,705.87
Eastern Terai	0.51	9.67	70.22	276.80	23.41	2,007.10
Central Mountain	0.50	5.59	68.81	423.57	8.33	1,306
Central Hills	0.60	6.50	79.80	340.84	17.69	1,595.28
Central Terai	0.55	6.52	74.98	293.08	23.36	1,538.65
Western Mountain	0.12	–8.64	26.72	678.16	–0.54	545.88
Western Hills	0.56	14.78	75.23	414.55	13.40	1,608.82
Western Terai	0.57	5.44	75.73	293.52	23.43	1,591.96
Mid-Western Mountain	0.27	4.48	44.50	525.22	3.11	954.05
Mid-Western Hills	0.59	8.64	78.60	339.43	15.89	1,431.84
Mid-Western Terai	0.58	6.28	77.47	319.13	22.92	1,482.38
Far-Western Mountain	0.41	4.06	60.21	438.06	9.13	1,379.08
Far-Western Hills	0.61	8.03	81.13	329.56	18.14	1,687.71
Far-Western Terai	0.59	6.81	78.74	331.79	23.63	1,803.00

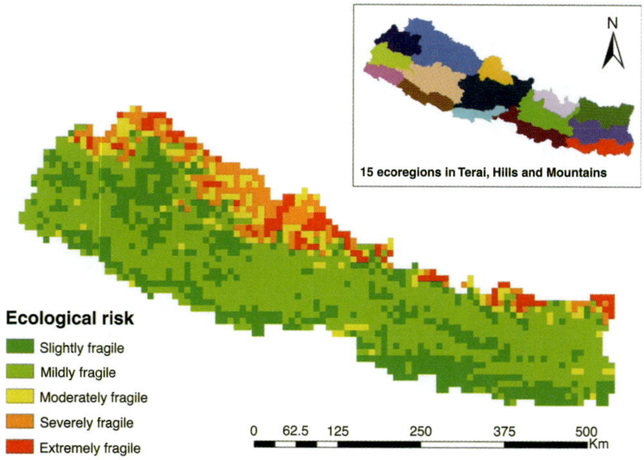

Figure 8.5 Ecological risk zone of Nepal based on annual SD of NDVI; the above right inset map shows the eco-regions in Nepal

observed in the Western and Mid-Western Mountains. The ecological risk map (Figure 8.5) was developed based on standard deviation of NDVI in each pixel from spatially averaged annual mean NDVI during 1982–2015. The standard deviation was classified as slightly fragile (152.85–307.69 × 10^{-4}), mildly fragile

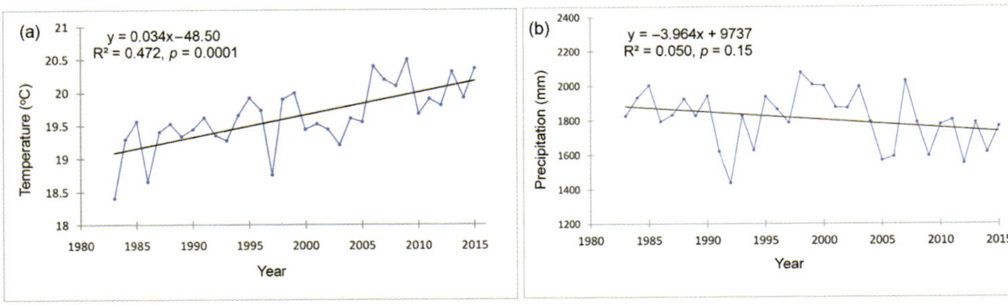

Figure 8.6 Average annual temperature and total precipitation trends; (a) temperature (°C yr^{-1}) and (b) precipitation (mm yr^{-1}) during 1982–2015 in Nepal

($307.69–445.32 \times 10^{-4}$), moderately fragile ($445.32–617.36 \times 10^{-4}$), severely fragile ($617.36–780.79 \times 10^{-4}$) and extremely fragile ($>780.79 \times 10^{-4}$). Higher deviations from the spatially averaged mean annual NDVI represent more a fragile environment and vice versa.

Nepal is a mountainous country with the Himalayas, a young mountain range, which exhibit various environmental risks. The NDVI variation showed that severe and extreme fragile ecological regions are located mainly in the Trans-Himalayan region. The inner Terai have shown a slightly fragile ecosystem but the mid hills showed a mild fragile ecological system. The grasslands which are found in the alpine regions have shown severe and moderate ecological system fragility in Nepal. At the same time, climate change is a dominant factor of pressure on the environment. In the last 30 years, temperature (average) has increased and precipitation has decreased. The annual average and maximum temperature significantly increased at the rate of 0.03°C yr^{-1} ($p = 0.0001$) and 0.04°C yr^{-1} ($p < 0.05$), respectively, during 1982–2015. In the same period, the total annual precipitation decreased at the rate of –3.96 mm yr^{-1}, although this is not significant at 5 per cent levels ($p = 0.15$) in Nepal (Figure 8.6). The mean and standard deviation of annual temperature are 19.5°C and 0.47°C, respectively. The increased temperature has significant effects on vegetation, ecosystem, water cycle, biodiversity and livelihood. The annual average total precipitation was about 1,810.14 millimetres during 1982–2015. The annual contribution of summer monsoon was approximately 80 per cent in Nepal. However, the summer monsoon decreased significantly at the rate of 4.29 mm yr^{-1}. Similarly, the average annual rainfall has a high standard deviation (i.e., 171.68 mm) in Nepal. A decrease in precipitation can increase drought, ecosystem deterioration and change in hydrological cycles and the agricultural system in Nepal.

The results show that, in Nepal, the annual NDVI and temperature were positively and significantly correlated, however NDVI and precipitation were negatively correlated during 1982–2015. The positive and significant correlation between NDVI and temperature was observed in 34.97 per cent of the investigated areas, while only 1.08 per cent of the study area showed a positive and significant correlation with precipitation. Most areas showed a negative correlation between NDVI and precipitation in Nepal (Baniya et al., 2018)

8.4 Discussion

The NOAA AVHRR NDVI and MODIS NDVI data showed greening in Nepal. The greening refers to statistically significant positive and browning refers to statistically significant negative NDVI trends. Generally, high NDVI values (indicating by an integrated NDVI or the annual maximum NDVI values) reflect a stable ecosystem. The positive NDVI trend observed in this study suggests that ecology and ecosystem conditions have improved over time. The ecological restorations and recent environmental changes have made a positive contribution to increase greening in Nepal. The previous studies of NDVI based on remote sensing tools showed similar positive trends in Nepal (Krakauer et al., 2017). The historical trend of the forest cover in Nepal exhibits fluctuations. Forest covered 43.5 per cent of areas in Nepal in 1960, but drastically decreased to reach 38.1 per cent in 1978 and 29 per cent in 1994 (DoFRS, 2015). However, forest has increased and recovered and reached 39.1 per cent in 2010 (Uddin et al., 2015), 40.36 per cent in 2015 (DoFRS, 2015) and 44.47 per cent in 2018 (MoFE, 2019). Overall, a recovery of the forest area suggests a reduction in the ecological risk. However, the observed decrease of environmentally sensitive lands such as wetlands (river, lakes, glacier and snow) in the Terai and mountain may have long-term environmental impacts. NDVI varies spatially, with less vegetation mainly in the northern parts such as the Trans-Himalayan regions and inner Terai. The NDVI trends were negative in some parts of the western regions, upper Terai and dense settlement areas, including the capital city in central Nepal, representing ecological risk zones. Negative NDVI trends are associated with a higher ecological risk exemplified by the extremely dry conditions experienced in the western mountains along with high vulnerability in Siwalik (i.e., upper Terai). The ecosystems in the Western Mountain and lands and livelihood at the downstream of Siwalik appear ecologically stable. The hills have shown relatively less ecological risk than Mountain and Terai. However, increased HFP have alarmed mid hills of Nepal. The human footprint pressures caused from built environments, croplands, pasture lands, land degradation, deforestation and population density have also increased in Nepal Himalaya over the past two-to-three decades. Identified increased human footprint pressure zones in Nepal during 2009 help to restore the degraded ecosystem and potential risk areas. The increased HFP alter the natural system and deteriorate environments. High human pressure on natural resource limits the carrying capacity of the environments and consequently deteriorates environments and creates ecological risk.

During 1977–1994, the annual maximum temperature increased at the rate of 0.06°C in most of the middle mountain and Himalayan regions and 0.03°C in Siwalik and Terai (Shrestha et al., 1999). Similarly, the average temperature increased at the rate of 0.04°C during 1975–2007(Sharma, 2009). In addition, the survey data of climate ($n = 5,060$) conducted by the Nepal government showed increased temperature. Similarly, gridded daily maximum and minimum temperatures obtained from NOAA Climate Prediction Centre (CPC) showed a significantly increased mean annual temperature of Nepal by 0.03°C yr^{-1} with an increase in maximum temperature by 0.02°C yr^{-1} and minimum temperature by 0.04°C yr^{-1} over the period from 1979 to 2016 (Shrestha et al., 2019). In line with these studies, our research showed that average temperature has increased at the rate of 0.03°C yr^{-1} during 1982–2015. The trends varied spatially, for instance, Mustang and Rara, which are located in the Himalayan region, have encountered an increase in the temperature at the rate of 0.06°C yr^{-1}. Several stations in the mid-hills and mountain have experienced an increase in the temperature higher than the average trends. Kathmandu, which is the capital of Nepal, showed an increase in the average temperature at the rate of 0.04°C yr^{-1} and in the maximum temperature at the rate of 0.06°C yr^{-1}. Conversely, the annual precipitation trend has significantly decreased during 1982–2015. The precipitation varied in space due to orographic influences in Nepal. The highest recorded annual average rainfall (5,471.28 millimetres) was found to be Lumle, located in Kaski district of Central Nepal (Index no: 814; altitude = 1,738 metres). Similarly, the lowest average rainfall was 209.28 millimetres found in Mustang (Index no: 610; altitude = 3,465 metres), which is located in the Trans-Himalayan region. The high temperature average values and low precipitation values are associated with the drought conditions observed in Nepal. An increase in temperature leads to more evaporation and transpiration and consequently provokes soil dryness and ice melting of the glaciers in the mountain.

Increasing temperatures have also impacted agriculture productivity in Nepal. The sensitivity of crop yields to temperature has been already observed (Lobell & Burke, 2008). The north western mountain experienced more severe drought in all time scales because it lies in the leeward side where the average rainfall is commonly very low (DHM, 2015) and the induced low moisture rate drastically limits plant growth (Tiwari et al., 2017a). The recent tree ring research also showed that the Trans-Himalayan region has experienced increasing drought (decreasing scPDSI and SPEI trends) in the past decades (Panthi et al., 2017). Three hundred and seven years of tree ring SPEI reconstruction indicate drought in the western Nepal Himalaya (Bhandari et al., 2019). These areas prone to drought experienced prolonged dry conditions during the nineteenth and twentieth centuries (Gaire et al., 2018; Baniya et al., 2019). Previous research works that have investigated drought suggested the presence of a high ecological risk in the Trans-Himalayan region including western and far-western region. This corroborates with our results showing negative NDVI values in the Trans-Himalayan region. Besides, the lowest precipitation and coldest temperature observed in the Trans-Himalayan region and the positive temperature trends in the mountains showed a fragile environment in this region in Nepal. However, some areas in this region exhibit positive NDVI trends which suggest slope stability because of increasing vegetation from herbs to shrubs or shrubs to trees. Thus, the growing vegetation in response to increasing temperature in some areas stabilize the soil and reduce the fragility of the ecosystems. The negative average temperature and lower average annual precipitation in the Western Mountain indicated unfavourable ecological conditions. The average temperature was also lower in the mountains which are considered as the limiting factor for vegetation growth. In cold and dry regions, plant's stomata cannot perform well for photosynthesis, resulting in dormant vegetation. The high population density and temperature observed in Terai can increase the ecological risk. Deforestation, ecosystem degradation and high temperature can increase plants and soil evaporation which leads to water scarcity and therefore increases the ecological risk in the Terai. Out of the total forest area, 37.80 per cent lies in Middle Mountain, 32.25 per cent in High Mountain and High Himal, 23.04 per cent in Siwalik and 6.9 per cent in the Terai (DoFRS, 2015). Relatively, the hills exhibit a stable ecosystem confirmed by the values observed for NDVI coverage, NDVI trends, climates and variance of NDVI. In Terai, the lower forest (only 6.9 per cent) coverage is very challenging to combat environmental changes because of the high population (CBS, 2011) and the land conversion from forest to agriculture and agriculture to built-up areas are rapidly increasing in the Terai and urban areas in Nepal (Uddin et al., 2015).

8.5 Conclusion

The study identified greening and browning through the analysis of spatial and temporal NDVI variation and trends during 1982–2015 and 2000–2017 using NOAA and MODIS NDVI, respectively. In both data sets, the spatial averaged NDVI showed a significant positive trend (i.e., greening revealing lower ecological risk at the annual scale). The average NDVI increased in the hills while exhibiting lower values in the plains and nearly less than 0.1 in the mountain. Spatially, the negative vegetation trends were mostly found in the Trans-Himalayan regions, Siwalik, urban areas and western parts of Nepal. The variance of NDVI, temperature and a reduction in the precipitation showed that mountain regions mainly in the Trans-Himalayas are associated with severe and extremely fragile ecosystems. For the majority of hills and Terai, the ecological risk was mild and slightly fragile. The increased human footprint pressure in lowlands and mid hills between 1993 and 2009 may have increased the ecological risk in the future. The spatially average temperature has increased but precipitation has decreased. The positive higher temperature trends and higher negative precipitation trends affect the ecosystem, water cycle, biodiversity, agriculture and people's livelihood. Mainly, the Western mountains, including the Trans-Himalayan region, need to be considered for ecological restoration where the less NDVI, negative NDVI changing ratio, high NDVI variance, high temperature trends and lower average precipitation were observed in Mid-Western Mountains. Similarly, higher temperature, lower forest coverage, higher land use dynamics and higher HFP in Terai may increase the environmental risk. Except in the Trans-Himalayan region, the average and low variance of NDVI showed a stable ecosystem due to humid climates. Growing vegetation in some parts of the mountain as a response to increased temperatures makes these areas more stable and less ecologically fragile.

References

Bajracharya, S., Maharjan, S., Shrestha, F., Bajracharya, O., & Baidya, S. (2014). *Glacier Status in Nepal and Decedal Changes from 1980 to 2010 based on Landsat data* (p. 88). Kathmandu, Nepal: ICIMOD.

Baniya, B., Tang, Q., Huang, Z., Sun, S., & Techato, K.-A. (2018) Spatial and temporal variation of NDVI in response to climate change and the implication for carbon dynamics in Nepal. *Forest* 9(6): 329.

Baniya, B., Tan, Q., Xu, X., Haile, G. G., & Chhipi-Shrestha, G. (2019). Spatial and temporal variation of drought based on satellite derived vegetation condition index in Nepal from 1982–2015. *Sensors* 19(2): 430.

Bhandari, S., Speer, J. H., Thapa, U. K., et al. (2019). A 307-year tree-ring SPEI reconstruction indicates modern drought in Western Nepal Himalayas. *Tree-Ring Research* 75(2): 73.

CBS (2011). *Population Monograph of Nepal*. Kathmandu, Nepal: National Planning Commission Secretariat, Central Bureau of Statistics (CBS), Government of Nepal Population Dynamics.

Chapagain, P. S., Rai, M. K., & Paudel, B. (2018). Land use land cover change and its pathways in Sidin VDC, Panchthar district, Nepal. *Geographical Journal of Nepal* 11: 77–94.

Chhetri, P. K., & Cairns, D. M. (2016) Dendroclimatic response of Abies spectabilis at treeline ecotone of Barun Valley, eastern Nepal Himalaya. *Journal of Forestry Research* 27(5): 1163–1170.

Dahal, P., Shrestha, M., Panthi, J., & Pradhanang, S. M. (2015). Drought risk assessment in central Nepal: Temporal and spatial analysis. *Natural Hazards* 80: 1913–1932.

DHM (2015). *Study of Climate and Climatic Variation over Nepal*. Kathmandu, Nepal: Department of Hydrology and Meteorology (DHM).

DHM (2017). *Observed Climate Trend Analysis in Nepal (1971–2014)*. Kathmandu, Nepal: Department of Hydrology and Meteorology (DHM).

Didan, K. (2015). MOD13Q1 MODIS/Terra Vegetation Indices 16-Day L3 Global 250m SIN V006 (Data set). *NASA EOSDIS LP DAAC*. doi:10.5067/MODIS/MOD13Q1.006

Dobremez, J. F. (1976). *Le Nepal Ecolgie et Biogeography (Ecology and Biogeography of Nepal)*. Paris: Editions du Centre National de la Researche Centifique.

DoFRS (2015). *State of Nepal's Forests, Forest Resource Assessment (FRA) Nepal*. Kathmandu, Nepal: Department of Forest Research and Survey (DoFRS).

DoFRS (2017). *Forest and Watershed Profile of Local Level (744) Structures of Nepal*. Kathmandu, Nepal: Department of Forest Research and Survey (DoFRS).

Dong, J. R., Kaufmann, R. K., Myneni, R. B., et al. (2003). Remote sensing estimates of boreal and temperate forest woody biomass: Carbon pools, sources, and sinks. *Remote Sensing of Environment* 84: 393–410.

EPA (1998). *Guidelines for Ecological Risk Assessment*. EPA/630/R-95/002F. Washington, DC: US Environmental Protection Agency.

Fensholt, R., & Proud, S. R. (2012). Evaluation of earth observation based global long term vegetation trends – Comparing GIMMS and MODIS global NDVI time series. *Remote Sensing of Environment* 119: 131–147.

Foley, J. A., Defries, R., Asner, G. P., et al. (2005). Global consequences of land use. *Science* 309(5734): 570–574.

Gaire, N. P., Bhuju, D. R., Koirala, M., et al. (2017a). Tree-ring based spring precipitation reconstruction in western Nepal Himalaya since AD 1840. *Dendrochronologia* 42: 21–30.

Gaire, N. P., Dhakal, Y. R., Shah, S. K., et al. (2018). Drought (scPDSI) reconstruction of trans-Himalayan region of central Himalaya using Pinus wallichiana tree-rings. *Palaeogeography, Palaeoclimatology, Palaeoecology* 514: 251–264.

Gaire, N. P., Koirala, M., Bhuju, D. R., & Carrer, M. (2017b). Site- and species-specific treeline responses to climatic variability in eastern Nepal Himalaya. *Dendrochronologia* 41: 44–56.

He, Y. Q., Lee, E., & Warner, T. A. (2017). A time series of annual land use and land cover maps of China from 1982 to 2013 generated using AVHRR GIMMS NDVI3g data. *Remote Sensing of Environment* 199: 201–217.

Holben, B. N. (1986). Characteristics of maximum value composite (MVC) images from temporal AVHRR data. *International Journal of Remote Sensing* 7(11): 1417–1434.

ICIMOD (2014a). *Land Cover of Nepal 1990*. Kathmandu, Nepal: International Center for Integrated Mountain Development (ICIMOD). Available from: http://rds.icimod.org (Last accessed 26 July 2021).

ICIMOD (2014b). *Land Cover of Nepal 2000*. Kathmandu, Nepal: International Center for Integrated Mountain Development (ICIMOD). Available from: http://rds.icimod.org (Last accessed 26 July 2021).

Karki, R., Hasson, S., Schickhoff, U., Scholten, T., & Bohner, J. (2017). Rising precipitation extremes across Nepal. *Climate* 5: 4.

Kendall, M. G. (1975). *Rank Correlation Methods*. London: Charles Griffin.

Kharal, D. K., Thapa, U. K., St George, S., et al. (2017). Tree-climate relations along an elevational transect in Manang Valley, central Nepal. *Dendrochronologia* 41: 57–64.

Klein Goldewijk, K., Beusen, A., Van Drecht, G., & De Vos, M. (2011). The HYDE 3.1 spatially explicit database of human induced global land use change over the past 12,000 years. *Global Ecology and Biogeography* 20(1): 73–86.

Kogan, F., & Sullivan, J. (1993). Development of Global Drought-Watch System Using NOAA AVHRR Data. *Advances in Space Research* 13(5): 219–222.

Krakauer, N. Y., Lakhankar, T., & Anadon, J. D. (2017). Mapping and attributing normalized difference vegetation index trends for Nepal. *Remote Sensing* 9(10): 986.

Li, L., Zhang, Y., Wu, J., et al. (2019). Increasing sensitivity of alpine grasslands to climate variability along an elevational gradient on the Qinghai-Tibet Plateau. *Science of the Total Environment* 678: 21–29.

Liang, L., Sun, Q., Luo, X., et al. (2017). Long-term spatial and temporal variations of vegetative drought based on vegetation condition index in China. *Ecosphere* 8(8): e01919.

Liu, X. F., Zhu, X. F., Li, S. S., Liu, Y. X., & Pan, Y. Z. (2015). Changes in growing season vegetation and their associated driving forces in China during 2001–2012. *Remote Sensing* 7(11): 15517–15535.

Lobell, D., & Burke, M. (2008). Why are agricultural impacts of climate change so uncertain? The importance of temperature relative to precipitation. *Environmental Research Letters* 3(034007).

LRMP (1986). *Land Resources Mapping Project*. Kathmandu, Nepal: Survey Department, HMGN and Kenting Earth Sciences.

Mann, H. B. (1945) Nonparametric tests against trend. *Econometrica* 13(3): 245–259.

MoFALD (2017) *Local Government Operative Act, 2017* (p. 84). Nepal, Kathmandu, Nepal: Ministry of Federal Affairs and Local Development (MoFALD).

MoFE (2019). *National Level Forests and Land Cover Analysis of Nepal using Google Earth Images*. Kathmandu, Nepal: Ministry of Forests and Environment, Forest Research and Training Centre.

Myneni, R. B., Dong, J., Tucker, C. J., et al. (2001). A large carbon sink in the woody biomass of Northern forests. *Proceedings of the National Academy of Sciences (USA)* 98(26): 14784–14789.

NAPA (2010). *National Adaptation Program in Action (NAPA) to Climate Change*. Kathmandu, Nepal: Ministry of Environment (MoE), Government of Nepal (GoN).

NRRC (2013). Nepal Risk Reduction Consortium; Flagship Programmes. *National Level Conference on Flagship Programmes*, Kathmandu, Nepal.

Panthi, S., Brauning, A., Zhou, Z. K., & Fan, Z. X. (2017). Tree rings reveal recent intensified spring drought in the central Himalaya, Nepal. *Global and Planetary Change* 157(17): 26–34.

Paudel, B., Zhang, Y., Li, S., et al. (2016). Review of studies on land use and land cover change in Nepal. *Journal of Mountain Science* 13(4): 643–660.

Paudel, B., Zhang, Y., Li, S., & Liu, L. (2018). Spatiotemporal changes in agricultural land cover in Nepal over the last 100 years. *Journal of Geographical Science* 28(10): 1519–1537.

Paudel, B., Zhang, Y., Raju, R., Li, L., & Wu, X. (2019). Farmer's understanding of climate change in Nepal Himalayas: Important determinants and implications for developing adaptation strategies. *Climate Change* 158(3–4): 485–502.

Pettorelli, N., Vik, J. O., Mysterud, A., et al. (2005). Using the satellite-derived NDVI to assess ecological responses to environmental change. *Trends in Ecology & Evolution* 20(9): 503–510.

Piao, S. L., Fang, J. Y., Zhou, L. M., et al. (2003). Interannual variations of monthly and seasonal normalized difference vegetation index (NDVI) in China from 1982 to 1999. *Journal of Geophysical Research-Atmospheres* 108(D14): 4401.

Piao, S. L., Fang, J. Y., Zhu, B., & Tan, K. (2005). Forest biomass carbon stocks in China over the past 2 decades: Estimation based on integrated inventory and satellite data. *Journal of Geophysical Research–Biogeosciences* 110: G01006.

Piao, S. L., Wang, X. H., Ciais, P., et al. (2011). Changes in satellite-derived vegetation growth trend in temperate and boreal Eurasia from 1982 to 2006. *Global Change Biology* 17(10): 3228–3239.

Rimal, B. (2013). Urbanization and the decline of agricultural land in Pokhara sub-metropolitan City, Nepal. *Journal of Agricultural Science* 5(1): 54–65.

Sen, P. K. (1968). Estimates of the regression coefficient based on Kendall's tau. *Journal of the American Statistical Association* 63(324): 1379–1389.

Sharma, K. P. (2009) *Maximum temperature trends in Nepal. An analysis based on temperature records from Nepal for the period 1975–2007.* Kathmandu, Nepal: Department of Hydrology and Meteorology (DHM), Babarmahal.

Shrestha, A. B., Wake, C. P., Mayewski, P. A., & Dibb, J. E. (1999). Maximum temperature trends in the Himalaya and its vicinity: An analysis based on temperature records from Nepal for the period 1971–94. *Journal of Climate* 12(9): 2775–2786.

Shrestha, K. B., Chhetri, P. K., & Bista, R. (2017). Growth responses of Abies spectabilis to climate variations along an elevational gradient in Langtang National Park in the central Himalaya, Nepal. *Journal of Forest Research* 22(5): 274–281.

Shrestha, U. B., Shrestha, A. M., Aryal, S., et al. (2019). Climate change in Nepal: A comprehensive analysis of instrumental data and people's perceptions. *Climate Change* 154(3): 315–334.

Sigdel, M., & Ikeda, M. (2012). Seasonal contrast in precipitation mechanisms over Nepal deduced from relationship with the large-scale climate patterns. *Nepal Journal of Science and Technology* 13(1): 115–123.

Sigdel, S. R., Wang, Y., Camarero, J. J., et al. (2018). Moisture-mediated responsiveness of treeline shifts to global warming in the Himalayas. *Global Change Biology* 24(11): 5549–5559.

Sigdyal, K. P. (1999) Save ecological balance: Environment problems and their solution in Nepal, insight in to diverse facets of topography, flora and ecology. *Nepal Nature Paradise* 1(1): 221–235.

Thapa, R. B., & Murayama, Y. (2012). Scenario based urban growth allocation in Kathmandu Valley, Nepal. *Landscape and Urban Planning* 105(1–2): 140–148.

Thapa, U. K., St George, S., Kharal, D. K., & Gaire, N. P. (2017). Tree growth across the Nepal Himalaya during the last four centuries. *Progress in Physical Geography* 41 (4): 478–495.

Tiwari, A., Fan, Z. X., Jump, A. S., Li, S. F., & Zhou, Z. K. (2017a). Gradual expansion of moisture sensitive Abies spectabilis forest in the Trans-Himalayan zone of central Nepal associated with climate change. *Dendrochronologia* 41: 34–43.

Tiwari, A., Fan, Z. X., Jump, A. S., & Zhou, Z. K. (2017b). Warming induced growth decline of Himalayan birch at its lower range edge in a semi-arid region of Trans-Himalaya, central Nepal. *Plant Ecology* 218: 621–633.

Tucker, C. J. (1979). Red and photographic infrared linear combinations for monitoring vegetation. *Remote Sensing of Environment* 8(2): 127–150.

Tucker, C. J., Pinzon, J. E., Brown, M. E., et al. (2005). An extended AVHRR 8-km NDVI dataset compatible with MODIS and SPOT vegetation NDVI data. *International Journal of Remote Sensing* 26(20): 4485–4498.

Uddin, K., Matin, M. A., & Maharjan, S. (2018). Assessment of land cover change and its impact on changes in soil erosion risk in Nepal. *Sustainability* 10(12): 4715.

Uddin, K., Shrestha, H. L., Murthy, M. S. R., et al. (2015). Development of 2010 national land cover database for the Nepal. *Journal of Environmental Management* 148: 82–90.

Venter, O., Sanderson, E. W., Magrach, A., et al. (2016). Global terrestrial Human Footprint maps for 1993 and 2009. *Scientific Data* 3: 160067.

Yengoh, G. T., Dent, D., Olsson, L., Tengberg, A. E., & Tucker, C. J. (2015). *Use of the Normalised Difference Vegetation Index (NDVI) to Assess Land Degradation at Multiple Scales; Current Status, Future Trends and Practical Considerations*. Springer Briefs in Environmental Science. Sweden: Springer.

Yuan, X. L., Li, L. H., Chen, X., & Shi, H. (2015). Effects of precipitation intensity and temperature on NDVI-based grass change over Northern China during the period from 1982 to 2011. *Remote Sensing* 7(8): 10164–10183.

Zhao, X., Tan, K., Zhao, S., & Fang, J. (2011). Changing climate affects vegetation growth in the arid region of the northwestern China. *Journal of Arid Environments* 75(10): 946–952.

Zhou, L. M., Tucker, C. J., Kaufmann, R. K., et al. (2001). Variations in northern vegetation activity inferred from satellite data of vegetation index during 1981 to 1999. *Journal of Geophysical Research-Atmospheres* 106(D17): 20069–20083.

9

Climate Warming Induced Frozen Soil Changes and the Corresponding Environmental Effect on the Tibetan Plateau

A Review

YUNYUN BAN, QIUHONG TANG AND XIMENG XU

9.1 Introduction

In the northern hemisphere, cold ground covered by permafrost and seasonally frozen soil cover approximately 23.9 and 56 per cent of the land surface, respectively (Zhang et al., 2008). With more than 3.0×10^6 km^2 and an average altitude around 4,500 metres above sea level (asl), the Qinghai–Tibetan Plateau (QTP) is the highest and largest plateau in the world (Royden et al., 2008). This region, shaped by seasonal frozen ground and permafrost, has been given the name of 'The third pole of the world'. Permafrost is commonly defined as the soil or rock ground remaining frozen for at least two continuous years (Widhalm et al., 2017). The thickness of permafrost and frozen soil in the QTP are distributed from 1 to 130 metres depending on the characteristics of slope gradient, altitude, landforms, soil features and water content. In those areas, the soil freeze–thaw (FT) cycle determines the hydrological cycle and thus has strong impacts on water and heat exchanges at the surface of the earth, hence on weather and climate (Zhao et al., 2019), which results in melted sea ice, snow and glacier and decreased permafrost area (Kurylyk et al., 2014). The soil FT cycle, connecting with climate warming, may substantially change the natural runoff characteristics and eventually increase meltwater erosion in cold regions (Immerzeel et al., 2009; Thorsteinsson et al., 2013).

Climate change resulted from greenhouse gas emissions is expected to increase territorial and/or global air temperatures and lead to variation in meteorological conditions, which may affect subsurface thermal conditions (Kurylyk et al., 2014). Rapid climate warming, which induces higher temperatures, has aroused considerable interest in recent years (Bloomfield et al., 2013; Menberg et al.,

This research is supported by the Second Tibetan Plateau Scientific Expedition and Research (2019QZKK0208), the Strategic Priority Research Program of Chinese Academy of Sciences (XDA20060402), and the Research Program of Newly Recruited Chinese Academy of Sciences (Y9V90214YZ).

2013). The literature shows a noticeable warming in the QTP that occurred from the middle 1950s to early 1960s, while corresponding increases in the average air temperature elsewhere in the world began about 20 years later (Houghton et al., 1996). The QTP is one of the most sensitive regions to climate change due to its high elevation and widely distributed permafrost.

Future potential air temperature changes in the QTP were predicted by coupling General Circulation Models such as the NCAR (Meehl et al., 2000), HADL (Mitchell et al., 1995) and GFDL (Mitchell et al., 1995) models. Air temperature is expected to increase more significantly in the QTP than elsewhere in China and should rise by 2.8–3.0°C until 2050, whereas the regions adjacent to the Qinghai–Tibetan Railway should be warmer by 3.8–4.8°C at the end of this century. The climate warming and increase in precipitation are expected to be more severe in winter. A progressive air temperature increase would considerably affect permafrost and frozen ground changes in the QTP, including temperature, distribution, thickness and FT cycles frequency.

Changes in frozen soil caused by increasing air temperature have an important impact on water and heat balances of the earth's surface (Subin et al., 2013; Schuur et al., 2015), such as soil moisture, runoff routes and stream flow regimes (Walvoord & Kurylyk, 2016), and therefore affect the ecosystems. This review helps understand the mechanisms behind frozen soil changes and its influences on ecosystems and is, therefore, important for water resources management and environment protection in the QTP regions.

9.2 Climate Warming Influence on Frozen Ground

The increase in air and land surface temperature has caused land surface changes, which have affected the surface and underground hydrology and meltwater amount in the QTP. In turn, this process could accelerate the rate of climate change by releasing greenhouse gases contained in the permafrost and seasonal frozen soil into the atmosphere (Kurylyk et al., 2014). In recent decades, the ground temperature changes caused by climate warming have been studied in the QTP. Under climate change, frozen ground has retreated severely in the QTP over the past 50 years. This includes a reduction of frozen ground areas, permafrost and frozen depth, frozen cycles and an early thawing period.

9.2.1 Ground Temperature Changes

The frozen ground in the QTP plays an important role in modulating the local climate. Much research has been conducted to observe and simulate the earth frozen ground conditions experiencing an increase in air temperature (Cheng &

Jin, 2013). Due to its special geographic location and high elevation, the frozen ground in the QTP is warm and shallow, which is sensitive to global warming. The area of frozen ground has decreased since the 1950s (Liu & Chen, 2000). The climate model simulations showed that the expected rise of the air temperature would result into a degradation of permafrost and frozen ground in the next 50 years (Zhu et al., 2013).

Ground temperature is a key parameter to describe the permafrost status. Permafrost is a type of frozen ground which is characterized by ground temperature and is thereby more sensitive to climate change. An increasing air temperature can induce a higher permafrost temperature and, in some cases, can lead to permafrost thaw. The increase in permafrost temperature along the QTP highway at depth of 6.0 metres ranged from 0.12 to 0.67°C from 1996 to 2006 (Wu & Zhang, 2008). Field experiments on the Tibetan Plateau have shown that the ground surface temperature near the frozen soil boundary areas in winter has increased by more than 2.5°C from 1980 to 2007 (Wu et al., 2013). From 1950 to 2001, the mean annual temperature in the QTP increased at the rate of 0.21–0.42°C every 10 years, and the diurnal temperature decreased at a rate of −0.08 to −0.51°C/decade (Ma & Hu, 2005). Pan and Li (1996) concluded that in QTP, the temperature has increased more and possibly sooner than elsewhere in the world.

The greatest air temperature warming happened in January, February, July, August and September in the Three River Source Region, located in the northeastern Tibetan Plateau. The mean annual increased temperatures in winter and summer were 0.06°C and 0.05°C, respectively, while those in spring or autumn were 0.04°C. The temperature of the frozen ground in the QTP has increased steadily according to the monitoring data collected between 2006 and 2010 in 27 sites. The average increasing rate of temperature on frozen ground above 6.0 metres was 0.02°C per year (Wu et al., 2012). However, the remote sensing data indicated that the earth surface temperature increased approximately 0.6°C from 1980 to 2005 (Jin et al., 2009). The change in soil temperature during the FT period at depth a from 0 to 320 centimetres has been investigated from 1980 to 2012, as shown in Table 9.1 (Luo et al., 2016). The average ground temperature was 4.7°C. The Changdu and Maduo stations collected the highest and the lowest temperature, which were 10.8°C and 1.6°C, respectively (Luo et al., 2016). Ground temperature is generally influenced by radiation, temperature and soil characteristics. The ground temperature usually increases from January to July. Beneath the surface, the increase of temperature underground occurs later than at the surface. However, at a depth of 80 centimetres, ground temperature increased from January or February to August.

The mean ground temperature increased from 1980 to 2014. The increasing trend was monitored at 1 per cent level with a mean rate of 0.07°C each year. The

Table 9.1. Increase of mean soil temperature at different depths in season and annual from year of 1980 to 2014 (data from meteorological stations listed)

	Spring	Summer	Autumn	Winter	Annual
Depth (0)					
Xinghai	4.46	2.80	4.16	3.20	5.57
Henan	4.46	5.02	4.45	4.28	5.34
Jiuzhi	4.19	4.53	4.04	4.72	5.37
Dari	3.32	2.33	2.88	3.00	3.90
Maduo	3.96	3.91	4.04	3.77	4.77
Qumalai	4.24	3.41	4.28	5.54	5.62
Yushu	3.35	2.79	5.09	4.50	5.72
Nangqian	3.75	3.62	4.22	5.29	4.83
Changdu	3.84	1.49	2.79	4.52	4.29
Mean	3.91	3.28	3.97	4.26	5.01
Depth (mean from 5 to 20 cm)					
Xinghai	2.97	2.69	3.77	3.42	4.74
Henan	4.12	4.80	4.03	2.89	5.18
Jiuzhi	4.95	5.32	4.42	4.51	5.93
Dari	3.66	3.83	2.58	3.37	4.46
Maduo	4.17	4.85	3.83	1.78	5.42
Qumalai	4.95	4.76	3.68	4.10	5.88
Yushu	0.91	3.75	4.40	4.75	5.49
Nangqian	3.72	3.95	3.21	5.32	4.87
Changdu	3.60	1.66	2.15	4.36	3.66
Mean	3.67	3.96	3.56	3.83	5.07
Depth (mean from 40 to 320 cm)					
Henan	4.81	4.93	4.85	4.43	5.45
Jiuzhi	4.45	5.72	3.47	1.45	5.24
Dari	3.65	4.68	3.75	3.87	4.88
Maduo	6.24	6.03	5.48	0.26	6.43
Qumalai	3.56	4.63	2.93	1.30	4.04
Changdu	2.31	1.90	2.82	3.94	2.71
Mean	4.17	4.65	3.88	2.49	4.79

monthly variation of ground temperature was observed in several months, including January, June, October, November and December. The largest warming tendency occurred in the winter months and the annual increase rate was 0.1°C each year, which was approximately 70 per cent higher than that in the summer months. The mean summer month increase rate was 0.06°C. The increase rates in spring and autumn were 0.07°C and 0.06°C each year, respectively. The increase rate in ground temperature is higher in winter than that in the other seasons (Yang et al., 2010). Model simulation results showed that the QTP region would increase by 0.6–0.9°C until 2050 (Zhu et al., 2013). Both daily minimum and maximum temperature would increase. The ground surface temperature in Xidatan and Tanggula along the QTP railway would increase by 0.43 and 0.52°C a^{-1} from 2012 to 2100, and the soil temperature at depths of 5, 15 and 30 metres would rise to 2.40, 0.52 and −0.30°C until 2100 (Sun et al., 2020). All in all, both the historical data and future simulations showed a warming trend of ground temperature due to a warmer climate.

Although scientists predicted that the rise of ground temperature caused by climate change would show regional variations, they reached consensus on the warming tendency of ground temperature. Permafrost is particularly sensitive to climate change and an increase in permafrost temperature is likely to become more serious by the end of this century.

9.2.2 Soil FT Cycles

Soil FT cycles of the active layer are significantly affected by ground temperature if daytime maximum ground temperature is above 0°C and the night minimum temperature is below 0°C. The seasonal soil FT cycles that occur at the surface with changes in soil temperature are sensitive to air temperature. The frequency of the FT cycles is strongly influenced by temperature fluctuations of the ground surface around 0°C. Research showed that soil FT cycles are more frequent with an increasing air temperature in winter because of a more frequent absorption and release of the active layer (Yang et al., 2007). From 1981 to 2010, the results show that the average ground freezing delayed its start by 3.4 days and ended 9.4 days earlier than normal. The shortened rate of frozen duration was approximately 0.3 days each year, and it decreased more in deeper layers (Guo & Wang, 2014). The frozen duration of the ground decreased by 19.2 days during 1981–2010. Higher frequency of FT cycles and shorter frozen duration are two typical characteristics of soil FT cycle in the QTP.

All in all, researchers have reported soil FT cycles shrinkage of about 12.8, 19.2 or 9 days in different periods. These FT cycles period reductions are expected to become more severe in the future.

9.2.3 Frozen Ground Distribution

It is estimated that the frozen areas in the QTP decreased by approximately 1.0×10^5 km^2 from 1970 to 1990. In the past decade, 10 per cent of the permafrost has been degraded (Qiu, 2014). The near-surface permafrost area exhibits a significant decreasing linear trend, with a decreasing rate of 9.9×10^3 km^2 yr^{-1}. The simulated deep permafrost area remains larger than the near-surface permafrost for the same period. In the following decades from the 1960s to

Table 9.2. Classification of permafrost by temperature (Ran et al., 2018)

Classification	Average permafrost temperature /°C	Average ground temperature /°C
Very cold	<−8.5	<−5.0
Cold	−6.5~−8.5	−3.0~−5.0
Cool	−5.0~−6.5	−1.5~−3.0
Warm	−4.0~−5.0	−0.5~−1.5
Very warm	−2.0~−4.0	0.5~−0.5
Likely thawing	−1.0~−2.0	>0.5

2010s, the simulated permafrost areas were estimated as 1.60, 1.49, 1.45, 1.36 and 1.27×10^6 km^2, respectively (Cheng et al., 2012). The mean air temperature continuously rose during 1960–2009, and the corresponding total decreased permafrost area was approximately 3.3×10^5 km^2.

The permafrost located in mountainous areas is significantly influenced by air warming. Wu et al. (2012) estimated that the area of permafrost in the QTP was 1.11×10^6 km^2 – this represents approximately 43.8 per cent of the QTP – and the frozen ground area was 1.24×10^6 km^2 in 2010. From the 1960s to 2000s, the permafrost occupied less area in the QTP. Ran et al. (2018) suggested a classification of permafrost based on temperature (Table 9.2).

The very cold, cold, cool and warm permafrost decreased by 72.79, 70.12, 27.24 and 4.77 per cent, respectively (Ran et al., 2018). Also 49.4 per cent of very warm and 95.95 per cent of likely thawing permafrost turned into seasonally frozen ground; the total area of degraded permafrost was 1.53×10^6 km^2, which accounted for 88 per cent of the total permafrost area in the 1960s. Temperature data on permafrost from 1975 to 1989 showed that some permafrost has disappeared in some regions in northern Xidatan (Cheng & Wu, 2007). Guo and Wang (2014) simulated present and future permafrost distribution through Community Land Model version 4 (CLM4) based on soil FT processes.

Zhao et al. (2019) simulated permafrost distribution of the Qilian Mountains located in the northeastern part of the QTP by remote sensing images from the 1990s to 2010s and predicted its future distribution from the 2010s to the 2040s through a logistical model. Results showed that the permafrost area would likely be reduced from 7.35×10^4 km^2 to 6.65×10^4 km^2. Nan et al. (2003) simulated the estimation of the permafrost area in the QTP (results in Table 9.3). The permafrost area in the QTP will continue to decrease since an increase of air temperature is expected according to model simulations. The permafrost area is expected to decrease by the rate of 9.2×10^3 km^2 yr^{-1}. The retreat of permafrost might reach 39 per cent in the 2050s and 81 per cent by 2100 (Guo et al., 2012).

Table 9.3. Model simulation results of permafrost area on the Qinghai–Tibetan Plateau (Nan et al., 2003)

Year	Air temperature increase rate /°C a^{-1}	Permafrost area /km^2	Decrease in area from 2001 /%
2001		1,202,140	
2050	0.02	1,094,010	8.8
	0.052	1,039,990	13.5
2100	0.02	1,040,970	13.4
	0.052	653,447	46

9.2.4 Active Layer Thickness

In the northeastern part of the QTP, the mean maximum depth of frozen ground gradually decreased over time at a rate of about 1.02 cm per year. The warmed tendency at different depths was not clear in October due to the buffer function associated with the freezing process. The deeper layer needs time to thaw during the warm season and shows a seasonal change in the temperature variability. The largest temperature increase began from April/May to September/October at depths of 40 cm and 80 cm, while important warming trends were observed from May/June to October/November at depths of 160 to 320 centimetres.

Research brought evidence that the frozen ground depths declined by 0.2 metre from 1980 to 2016 (Jin et al., 2009). The active layer has thickened by the rate of 3.6~7.5 cm a^{-1} (Wu et al., 2010; Li et al., 2012). The increase in air temperature warms permafrost by energy exchange and shifts permafrost into seasonally frozen soil. The increase in warm days and nights between 1961 and 2005 resulted in the increased thawing days for frozen ground, permafrost and the deeper thawed layer. Cheng and Wu (2007) did research on soil temperature using boreholes in lower altitude regions of the Qinghai Tibetan Plateau and found that the active layer thickness of frozen soil deepened by 0.15 to 0.50 metre from 1996 to 2001. An increase in permafrost in the northern QTP was 25 metres from 1976 to 2006, and the increase was 50–80 metres in the south from 1986 to 2006. However, Wu et al. (2010) reported that the increase in active layer in higher altitude areas was 7.5 centimetres each year from 1995 to 2007 which was lower than the results from Cheng and Wu (2007). Zhao et al. (2004) observed an increasing trend of thawing depth in the seasonally frozen soil in the QTP. However, the temporal and spatial changes in frozen soil remain unclear.

To simulate soil freezing and thawing processes, Gao et al. (2018) divided frozen soil into 39 layers. The results showed that the trend of ground temperature increased more significantly in winter than other seasons. The deepest frozen depth of seasonally frozen soil decreased by 0.032 centimetres each decade. The active layer became deeper by the rate of 0.043 cm/decade. In addition, the permafrost thickness decrease resulted in some permafrost turned into seasonally frozen soil. The mean permafrost thickness around Beilu River decreased by 42 centimetres during 2003–2016 (Yin et al., 2017). A model for active layer depth was developed by Li et al. (2012). The results showed that from 1981 to 2010, the active layer thickness increased by the rate of 1.33 centimetres per year. On the southern Tanggula Mountains, the active layer thickness increased by 0.29 and 0.41 metres in undisturbed areas and beneath the embankment from 2006 to 2012 (Sun et al., 2014). The lower edge of permafrost moved up by 25 metres from 1975 to 2002 in the southern QTP (Nan et al., 2003). From the 1990s to 2010s, the elevation edge of permafrost in northern and southern hillslopes of Bayan Har were approximately 90 and 100 metres, respectively (Luo et al., 2013).

All the above research demonstrated that the permafrost degradation happened and active layer thickness deepened in the past several decades. The active layer thickness of 0.5–1.5 metres found in the present day may increase to approximately 1.5–2.0 metres by 2030–2050. This increase is expected to continue and reach a level of 2.0–3.5 metres by 2080–2100. The permafrost table should decline slowly from 2012 to 2060 and sharply from 2060 to 2100 by 0.94 and 11.38 centimetres per year under the RCP6.0 (Sun et al., 2020). A reduction of the permafrost is expected to continue and permafrost residuals could disappear by the end of this century (Guo et al., 2012).

9.3 Environmental Effects

Frozen ground and permafrost are sensitive to air temperature and closely related to the alpine ecosystem which may substantially affect the environment in cold areas (Immerzeel et al., 2012). Although the influence of global warming is less pronounced on weather, it has already considerably impacted our environment, society and economy. Under climate change, the QTP region is experiencing serious degradation of its ecosystem.

9.3.1 Vegetation Degradation

The water content in the active layer is determined by the thawed depth of frozen ground. Therefore, the FT status of frozen ground severely affects the hydrological and thermal processes at the soil surface and in the ecosystem. Vegetation in the QTP region is the major component of its ecosystem and is very sensitive to soil temperature, water content and FT status. The vegetation status is closely associated with permafrost conditions (Wang et al., 2006). Permafrost degradation results in a lower water table, with degradation of wetland and grassland (Cheng & Zhao, 2000). The thawed permafrost releases a large amount of water vapor in the air through evaporation of frozen water in the soil. The heat and water exchange in frozen soils affect plant growth. The QTP region is expected to become drier under climate warming, which should result in smaller vegetation cover (Yang et al., 2010). Moreover, more water at the soil surface should infiltrate into deep soil, because of the decrease of infiltration ability and the increased hydraulic conductivity of thawed soil. A decline in the precipitation in the QTP regions should lead to a vegetation degradation in the region.

Alpine meadows and steppes are the most common plant formations in the QTP regions. In the alpine meadows and steppes, the soil thawed date advanced by 20 and 18 days from 2002 to 2012 caused by permafrost degradation (Jin et al., 2015). Another prediction model by Peng et al. (2012) showed that 82.4 per cent of the vegetation increased and 50 per cent of the broadleaf forest decreased from 1982 to 2003. According to the research in Kunlun to the south slope of Tanggula by Wang et al. (2012), swamp meadow showed a tendency of changing into meadow, caused by the retreat of permafrost. From 1980 to 2000, the mean increase in permafrost temperature was 0.2°C and the area decreases in alpine meadow and swamp meadow were 7.98 per cent and 28.11 per cent, respectively. However, the increase in alpine meadows and steppes coverage area along the QTP railway from 2000 to 2016 was 42.8 per cent and 43.4 per cent, respectively. Since 2007, a rapid vegetation degradation occurred (Song et al., 2018).

Zhang et al. (2010) employed an empirical model to simulate variation in annual biomass production by 2050. Alpine meadow is sensitive to temperature, whereas alpine steppe is sensitive to precipitation. The authors found that with an annual increase of temperature of 0.04°C and annual decrease of precipitation of 0.8 millimetres, the alpine meadow and steppe should keep unchanged. However, if the precipitation remains constant, the biomass production should decrease by 0.27 per cent and 0.24 per cent. Research results showed that vegetation in the QTP regions are likely to face further degradation under a future climate change.

Permafrost and frozen ground have significantly affected the vegetation distribution in the QTP regions. Alpine meadows and steppes showed inconsistent changes tendency, and swamp meadow changed into meadow during the last 20 years of the last century. Predicting the vegetation tendency based on single factors such as the temperature, precipitation and degradation of permafrost remains challenging.

9.3.2 Desertification

A large amount of literature has investigated the relationship among temperature, permafrost and desertification based on soil FT depth and sandstorm data. Permafrost degradation has induced desertification in some regions of the QTP regions since 2000 (Xue et al., 2009). Comparing with sandstorms in other areas, permafrost degradation occurred more frequently in the QTP regions (Han et al., 2004). The main characteristics in this area were strong wind, low temperature and FT cycles. Zhao et al. (2005) stated that the sandstorms occurrence was closely related to minimum FT depth. Sandstorms often appear in the northwest of China (Yang et al., 2006). Field observations from 1964 to 2004 indicated that sandstorms occur more frequently in the QTP regions and in the arid and semi-arid areas of northern China (Song et al., 2004). One of the largest sandstorm centres was situated in the QTP regions and has gradually moved northward toward Korea and Japan from December to March each year. Dust migrated with wind vortex at lower atmosphere level and with the global westerlies at higher level, giving rise to a multilayer structure of the wind.

The northern part of the QTP regions is a sandy desertification district that has attracted broad research on the desertification process and mechanisms. Data analysis showed that the characteristic of sandy desertification in the northern QTP regions is strongly associated with natural and anthropogenic factors. It is related to climate warming and a precipitation decrease that has occurred in the past several decades (Dong, 2001). These climate changes have caused glacial melting, rising snow lines, and provoked permafrost thaw and lake disappearance. These phenomena have also contributed to a rise in farmland salinity and wind and meltwater erosion, while overgrazing and lack of management have led to a degradation of the vegetation cover and soil. The land surface covered by grassland has become a desert and this type of desertification process should extend and become more severe in the future (Yang et al., 2004). The annual degradation of grassland has had an irreversible influence on the environment and economy in those districts. Former research concluded that the region covered by desert increased by 1.8 per cent each year on the Tibetan Plateau during the research period between 1957 and 1977 (Shi, 1992). In 1993, the desert area was 1.9×10^3 km^2 in the area of the Yaluzangbu, Lahsa and Nyingchi River in the QTP regions. Desertification leads to more serious soil erosion and the degradation of permafrost and frozen soil (Yang et al., 2004).

Permafrost degradation has caused desertification in some regions of the QTP since the beginning of this century. This phenomenon is closely related to the air warming and precipitation decrease that has occurred in the past several decades and has contributed to farmland salinity, erosion and grazing.

9.3.3 Soil Erosion

Soil erosion is a common phenomenon on cultivated slopes in many regions across the world. Soil erosion in the QTP regions is susceptible to rising temperature, vegetation and soil water content (Nearing et al., 2004). Frozen ground degradation affects soil moisture and groundwater level, which leads to increase in meltwater runoff and serious soil erosion.

Repeated FT cycles destroy the soil aggregate structure, undermine soil strength and affect soil bulk density, thereby causing high soil erodibility and low critical shear force. With more frequent FT cycles, FT erosion become more severe. During the soil freezing process, temperature gradually decreases below 0°C and soil gradually freezes from the surface to greater depths when the effect of water surface accumulation takes place at the soil surface (Xie et al., 2011). Water gathering at the upper layer resulting from the migration of water is driven by temperature gradient. In the freezing period, moisture in soil changes from liquid to solid, while moisture in a soil pore freezes as an ice crystal. The higher the water content in the soil pore is, the larger the soil swelling volume during freezing will be, possibly resulting in the breaking up of soil aggregates. Surface runoff decreases but subsurface runoff increases, both relative to the difference between precipitation and evapotranspiration (Guo et al., 2012). These phenomena result in fluctuations in the runoff intensities of snow and glacier-generated water flows and the thawed depth of a frozen soil layer, leading to serious soil erosion (Kurylyk et al., 2014).

The climate warming has contributed to make a deeper thawed thickness of the land surface soil. With a thicker active layer, the organic matter content of the soil in cold areas reduces exponentially and the surface soil disposition becomes coarse and gravelly (Wang et al., 2006). Loose sediment on the surface of permafrost and frozen soil was plentiful during thawing seasons compared with bedrock (Zhang et al., 2016). Frequent FT cycles caused by climate warming have led to the rearrangement of the soil structure (Liu et al., 2009), which has affected the permeability (Kurylyk & Watanabe, 2013), soil stability (Oztas & Fayetorbay, 2003), density (Zhang et al., 2007) and strength (Jamshidi & Lake, 2015) and soil erodibility (Fang et al., 2015). The area of FT erosion in 2015 was severer than that of 2000 (Guo et al., 2017). The area of erosion increased from 10.55×10^4 km^2 to 11.26×10^4 km^2 between 2000 and 2015. However, large scale erosion data to assess the conditions of soil degradation caused by increasing air temperature was difficult to obtain. In this context, model predictions are suitable tools to evaluate the response of soil erosion to climate change. It is reported that the soil erosion area increased during 1961–2012 (Wang et al., 2020) due to the higher precipitation in the south QTP. The average annual rainfall erosivity value of the mid-Yarlung Tsangpo River region located in the southern part of the QTP was 867.6 MJ mm ha^{-1} h^{-1} a^{-1} during 2001–2015 (Teng et al., 2018), which was positively related to precipitation. The mean K factor decreased from 0.034 to 0.022 t ha ha^{-1} MJ^{-1} mm^{-1}, indicating the soil erosion capacity was affected by soil FT effect (Wang et al., 2020). The soil erosion rate of this region was 29.1 t ha^{-1} a^{-1} and the most severe soil erosion occurred in low temperature and high precipitation years. A downward tendency of soil erosion caused by climate change of about 1.14 per cent per year was identified between 2001 and 2015. The model predictions by Wang et al. (2020) also showed that the increased precipitation and temperature would result in a higher eroded soil amount until 2050.

Frozen ground degradation increases meltwater runoff and may lead to serious soil erosion. The soil organic matter content in cold areas is reduced and the surface soil disposition may become coarse and gravelly. The soil erosion area increased caused by meltwater in the past decades, however, it had a decreasing tendency from 2001 to 2015. The model predictions suggest that soil erosion would continue to increase until 2050.

9.3.4 Carbon Emission

Large amounts of carbon are widely distributed in the wetland and peatland in the QTP regions, and much of them were stored in permafrost (Schuur et al., 2015). The degradation of permafrost has accelerated the biogeochemical cycle, converting the wetland in the cold areas into a source of carbon dioxide, which would certainly contribute to accelerate global warming.

Initial works on the release of greenhouse gases from thawed permafrost have been focused on the QTP regions. The release of greenhouse gases from the thawed permafrost layer in the Tibetan Plateau was initially found in the form of CO_2. Grasslands may release a small amount of CH_4 to the atmosphere in the winter. Findings showed that the amount of CH_4 emission was closely related to the soil temperature at different depths and air temperature. The emission amount of CH_4 in the permafrost regions of the Tibetan Plateau was relatively much higher than that released from dry land or farmland (Yang et al., 2010). The excess amount of CH_4 stored in the permafrost on the Tibetan Plateau was in the form of hydrates. Research showed that the potential reserves in the permafrost in the form of hydrates ranged approximately from 1.2×10^{11} to 2.4×10^{14} m^3 on the Tibetan Plateau (Chen et al., 2005). Permafrost degradation should increase the emissions of CH_4 and CO_2. The CH_4 emissions from the wetland ranged from 0.7 to 1.0 T_g a^{-1}. The CH_4 hydrates may be released from a depth between 27 and 2,070 metres, in frozen regions with a low thermal gradient. The variation of temperature generally infiltrates until 10 metres in the soil of the Tibetan Plateau but cannot influence the CH_4 hydrate emission. Research conducted in the permafrost regions along the railway in the QTP regions showed that the increasing air temperature in the last 30 years has had a significant influence on permafrost temperature above the depth of 40 metres (Pan et al., 2002).

The air temperature is expected to continue rising in the future, which is likely to trigger the deeper permafrost thaw. These changes increase the risk of releasing carbon hydrates from the permafrost to the air as greenhouse gases and, in turn, increase the air temperature.

9.3.5 Infrastructure Project

The increase in the economic development on the Tibetan Plateau has led to infrastructures beneficial for the local population, such as highway, railway, oil and gas product pipeline. The QTP railway is the highest in the world, with a length of 960 kilometres and the highest section is on a mountain at the altitude of 5,072 metres. Most destructions of projects in these areas, such as ground heave and thawed frozen soil, were provoked by the degradation of permafrost. A lot of engineering problems in the QTP regions related to climate changes were all associated with an increasing temperature of the surface of the permafrost, deeper thaw depth of seasonal frozen soil and the degradation of permafrost. These changes have resulted in the decrease of soil bearing strength, increase in soil infiltration, increased potential damage infrastructure, destroyed aggregates, and contributed to expand melt ground. Each of these damages has the potential to destroy various human infrastructures.

The warming tendency in the QTP regions during the past years has led to building collapse (Lu et al., 2018). The first and foremost damage in the Tibetan regions was a highway built on the permafrost and affected by the thawed settlement. The infrastructure projects in the QTP areas would make the nearby permafrost thaw further. The soil FT cycle changes soil structure, which would further cause a soil deformation process and even substantially change the engineering facility (Özgan et al., 2015). In the QTP, more than half of the transportation structures were built on the permafrost with a temperature above -1.0°C (Cheng, 2005). The main forms of damage to the infrastructure were permafrost thaw and soil deformation. Water migration and redistribution were the main reasons for the deformation; water in the soil transforms into ice during freezing periods, and the ice volume becomes bigger than the water, which would affect the soil structure. Also, during the thawing period, the soil would deform again and contain less water volume (Zhang et al., 2016). Large ground elevations (>8 mm) were reported to occur on the central flat terrain of the basin and some small surface deformations (2.5–8 millimetres) were found on the slope because of a reduction of water in the soil. In addition, the annual thaw settlement of 3 mm was observed during the past several years which showed an increasing tendency in some regions due to long-term variations in soil water content (Daout et al., 2017). Changes of the physical and mechanical characteristics caused by the FT cycles could have negative influences in cold region engineering. This kind of change in those areas can damage the substrate bearing strength, soil permeability, mass movements and thermokarst terrain. All of the above phenomena have a negative influence on human infrastructure, such as railways, buildings and bridges.

A total of 83.5 per cent of the damage to the highway in the QTP in the frozen area resulted from subsidence, and the remaining part was caused by soil elevation (Tong & Wu, 1996). The Qinghai–Tibetan Railway was constructed from 2001 to 2006. More than half of the railway (632 kilometres of a total of1,118 kilometres) was built in a permafrost area. A 'roadbed cooling' approach was suggested for road constructions in 'warm' permafrost (Cheng, 2005). The influence of infrastructure built on the permafrost, vegetation and soil in the QTP has been known as the leading factor that deepened the permafrost thawed depth. The engineering behaviour has had negative effects on the vegetation and has affected the depth of frozen ground, permafrost, soil, ice, snow and other natural processes.

The influence of permafrost degradation on hydrothermal migration processes between the earth surface and the atmosphere require long-term monitoring and further research on the environment in the QTP regions. To protect the permafrost, glacier, ice and snow from retreating in QTP, appropriate laws and regulations must be adopted such as controlling overgrazing and strengthening management.

9.4 Conclusion

We reviewed previous literature that has considered climate change as the primary factors caused by territorial and/or global air temperature, and thereby affecting FT cycles, frozen ground distribution and depth changes in the QTP. From 2000 to 2015, the average freezing days of the ground reduced by 12.8 days and the total area decreased by 3.3×10^5 km^2. In addition, the active layer became thicker by a rate of about 3.6–7.5 centimetres per year. Model simulations suggested that the frozen ground would deepen by 2.0–3.5 metres by 2080–2100. These changes have had serious impacts on the vegetation, soil erosion, sandstorms, ecohydrology and infrastructure. Much of the infrastructure project in the QTP has mainly focused on highway and railway. These changes in the QTP would continue and will likely worsen by the end of 2100. Knowledge about the QTP remains limited and the future impacts of global warming in the QTP are still unclear. In those areas, the monitoring stations and observation programmes have not been carried out because of the difficult weather and terrain conditions. An increasing number and density of monitoring stations should give enough support to meet the urgent need of research and contribute to improve the life quality of the local residents in the QTP. Interactions between atmosphere and vegetation are another topic that deserves further attention. In addition, more research on its relationship with a warming climate is required. More scientists need to work together to protect the frozen ground resources and environmental values in this critical third pole region threatened by future climate changes.

References

Bloomfield, J. P., Jackson, C. R., & Stuart, M. E. (2013). Changes in groundwater levels, temperature, and quality in the UK over the 20th century: An assessment of evidence of impacts from climate change. *Living With Environmental Change Report, UK*, 14 pp., Available from http://nora.nerc.ac.uk/503271/ (Last accessed 19 September 2013).

Chen, D. F., Wang, M. C., & Xia, B. (2005). Formation condition and distribution prediction of gas hydrate in Qinghai–Tibet Plateau permafrost. *Chinese Journal of Geophysics* 48(1): 166–172.

Cheng, G. D. (2005). A roadbed cooling approach for the construction of Qinghai-Tibet Railway. *Cold Region Science and Technology* 42(2): 169–176.

Cheng, G. D., & Wu, T. (2007). Responses of permafrost to climate change and their environmental significance, Qinghai-Tibet Plateau. *Journal of Geophysical Research* 112(F02S03): 1–10.

Cheng, G., & Zhao, L. (2000), The problems associated with permafrost in the development of the Qinghai-Xizang Plateau (in Chinese with English abstract). *Quaternary Sciences*, 20(6): 521–531.

Cheng, W., Zhao, S., Zhou, C., & Chen, X. (2012). Simulation of the decadal permafrost distribution on the Qinghai-Tibet plateau (China) over the past 50 years. *Permafrost and Periglacial Processes* 23(4): 292–300.

Daout, S., Doin, M. P., Peltzer, G., Socquet, A., & Lasserre, C. (2017). Large-scale InSAR monitoring of permafrost freeze-thaw cycles on the Tibetan plateau. *Geophysical Research Letters* 44(2): 901–909.

Dong, Y. (2001). Driving mechanism and status of sandy desertification in the northern Tibet Plateau. *Journal of Mountain Science*, 19(5): 385–391 (in Chinese, English abstract).

Fang, H. B., Zhao, L., Wu, X. D., et al. (2015). Soil taxonomy and distribution characteristics of the permafrost region in the Qinghai-Tibet Plateau, *China. Journal of Mountain Science* 12(6): 1448–1459.

Gao, B., Yang, D., Qin, Y., et al. (2018). Change in frozen soils and its effect on regional hydrology, upper Heihe basin, northeastern Qinghai–Tibetan Plateau. *The Cryosphere* 12(2): 657–673.

Guo, B., Luo, W., Wang, D., & Jiang, L. (2017). Spatial and temporal change patterns of freeze–thaw erosion in the three-river source region under the stress of climate warming. *Journal of Mountain Science*, 14(6): 86–99.

Guo, D., & Wang, H. (2014). Simulated change in the near-surface soil freeze/thaw cycle on the Tibetan plateau from 1981 to 2010. *Chinese Science Bulletin* 59(20): 2439–2448.

Guo, D., Wang, H., & Li, D. (2012). A projection of permafrost degradation on the Tibetan plateau during the 21st century. *Journal of Geophysical Research: Atmospheres* 117 (D5): 1–15.

Han, Y., Xi, X., Song, L., Ye, Y., & Li, Y. (2004), Spatio-temporal sand-dust distribution in Qinghai-Tibet Plateau and its climatic significance. *Journal of Desert Research* 24 (5): 72–76 (in Chinese with English abstract).

Houghton, J., Filho, L., Callander, B., et al. (eds.) (1996). *Climate Change 1995: The Science of Climate Change*. Cambridge: Cambridge University Press.

Immerzeel, W. W., Droogers, P., De Jong, S. M., & Bierkens, M. F. P. (2009). Large-scale monitoring of snow cover and runoff simulation in Himalayan river basins using remote sensing. *Remote Sensing of Environment* 113(1): 40–49.

Immerzeel, W. W., van Beek, L. P. H., Konz, M., Shrestha, A. B., & Bierkens, M. F. P. (2012). Hydrological response to climate change in a glacierized catchment in the Himalayas. *Climatic Change* 110: 721–736.

Jamshidi, R. J., & Lake, C. B. (2015). Hydraulic and strength properties of unexposed and freeze–thaw exposed cement-stabilized soils. *Canadian Geotechnical Journal* 52(3): 283–294.

Jin, H., He, R., Cheng, G., et al. (2009). Changes in frozen ground in the source area of the Yellow River on the Qinghai-Tibet Plateau, China, and their eco-environmental impacts. *Environmental Research Letters* 4(4): 045206.

Jin, R., Zhang, T. J., Li, X., Yang, X. G., & Ran, Y. H. (2015). Mapping surface soil freeze–thaw cycles in China based on SMMR and SSM/I brightness temperatures from 1978–2008. *Arctic, Antarctic, and Alpine Research* 47(2): 213–229.

Kurylyk, B. L., MacQuarrie, K. T. B., & Voss, C. I. (2014). Climate change impacts on the temperature and magnitude of groundwater discharge from shallow, unconfined aquifers. *Water Resources Research* 50(4): 3253–3274.

Kurylyk, B. L., & Watanabe, K. (2013). The mathematical representation of freezing and thawing processes in variably saturated, non-deformable soils. *Advances in Water Resources* 60: 160–177.

Li, R., Zhao, L., Ding, Y., et al. (2012). Temporal and spatial variations of the active layer alone the Qinghai–Tibetan Highway in a permafrost region. *Chinese Science Bulletin* 57(35): 4609–4619.

Liu, X., & Chen, B. (2000). Climatic warming in the Tibetan Plateau during recent decades. *International Journal of Climatology* 20(14): 1729–1742.

Liu, X. D., Cheng, Z. G., & Yan, L. B. (2009). Elevation dependency of recent and future minimum surface air temperature trends in the Tibetan Plateau and its surroundings. *Global and Planetary Change* 68(3): 164–174.

Lu, J., Zhang, M., Zhang, X., Pei, W., & Bi, J. (2018). Experimental study on the freezing–thawing deformation of a silty clay. *Cold Regions Science and Technology* 151: 19–27.

Luo, D. L., Jin, H. J., Lin, L., et al. (2013). Distributive features and controlling factors of permafrost and the active layer thickness in the Bayan Har Mountains along the Qinghai-Kangding highway on Northeastern Qinghai-Tibet plateau. *Scientia Geographica Sinica* 33(5): 635–640 (In Chinese, English abstract).

Luo, S., Fang, X., Lu, S., Ma, D., & Chen, H. (2016). Frozen ground temperature trends associated with climate change in the Tibetan plateau three river source region from 1980 to 2014. *Climate Research* 67(3): 241–255.

Ma, X. B., & Hu, Z. Y. (2005). Precipitation variation characteristics and abrupt change-over Qinghai–Xizang Plateau in recent 40 year. *Journal of Desert Research* 25(1): 137–139 (in Chinese, English abstract).

Meehl, G. A., Washington, W. M., Arblaster, J. M., Bettge, T. W., & Strand, W. G. (2000). Anthropogenic forcing and climate system response in simulations of 20th and 21st century climate. *Journal of Climate* 13(21): 3728–3744.

Menberg, K., Blum, P., Schaffitel, A., & Bayer, P. (2013). Long-term evolution of anthropogenic heat fluxes into a subsurface urban heat island. *Environment Science and Technology* 47(17): 9747–9755.

Mitchell, J. F. B., Johns, T. J., Gregory, J. M., & Tett, S. F. B. (1995). Climate response to increasing level of greenhouse gases and sulphate aerosols. *Nature* 376: 501–504.

Nan, Z., Gao, Z., Li, S., & Wu, T. (2003). Permafrost changes in the northern limit of permafrost on the Qinghai-Tibet Plateau in the last 30 years, *Acta Geographica Sinica* 58(6): 817–823 (In Chinese, English abstract).

Nearing, M. A., Pruski, F. F., & O'Neal, M. R. (2004) Expected climate change impacts on soil erosion rates: A review. *Journal of Soil and Water Conservation* 59(1): 43–50.

Özgan, E., Serin, S., Ertürk, S., & Vural, I. (2015). Effects of freezing and thawing cycles on the engineering properties of soils. *Soil Mechanics and Foundation Engineering* 52(2): 95–99.

Oztas, T., & Fayetorbay, F. (2003). Effect of freezing and thawing processes on soil aggregate stability. *Catena* 52(1): 1–8.

Pan, B., & Li, J. (1996). Qinghai–Tibetan Plateau: A driver and amplifier of the global climatic change. *Journal of Lanzhou University (Natural Sciences)* 32: 108–115 (in Chinese, English abstract).

Pan, W., Yu, S., Jia, H., & Liu, D. (2002). Variation of the ground temperature field in permafrost regions along the Qinghai–Tibetan railway. *Journal of Glaciology and Geocryology* 24(6): 774–779 (in Chinese, English abstract).

Peng, J., Liu, Z., Liu, Y., Wu, J., & Han, Y. (2012). Trend analysis of vegetation dynamics in Qinghai–Tibet Plateau using Hurst Exponent. *Ecological Indicators* 14(1), 28–39.

Qiu, J. (2014). Double threat for Tibet: Climate change and human development are jeopardizing the platear's fragile environment. *Nature* 512: 240–241.

Ran, Y., Li, X., & Cheng, G. (2018). Climate warming over the past half century has led to thermal degradation of permafrost on the Qinghai–Tibet Plateau. *Cryosphere* 12(2): 595–608.

Royden, L. H., Burchfiel, B. C., & van der Hilst, R. D. (2008). The geological evolution of the Tibetan Plateau. *Science* 321(5892): 1054–1058.

Schuur, E. A. G., McGuire, A. D., Schädel, C., et al. (2015). Climate change and the permafrost carbon feedback. *Nature* 520: 171–179.

Shi, Y. (1992). The discussion of the grassland ecological environment maladjustment and control strategy. *Qinghai Environment* 1(2): 7–13 (in Chinese, English abstract).

Song, L. C., Han, Y. X., Zhang, Q., Xi, X. X., & Ye, Y. H. (2004). Monthly temporal–spatial distribution of sandstorms in China as well as the origin of Kosa in Japan and Korea. *Chinese Journal of Atmospheric Sciences* 28(6): 820–828 (in Chinese, English abstract).

Song, Y., Jin, L., & Wang, H. (2018). Vegetation changes along the Qinghai–Tibet Plateau engineering corridor since 2000 induced by climate change and human activities. *Remote Sensing* 10(1): 95–115.

Subin, Z. M., Koven, C. D., Riley, W. J., et al. (2013). Effects of soil moisture on the responses of soil temperatures to climate change in cold regions. *Journal of Climate* 26(10): 3139–3158.

Sun, Z. Z., Wu, G. L., Yun, H. B., Liu, G. J., & Rui, P. F. (2014). Permafrost degradation under an embankment of the Qinghai–Tibet Railway in the southern limit of permafrost. *Journal of Glaciology and Geocryology* 36(4): 767–771 (in Chinese, English abstract).

Sun, Z. Z., Zhao, L., Hu, G., et al. (2020). Modeling permafrost changes on the Qinghai–Tibetan plateau from 1966 to 2100: A case study from two boreholes along the Qinghai–Tibet engineering corridor. *Permafrost and Periglacial Processes* 31(1): 156–170.

Teng, H., Liang, Z., Chen, S., et al. (2018). Current and future assessments of soil erosion by water on the Tibetan Plateau based on RUSLE and CMIP5 climate models. *Science of the Total Environment*, 635: 673–686.
Thorsteinsson, T., Johannesson, T., & Snorrason, A. (2013). Glaciers and ice caps: Vulnerable water resources in a warming climate. *Current Opinion in Environmental Sustainability* 5(6): 590–598.
Tong, C. J., & Wu, Q. B. (1996). The effect of climate warming on the Qinghai–Tibet Highway. *Cold Region Science and Technology* 24(1): 101–106.
Walvoord, M. A., & Kurylyk, B. L. (2016). Hydrologic impacts of thawing permafrost – A review. *Vadose Zone Journal* 15(6): 1–20.
Wang, G. X., Li, Y. S., Wu, Q. B., & Wang, Y. B. (2006). Impacts of permafrost changes on alpine ecosystem in Qinghai–Tibet Plateau. *Science in China (series D)* 49: 1156–1169.
Wang, G. X., Liu, G., Li, C., & Yang, Y.. (2012). The variability of soil thermal and hydrological dynamics with vegetation cover in a permafrost region. *Agricultural & Forest Meteorology* 162–163, 44–57.
Wang, L., Zhang, F., Fu, S., et al. (2020). Assessment of soil erosion risk and its response to climate change in the mid-Yarlung Tsangpo river region. *Environmental Science and Pollution Research* 27(1): 607–621.
Widhalm, B., Bartsch, A., Leibman, M., & Khomutov, A. (2017). Active-layer thickness estimation from x-band SAR backscatter intensity. *Cryosphere* 11(1): 483–496.
Wu, Q., & Zhang, T. (2008). Recent permafrost warming on the Qinghai-Tibetan Plateau. *Journal of Geophysics Research* 113(D13): D13108.
Wu, Q., Zhang, T., & Liu, Y. (2010). Permafrost temperatures and thickness on the Qinghai-Tibet Plateau. *Global and Planetary Change* 72(1–2): 32–38.
Wu, Q., Zhang, T., & Liu, Y. (2012). Thermal state of the active layer and permafrost along the Qinghai-Xizang (Tibet) railway from 2006 to 2010. *The Cryosphere* 6(3): 607–612.
Wu, T., Zhao, L., Li, R., et al. (2013). Recent ground surface warming and its effects on permafrost on the Central Qinghai-Tibet Plateau. *International Journal of Climatology* 33(4): 920–930.
Xie, H., Nkonya, E., & Wielgosz, B. (2011). Technical Note: Assessing the risks of soil erosion and small reservoir siltation in a tropical river basin in Mali using the SWAT model under limited data condition. *Applied Engineering in Agriculture* 27(6): 895–904.
Xue, X., Guo, J., Han, B. S., Sun, Q. W., & Liu, L. C. (2009). The effect of climate warming and permafrost thaw on desertification in the Qing–Tibetan Plateau. *Geomorphology* 108(3–4): 182–190.
Yang, M. X., Nelson, F., Shiklomanov, N., Guo, D., & Wan, G. (2010). Permafrost degradation and its environmental effects on the Tibetan Plateau: A review of recent research. *Earth Science Review* 103(1–2): 31–44.
Yang, M. X., Wang, S. L., Yao, T. D., et al. (2004). Desertification and its relationship with permafrost degradation in Qinghai–Xizang (Tibet) Plateau. *Cold Region Science and Technology* 39(1): 47–53.
Yang, M. X., Yao, T. D., Gou, X. H., & Wang, H. J. (2006). Effect of heavy snowfall on ground temperature, Northern Tibetan Plateau. *Annals of Glaciology* 43(1): 317–322.
Yang, M. X., Yao, T. D., Gou, X. H., et al. (2007). The diurnal thaw/freeze cycles of the surface ground on the Tibetan Plateau. *Chinese Science Bulletin* 52(1): 136–139.
Yin, G. A., Niu, F. J., Li, Z. J., Luo, J., & Liu, M. H. (2017). Effects of local factors and climate on permafrost conditions and distribution in Beiluhe basin, Qinghai-Tibet Plateau, China. *Science of the Total Environment* 581–582: 472–485.

Zhang, M. L., Wen, Z., & Xue, K. (2016). Temperature and deformation analysis on slope subgrade with rich moisture of Qinghai–Tibet railway in permafrost regions. *Chinese Journal Rock Mechanic Engineering* 35(8): 1677–1687.

Zhang, S. J., Lai, Y. M., Sun, Z. Z., & Gao, Z. H. (2007). Volumetric strain and strength behavior of frozen soils under confinement. *Cold Regions Science and Technology* 47 (3): 263–270.

Zhang, T., Barry, R. G., Knowles, K., Heginbottom, J. A., & Brown, J. (2008). Statistics and characteristics of permafrost and ground-ice distribution in the northern hemisphere. *Polar Geography* 31(1–2): 47–68.

Zhang, Y., Wang, G., & Wang, Y. (2010). Response of biomass spatial pattern of alpine vegetation to climate change in permafrost region of the Qinghai-Tibet plateau, China. *Journal of Mountain Science* 7(4): 2–15.

Zhao, J., Yu, Y., & Sun, G. (2005). Influence of frozen soil on sandstorms. *Journal of Desert Research* 25(5): 658–662 (in Chinese, English abstract).

Zhao, L., Ping, C. L., Yang, D. Q., et al. (2004). Changes of climate and seasonally frozen ground over the past 30 years in Qinghai-Xizang (Tibetan) Plateau, China, Global Planet. *Change* 43(1/2): 19–31.

Zhao, S., Zhang, S., Cheng, W., & Zhou, C. (2019). Model simulation and prediction of decadal mountain permafrost distribution based on remote sensing data in the Qilian mountains from the 1990s to the 2040s. *Remote Sensing* 11(2): 183.

Zhu, X., Wang, W., & Fraedrich, K. (2013). Future climate in the Tibetan Plateau from a statistical regional climate model. *Journal of Climate* 26(24): 10125–10138.

10

A Review of the Effects of Climate Extremes on Agriculture Production

XIAOMENG YIN AND GUOYONG LENG

10.1 Introduction

Large growth rates of crop production have been observed in the past half-century including wheat, rice, maize and soybean, which provide two-thirds of human caloric intake (Zhao et al., 2017) and have saved numerous people from hunger by supplying sufficient protein and energy from their diet (Mondiale, 2008; Godfray et al., 2010; Zhao et al., 2017). Other food sources, such as livestock, fish, vegetables and fruits, have been extensively studied, but typically bring fewer calories than crops in human diets (Varma & Bebber, 2019). To feed the expected nine billion people by 2050, increasing demand is expected, especially in developing countries, which emphasizes the urgent need for increasing crop production (Godfray et al., 2010).

Clusters of population are mainly located in China, India, the United States, Indonesia, Pakistan, Brazil, Nigeria, Bangladesh, Russia and Japan, which represent 58 per cent of the total population in the world.[1] The largest increase in population growth occurs in the developing and poverty-stricken countries, which affects disproportionally African countries (Pretty et al., 2006; Fess et al., 2011). Most of the countries that show a large increase of population are developing countries, except for the United States and Japan, and a significant correlation between hunger and poverty is observed (Pretty et al., 2003, 2006; Roberts, 2011). The demand for crops and livestock is expected to increase by 40 per cent by 2030 and 70 percent by 2050, according to the UN Food and Agricultural Organization (Tilman et al., 2011). Low-income countries are more vulnerable to the availability and price of crops, for example, some farmers in remote areas do not have access to new technologies available on the market to increase crop production and trade-off market conditions (Godfray et al., 2010). For example, farmers may face food shortage agriculture in sub-Saharan Africa, which is considered as one of the most vulnerable regions and makes up 20–40 per cent of the gross domestic product. People account the high investments to varieties, irrigation, fertilizer, soil conservation and transportation and low returns, however, which finally leads to a high price of crops and enhances the challenges of food supply (Godfray et al., 2010; Hwalla et al., 2016). Also, the highest proportion of people affected by food security live in sub-Saharan Africa, which recorded 26.8 per cent of people as being undernourished in 2010–2012. The largest number of people affected by food security are in south Asia. Bangladesh imported 11.5 per cent of the total available food grains in 2007–2008 because the demand for staple food was higher than the domestic production (Amin et al., 2015). Because of the versatility of the food prices on the market, which happened to rise rapidly in mid-2008, it was encouraged to improve food production rather than relying on market trade (Godfray et al., 2010).

To satisfy a strong demand for food, crop production can be increased by extending the lands dedicated to plant crop. However, crop land has increased by only 9 per cent worldwide since 1960s, while grain production doubled over the same period. Since priority is often given to urbanization and the development of human activities, the expansion of agricultural land is difficult (Godfray et al., 2010; Smith, 2013; Yu et al., 2018). In the United States, cultivated areas suited to dryland winter-soybeans increased by up to 28 per cent from 1988 to 2012 (Seifert & Lobell, 2015). Crop production always depends on the harvested area and yield; therefore, the crop yield is required to improve in order to devote more to crop production. However, there is a gap between the actual and the best yield, which results from the discrepancy in use of seeds, water, nutrients, pest management, soils, biodiversity on condition under consistent genetic material, available technologies and management (Pretty et al., 2003; Godfray et al., 2010; Fess et al., 2011). About 60 per cent of the best yield is found in southeast and central Asia, Argentina and Brazil, which has probably decreased due to greenhouse gases, nutrient run-off, water shortages, soil degradation and inappropriate management (Stern, 2008). The best yield in southeast Asia is estimated to almost double current rice yields, and is derived from the different use of seeds, water, nutrients, pest management, soils, biodiversity and knowledge (Godfray et al., 2010). More efficient use of water may increase crop yields from 400 to 800–1,000 kg ha^{-1}, which includes the production of wheat, pigeonpeas and sorghum. For example, velvetbean yield is projected to increase substantially with the incorporation of nitrogen-fixing legumes in the Central United States (Pretty et al., 2003). The increase in crop production varies across specific sites and countries (Godfray et al., 2010). Grain production has more than doubled over the past 50 years, which can be ascribed to the increase in crop yield and the extension of land dedicated to agricultural use. Projections for future global crop production suggest an increase of more than 70 per cent by 2050 (Wiebe, 2009; Knox et al., 2012).

In summary, the increase of crop production – crop production is calculated by multiplying the harvested area with yield – is associated with the increase of crop yield and expansion of agricultural area. There is a large gap between potential and actual harvested area and between potential and actual yield (Ramankutty et al., 2008; Godfray et al., 2010; Foley et al., 2011; Iizumi & Ramankutty, 2015; Yu et al., 2017a). Cropland areas represent 12 per cent of the Earth's ice-free land, at 1.53 billion hectares. Pastures cover 26 per cent of the Earth's ice-free land, with 3.38 billion hectares. Cropland and pasture together represent the largest use of land on the earth, according to the Food and Agriculture Organization (FAO) (Ramankutty et al., 2008; Foley et al., 2011). Between 1985 and 2005, common crops production increased by 28 per cent, which is attributed to an increase in harvested area by 7 per cent and yield by 20 per cent. This increase is smaller compared to the yield increase of 65 per cent that was observed between 1965 and 1985. The contribution of yield increase is about 10-times that of the increase in the harvested area in the increase of production. This indicates that yield increase plays a dominant role in the production increase (Foley et al., 2011). Studies conclude that maize production increases by 2.1 per cent per year during 1980–2010, which can be mainly attributed to yield improvements in the United States (Leng et al., 2019).

Despite its minor role in crop production compared to yield improvements, the harvested area is a determining factor of production growth. The harvested area is contained in the cropping area, which is greatly affected by climate and weather, especially by unfavourable climate and extreme events (Lobell et al., 2008; Koide et al., 2013; Kotera et al., 2013; Amin et al., 2015; Iizumi & Ramankutty, 2015; Cohn et al., 2016). The effects of climate change on cropping area of rice have been reported by Amin et al. (2015) and Koide et al. (2013). They concluded that the maximum temperature displays a negative effect on cropping area, while both the minimum temperature and precipitation make a positive impact on cropping area. Furthermore, extreme events, such as flood, monsoon, and heat waves, may provoke immediate and unanticipated damage to harvested areas. An example is the severe flood that occurred in Thailand in 2011 which ruined 14 per cent of rice production (Koide et al., 2013).

Yield change remains the most important factor that impacts crop productivity. The increasing rate of crop yield has slowed down in the United States by 3–5 per cent since a peak observed in the early 1960s. A capped increasing rate of crop yield might not suffice to satisfy the future demand for food from a growing population (Kucharik & Ramankutty, 2005). The yields of maize, rice, wheat and soybean increased by 1.6, 1.0, 0.9 and 1.3 per cent per year, respectively, which are far lower than the 2.4 per cent per year required to feed the world population – expected to double by 2050 (Ray et al., 2012, 2013). Specifically, at about 26 per cent of world maize-harvested areas, maize yield improvements stagnated, especially in Western Europe and Africa, and the stagnated areas occupied 52 per cent of maize-growing areas in China. The proportion of rice stagnated area is 35 per cent of the rice-harvested areas globally, affecting 79 per cent of the rice-growing areas in China and 81 per cent in Indonesia. Thirty-seven per cent of global wheat-harvested areas exhibit yield stagnation, mainly in China (with 56 per cent of wheat-growing areas) and India (with 70 per cent of wheat-growing areas). Following these, China has the world's largest area affected by yield stagnation (Ray et al., 2012, 2013). Climate change may affect cultivar, soil, crop management and technology. In particular, climate extremes can have impacts on various stages of crop growth and crop conditions (Rosenzweig et al., 2001; Porter & Semenov, 2005; Lobell & Field, 2007; Moriondo et al., 2010; Lobell et al., 2011, 2013; Maltais-Landry & Lobell, 2012; Challinor et al., 2014; Kassie et al., 2014; Ray et al., 2015; Troy et al., 2015; Lesk et al., 2016; Leng, 2017; Leng & Huang, 2017; Leng & Hall, 2019). Globally, a third of crop yield variability can be explained by climate variability, including temperature and precipitation and their interactions (Ray et al., 2015). Particular climate extremes, such as drought and extreme heat, negatively affect yield variation. The projected reduction of maize yield could double by the 2080s (relative to the 1980s), when taking into account the effects of extreme heat stress (Lobell et al., 2012, 2014; Deryng et al., 2014).

10.2 Climate Change and Climate Extremes

The increase in temperature and reduction of precipitation under climate change poses a threat to crop production. Climate changes are described as average changes in temperature or precipitation (Easterling et al., 2000), while climate extremes and weather include high winds, flooding, heavy snowfall, wildfires, as well as heat waves and droughts (Easterling et al., 2000). Temperatures have increased by 0.78°C between 1850–1900 and 2003–2012. The GCMs (General Circulation Models) projected that mean global warming should range from 1.5 to 5.8°C, and precipitation should increase from 5 to 15 per cent globally by the end of the century, depending on changes in population growth, economic development, energy use and land use (Rosenzweig et al., 2001; Rogelj et al., 2015). The period 1980–2010: (1) was the warmest since 1880; (2) was associated with the largest number of natural disasters, with peaks in 2007 and 2010; (3) was responsible for most social and economic costs and caused 65 per cent of economic damage (Easterling et al., 2000; Luber & McGeehin, 2008). Therefore, understanding climate change and climate extremes in the past and future is key to assess the effects of climate extremes on agriculture production.

10.2.1 Heat Extremes

In recent decades, a record-breaking number of heat waves associated with global warming have affected various parts of the world. The hottest summer during the period of 1500–2010 was recorded in 2003 in Europe, with temperatures in Switzerland exceeding the previous record by 2.4°C (Ilani et al., 2004; Robine et al., 2008). In 2007, Athens, Greece showed the maximum temperature record at 44.8°C, higher than the previous record of 43°C in 1891–2006 (Founda & Giannakopoulos, 2009). In July 2010, temperatures in Moscow exceeded the previous record by 2.5°C. The US Southern Plains suffered its worst heatwave in July 2011 (Coumou & Rahmstorf, 2012). The warmest year was 2005, with an extreme of 53.3°C recorded on 26 May in Mohenjo-daro in Pakistan (Huber & Gulledge, 2011).

Heat extremes have increased worldwide in recent decades and are expected to increase in the future. In western Europe, the number of summer heatwaves has doubled from 1880 to 2005. Since the 1960s, the intensity, duration and number of heatwaves have increased in the eastern Mediterranean (Coumou & Rahmstorf, 2012). The global warming of about an average 0.5°C observed since the 1970s has increased the probability of occurrence of heat extremes. The CMIP5 (Coupled Model Intercomparison Project) predicted that heat extremes will occur more frequently by 2040, regardless of the emission scenario. Predictions show that by 2100 the coldest summer months will be hotter than the hottest months in 2000–2012 (Coumou & Robinson, 2013). Likewise, the IPCC (Intergovernmental Panel on Climate Change) projected that extreme heat events would intensify their magnitude and duration in the United States (Luber & McGeehin, 2008). GCMs projected that the number of heatwaves would increase by 25 per cent in Chicago, and annual heatwave days would increase from 12 to 44–95 in Los Angeles for the period 2070–2099 (Hayhoe et al., 2004; Meehl & Tebaldi, 2004). Given the effects of global warming on the occurrence of heat extremes, it is urgent to better understand the mechanisms behind heat extremes and develop mitigating measures.

10.2.2 Droughts

Droughts affect various regions worldwide (Easterling et al., 2000; Mishra & Singh, 2010; Dai, 2012; Lyon & DeWitt, 2012; Sun & Yang, 2012; van Dijk et al., 2013; Gleick, 2014; Griffin & Anchukaitis, 2014). For example, during the period of 2001–2009, the worst drought on record occurred in southeast Australia. And in this time, southeast Australia showed a precipitation level below the median level since 1900 (van Dijk et al., 2013). The poor short rains and long rains in 2010–2011, a typically bimodal rainfall season across East Africa, induced severe drought in many regions in East Africa (Lyon & DeWitt, 2012). In Syria, six significant droughts that occurred from 1900 to 2005 decreased the average monthly precipitation in the main rainfall season by one-third of normal (Gleick, 2014). The impacts of droughts on agriculture and social economy are detrimental. In 2011, a severe drought occurred in southern China, which affected the middle-lower reaches of the Yangtze River basin in particular and destroyed about 98.9 million hectares of crops (Sun & Yang, 2012). In 1988, drought resulted in about 40 billion dollars of economic loss in the United States. The drought in 2011–2012 was one of the most severe on record in California (Seneviratne, 2012; Griffin & Anchukaitis, 2014).

10.2.3 Climate Models

The accuracy of climate change and climate extremes predictions has improved through the development of complex climate models, including the Atmosphere-Ocean General Circulation Models (AOGCMs), Earth System Models (ESMs), Earth System Models of Intermediate Complexity (EMICs) and Regional Climate Models (RCMs) (Dickinson et al., 1989; Gordon et al., 2000; Burke et al., 2006; Forster & Taylor, 2006; Beniston et al., 2007; Kiehl, 2007; Moss et al., 2010; Knutti & Sedláček, 2012; Taylor et al., 2012; Yukimoto et al., 2012; Flato et al., 2014). AOGCMs were considered as the standard climate models based on the dynamics of the physical components of the climate system, accounting for future greenhouse gas and aerosol forcing, which also applied to specific regions with a large spectrum of spatial resolution, including high resolution. Expanding on AOGCMs, ESMs account for various biogeochemical cycles and are currently considered as state-of-the-art models. In addition, EMICs are usually employed to answer specific scientific questions associated with relevant components of the Earth system. RCMs are applied to simulate dynamic processes using downscaling processes applied to restricted areas (Flato et al., 2014).

The World Climate Research Programme's (WCRP's) Coupled Model Intercomparison Project Phase 3 (CMIP3) contains over 20 AOGCMs. Additional climate models contained in the Coupled Model Intercomparison Project Phase 5 (CMIP5) show similar sensitivity ranging from 2.1 to 4.7 used in the evaluation from the Intergovernmental Panel on Climate Change (IPCC) Fourth Assessment Report (AR4). These models have been recently updated to the CMIP6 generation (Eyring et al., 2016). Climate models are based on historical data from the period

1850–2005 under four Representative Concentration Pathway (RCP) scenarios (RCP2.6, RCP4.5, RCP6.0, RCP8.5) and are used to predict future climate in 2006–2100 (Flato et al., 2014). The literature showed evidence that CMIP3 and CMIP5 has well captured the observed trends in temperature extremes since 1950, and ensembled CMIP3 have well simulated large-scale droughts in the twentieth century, whose results correlate with various definitions of drought (Sheffield & Wood, 2007; Sillmann et al., 2013; Flato et al., 2014).

10.2.4 Climate Data Sets

Climate change and climate extreme models rely on climate data. Common climate data sets include the Climate Research Unit (CRU) and HadEX with their various versions (New et al., 2002; Donat et al., 2013; Harris et al., 2014; Vogel et al., 2019). The latest version of CRU TS is CRU TS 4.03 covering the period 1901–2018, with monthly climate data based on more than 5,000 stations globally, which is obtained both at global and country scale with the resolution $0.5^\circ \times 0.5^\circ$ (New et al., 2002; Harris et al., 2014). Compared to CRU emphasizing on mean climate, the second version of HadEX is usually used to study climate extremes; it includes 17 temperature and 12 precipitation extreme indices from 1901 to 2010 and offers daily climate extreme data with the resolution $3.75^\circ \times 2.5^\circ$ (longitude–latitude) from 7,400 temperature stations and 11,600 precipitation stations worldwide (Donat et al., 2013; Vogel et al., 2019).

10.3 The Effects of Climate Extremes on Yields

Crop growth is impacted by high temperature and drought from sowing to grain maturity. In the early stage of crop growth, the plant organs are well protected by vegetative tissues. However, the impacts of high temperature and drought generate a delay in anthesis in the phase of vegetative generative transition in the shoot apical meristems, flower initiation, because of the sensitivity of apical morphogenesis to water (Barnabas et al., 2008). The flowering process is critical in the determination of the final crop yields. Generally, the number of grains per land area is determined during the flowering process and the grain filling stage gets stable grain weights, which determine the final yields.

10.3.1 The Effects of High Temperature on Crop Growth

Warm periods are essential throughout the entire development of crop growth and help grains bear more fruits. However, temperature extremes cause adverse impacts on crops. Specifically, during the flower initiation, an increase of 1°C in mean temperature reduces the kernel number per unit area by 4 per cent before anthesis (Fischer, 1985; Boyer, 2004). A decrease of kernel number also occurs in early development, including meiosis, anthesis, pollen fertility, pollination, female fertility and early zygote developments (Boyer, 2004). During the reproductive phase, high temperature would lead to pollen sterility, tissue dehydration and lower CO_2 assimilation, and temperatures over 30°C cause complete sterility during floret formation (Saini & Aspinall, 1982). The number and weight of grains are sensitive to temperature. Temperatures over 20°C reduce the grain number per spike between spike initiation and anthesis, as well as spikelet initiation, pollination and fertilization. The most sensitive parts of the crop to temperatures are between the appearance of doubled ridges on the shoot apex and flag leaf. In the preceding anthesis, an increase of 1°C between 15°C and 22°C reduces the grain number by 4 per cent per spike (Farooq et al., 2011). In ovary and embryo sac development, high temperatures significantly affect pre-anthesis stem reserve accumulation, reducing the stem storage capacity and crop growth in pollen development. Impacts on the anthesis and grain filling stages affect the final crop yields. Since the photosynthesis rate is sensitive to high temperatures, temperature extremes beyond the tolerated range cause detrimental impacts, including the reduction of kernel number and weights, even early senescence and reduction of duration of grain filling. During flowering and fertilization, high temperatures reduce the spikelet fertility for rice and temperatures over 38°C reduce maize seed set. Water deficit reduces 60 per cent of the harvest index (Barnabas et al., 2008). During grain filling – the final stage of crop growth – high temperatures increase the grain-filling rate at the beginning of the process, but decrease it at the end of the process, which eventually cause yield losses, and the senescence of leaves is accelerated by large diurnal temperature differences (Zhao et al., 2007; Barnabas et al., 2008). The grain filling duration of wheat is reduced by 12 days with an increase of 5°C above 20°C, while grain growth accelerates during grain filling, depending on the number of grains per spike (Farooq et al., 2011). Similarly, temperatures above 31°C reduce the rate of wheat grain filling (Chaves et al., 2009). Besides, leaf senescence usually occurs along with a progressive reduction of green leaf area and high temperatures inhibit chlorophyll biosynthesis, which severely affect photosynthesis. Photosynthesis is affected most severely by high temperatures through disruptions in the structure and function of chloroplasts and reductions in the chlorophyll content (Farooq et al., 2011).

10.3.2 The Effects of Drought on Crop Growth

Plant growth and grain maturity need sufficient amounts of water. However, water deficits affect various stages and processes associated with transpiration and photosynthesis, which results in yield losses. Specifically, drought reduces the carbon assimilation of wheat, which decreases the stems storage and flower and grain development, resulting in low yield. Drought inhibits female organ development and the photosynthesis process of maize, reducing the nutrient supply to generative organs (Fischer, 1985; Boyer, 2004; Barnabas et al., 2008). Besides, 15–45 per cent of the ovules develop abnormally from droughts of various magnitude and duration. In ovary and embryo sac development, drought can also reduce the stem storage capacity and crop growth in pollen development. Drought affects plant transpiration by closing their stomata, which prevents water loss from transpiration, as well as the reduction of stomata's conductance (Mahajan & Tuteja, 2005; Chaves et al., 2009).

Studies have shown that high temperature and drought result in higher yield losses in anthesis than that at the end of spikelet initiation. Yield appears to be more affected by high temperatures than by drought (Zhang et al., 2013). Also, high temperature and drought reduce grain yield at any growth stage and shorten the grain filling period and grain yield before the end of grain filling, especially between days 1–14 after anthesis, with the most severe impacts in grain filling (Ferris et al., 1998; Saini & Westgate, 1999; Gooding et al., 2003; Zhang et al., 2013).

10.3.3 Recent Research on the Effects of Climate Extremes on Yield

In view of the substantial production losses induced by heatwaves and droughts, many research works have investigated the role of climate extremes on crop production (Zhang, 2004; Li et al., 2009, 2015; Mohammed & Tarpley, 2009; Simelton, 2011; Shi & Tao, 2014; Lesk et al., 2016; Zhang et al., 2016; Zipper et al., 2016; Zampieri et al., 2017). At the global scale, Vogel et al. (2019) assessed the impacts of climate change and climate extremes on crop yield variability using subnational crop yield data and climate data. Climate extreme indices are composed of growing-season maximum and minimum temperatures, warm day frequency, cold night frequency, maximum 5-day precipitation intensity and SPI-6. Flexible models such as random forest can account for and quantify the relative importance of each feature in the model (independent variables) and accommodate complex relationships with the label (dependent variable). Results indicate that climate extremes explain 18–43 per cent of crop yield variability for maize, rice, soybean and spring wheat, compared to 20–49 per cent by mean climate variability, and temperature-related extremes are more sensitive to crop yield than precipitation-related extremes. The models identified regions that are more vulnerable to climate extremes, which include North America for maize, spring wheat and soybean, Asia for maize and rice, as well as Europe in the case of spring wheat. The use of national crop yield data is common given the difficulty to obtain subnational crop yield data on a global scale. Besides crop yield, 9–10 per cent of the reduction in national crop production is caused by heat stress and drought, with a peak of damage from drought that reached 7 per cent in 1964–2007. Developed countries experience 7–11 per cent more damage than developing countries. Crop production in North America, Europe and Australia reduced by an average of 19.9 per cent, compared to 12.1 per cent in Asia and 9.2 per cent in Africa from 1964 to 2007 (Lesk et al., 2016). Latin America and the Caribbean have not shown significant effects (Zampieri et al., 2017). Furthermore, we observed that the effect of heat stress on crop production of wheat increased significantly after the mid-1990s (Zampieri et al., 2017).

Meanwhile, the production losses induced by drought show discrepancies in different periods. Recent droughts during 1985–2007 reduced production by about 13.7 per cent on average, which corresponds to about twice the average production reduction (6.7 per cent) observed in 1964–1984 (Lesk et al., 2016). The effects of future climate extremes on crop production are assessed. Based on 20 GCMs, 120 drought disasters are expected to occur in 2100 and areas affected by drought – drought is measured through a revised PDSI (the Palmer drought severity index) – should increase from 15.4 per cent to 44.0 per cent by 2100. Furthermore, the global cropland drought risk measured as DRI (Drought Risk Index) should double in 2050 and is projected to increase from 52.45 to 129.40 globally, with the highest increase, from 95.77 to 205.46, expected to occur in Africa (Li et al., 2009). Projections on the effects of heat stress on crop production using process-based model PEGASUS and climate models suggest that maize and spring wheat production should progressively be reduced by half by 2080 (Deryng et al., 2014). These model projections can help design and implement relevant measures to alleviate the risk of drought on crop production.

At the regional scale, crop yield shows various responses to climate extremes. Rice and winter wheat display little sensitivity to temperature extremes in the United States, but spring wheat yield losses are strongly affected by hot days and higher maximum temperatures. Similarly, maize and soybean yield losses are caused by long-lasting heat stress in 1948–2010 (Troy et al., 2015). High night-time temperatures also affect crop yield by inhibiting crop growth. Rice yield may decrease by 90 per cent with night-time temperatures of 32°C compared to those under 27°C (Mohammed & Tarpley, 2009).

The sensitivity of crop yield anomaly to drought also varies across regions. Zipper et al. (2016) quantified meteorological drought effects on maize and soybean yield with the standardized precipitation evapotranspiration index (SPEI) and found that crop yields were highly sensitive to short-term droughts, showing 13 per cent of crop yield variability in the central and southeastern United States.

Maize and soybean yields were reduced by a longer duration of dry spells. However, due to the irrigation systems in place, rice yield appeared to not be affected by variations in precipitation in the United States. In China, Shi and Tao (2014) concluded that droughts predominantly affected northern China in 1983–2008, a peak of wheat yield loss of about 35 per cent induced by drought events in Shanxi in the 2000s (Simelton, 2011). In Songliao Plain, maize yield losses caused by drought accounted for 60 per cent in yield loss induced by meteorological disasters from 1949 to 1990 (Zhang, 2004). Maize yield decrease induced by drought was about 9.94 per cent in Jilin, compared to 6.91 per cent in Henan during 1978–2010. The estimation of the effect was based on a composite drought disaster risk index based on SPEI, including drought intensity and frequency (Zhang & Zhang, 2016; Zhang et al., 2016). Furthermore, the impacts of drought on crop yield were estimated among several regions. Under the RCP8.5 scenario, the worst drought effects on winter wheat should occur around 2030 during the period of 2010–2050. Beijing, Tianjin, Hebei and Shandong are expected to be more sensitive to drought than other Chinese regions due to their particular characteristics that affect their exposure, sensitivity and adaptive capacity induced by precipitation and evapotranspiration (Li et al., 2015). The analysis of past and future impacts of heat stress and drought on crop production have also shown substantial yield losses in other regions, including Europe, Canada and India (Quiring & Papakryiakou, 2003; Ciais et al., 2005; Mkhabela et al., 2010; Siebert et al., 2014; Chen et al., 2018; Song et al., 2018).

10.3.4 The Indices of Drought

The effects of climate extremes on crop yield are measured through various indices, such as maximum temperature, warm day frequency and PDSI. The indices highlight different aspects of climate extremes and are chosen according to different research contexts. Here, we review the main drought indices that have been used in recent literature.

Drought is grouped into meteorological, agricultural and hydrological types, quantified by precipitation deficit, soil moisture and low river flow, caused by below-normal precipitation measured over a time period that ranges from months to years (Heim Jr, 2002; Mishra & Singh, 2010; Dai, 2011; Leng et al., 2015; Huang et al., 2016). Drought is composed of three characteristics: (1) frequency: the number of droughts over a defined period; (2) duration: length of time of a drought event; and (3) intensity: average drought index value associated with a drought event, which can be defined by various drought indices (Leng et al., 2015). Common drought indices used to quantify drought include the Palmer drought severity index (PDSI), rainfall index (RAI), deciles, crop moisture index

(CMI), Bhalme and Mooly drought index (BMDI), surface water supply index (SWSI), national rainfall index (NRI), reclamation drought index (RDI), standardized precipitation index (SPI), standardized precipitation evapotranspiration index (SPEI), standardized soil moisture index (SSWI) and standardized runoff index (SRI) (Heddinghaus & Sabol, 1991; Guttman, 1998; Shukla & Wood, 2008; Mishra & Singh, 2010; Vicente-Serrano et al., 2010; Dai, 2011; AghaKouchak, 2014; Leng et al., 2015; Costa & Rodrigues, 2017). Generated from CMI, the crop-specific drought index (CSDI) is also divided into corn drought index (CDI), soybean drought index (SDI) and vegetation condition index (VCI) based on the specific crop. Furthermore, all these drought indices are based on precipitation only or in combination with other elements (Mishra & Singh, 2010). Each drought index has its own strengths and weaknesses and is chosen according to the context of the study (Morid et al., 2006; Mishra & Singh, 2010).

PDSI is the most widely used regional drought index and can be used to explain the spatio-temporal distribution of drought, as well as drought magnitude. PDSI has been used to assess the impacts of the most severe drought in California from the last 1,200 years, which occurred in 2014 (Mishra & Singh, 2010; Dai, 2011, 2012; Coumou & Robinson, 2013; Griffin & Anchukaitis, 2014). SPI is used to assess the probability distribution based on precipitation at varying time scales and can accommodate gamma, Pearson Type III, lognormal, extreme value and exponential distributions. Based on the joint probability distribution of drought properties using the SPI index, the Guanzhong Plain and the surrounding areas are at high drought risk (Mishra & Singh, 2010; Huang et al., 2014). Like SPI, SPEI is based on precipitation. However, SPEI also accounts for the effects of temperature on drought and potential evapotranspiration (Dai, 2011; Zhang & Zhang, 2016). The impacts of climate change on meteorological, agricultural and hydrological drought are estimated with SPI, SSWI and SRI, respectively. Under global warming, drought frequency, duration and intensity are expected to increase (Mishra & Singh, 2010; Leng et al., 2015). Using similar drought indices and methods, the literature suggests that global warming is expected to increase the frequency and magnitude of agricultural and hydrological droughts to a larger extent compared to meteorological droughts (Huang et al., 2017).

Drought indices are particularly well suited to assess the effects of droughts on crop yields. One common index, the SPI, represents the probability density distribution of precipitation over time, which is usually estimated within a time window from 1 to 48 months (Gebrehiwot et al., 2011; Huang et al., 2017). Except for single drought indices which reflect one aspect of water deficit, simultaneous and integrated drought indexes are also commonly used to assess the impacts of droughts (Huang et al., 2017). For example, Antwi-Agyei et al. (2012) used a vulnerability index to assess the vulnerability of crop production to drought, which was obtained by a function of exposure index, sensitivity index and adaptive capacity index. A similar drought vulnerability index was employed to estimate the risk of winter wheat production losses in the future (Li et al., 2015). Mkhabela et al. (2010) adopted four types of drought indices, considering water supply, water demand, water balance and water use, to assess the impact of drought on spring wheat yield and quality. And the most accurate assessments based on water demand and water balance were concluded. To account for heat stress, the combined stress index (CSI) was developed. This combines SPEI with heat magnitude day (HMD) (Zampieri et al., 2017). The Integrated Drought Index (IDI) represents a suitable approach to reflect meteorological, hydrological and agricultural drought, which combines SPI, runoff and soil moisture (Huang et al., 2015). Zhang (2004) developed a risk assessment model combining occurrence frequency, duration and intensity of drought, spatial extent of damage and regional production level of maize to quantify risk of drought. A similar combination of drought characteristics, frequency, and intensity accounted by SPEI were combined into the drought hazard (DH). DH is based on the probability density functions of SPEI, which is a more accurate measure than drought grades (Zhang & Zhang, 2016). A comprehensive analysis of the hazard and vulnerability of drought risk was employed based on frequency, intensity and crop yield sensitivity and drought recoverability (Zhang et al., 2016). Zipper et al. (2016) applied SPEI at 14 different timescales to study drought impacts on maize and soybean yield variability, and concluded that short-time (one-to-three months) droughts showed the highest levels of sensitivity.

Choosing suitable drought indices for assessing drought impacts on crop yield is important. Each index may reflect a particular aspect of the impact of drought on crop yield. Leng and Hall (2019) used both SPI and SPEI to estimate drought impacts on crop yield and found that the impacts of drought on crop yield differed according to the index used to measure drought.

10.3.5 Approaches to Assess the Impacts of Climate Extremes

The impacts of climate extremes are commonly analysed using simple linear regression models (Zipper et al., 2016). More flexible approaches based on machine learning algorithms may offer higher predictive accuracy while quantifying the importance of the variables in the model. Random forest algorithms have been applied to analyse the effects of climate extreme on crop yields (Fukuda et al., 2013; Jeong et al., 2016; Hoffman et al., 2018). Copula are an alternative reliable approach that has been used to assess and predict the probability of risk loss of crop yield induced by drought (Renard & Lang, 2007; Tebaldi & Lobell, 2008; Sadegh et al., 2017). Furthermore, experiments were carried out to assess the effects of night-time temperature on rice productivity, which suggested that high night-time temperatures impact growth, development, physical appearance and yield of crop and eventually lead to a loss in the final yield (Mohammed & Tarpley, 2009; Mkhabela et al., 2010).

10.4 Adaptation Measures to Mitigate Yield Loss

Along with climate conditions, crop yields rely on agricultural management practices, including the use of seeds, water, nutrients, pest managements, soils, biodiversity and knowledge. Agricultural management can mitigate crop yield reduction impacted by climate extremes (Godfray et al., 2010). Conservation tillage, soil mulching, multiple cropping, irrigation and fertilization are usually applied to enhance crop yield, as well as the application of technology and genetic engineering.

For example, Troy et al. (2015) demonstrated that irrigation eliminated yield reduction by climate extremes (Butler & Huybers, 2012). Since soil water availability is affected by irrigation amount and frequency, the impacts of irrigation on yield reduction were distinct at the global and country scale (Deryng et al., 2011; Butler & Huybers, 2012; Leng et al., 2016). Deryng et al. (2011) concluded that under climate change global cereal production would decrease by 1–8 per cent using CO_2 fertilization compared to 11–20 per cent without its usage, and the largest change in yield loss with irrigation ranges from –19 per cent to –38 per cent for soybean.

In addition, 7–18 per cent of the crop losses could be avoided by adapting planting dates and cultivar choices based on the mechanisms of PEGASUS, which pointed to the necessity of irrigation, fertilization and adaptive planting and harvested data, and cultivar. McGrath and Lobell (2011) used elevated levels of carbon dioxide (CO_2) to increase maize and soybean yield by 9 per cent and 14 per cent, respectively. Karlen et al. (2013) concluded that corn–soybean rotation would be more profitable than continuous corn, and good nutrient management and crop rotation minimize the differences between no-tillage and more intensive tillage systems. Seifert and Lobell (2015) proposed the double crop (DC) system that contains dryland winter wheat and soybeans, and their yields may be reduced by 0.4–0.75 per cent per decade under climate change, which is far below the expected yield reduction without the DC system. Ruffo et al. (2015) showed that high technology and the specific level of agronomic management factors, which were different from current practice, can increase maize yields by 28 per cent compared to standard technology. Moreover, genetic modifications improve the stress tolerance of crops, including the tolerance to herbicide and pathogens. And the use of a NAC-type transcription factor in rice showed enhanced resistance to drought, with seed settings with a level 22–34 per cent higher than under severe drought conditions (Barnabas et al., 2008).

Understanding the effects of climate extremes on crop production is necessary to help policymakers develop targeted measures and alleviate the damage induced by climate and climate extremes.

10.5 Conclusion

Large populations rely on a substantial supply of crop food and increasing heat extremes and droughts inhibit crop growth to different extents, during flowering and grain filling stages, which determine final crop yields. The role of climate extremes effects, especially heat extremes and droughts, on crop yields has been assessed at global and regional research scales. The sensitivity of crop yield to variations of temperature and precipitation induced by heat extremes and droughts varies across crops, regions and periods. Various climate extreme indices can be used to assess the impacts on crop production. Each drought index reflects different aspects of drought, therefore integrated indices may encompass a wider range of drought specificities. A better understanding of the role of climate extremes on crop production aims at improving adaptation measures to offset yield loss and enhance food security.

Note

1 FAOSTAT (2018). Available from www.fao.org/faostat/en/#data/OA (Last accessed 31 August 2021).

References

AghaKouchak, A. (2014). A baseline probabilistic drought forecasting framework using standardized soil moisture index: Application to the 2012 United States drought. *Hydrology and Earth System Sciences* 18(7): 2485–2492.

Amin, M., Zhang, J., & Yang, M. (2015). Effects of climate change on the yield and cropping area of major food crops: A case of Bangladesh. *Sustainability* 7(1): 898–915.

Antwi-Agyei, P., Fraser, E. D. G., Dougill, A. J., Stringer, L. C., & Simelton, E. (2012). Mapping the vulnerability of crop production to drought in Ghana using rainfall, yield and socioeconomic data. *Applied Geography* 32(2): 324–334.

Barnabas, B., Jager, K., & Feher, A. (2008). The effect of drought and heat stress on reproductive processes in cereals. *Plant, Cell & Environment* 31(1): 11–38.

Beniston, M., Stephenson, D. B., Christensen, O. B., et al. (2007). Future extreme events in European climate: An exploration of regional climate model projections. *Climatic Change* 81(S1): 71–95.

Boyer, J. S. (2004). Grain yields with limited water. *Journal of Experimental Botany* 55 (407): 2385–2394.

Burke, E. J., Brown, S. J., & Christidis, N. (2006). Modeling the recent evolution of global drought and projections for the twenty-first century with the Hadley centre climate model. *Journal of Hydrometeorology* 7(5): 1113–1125.

Butler, E. E., & Huybers, P. (2012). Adaptation of US maize to temperature variations. *Nature Climate Change* 3(1): 68–72.

Challinor, A. J., Watson, J., Lobell, D. B., et al. (2014). A meta-analysis of crop yield under climate change and adaptation. *Nature Climate Change* 4(4): 287–291.

Chaves, M. M., Flexas, J., & Pinheiro, C. (2009). Photosynthesis under drought and salt stress: Regulation mechanisms from whole plant to cell. *Annals of Botany* 103(4): 551–560.

Chen, Y., Zhang, Z., & Tao, F. (2018). Impacts of climate change and climate extremes on major crops productivity in China at a global warming of 1.5 and 2.0°C. *Earth System Dynamics* 9(2): 543–562.

Ciais, P., Reichstein, M., Viovy, N., et al. (2005). Europe-wide reduction in primary productivity caused by the heat and drought in 2003. *Nature* 437(7058): 529–533.

Cohn, A. S., VanWey, L. K., Spera, S. A., & Mustard, J. F. (2016). Cropping frequency and area response to climate variability can exceed yield response. *Nature Climate Change* 6(6): 601–604.

Costa, J. A., & Rodrigues, G. P. (2017). Space-time distribution of rainfall anomaly index (Rai) for the Salgado Basin, Ceará State – Brazil. *Ciência e Natura* 39(3): 627–634.

Coumou, D., & Rahmstorf, S. (2012). A decade of weather extremes. *Nature Climate Change* 2(7): 491–496.

Coumou, D., & Robinson, A. (2013). Historic and future increase in the global land area affected by monthly heat extremes. *Environmental Research Letters* 8(3): 1–6.

Dai, A. (2011). Drought under global warming: A review. *Wiley Interdisciplinary Reviews: Climate Change* 2(1): 45–65.

Dai, A. (2012). Increasing drought under global warming in observations and models. *Nature Climate Change* 3(1): 52–58.

Deryng, D., Conway, D., Ramankutty, N., Price, J., & Warren, R. (2014). Global crop yield response to extreme heat stress under multiple climate change futures. *Environmental Research Letters* 9(3): 1–13.

Deryng, D., Sacks, W. J., Barford, C. C., & Ramankutty, N. (2011). Simulating the effects of climate and agricultural management practices on global crop yield. *Global Biogeochemical Cycles* 25(2): GB2006.

Dickinson, R. E., Errico, R. M., Giorgi, F., & Bates, G. T. (1989). A regional climate model for the Western United States. *Climatic Change* 15(3): 383–422.

van Dijk, A. I. J. M., Beck, H. E., Crosbie, R. S., et al. (2013). The Millennium Drought in southeast Australia (2001–2009): Natural and human causes and implications for water resources, ecosystems, economy, and society. *Water Resources Research* 49 (2): 1040–1057.

Donat, M. G., Alexander, L. V., Yang, H., et al. (2013). Updated analyses of temperature and precipitation extreme indices since the beginning of the twentieth century: The HadEX2 dataset. *Journal of Geophysical Research: Atmospheres* 118(5): 2098–2118.

Easterling, D. R., Meehl, G. A., Parmesan, C., et al. (2000). Climate extremes: Observations, modeling, and impacts. *Science* 289(5487): 2068–2074.

Eyring, V., Bony, S., Meehl, G. A., et al. (2016). Overview of the Coupled Model Intercomparison Project Phase 6 (CMIP6) experimental design and organization. *Geoscientific Model Development* 9(5): 1937–1958.

Farooq, M., Bramley, H., Palta, J. A., & Siddique, K. H. M. (2011). Heat stress in wheat during reproductive and grain-filling phases. *Critical Reviews in Plant Sciences* 30 (6): 491–507.

Ferris, R., Ellis, R. H., Wheeler, T. R., & Hadley, P. (1998). Effect of high temperature stress at anthesis on grain yield and biomass of field-grown crops of wheat. *Annals of Botany* 82(5): 631–639.

Fess, T. L., Kotcon, J. B., & Benedito, V. A. (2011). Crop breeding for low input agriculture: A sustainable response to feed a growing world population. *Sustainability* 3(10): 1742–1772.

Fischer, R. A. (1985). Number of kernels in wheat crops and the influence of solar-radiation and temperature. *Journal of Agricultural Science* 105: 447–461.

Flato, G., Marotzke, J., Abiodun, B., et al. (2014). Evaluation of climate models. Climate change 2013: The physical science basis. *Contribution of Working Group I to the Fifth Assessment Report of the Intergovernmental Panel on Climate Change* (pp. 741–866). Cambridge: Cambridge University Press.

Foley, J. A., Ramankutty, N., Brauman, K. A., et al. (2011). Solutions for a cultivated planet. *Nature* 478(7369): 337–342.

Forster, P. M. D., & Taylor, K. E. (2006). Climate forcings and climate sensitivities diagnosed from coupled climate model integrations. *Journal of Climate* 19(23): 6181–6194.

Founda, D., & Giannakopoulos, C. (2009). The exceptionally hot summer of 2007 in Athens, Greece – A typical summer in the future climate? *Global and Planetary Change* 67(3–4): 227–236.

Fukuda, S., Spreer, W., Yasunaga, E., et al. (2013). Random forests modelling for the estimation of mango (*Mangifera indica* L. cv. Chok Anan) fruit yields under different irrigation regimes. *Agricultural Water Management* 116: 142–150.

Gebrehiwot, T., van der Veen, A., & Maathuis, B. (2011). Spatial and temporal assessment of drought in the Northern highlands of Ethiopia. *International Journal of Applied Earth Observation and Geoinformation* 13(3): 309–321.

Gleick, P. H. (2014). Water, drought, climate change, and conflict in Syria. *Weather, Climate, and Society* 6(3): 331–340.

Godfray, H. C. J., Beddington, J. R., Crute, I. R., et al. (2010). Food security: The challenge of feeding 9 billion people. *Science* 327(5967): 812–818.

Gooding, M. J., Ellis, R. H., Shewry, P. R., & Schofield, J. D. (2003). Effects of restricted water availability and increased temperature on the grain filling, drying and quality of winter wheat. *Journal of Cereal Science* 37(3): 295–309.

Gordon, C., Cooper, C., Senior, C. A., et al. (2000). The simulation of SST, sea ice extents and ocean heat transports in a version of the Hadley Centre coupled model without flux adjustments. *Climate Dynamics* 16(2–3): 147–168.

Griffin, D., & Anchukaitis, K. J. (2014). How unusual is the 2012–2014 California drought? *Geophysical Research Letters* 41(24): 9017–9023.

Guttman, N. B. (1998). Comparing the Palmer Drought Index and the standardized precipitation index. *Journal of the American Water Resources Association* 34(1): 113–121.

Harris, I., Jones, P. D., Osborn, T. J., & Lister, D. H. (2014). Updated high-resolution grids of monthly climatic observations – The CRU TS3.10 dataset. *International Journal of Climatology* 34(3): 623–642.

Hayhoe, K., Cayan, D., Field, C. B., et al. (2004). Emissions pathways, climate change, and impacts on California. *Proceedings of the National Academy of Sciences (USA)* 101 (34): 12422–12427.

Heddinghaus, T. R., & Sabol, P. (1991). A review of the Palmer Drought Severity Index and where do we go from here. *Proceedings of the 7th Conference on Applied Climatology*. American Meteorological Society Boston, MA, pp. 242–246.

Heim, Jr., R. R. (2002). A review of twentieth-century drought indices used in the United States. *Bulletin of the American Meteorological Society* 83(8): 1149–1166.

Hoffman, A. L., Kemanian, A. R., & Forest, C. E. (2018). Analysis of climate signals in the crop yield record of sub-Saharan Africa. *Global Change Biology* 24(1): 143–157.

Huang, S., Chang, J., Huang, Q., & Chen, Y. (2014). Spatio-temporal changes and frequency analysis of drought in the Wei River Basin, China. *Water Resources Management* 28(10): 3095–3110.

Huang, S., Chang, J., Leng, G., & Huang, Q. (2015). Integrated index for drought assessment based on variable fuzzy set theory: A case study in the Yellow River basin, China. *Journal of Hydrology* 527: 608–618.

Huang, S., Huang, Q., Leng, G., & Liu, S. (2016). A nonparametric multivariate standardized drought index for characterizing socioeconomic drought: A case study in the Heihe River Basin. *Journal of Hydrology* 542: 875–883.

Huang, S., Leng, G., Huang, Q., et al. (2017). The asymmetric impact of global warming on US drought types and distributions in a large ensemble of 97 hydro-climatic simulations. *Science Reports* 7(1): 5891.

Huber, D. G., & Gulledge, J. (2011). *Extreme Weather and Climate Change: Understanding the Link, Managing the Risk*. Arlington: Pew Center on Global Climate Change.

Hwalla, N., El Labban, S., & Bahn, R. A. (2016). Nutrition security is an integral component of food security. *Frontiers in Life Science* 9(3): 167–172.

Iizumi, T., & Ramankutty, N. (2015). How do weather and climate influence cropping area and intensity? *Global Food Security* 4: 46–50.

Ilani, S., Martin, J., Teitelbaum, E., et al. (2004). The microscopic nature of localization in the quantum Hall effect. *Nature* 427(6972): 328–332.

Jeong, J. H., Resop, J. P., Mueller, N. D., et al. (2016). Random forests for global and regional crop yield predictions. *PLoS ONE* 11(6): e0156571.

Karlen, D. L., Kovar, J. L., Cambardella, C. A., & Colvin, T. S. (2013). Thirty-year tillage effects on crop yield and soil fertility indicators. *Soil and Tillage Research* 130: 24–41.

Kassie, B. T., Van Ittersum, M. K., Hengsdijk, H., et al. (2014). Climate-induced yield variability and yield gaps of maize (*Zea mays* L.) in the Central Rift Valley of Ethiopia. *Field Crops Research* 160: 41–53.

Kiehl, J. T. (2007). Twentieth century climate model response and climate sensitivity. *Geophysical Research Letters* 34(22): 1–4.

Knox, J., Hess, T., Daccache, A., & Wheeler, T. (2012). Climate change impacts on crop productivity in Africa and South Asia. *Environmental Research Letters* 7(3): 1–8.

Knutti, R., & Sedláček, J. (2012). Robustness and uncertainties in the new CMIP5 climate model projections. *Nature Climate Change* 3(4): 369–373.

Koide, N., Robertson, A. W., Ines, A. V. M., et al. (2013). Prediction of rice production in the Philippines using seasonal climate forecasts. *Journal of Applied Meteorology and Climatology* 52(3): 552–569.

Kotera, A., Nguyen, K. D., Sakamoto, T., Iizumi, T., & Yokozawa, M. (2013) A modeling approach for assessing rice cropping cycle affected by flooding, salinity intrusion, and monsoon rains in the Mekong Delta, Vietnam. *Paddy and Water Environment* 12(3): 343–354.

Kucharik, C. J., & Ramankutty, N. (2005). Trends and variability in U.S. corn yields over the twentieth century. *Earth Interactions* 9(1): 1–29.
Leng, G. (2017). Recent changes in county-level corn yield variability in the United States from observations and crop models. *Science of the Total Environment* 607–608: 683–690.
Leng, G., & Hall, J. (2019). Crop yield sensitivity of global major agricultural countries to droughts and the projected changes in the future. *Science of the Total Environment* 654: 811–821.
Leng, G., & Huang, M. (2017). Crop yield response to climate change varies with crop spatial distribution pattern. *Science Reports* 7(1): 1463.
Leng, G., Peng, J., & Huang, S. (2019). Recent changes in county-level maize production in the United States: Spatial-temporal patterns, climatic drivers and the implications for crop modelling. *Science of the Total Environment* 686: 819–827.
Leng, G., Tang, Q., & Rayburg, S. (2015). Climate change impacts on meteorological, agricultural and hydrological droughts in China. *Global and Planetary Change* 126: 23–34.
Leng, G., Zhang, X., Huang, M., et al. (2016). Simulating county-level crop yields in the conterminous United States using the Community Land Model: The effects of optimizing irrigation and fertilization. *Journal of Advances in Modeling Earth Systems* 8(4): 1912–1931.
Lesk, C., Rowhani, P., & Ramankutty, N. (2016). Influence of extreme weather disasters on global crop production. *Nature* 529(7584): 84.
Li, Y., Huang, H., Ju, H., et al. (2015). Assessing vulnerability and adaptive capacity to potential drought for winter-wheat under the RCP 8.5 scenario in the Huang-Huai-Hai Plain. *Agriculture, Ecosystems & Environment* 209: 125–131.
Li, Y., Ye, W., Wang, M., & Yan, X. (2009). Climate change and drought: A risk assessment of crop-yield impacts. *Climate Research* 39: 31–46.
Lobell, D. B., Burke, M. B., Tebaldi, C., et al. (2008). Prioritizing climate change adaptation needs for food security in 2030. *Science* 319(5863): 607–610.
Lobell, D. B., & Field, C. B. (2007). Global scale climate–crop yield relationships and the impacts of recent warming. *Environmental Research Letters* 2(1): 1–7.
Lobell, D. B., Hammer, G. L., McLean, G., et al. (2013). The critical role of extreme heat for maize production in the United States. *Nature Climate Change* 3(5): 497–501.
Lobell, D. B., Roberts, M. J., Schlenker, W., et al. (2014). Greater sensitivity to drought accompanies maize yield increase in the US Midwest. *Science* 344(6183): 516–519.
Lobell, D. B., Schlenker, W., & Costa-Roberts, J. (2011). Climate trends and global crop production since 1980. *Science* 333(6042): 616–620.
Lobell, D. B., Sibley, A., & Ivan Ortiz-Monasterio, J. (2012). Extreme heat effects on wheat senescence in India. *Nature Climate Change* 2(3): 186–189.
Luber, G., & McGeehin, M. (2008). Climate change and extreme heat events. *American Journal of Preventive Medicine* 35(5): 429–435.
Lyon, B., & DeWitt, D. G. (2012). A recent and abrupt decline in the East African long rains. *Geophysical Research Letters* 39(2): 2702.
Mahajan, S., & Tuteja, N. (2005). Cold, salinity and drought stresses: An overview. *Archives of Biochemistry and Biophysics* 444(2): 139–158.
Maltais-Landry, G., & Lobell, D. B. (2012). Evaluating the contribution of weather to maize and wheat yield trends in 12 U.S. counties. *Agronomy Journal* 104(2): 301–311.
McGrath, J. M., & Lobell, D. B. (2011). An independent method of deriving the carbon dioxide fertilization effect in dry conditions using historical yield data from wet and dry years. *Global Change Biology* 17(8): 2689–2696.

Meehl, G. A., & Tebaldi, C. (2004). More intense, more frequent, and longer lasting heat waves in the 21st century. *Science* 305(5686): 994–997.

Mishra, A. K., & Singh, V. P. (2010) A review of drought concepts. *Journal of Hydrology* 391(1–2): 202–216.

Mkhabela, M., Bullock, P., Gervais, M., Finlay, G., & Sapirstein, H. (2010). Assessing indicators of agricultural drought impacts on spring wheat yield and quality on the Canadian prairies. *Agricultural and Forest Meteorology* 150(3): 399–410.

Mohammed, A. R., & Tarpley, L. (2009). High nighttime temperatures affect rice productivity through altered pollen germination and spikelet fertility. *Agricultural and Forest Meteorology* 149(6–7): 999–1008.

Mondiale, B. (2008) *World Development Report 2008*. Agriculture for Development. Washington, DC: World Bank.

Morid, S., Smakhtin, V., & Moghaddasi, M. (2006). Comparison of seven meteorological indices for drought monitoring in Iran. *International Journal of Climatology* 26(7): 971–985.

Moriondo, M., Giannakopoulos, C., & Bindi, M. (2010). Climate change impact assessment: The role of climate extremes in crop yield simulation. *Climatic Change* 104 (3–4): 679–701.

Moss, R. H., Edmonds, J. A., Hibbard, K. A., et al. (2010). The next generation of scenarios for climate change research and assessment. *Nature* 463(7282): 747–756.

New, M., Lister, D., Hulme, M., & Makin, I. (2002). A high-resolution data set of surface climate over global land areas. *Climate Research* 21(1): 1–25.

Porter, J. R., & Semenov, M. A. (2005). Crop responses to climatic variation. *Philosophical Transactions of the Royal Society of London, B: Biological Sciences* 360(1463): 2021–2035.

Pretty, J. N., Morison, J. I. L., & Hine, R. E. (2003). Reducing food poverty by increasing agricultural sustainability in developing countries. *Agriculture, Ecosystems & Environment* 95(1): 217–234.

Pretty, J. N., Noble, A. D., Bossio, D., et al. (2006). Resource-conserving agriculture increases yields in developing countries. *Environmental Science & Technology* 40(4): 1114–1119.

Quiring, S. M., & Papakryiakou, T. N. (2003). An evaluation of agricultural drought indices for the Canadian prairies. *Agricultural and Forest Meteorology* 118(1–2): 49–62.

Ramankutty, N., Evan, A. T., Monfreda, C., & Foley, J. A. (2008). Farming the planet: 1. Geographic distribution of global agricultural lands in the year 2000. *Global Biogeochemical Cycles* 22(1): GB1003.

Ray, D. K., Gerber, J. S., MacDonald, G. K., & West, P. C. (2015). Climate variation explains a third of global crop yield variability. *Nature Communications* 6(5989): 1–9.

Ray, D. K., Mueller, N. D., West, P. C., & Foley, J. A. (2013). Yield trends are insufficient to double global crop production by 2050. *PLoS ONE* 8(6): e66428.

Ray, D. K., Ramankutty, N., Mueller, N. D., West, P. C., & Foley, J. A. (2012). Recent patterns of crop yield growth and stagnation. *Nature Communications* 3 (1293): 1–7.

Renard, B., & Lang, M. (2007). Use of a Gaussian copula for multivariate extreme value analysis: Some case studies in hydrology. *Advances in Water Resources* 30(4): 897–912.

Roberts, L. (2011). 9 Billion? *Science* 333(6042): 540–543.

Robine, J. M., Cheung, S. L., Le Roy, S., et al. (2008). Death toll exceeded 70,000 in Europe during the summer of 2003. *Comptes Rendus Biologies* 331(2): 171–178.

Rogelj, J., Luderer, G., Pietzcker, R. C., et al. Energy system transformations for limiting end-of-century warming to below 1.5°C. *Nature Climate Change* 2015; 5(6): 519–527.

Rosenzweig, C., Iglesius, A., Yang, X. -B., Epstein, P. R., & Chivian, E. (2001). Climate change and extreme weather events – Implications for food production, plant diseases, and pests. *Global Change Human Health* 2(2): 90–104.

Ruffo, M. L., Gentry, L. F., Henninger, A. S., Seebauer, J. R., & Below, F. E. (2015). Evaluating management factor contributions to reduce corn yield gaps. *Agronomy Journal* 107(2): 495–505.

Sadegh, M., Ragno, E., & AghaKouchak, A. (2017). Multivariate copula analysis toolbox (MvCAT): Describing dependence and underlying uncertainty using a Bayesian framework. *Water Resources Research* 53(6): 5166–5183.

Saini, H. S., & Aspinall, D. (1982). Abnormal sporogenesis in wheat (*Triticum-aestivum* L) induced by short periods of high-temperature. *Annals of Botany* 49(6): 835–846.

Saini, H. S., & Westgate, M. E. (1999). Reproductive development in grain crops during drought. In D. L. Spartes (ed.), *Advances in Agronomy* (vol. 68, pp. 59–96). San Diego, CA: Academic Press.

Seifert, C. A., & Lobell, D. B. (2015). Response of double cropping suitability to climate change in the United States. *Environmental Research Letters* 10(2): 1–6.

Seneviratne, S. I. (2012). Climate science: Historical drought trends revisited. *Nature* 491 (7424): 338–339.

Sheffield, J., & Wood, E. F. (2007). Projected changes in drought occurrence under future global warming from multi-model, multi-scenario, IPCC AR4 simulations. *Climate Dynamics* 31(1): 79–105.

Shi, W., & Tao, F. (2014). Spatio-temporal distributions of climate disasters and the response of wheat yields in China from 1983 to 2008. *Natural Hazards* 74(2): 569–583.

Shukla, S., & Wood, A. W. (2008). Use of a standardized runoff index for characterizing hydrologic drought. *Geophysical Research Letters* 35(2): 1–7.

Siebert, S., Ewert, F., Eyshi Rezaei, E., Kage, H., & Graß, R. (2014). Impact of heat stress on crop yield – On the importance of considering canopy temperature. *Environmental Research Letters* 9(4): 1–8.

Sillmann, J., Kharin, V. V., Zhang, X., Zwiers, F. W., & Bronaugh, D. (2013). Climate extremes indices in the CMIP5 multimodel ensemble: Part 1. Model evaluation in the present climate. *Journal of Geophysical Research: Atmospheres* 118(4): 1716–1733.

Simelton, E. (2011). Food self-sufficiency and natural hazards in China. *Food Security* 3 (1): 35–52.

Smith, P. (2013) Delivering food security without increasing pressure on land. *Global Food Security* 2(1): 18–23.

Song, L., Guanter, L., Guan, K., et al. (2018). Satellite sun-induced chlorophyll fluorescence detects early response of winter wheat to heat stress in the Indian Indo-Gangetic Plains. *Global Change Biology* 24(9): 4023–4037.

Stern, N. (2008). The economics of climate change. *American Economic Review* 98(2): 1–37.

Sun, C., & Yang, S. (2012). Persistent severe drought in southern China during winter–spring 2011: Large-scale circulation patterns and possible impacting factors. *Journal of Geophysical Research: Atmospheres* 117(D10).

Taylor, K. E., Stouffer, R. J., & Meehl, G. A. (2012). An overview of CMIP5 and the experiment design. *Bulletin of the American Meteorological Society* 93(4): 485–498.

Tebaldi, C., & Lobell, D. B. (2008). Towards probabilistic projections of climate change impacts on global crop yields. *Geophysical Research Letters* 35(8): 1–8.
Tilman, D., Balzer, C., Hill, J., & Befort, B. L. (2011). Global food demand and the sustainable intensification of agriculture. *Proceedings of the National Academy of Sciences (USA)* 108(50): 20260–20264.
Troy, T. J., Kipgen, C., & Pal, I. (2015). The impact of climate extremes and irrigation on US crop yields. *Environmental Research Letters* 10(5): 1–10.
Varma, V., & Bebber, D. P. (2019). Climate change impacts on banana yields around the world. *Nature Climate Change* 9(10): 752–757.
Vicente-Serrano, S. M., Beguería, S., & López-Moreno, J. I. (2010). A multiscalar drought index sensitive to global warming: The standardized precipitation evapotranspiration index. *Journal of Climate* 23(7): 1696–1718.
Vogel, E., Donat, M. G., Alexander, L. V., et al. (2019). The effects of climate extremes on global agricultural yields. *Environmental Research Letters* 14(5): 1–12.
Wiebe, K. (2009). How to feed the world in 2050. *Insights from an Expert Meeting at FAO*, pp. 24–26.
Yu, Q., Shi, Y., Tang, H., et al. (2017a). eFarm: A tool for better observing agricultural land systems. *Sensors (Basel)* 17(3): 453.
Yu, Q., van Vliet, J., Verburg, P. H., et al. (2018). Harvested area gaps in China between 1981 and 2010: Effects of climatic and land management factors. *Environmental Research Letters* 13(4): 1–10.
Yukimoto, S., Adachi, Y., Hosaka, M., et al. (2012). A new global climate model of the meteorological research Institute: MRI-CGCM3 – Model description and basic performance. *Journal of the Meteorological Society of Japan* 90A: 23–64.
Zampieri, M., Ceglar, A., Dentener, F., & Toreti, A. (2017). Wheat yield loss attributable to heat waves, drought and water excess at the global, national and subnational scales. *Environmental Research Letters* 12(6): 1–11.
Zhang, J. (2004). Risk assessment of drought disaster in the maize-growing region of Songliao Plain, China. *Agriculture, Ecosystems & Environment* 102(2): 133–153.
Zhang, Q., & Zhang, J. (2016). Drought hazard assessment in typical corn cultivated areas of China at present and potential climate change. *Natural Hazards* 81(2): 1323–1331.
Zhang, Q., Zhang, J., & Wang, C. (2016). Risk assessment of drought disaster in typical area of corn cultivation in China. *Theoretical and Applied Climatology* 128(3–4): 533–540.
Zhang, X., Cai, J., Wollenweber, B., et al. (2013). Multiple heat and drought events affect grain yield and accumulations of high molecular weight glutenin subunits and glutenin macropolymers in wheat. *Journal of Cereal Science* 57(1): 134–140.
Zhao, C., Liu, B., Piao, S., et al. (2017). Temperature increase reduces global yields of major crops in four independent estimates. *Proceedings of the National Academy of Sciences (USA)* 114(35): 9326–9331.
Zhao, H., Dai, T., Jing, Q., Jiang, D., & Cao, W. (2007). Leaf senescence and grain filling affected by post-anthesis high temperatures in two different wheat cultivars. *Plant Growth Regulation* 51(2): 149–158.
Zipper, S. C., Qiu, J., & Kucharik, C. J. (2016). Drought effects on US maize and soybean production: Spatiotemporal patterns and historical changes. *Environmental Research Letters* 11(9): 1–11.

11

Agricultural Water Use Estimation and Impact Assessment on the Water System in China

MENGFEI MU, QIUHONG TANG, XIMING CAI, SIAO SUN
AND HUIJUAN CUI

11.1 Introduction

The global hydrological cycle has been considerably altered by cropland expansion, irrigation water extraction (Rost et al., 2008) and other anthropogenic activities such as urbanization, deforestation and water withdrawal for domestic and industrial use. During the last three decades, China has expanded irrigated cropland to satisfy increasing food demand induced by fast population growth and dietary structure change, although rainfed cropland areas decreased. While 22 per cent of the population of the world live in China, the country has only 7 per cent of the world's arable lands (NBSC, 2015). In China, the arable lands are mainly located in the north, where crop growth needs intense irrigation, which is commonly performed using blue water (Falkenmark & Rockström, 2004).

Nearly 70 per cent of water withdrawals in the world are used as blue water for irrigation to increase crop yield (Molden et al., 2007). In developing countries, blue water used for irrigation represents about 90 pe cent of water withdrawals (Rosegrant et al., 2002b). Blue water use in cropland has higher opportunity cost than green water because it is in direct competition with domestic and industrial water uses. When surface water resources cannot meet the demand for irrigation, groundwater has been exploited to increase crop yields, which has caused the groundwater level to drop (Tang et al., 2013), land subsidence (Tilman et al., 2001) and salt intrusion in deltaic areas (Wada et al., 2010). Water is often collected from basins in the neighbourhood of irrigation areas when the availability of local water does not satisfy the needs. The exploitation of deep fossil groundwater and inter-basin transfers of water, which are non-renewable and non-local blue water resources (NNBW) (Vörösmarty et al., 2005), may provoke considerable impacts on the environment.

Funding for this research was provided by the National Natural Science Foundation of China (41730645, 41425002).

However, agriculture is also largely dependent on green water use, which is part of precipitation stored in the soil and naturally absorbed by crops (Falkenmark & Rockström, 2004). It is estimated that around 60 per cent (Rosegrant et al., 2002a; Rost et al., 2008) of the world's food is produced on rainfed land where only green water is provided. Even on irrigated cropland, green water comprises 20–35 per cent of the total water consumption as a global average (Rost et al., 2008). As irrigation-related problems emerge and blue water demands by other sectors increase, ever growing attention has since been paid to green water use, given its invisible property and low opportunity cost compared to blue water. Rainfed agriculture is expected to play a more important role than before in the foreseeable future, which should help reduce blue water use (Rosegrant et al., 2002b) and increase blue water availability for other water usages. For this purpose, more green water should be shifted from unproductive to productive evapotranspiration (Hoff et al., 2010). Green water management measures such as rainwater harvesting (Bruins et al., 1986; Wisser et al., 2010) may enhance soil infiltration. Water productivity of crops can be improved by increasing the harvest index (the ratio of yield to biomass), improving soil fertility, and reducing evaporation while increasing productive transpiration (Molden & Sakthivadivel 1999; Molden et al., 2007). Additional policy reforms and infrastructure investments in both less-favoured and high-potential rainfed areas are needed to close the considerable gap between potential and current yields (Rosegrant et al., 2002b). Meanwhile, increasing water use efficiency (WUE), especially for blue water, remains an important goal in developing countries with traditional irrigation systems.

Both green and blue water are essential to satisfy crop water demand (CWD). CWD varies by crop types, local climates and crop management practices. Quantifying and partitioning the green and blue water components in CWD help to better understand and manage the use of agricultural water. Global assessments using various models have been carried out by Döll and Siebert (2002), Rost et al. (2008), Wisser et al. (2008), Siebert and Döll (2010), Liu and Yang (2010) and Wada et al. (2012). Recent studies have assessed green and blue water uses in various parts of China (Sun et al., 2012; Zang et al., 2012; Zeng et al., 2012; Cao et al., 2014; Zhuo et al., 2016). However, they have mainly focused on some specific basins in China or on irrigated croplands and therefore have failed to provide a whole picture of China's agricultural water use over a long period. Moreover, these regional studies have focused on consumptive water use and have seldom analysed water withdrawal, although the latter one is of greater practical significance for both water consumers and suppliers. In sum, the local water resource availability can be larger than the consumptive water use but access to water withdrawal may still be limited, which can result in an underestimation of water scarcity and risks. For this reason, irrigation efficiency evolution is included in our study to account for water withdrawal and its inter-annual changes.

To these ends, the objectives and significance of this study are to identify and quantify: (1) temporal variability (at both annual and monthly level) of agricultural green and blue water uses in China at national and basin levels; (2) spatial patterns of agricultural green and blue water uses by basin; and (3) water scarcity or water stress of renewable agricultural water use on the water system in China. The scientific innovations of this study include: (1) the use of a dynamic cropland areas distribution dataset (MIRCA); (2) accounting water withdrawal and its inter-annual changes in addition to water consumption dynamic annual irrigation efficiency. Most global studies have ignored the dynamic aspect of irrigation efficiency using one year (2000) as a benchmark throughout the study period or have only accounted for water consumption; (3) by partitioning NNBW into non-renewable and non-local blue water, we make the distinction between renewable and non-renewable water uses. None of previous studies except Wada et al. (2010) have made this kind of partition; and (4) comparing basins between different regions across China with various climate and water resources conditions.

This systematic study aims to provide a global picture of the dynamics of agricultural water use over the past three decades in China and help develop an integrated agricultural water management strategy to mitigate the environmental risk. The results of this study may benefit other countries or regions that encounter similar issues associated with the use and management of agricultural water.

11.2 Methodology

11.2.1 The Model

The integrated global hydrological model H08 (Hanasaki et al., 2008) is utilized to simulate the hydrological cycle in China during 1981–2010. It consists of six modules: (1) land surface hydrology; (2) river routing; (3) crop growth; (4) reservoir operation; (5) environmental flow requirements; and (6) anthropogenic water withdrawals. Its model output, i.e., simulated runoff, was validated with observational global runoff data sets and streamflow records at gauging stations of 32 major rivers around the world.

We ran the H08 model from 1960 to 2010 with a daily time step, and used the former two decades, that is, 1960–1980, as the 'warm-up period'. The analysis focuses on the latter three decades, that is, 1981–2010. The simulation was done using a fine spatial resolution of $0.5° \times 0.5°$.

11.2.2 Green and Blue Water Uses Estimation

The estimation of green and blue water uses follows the conceptual framework developed by Molden and Sakthivadivel (1999). In this framework, both water consumption and water withdrawal accounted for irrigation in the agricultural sector.

Crop water demand (CWD) is determined by actual evapotranspiration (ET_a) calculated by the energy balance method (Hanasaki et al., 2008) as:

$$\mathrm{CWD} = ET_a. \tag{11.1}$$

Blue water consumption ($Blue_{consumption}$, i.e., water deficit between total crop water demand and green water supply) is determined as:

$$Blue_{consumption} = \mathrm{CWD} - Green \tag{11.2}$$

Green water is supplied by soil moisture, i.e., effective precipitation.

Gross blue water use (*Blue*, i.e., water withdrawal for irrigation; if not specified, blue water use in this study represents the blue water withdrawal) is defined as the ratio between blue water consumption and irrigation efficiency (E_i):

$$Blue = Blue_{consumption}/E_i \tag{11.3}$$

Blue water use is provided in sequence via three sources. The first source is water withdrawal from river channels $(RIVER)$. When river discharge is not sufficient, withdrawal from reservoirs $(RESERVOIR)$ is activated. $NNBW$ is the last choice to fill the gap between water demand and supply in the H08 model. Water balance in the blue water system can be described as:

$$Blue = RIVER + RESERVOIR + NNBW \tag{11.4}$$

We further partition $NNBW$ into non-renewable and non-local blue water, making non-renewable blue water in the year 2000 consistent with the 26 ± 5 km^3 estimated by Wada et al. (2010) at national level. We suggest sensible partitioning ratios (R_{part}: non-renewable blue water use to $NNBW$) among the ten basins in China based on our experience and knowledge. The ratios are 5 per cent in Songhua, Liao, Northwest and Yellow River basins, 20 per cent in Hai and Huai River basins and 0.1 per cent in Yangtze, Southeast, Southwest and Pearl River basins. Others are non-local water through water diversions within each basin. Inter-basin transfers of non-local water from remote areas, for example, the South to North Water Transfer (SNWT) project, are not included in the simulation due to the limitations of the model.

Renewable blue water is defined as the sum of withdrawals from rivers, reservoirs and non-local sources:

$$Blue_{renew} = RIVER + RESERVOIR + NNBW * (1 - R_{part}) \tag{11.5}$$

Non-renewable blue water use is expressed as:

$$Blue_{nonrenew} = NNBW * R_{part} \tag{11.6}$$

Total agricultural water use (SUM) is defined as:

$$SUM = Green + Blue \tag{11.7}$$

11.2.3 Temporal Variabilities and Spatial Patterns

The Mann-Kendall (MK) trend test method (Mann, 1945) is used to analyse the trends of green and blue water uses, as well as crop areas, irrigated areas and precipitation both during 1981–2000 and 2001–2010. Spatial patterns of green and blue water uses are identified by mapping and comparison.

11.2.4 Impact Assessment

The total natural runoff (Q_{nat}) on natural condition quantifies the availability of renewable blue water resources in each basin. The total withdrawal of renewable blue water for irrigation as agricultural blue water use (Q_{agr}) is equal to $Blue_{renew}$:

$$Q_{agr} = Blue_{renew} \tag{11.8}$$

The withdrawal ratio (R_{agr}) is used to quantify the impact of renewable agricultural blue water withdrawal on natural runoff:

$$R_{agr} = Q_{agr}/Q_{nat} \tag{11.9}$$

11.3 Study Area and Data Sets

11.3.1 Study Area

We focus our study area on the mainland of China, which is composed of 10 major basins. There are six basins in northern China: Songhua, Liao, Northwest, Hai, Huai and Yellow River basins; and four basins in southern China: Yangtze, Southeast, Southwest and Pearl River basins.

11.3.2 Data Sets

The re-analysis data set WFDEI (Weedon et al., 2011, 2014) is used as meteorological forcing data. The global river network map is taken from the global drainage direction map DDM30 provided by Döll and Lehner (2002) and compiled by Haddeland et al. (2011).

The global data set of monthly irrigated and rainfed crop areas (MIRCA) provides the annual fractional coverage in all grid cells, with irrigated and rainfed areas of 13 crops separated (Fader et al., 2010). The data set was developed following the MIRCA2000 program (Portmann et al., 2010).

The national-scale irrigation efficiency (E_i) is considered as 0.35 referring to Rohwer et al. (2007) for the year 2000. We assume that E_i in China has improved linearly from 1981 to 2000 and from 2001 to 2010 as well. E_i takes the values 0.3, 0.35 and 0.5 for the years 1981, 2000 and 2010, respectively.

National industrial and domestic water withdrawal data are taken from AQUASTAT (FAO, 2016), and downscaled from national level to grid level using the H08 model. National industrial and domestic water uses compete with blue water for irrigation and usually have priority over agricultural blue water use in most regions. Accordingly, the H08 model puts priority on domestic and industrial sectors for water demands.

Reservoirs across China with a storage capacity above 1 km^3 are included and geo-referenced in the reservoir module. They are operated according to general rules (Hanasaki et al., 2006) aiming at reducing the magnitude and frequency of annual and seasonal fluctuations in streamflow and downstream water demand. Considering that rivers in northern China have mostly run dry during the past years, environmental flows are set as relatively small values, which are 10 per cent of the normal environmental flows' requirement computed in H08.

11.4 Results

11.4.1 Temporal Variability of Green and Blue Water Uses

The 30-year averaged amount of green and blue water uses in China are 735 and 360 $km^3\ yr^{-1}$ (Table 11.1), respectively. Both green and blue water uses in China's irrigated cropland around the year 2000 are compared with those from previous studies (Döll & Siebert, 2002; Rost et al., 2008; Wisser et al., 2008; Liu & Yang, 2010; Siebert & Döll, 2010; Wada et al., 2012). Our results lie in the range and agree well with the average values of these studies.

At national level, Table 11.2 shows that green water use in cropland decreased significantly during 1981–2000 and increased significantly during 2001–2010. The decrease observed during 1981–2000 might be the result of a significant cropland reduction and precipitation decrease in most investigated basins (Table 11.2). The Yangtze River basin contributed most to the decrease; it has the largest share of cropland area and shows the second highest annual precipitation among the 10 investigated basins. Despite the cropland area decreasing, the increase of green water use during 2001–2010 might be attributed to a slight precipitation increase in

Table 11.1. Thirty-year averaged green and blue water uses on 10 river basins' croplands in China and ratios of blue water use to the sum of green and blue water uses, the ratios of non-renewable blue water use to blue water use (values in the tables and figures hereinafter are from the model results)

		Blue				
Basin	Green (km^3)	Total (km^3)	Renewable (km^3)	Non-renewable (km^3)	Blue/ (Blue+Green) (100%)	Non-renewable/ Blue (100%)
Songhua	82.21	29.28	28.47	0.81	26%	3%
Liao	37.80	31.91	30.65	1.27	46%	4%
Northwest	16.80	49.13	47.73	1.39	75%	3%
Hai	51.47	65.68	53.93	11.75	56%	18%
Yellow	67.41	53.95	52.17	1.78	44%	3%
Yangtze	250.08	49.77	49.74	0.02	17%	0.05%
Huai	106.02	61.29	50.80	10.49	37%	17%
Southeast	26.19	4.83	4.82	0.003	16%	0.05%
Southwest	16.99	3.08	3.08	0.001	15%	0.05%
Pearl	79.95	11.36	11.36	0.006	12%	0.06%
China	734.93	360.28	332.75	27.53	33%	8%

most basins, especially in those with large cropland areas, that is Yangtze, Huai and Songhua.

In contrast, blue water use increased slightly (Table 11.2) during 1981–2000 and decreased significantly during 2001–2010. During 1981–2000, there was a significant irrigated cropland expansion and a precipitation decrease both in major food-producing basins and in the whole country. As a result, agricultural blue water consumption increased significantly. Offset to some extent by irrigation efficiency improvement, eventually the blue water withdrawal increased slightly. During 2001–2010, irrigated cropland expansion was slower than that during 1981–2000 and precipitation increased in nearly all basins, especially in all arid and semi-arid basins. These together resulted in a significant increase in green water use and therefore a slight decrease in blue water consumption. The joint effects of factors such as irrigation efficiency improvement, increasing precipitation, and water transfer from agricultural sector to non-agricultural sectors (Cai, 2008) decreased agricultural blue water use significantly in this period, although irrigation efficiency improvement was the main contributor.

Compared to blue water consumption, green water use considerably increased from 2001 to 2010 (Table 11.2), probably because more favoured rainfed cropland with higher precipitation, where a greater proportion of water use was green, was converted into irrigated cropland. Despite the decrease of the total cropland area,

Table 11.2. The results of the Mann-Kendall test for 10 river basins and China at the nation level

	Green water use		Blue water use		Blue water consumption		Cropland areas		Irrigated areas		Precipitation	
Basins	1981–2000	2001–2010	1981–2000	2001–2010	1981–2000	2001–2010	1981–2000	2001–2010	1981–2000	2001–2010	1981–2000	2001–2010
Songhua	x ↓	** ↑	x ↑	x ↓	** ↑	x ↓	***↑	** ↑	***↑	**↑	x ↓	x ↑
Liao	* ↓	x ↑	x ↑	x ↓	* ↑	x	***↓	** ↓	***↑	**↑	x ↓	x ↑
Northwest	x ↑	x ↓	x ↑	x ↓	* ↑	* ↑	x ↑	** ↑	***↑	**↑	x ↑	x ↑
Hai	x ↓	x ↑	x ↑	* ↓	x ↑	x ↓	***↓	** ↓	***↑	**↑	x ↑	x ↑
Yellow	** ↓	x ↑	x ↑	* ↓	* ↑	x ↑	***↓	** ↓	***↑	**↑	* ↓	x ↑
Yangtze	** ↓	x ↑	x ↑	x ↓	x ↑	x	***↓	** ↓	***↑	**↑	x ↓	x ↑
Huai	x ↓	x ↑	x ↓	x ↓	x ↑	x	***↓	** ↓	***↑	**↑	x ↑	x ↑
Southeast	x ↓	x ↑	x ↑	x ↓	x ↑	x ↓	***↓	** ↓	***↑	**↑	x ↑	x ↑
Southwest	x ↓	x ↑	x ↑	x ↑	* ↑	x ↑	***↓	x ↑	***↑	**↑	* ↑	x ↓
Pearl	** ↓	x ↑	* ↑	x ↓	** ↑	x ↓	***↓	** ↓	***↑	**↑	x ↓	x ↓
China	* ↓	* ↑	x ↑	* ↓	* ↑	x ↓	***↓	** ↓	***↑	**↑	x ↓	x ↑

Notation: x, not significant; *, significant at $p < 0.05$; **, significant at $p < 0.01$; ***, significant at $p < 0.001$; ↑, increasing trend; ↓, decreasing trend.

the overall green water use increased during 2001–2010 due to the increase of precipitation.

Significant spatial heterogeneity is observed for both green and blue water uses. Table 11.2 shows that from 1981 to 2000 the green water use decreased while the blue water use increased in six basins (i.e., Songhua, Liao, Hai, Yellow, Yangtze and Pearl). Yangtze experienced the largest decrease in green water use. Songhua, Hai and Yellow exhibited the largest three increases in blue water use. There was a slight increase in both green and blue water uses in the Northwest. No obvious change is found in either green or blue water uses in Huai, Southeast and Southwest during 1981–2000. For the period 2001–2010, green water use experienced an increase in nearly all 10 basins and showed a considerable increase in Songhua, Yangtze and Huai. Blue water use decreased at a high rate in Songhua, Liao, Northwest, Hai, Yellow, Yangtze and Huai, while changes of rate were not observed in the Southeast, Southwest and Pearl – in these areas, the blue water use decrease caused by irrigation efficiency improvement was offset by an increase of blue water consumption requirement due to irrigated cropland expansion.

We further examine non-renewable blue water use, which accounts for 8 per cent (Table 11.1) of total blue water use in China. More than 10 $km^3\ yr^{-1}$ non-renewable blue water was used in Hai and Huai. Only the last decade encountered a decreasing tendency observed in both basins and in China. Around 1.8, 1.4 and 1.3 $km^3\ yr^{-1}$ non-renewable blue water was used, respectively, in the Yellow, Northwest and Liao basins. A small amount of non-renewable blue water was used in southern China basins, since renewable water is usually in sufficient quantity in these areas.

Monthly mean green and blue water uses are essential for sub-annual analysis. The phase of the annual cycle of blue water is shifted with regard to the green water use phase. At the national level, Figure 11.1 shows that blue water use reaches a peak in

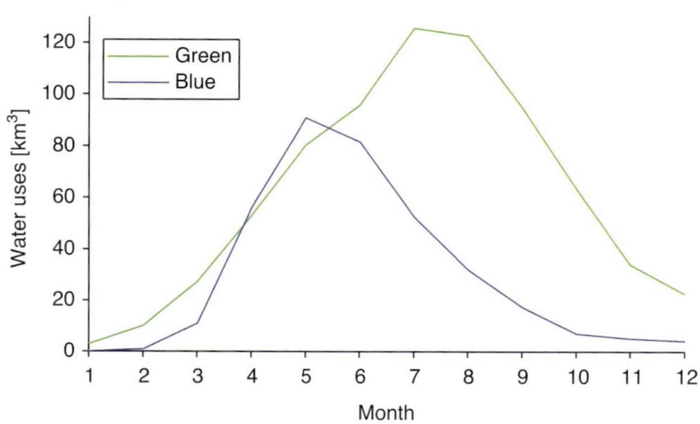

Figure 11.1 Monthly mean green and blue water uses in China's cropland during 1981–2010

May. Before May, a large amount of blue water is required since green water is limited due to low precipitation. After May, a larger amount of green water is available in the raining season for crop growth, while the use of blue water decreases.

Figure 11.2 shows the monthly mean green and blue water uses per area in each basin during 1981–2010 to compare the type of water uses among basins. The phase shift of blue water use with respect to green water use is more pronounced in northern China. In the Northwest, blue water is used more than green water during almost the entire crop growth season. From May to August, blue water use decreases when green water becomes more available. Other basins in northern China, including Songhua, Liao, Hai, Yellow and Huai, share a similar temporal distribution pattern. Peaks of blue water use (in May) appear earlier than those of green water use (in August). The turning points of blue water use in the Yellow and Hai basins come in June. This may partly explain why a large amount of non-local water has to be diverted and non-renewable water has to be withdrawn from these basins. Moreover, precipitation (dashed line in Figure 11.2) cannot cover the total of green and blue water use (SUM) associated with the croplands in these basins. Diversions and irrigation systems are needed to redistribute water to croplands. In southern China basins such as the Yangtze, Southeast, Southwest and Pearl, the blue and green water uses are basically synchronized over the year and blue water plays only a minor role, complementary to green water. This phenological difference observed in northern and southern China mainly results from different climatic environments, especially different precipitation patterns (see dashed line in Figure 11.2). The raining season comes about 1 month earlier in southern China, with higher levels of precipitation and, therefore, larger quantities of green water available than that in northern China. Southern basins usually have multiple cropping over the year, which explains the presence of several peaks (two or three) of blue water use in these regions.

11.4.2 Spatial Patterns of Green and Blue Water Uses

Figure 11.3(a) and (b) show the spatial pattern of green and blue water uses, respectively, with their amount provided in Table 11.1. Green water use (Figure 11.3(a)) is concentrated in Yangtze, Huai, Songhua and Pearl river basins. The total amount of green water use in these four basins is 518 $km^3\ yr^{-1}$, representing 70 per cent of the green water use in the whole country; the green water use in the largest two basins, Yangtze and Huai, is 356 $km^3\ yr^{-1}$, representing nearly half (48%) of the green water use in China. Blue water use (Figure 11.3(b)) is concentrated in Hai, Huai and Yellow river basins. The total blue water use in these three basins is 181 $km^3\ yr^{-1}$, which is about half the national total blue water use.

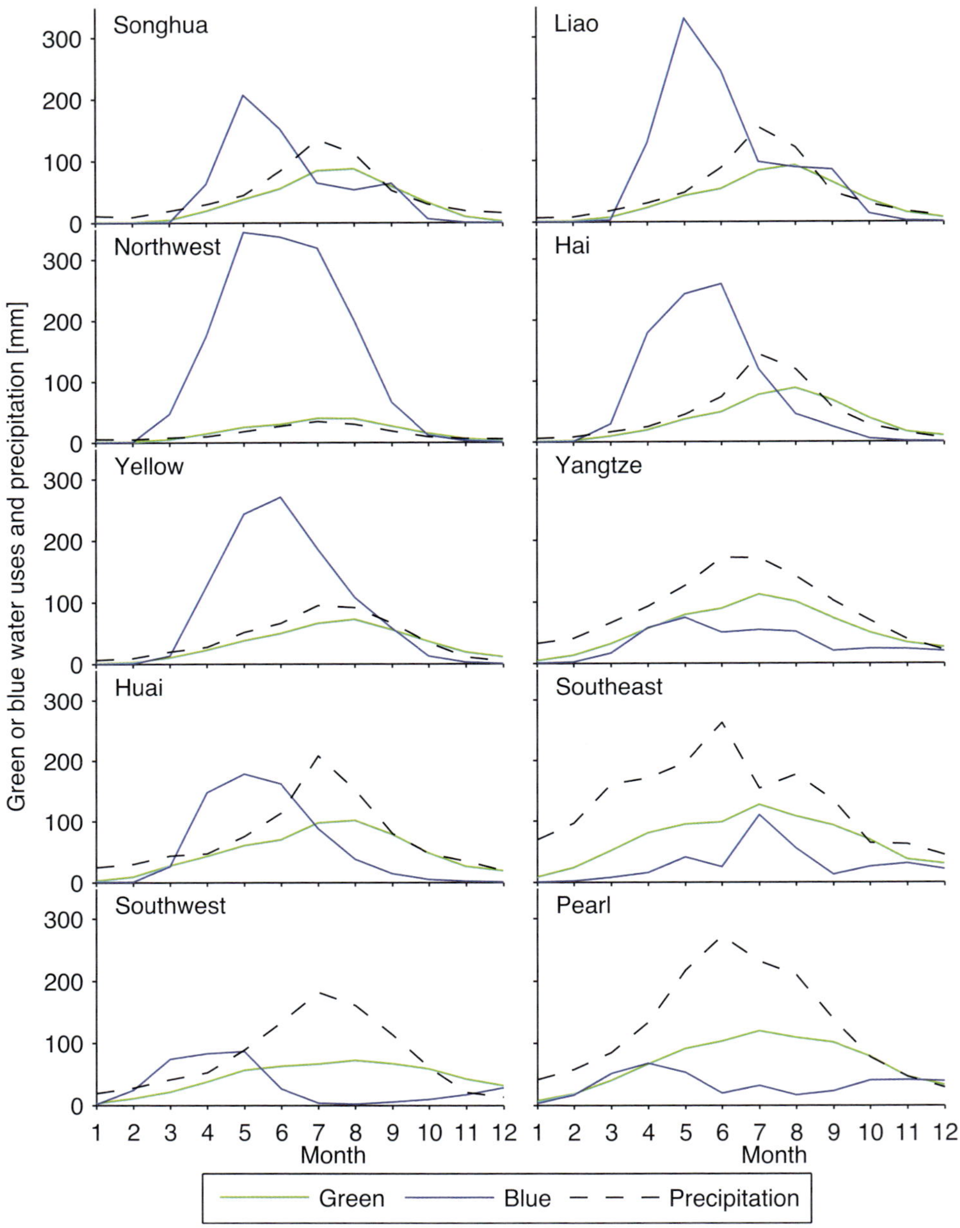

Figure 11.2 Monthly mean green (normalized by total cropland area) or blue (normalized by irrigated cropland area) water uses and precipitation in each basin during 1981–2010

Figure 11.3(c) shows the ratio of blue water use to the SUM of both blue and green water uses, which indicates to what extent the blue water contributes to crop growth. As shown in Figure 11.3(c) (also in Table 11.1), blue water use dominates the agricultural water use in northern China (i.e., in Liao, Northwest, Hai, Yellow

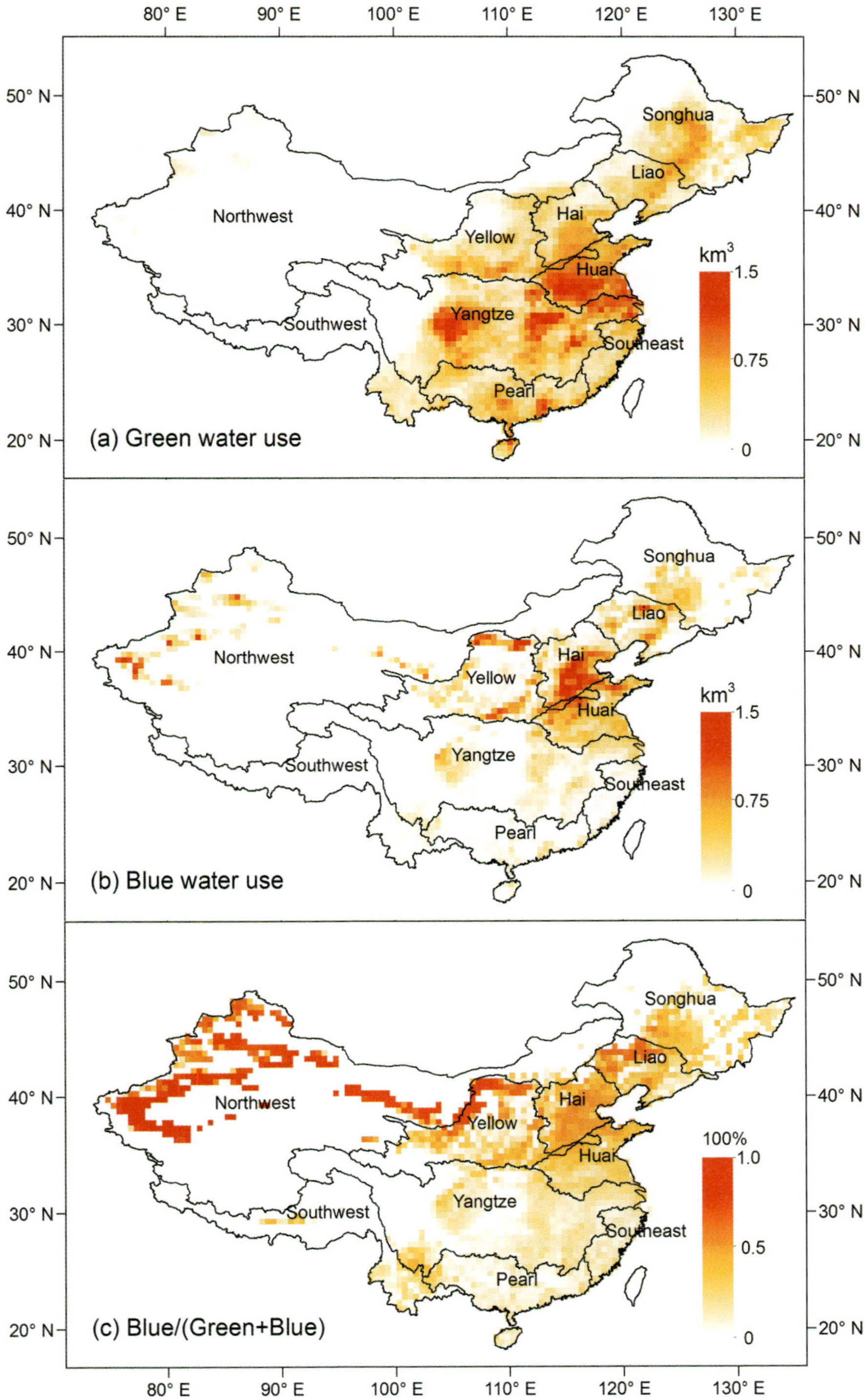

Figure 11.3 (a) 30-year averaged green water use over cropland; (b) 30-year averaged blue water use over cropland; (c) the ratio (defined as Blue/(Blue +Green)) of averaged blue water use for 1981–2010

and northern parts of Huai); while green water plays a major role in southern China (i.e., Yangtze, Southwest, Southeast and Pearl). The largest ratio of blue water (75%) is found in the Northwest, because of its concentrated irrigated cropland and low green water availability, resulting from low precipitation and its uneven temporal distribution. The Hai River basin takes the second place with a bit more green water available, due to higher precipitation. However, a higher percentage of non-renewable blue water, nearly 18 per cent, is used in intensive irrigated cropland. A large amount of non-local blue water is transferred by waterway diversions within the basin. The Yellow River basin also faces renewable blue water stresses. The ratio of blue water use is 44 per cent, and 1.8 km^3 comes from non-renewable water. The ratio of blue water use in southern China is lower than 20 per cent, which means that blue water plays a minor role in southern China.

11.4.3 Impact Assessment of Agricultural Water Use on the Water System

We further assess the impact of agricultural renewable blue water use, which is separated from those of domestic and industrial water uses, on runoff at both the national and basin levels in Table 11.3.

Over the entire year, 330 km^3 of agricultural blue water use are withdrawn for irrigation in China, comprising 16 per cent of natural runoff, while the global runoff decrease caused by irrigation is 1.2 per cent in Rost et al. (2008). The impact at the national level may seem insignificant. However, there is high spatial

Table 11.3. Total natural runoff (Q_{nat}), withdrawal for irrigation as agricultural renewable blue water use (Q_{agr}) and withdrawal ratio (R_{agr}) in each basin and in China as a whole

	Over the year			During crop season		
Basins	Q_{nat} (km^3)	Q_{agr} (km^3)	R_{agr} (100%)	Q_{nat} (km^3)	Q_{agr} (km^3)	R_{agr} (100%)
Songhua	112.09	28.12	25%	68.84	24.77	36%
Liao	35.97	30.16	84%	26.38	26.99	102%
Northwest	61.59	47.37	77%	33.35	44.65	134%
Hai	25.12	53.56	213%	16.87	51.51	305%
Yellow	51.33	52.20	102%	23.53	48.20	205%
Yangtze	732.61	48.94	7%	521.44	37.69	7%
Huai	71.97	50.51	70%	50.70	48.55	96%
Southeast	155.48	4.82	3%	119.80	3.51	3%
Southwest	371.27	3.08	1%	258.42	2.34	1%
Pearl	399.79	11.21	3%	322.55	6.62	2%
China	2,017.21	329.97	16%	1,441.88	294.82	20%

heterogeneity among the 10 river basins. Although agricultural withdrawals in the Northwest, Hai, Yellow, Yangtze and Huai are similar, around 50 km^3, withdrawal ratios differ among basins. The withdrawal ratio in the Yangtze basin is the smallest due to its largest natural runoff (732 km^3), followed by the Huai and Northwest basins. In the Yellow basin, the withdrawal ratio is above 100 per cent, which means that almost all renewable blue water resources need to be withdrawn if irrigation demand is fully satisfied, although most of the water will return to the river channel. The withdrawal ratio in the Hai River basin is higher than 200 per cent. This means that a large proportion of water is reused or extra blue water resources are transferred from other basins or exploited from non-renewable sources; the latter process is not investigated here and its analysis would go beyond the scope of our study. The impact on Songhua seems moderate, with the withdrawal ratio being 25 per cent. The impact on Liao is significant with a small natural runoff and large agricultural withdrawal. In the Southeast, Southwest and Pearl basins, agricultural blue water withdrawals are small compared to their large natural runoffs, resulting in small withdrawal ratios. Therefore, the impact of agricultural blue water use on these basins is relatively small.

During the crop season – for most regions in China, crop growing season is from March to August – when a large amount of blue water is withdrawn to meet crop water demand, the impact seems more significant than that over the entire year (Table 11.3). Agricultural withdrawal is around 295 km^3 at the national level, comprising 20 per cent of natural runoff during the crop season. Withdrawals in the northern basins (i.e., Songhua, Liao, Northwest, Hai, Huai and Yellow) are slightly smaller than those over the entire year, whereas natural runoffs in these basins reduce by almost half. This results in higher withdrawal ratios over 100 per cent, which indicates that the temporal distribution of runoff does not align with the agricultural withdrawal and blue water that is captured in the raining season for irrigation during the crop season. Otherwise, the irrigation demand would not be fully satisfied. The withdrawal ratios in southern China remain small due to their large runoffs observed during this period and over the year. These areas face a little amount of stress associated with blue water supply, despite issues of water quality.

11.5 Discussion

11.5.1 Uncertainty

Similar to previous studies, there is uncertainty associated with the results of our study. Uncertainty is partly attributed to input data, such as reanalysis meteorological forcing data, cropland extent and irrigated areas. The largest uncertainty lies in irrigated areas and irrigation efficiency settings.

Additional uncertainty stems from the H08 model settings. (1) Surface runoff and subsurface runoff are not well partitioned. (2) Soil information and its heterogeneity are not well considered. (3) Crop areas and non-crop areas share the same evapotranspiration capability, which may underestimate the relative amount of green and blue water uses in cropland. (4) Irrigation increases soil moisture in local grids and, as a result, additional runoff is expected. Consequently, less NNBW is needed to fully satisfy the crop water demand, and the initial NNBW is therefore overestimated. To this end, iteration should be done to reach a model equilibrium. However, the current H08 settings do not consider this iteration nor examine model equilibrium. (5) NNBW is not well partitioned or traced. Non-local blue water, one part of NNBW, is a relative concept, which is sensitive to spatial resolution. Non-local blue water in a 0.5-degree simulation may turn into local blue water in a 1-degree simulation. NNBW is well partitioned according to Wada et al. (2012), however, as far as we know, withdrawal rules for non-local blue water has not been introduced into hydrological models and H08. Future enhancements in the H08 model should address the aforementioned problems.

11.5.2 Implications for Agricultural Water Resources Management

With urbanization and population growth, industrial and domestic water uses in China have increased dramatically from 1981 to 2010. Industrial water use nearly tripled from 45 km^3 yr^{-1} to 130 km^3 yr^{-1}, while domestic water use increased more than 10-times, from 6.7 km^3 yr^{-1} to 68 km^3 yr^{-1} (FAO, 2016). The total blue water use by industrial, domestic, and agricultural sectors increased during 1981–2000 and showed a decrease during 2001–2010. The non-agricultural blue water use is in direct competition with the agricultural sector, which poses challenges to both water and food security in China. Over the past decades, irrigation efficiency has improved and contributed to mitigate the trade-off. However, the agricultural sector remains vulnerable to water transfer to non-agricultural sectors. Continuous efforts to improve water use efficiency are needed to save more water and ensure food security to support the economic and social transformations that have occurred in China. There is still potential to save more water with the current irrigation efficiency at the national level being about 0.5, in comparison to developed countries such as Canada which had an efficiency of about 0.7 in the year 2000. Policies such as reducing water and electricity prices for water pumping or rural households are likely to foster the use of blue water (Nature Agency, 2013). A successful economic-based project, ‘Water Licensing and Water Resources Fee Collection and Management Regulations’, has been implemented since 2006 by the Ministry of Water Resources (MWR). The China 2011 NO.1 Document introduced the Three Red Lines Strategy that has deployed a strict water resources management system, with an emphasis on agricultural water use. Additional biological and water-saving irrigation technologies are promising tools to improve the agricultural water use efficiency.

In some arid and semi-arid basins in northern China, small distributed rainwater harvesting engineering could store extra rainwater and reduce water withdrawal from river channels and aquifers. In addition, it would be reasonable to transform irrigated cropland into grassland, forest or other natural landscapes, as long as food supply is ensured by means of food import and social equity is maintained by means of subsidies to farmers who give back their croplands. If irrigated cropland in both Inner Mongolia and the greater Beijing area would decrease by 50 per cent, national corn and wheat production should decrease by only 4.3 per cent and 4.5 per cent, respectively, while national agricultural blue water use should decrease by almost 14 per cent (Dalin et al., 2015). In humid and semi-humid basins in southern China, more work could be done to at least ensure food self-sufficiency instead of the current situation that relies on the import of large amount of food from northern China. Uncommitted outflow (Molden & Sakthivadivel, 1999), such as non-productive evaporation from fallow lands or water surfaces and excess outflow to the sea, is noticeable in southern China. The engineering project SNWT allows a committed utilization of the excess outflow. However, a large proportion of river flow remains unutilized. Management strategies that combine green and blue water should be developed to make these basins more committed. Water and food security must be ensured by the holistic view from a national management system rather than a basin-scale one.

Favourable climate conditions, such as the increasing precipitation observed in the last decade (2001–2010), contribute to increasing green water availability and therefore may offset, to some extent, the increasing blue water demand due to an expansion of irrigated cropland. A future reduction in the increase of precipitation or drier conditions should increase the use of blue water and threaten water security. The effects of climate change on the expected amount of precipitation in China remains unclear. Therefore, we suggest reducing the current dependency on blue water by increasing the use of uncommitted green water and developing systems to improve the efficiency of irrigation systems.

A large amount of non-renewable blue water is used in northern China and such an irrigation mode is unsustainable, which leaves a serious impact on the hydrological cycle and environment in the long run. Therefore, efforts should be made to reduce that in the future if sustainable agriculture development is desired. Irrigated agriculture locations and extents should be redesigned by taking advantage of local available water resources instead of heavily relying on non-local or non-renewable water, which is beyond the hydrological scope and is related to broader social, economic and political considerations. Advances in rainfed agriculture should also contribute to reducing NNBW and thus blue water use.

Regions with under-developed irrigation systems that encounter agricultural and social development levels similar to China may benefit from the results of our study. Utilizing uncommitted green water and improving irrigation efficiency can be effective and are key tools to mitigate risks associated with water and agriculture.

11.6 Conclusions

In this study, dynamic cropland areas and irrigation efficiency are utilized to: (1) provide spatially and temporally explicit estimations of agricultural uses of green and blue water, and (2) assess their impact on the water system in China. The findings are summarized here.

Green and blue water uses play different roles in agriculture across 10 river basins in China and show different trends in 1981–2000 and 2001–2010. Irrigated cropland expansion and decreasing precipitation increase the blue water use during 1981–2000, while irrigation efficiency improvement and increasing precipitation are the main drivers of the decrease of blue water use observed during 2001–2010. Future agricultural water management in China faces two challenges: (1) the increase in precipitation observed during 2001–2010 might not continue to increase or, worse, might decrease under climate change and (2) the current blue water use includes a large amount of non-renewable water in northern China. A continuous use of blue water is not sustainable. These challenges may represent an important threat on water and food security. Therefore, continuous irrigation efficiency improvement and additional green water management remain critical to sustain water and food security. Moreover, because the phase of the annual cycle of blue water use is not consistent with the phase of precipitation in most of the basins in northern China, utilization of rainwater harvesting besides the existing large reservoirs during the non-crop season may help provide irrigation water for crops and mitigate water stress during the crop season.

Intensive agricultural water use greatly affects the water system in China. At the national level, agricultural renewable blue water withdrawal ratios are 16 per cent over the year and 20 per cent during the crop season. Large amounts of agricultural blue water are withdrawn from a vulnerable water system in northern China, while relatively small amounts of that are withdrawn from a robust blue water system in southern China. This suggests the presence of higher levels of water stress in northern China and a potential for water development in southern China.

Considerable spatial heterogeneity is observed between green and blue water use, as well as in the impact on the water system. This suggests that the water resources management strategies should adapt to the specific properties of the basins. Despite the uncertainty in our results, our study provides insight and recommendations for large-scale agricultural water resources management that can be applied in China and in other parts of the world.

References

Bruins, H. J., Evenari, M., & Nessler, U. (1986). Rainwater-harvesting agriculture for food production in arid zones: The challenge of the African famine. *Applied Geography* 6 (1): 13–32.

Cai, X. (2008). Water stress, water transfer and social equity in Northern China – Implications for policy reforms. *Journal of Environmental Management* 87(1): 14–25.

Cao, X., Wu, P., Wang, Y., & Zhao, X. (2014). Water footprint of grain product in irrigated farmland of China. *Water Resources Management* 28, 2213–2227.

Dalin, C., Qiu, H., Hanasaki, N., Mauzerall, D. L., & Rodriguez-Iturbe, I. (2015). Balancing water resource conservation and food security in China. *Proceedings of the National Academy of Sciences (USA)* 112(15): 4588–4593.

Döll, P., & Lehner, B. (2002). Validation of a new global 30-min drainage direction map. *Journal of Hydrology* 258(1–4), 214–231.

Döll, P., & Siebert, S. (2002). Global modeling of irrigation water requirements. *Water Resources Research 38*(4): 1–10.

Fader, M., Rost, S., Müller, C., Bondeau, A., & Gerten, D. (2010). Virtual water content of temperate cereals and maize: Present and potential future patterns. *Journal of Hydrology* 384(3–4), 218–231.

Falkenmark, M., & Rockström, J. (2004). *Balancing Water for Humans and Nature: The New Approach in Ecohydrology*. Oxford: Routledge.

FAO: AQUASTAT, Food and Agriculture Organization of the United Nations. Available from www.fao.org/nr/water/aquastat/main/index.stm (Last accessed 10 March 2016).

Haddeland, I., Clark, D. B., Franssen, W., et al. (2011). Multimodel estimate of the global terrestrial water balance: Setup and first results. *Journal of Hydrometeorology* 12(5): 869–884.

Hanasaki, N., Kanae, S., & Oki, T. (2006). A reservoir operation scheme for global river routing models. *Journal of Hydrology* 327(1–2): 22–41.

Hanasaki, N., Kanae, S., Oki, T., et al. (2008). An integrated model for the assessment of global water resources – Part 1: Model description and input meteorological forcing. *Hydrology and Earth System Sciences* 12(4): 1007–1025.

Hoff, H., Falkenmark, M., Gerten, D., Gordon, L., Karlberg, L., & Rockström, J. (2010). Greening the global water system. *Journal of Hydrology* 384(3–4): 177–186.

Liu, J., & Yang, H. (2010). Spatially explicit assessment of global consumptive water uses in cropland: Green and blue water. *Journal of Hydrology* 384(3–4): 187–197.

Mann, H. B. (1945). Nonparametric tests against trend. *Mann Source: Econometrica* 13(3): 245–259.

Molden, D., Oweis, T., Steduto, P., Kijne, J. W., Hanjra, M. A., & Bindraban, P. S. (2007). Pathways for increasing agricultural water productivity. In D. Molden (ed.), *Water for Food, Water for Life: A Comprehensive Assessment of Water Management in Agriculture* (pp. 278–310). London: Earthscan.

Molden, D., & Sakthivadivel, R. (1999). Water accounting to assess use and productivity of water. *International Journal of Water Resources Development* 15(1–2), 55–71.
Nature Agency (2013). *Groundwater in China – Part 1: Occurrence and Use*. Beijing: Ministry of the Environment.
NBSC (2015). *China Statistical Year Book 2015 (in Chinese)*. National Bureau of Statistics of China. Beijing: China Statistics Press.
Portmann, F. T., Siebert, S., & Döll, P. (2010). MIRCA2000 – Global monthly irrigated and rainfed crop areas around the year 2000: A new high-resolution data set for agricultural and hydrological modeling. *Global Biogeochemical Cycles* 24(1): 1–24.
Rohwer, J., Gerten, D., & Lucht, W. (2007). Development of functional irrigation types for improved global crop modelling. *PIK Report* 104: 1–61.
Rosegrant, M. W., Cai, X., & Cline, S. (2002a). *World Water and Food to 2025: Dealing with Scarcity*. Washington, DC: International Food Policy Research Institute (IFPRI).
Rosegrant, M. W., Cai, X., Cline, S., & Nakagawa, N. (2002b). *The Role of Rainfed Agriculture in the Future of Global Food Production. EPTD discussion papers 90*. Washington, DC: International Food Policy Research Institute (IFPRI).
Rost, S., Gerten, D., Bondeau, A., Lucht, W., Rohwer, J., & Schaphoff, S. (2008). Agricultural green and blue water consumption and its influence on the global water system. *Water Resources Research* 44(9): 1–17.
Siebert, S., & Döll, P. (2010). Quantifying blue and green virtual water contents in global crop production as well as potential production losses without irrigation. *Journal of Hydrology* 384(3–4): 198–217.
Sun, S., Wu, P., Wang, Y., Zhao, X., Liu, J., & Zhang, X. (2012). The temporal and spatial variability of water footprint of grain: A case study of an irrigation district in China from 1960 to 2008. *Journal of Food, Agriculture and Environment* 10(3–4): 1246–1251.
Tang, Q., Zhang, X., & Tang, Y. (2013). Anthropogenic impacts on mass change in North China. *Geophysical Research Letters* 40(15): 3924–3928.
Tilman, D., Fargione, J., Wolff, B., et al. (2001). Forecasting agriculturally driven global environmental change. *Science* 292(5515): 281–284.
Vörösmarty, C. J., Bos, R., & Balvanera, P. (2005). Fresh water. In R. Hassan, R. Scholes, & N. Ash (eds.), *Ecosystems and Human Well-Being: Current States and Trends, Millennium Ecosystem Assessment Report* (pp. 165–207). Washington, DC: Island Press.
Wada, Y., Van Beek, L. P. H., & Bierkens, M. F. P. (2012). Nonsustainable groundwater sustaining irrigation: A global assessment. *Water Resources Research* 48(6): 2055.
Wada, Y., Van Beek, L. P. H., Van Kempen, C. M., Reckman, J. W. T. M., Vasak, S., & Bierkens, M. F. P. (2010). Global depletion of groundwater resources. *Geophysical Research Letters* 37(20): 1–5.
Weedon, G. P., Balsamo, G., Bellouin, N., Gomes, S., Best, M. J., & Viterbo, P. (2014). Data methodology applied to ERA-Interim reanalysis data. *Water Resources Research* 50(9): 7505–7514.
Weedon, G. P., Gomes, S., Viterbo, P., et al. (2011). Creation of the WATCH forcing data and its use to assess global and regional reference crop evaporation over land during the twentieth century. *Journal of Hydrometeorology* 12(5): 823–848.
Wisser, D., Frolking, S., Douglas, E. M., Fekete, B. M., Schumann, A. H., & Vörösmarty, C. J. (2010). The significance of local water resources captured in small reservoirs for crop production – A global-scale analysis. *Journal of Hydrology* 384(3–4), 264–275.
Wisser, D., Frolking, S., Douglas, E. M., Fekete, B. M., Vörösmarty, C. J., & Schumann, A. H. (2008). Global irrigation water demand: Variability and uncertainties arising from agricultural and climate data sets. *Geophysical Research Letters* 35(24), 1–5.

Zang, C. F., Liu, J., Van Der Velde, M., & Kraxner, F. (2012). Assessment of spatial and temporal patterns of green and blue water flows under natural conditions in inland river basins in Northwest China. *Hydrology and Earth System Sciences* 16(8): 2859–2870.

Zeng, Z., Liu, J., Koeneman, P. H., Zarate, E., & Hoekstra, A. Y. (2012). Assessing water footprint at river basin level: A case study for the Heihe River Basin in northwest China. *Hydrology and Earth System Sciences* 16(8): 2771–2781.

Zhuo, L., Mekonnen, M. M., Hoekstra, A. Y., & Wada, Y. (2016). Inter- and intra-annual variation of water footprint of crops and blue water scarcity in the Yellow River basin (1961–2009). *Advances in Water Resources* 87: 29–41.

12

Impact of Inter-Basin Water Transfer on Water Scarcity in Water-Receiving Area under Global Warming

A Case Study of the South-to-North Water Diversion Project

YUANYUAN YIN, QIUHONG TANG AND LEI WANG

12.1 Introduction

Water scarcity is increasingly perceived as a risk that threatens sustainable development in many areas of the world, especially in semi-arid and arid regions (Hoekstra, 2014; Cosgrove & Loucks, 2015). Nowadays, approximately 36 per cent of the world's population (2.5 billion people) live in water-scarce regions where more than 20 per cent of global GDP is produced (Kummu et al., 2010; High-Level Panel on Water, 2018). The growing impacts of climate change are exacerbating the problem of water scarcity in key regions of the world (Haddeland et al., 2014; Schewe et al., 2014), making the mitigation of water scarcity a critical task in achieving the goals of sustainable development (Gain et al., 2016). Therefore, within the context of climate change in which the nature of stream flows can no longer be regarded as stationary (Milly et al., 2008), water resource management faces a difficult dilemma of securing access to adequate water resources for demographic growth, economic development, and expansion of irrigated areas, whilst maintaining healthy freshwater ecosystems and the vital services they provide (WWAP, 2009).

Inter-basin water transfer (IBT), which transfers water from areas with perceived surpluses to those with shortages, is both widely considered and broadly controversial as a climate adaptation strategy to minimize water scarcity and to distribute water as per requirements. IBT started with the Roman civilization (Verma et al., 2009). There are 34 existing mega inter-basin water transfer projections (MWTP) worldwide that meet one of the following criteria: construction costs larger than 1 billion US dollars, distance of transfer larger than 190 km, or volume of water transferred exceeding 0.23 km^3 yr^{-1}. A total of 76 MWTP are either under construction or in the planning phase (Shumilova et al., 2018). Worldwide, existing IBT projects divert an estimated 500 billion m^3 of water annually (Shiklomanov,

This research is supported by the National Natural Science Foundation of China (41790424, 41730645).

2000), the equivalent of about 14 per cent of the global water withdrawals (Shumilova et al., 2018). Global water withdrawals through transfer schemes are expected to increase by 25 per cent until 2025 (Gupta & van der Zaag, 2008), primarily through an expansion of water transfer schemes. As droughts continue to affect semi-arid and arid regions across the world, IBT will play an increasing role in socioeconomic expansion. Given the impacts of ongoing droughts on water supplies and the predicted impacts of climate change on water resources (Rajagopalan et al., 2009; IPCC, 2014), it is important to understand the effectiveness of transfers as a strategy to mitigate climate-driven water shortages.

The Huang–Huai–Hai (HHH) region is one of the most water-scarce but densely populated regions in the world (Xia et al., 2012; Haddeland et al., 2014). Over the past 20 years, mainstream water flows have declined by 41 per cent in the Hai River Basin (HaiRB), 15 per cent in the Yellow (Huang) River Basin (YRB), and 15 per cent in the Huai River Basin (HuaiRB) (NDRC, 2007). The region is experiencing growing pressure to divert water from the agricultural sector to the municipal and industrial sectors. To alleviate severe water shortages in this region, China is building the world's largest inter-basin water diversion project – the South-to-North Water Diversion (SNWD) project – to transfer 45 billion m^3 of water annually from the Yangtze River to this region (Liu & Zhang, 2002; Zhang et al., 2009). Nonetheless, in the context of climate change, the effectiveness of the SNWD project as a strategy to mitigate climate-driven water shortages is under debate. Therefore, this chapter took the SNWD project as an example to quantitatively investigate the impact of the large-scale inter-basin water transfer on water scarcity in the HHH region within the context of climate change. The HHH region is often referred to as China's breadbasket and irrigation is its dominant water use sector. 'Three Red Lines' of water resources in China proposed in the No.1 Document in 2009 lead water resources management to gradually orient its scheme toward demand-oriented measures. We discuss the effectiveness of SNWD within the context of improving irrigation water use in Section 12.6.

12.2 South-to-North Water Diversion Project: A Primer

The South-to-North Water Diversion (SNWD) project is thus far the world's largest inter-basin water diversion project with the longest waterway aimed at alleviating severe water shortages in the HHH Region (Office of the South-to-North Water Diversion Project Construction Committee, State Council, PRC, 2016). The SNWD encompasses three water diversion routes – the Eastern, Central and Western (i.e. vertically) – diverting water from the lower, middle and upper reaches of the Yangtze River, respectively. It connects four major rivers – the Yangtze River, Huai River, Yellow River and Hai River – which flow west to east

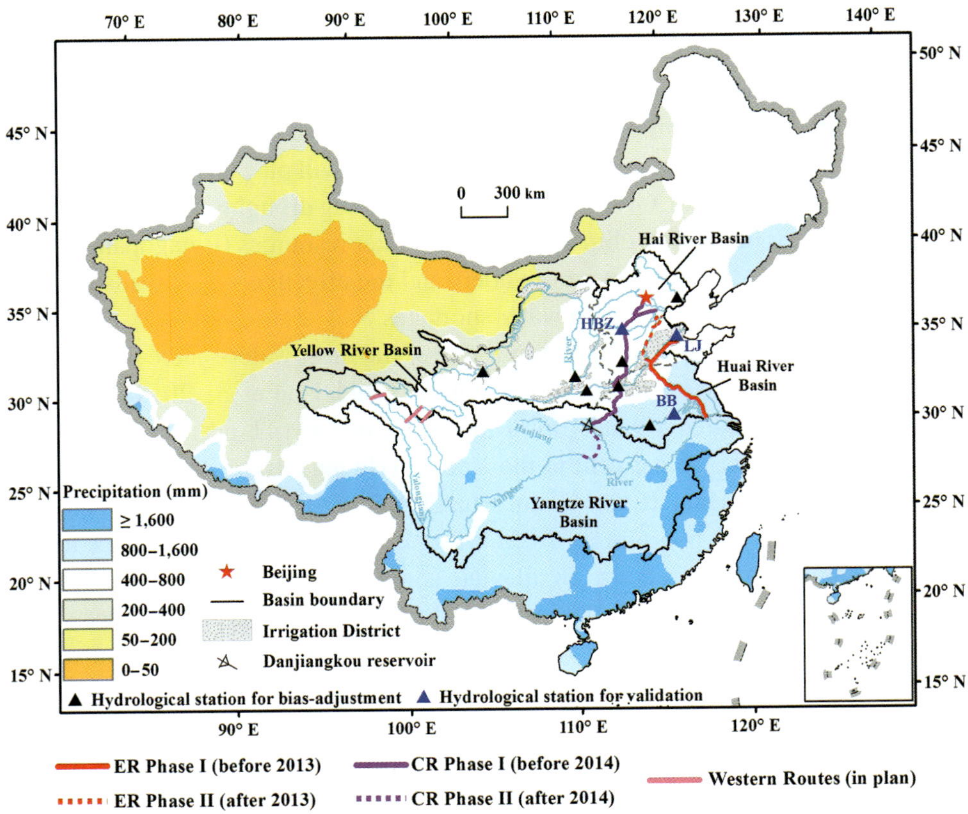

Figure 12.1 Sketch of the South-to-North Water Diversion (SNWD) project in China (CAB-SNWDP, 2003), and locations of major hydrological stations for bias-adjustment of the simulated runoff. Ultimately, the project aims to channel freshwater annually from the Yangtze River in southern China to the HHH (short for Huang–Huai–Hai) region in the arid northern China. LJ, HBZ and BB are short for Lijin, Huangbizhuang and Bengbu hydrological stations, respectively. ER and MR are the abbreviations of Eastern Route and Middle Route, respectively. The colour map shows the annual mean precipitation in China (1980–2009)

(i.e., horizontally). Thus, the pattern of water resources allocation in China associated with the SNWD is referred to as a 'four horizontals and three verticals' layout (Ministry of Water Resources (MWR), 2002) (Figure 12.1).

The idea of the SNWD began in 1952, when Chairman Mao Zedong said that 'the south has a lot of water, the north little. If possible, it is ok to lend a little water' (Ma, 2004). The planning and design of the project were overseen by the Planning, Design and Management Bureau of the SNWD, Ministry of Water Resources (MWR), established in 1999, and the masterplan was completed in 2002. The construction of the project officially commenced in December 2002 (Xia, 2012). According to the Master-plan, the SNWD project is estimated be as

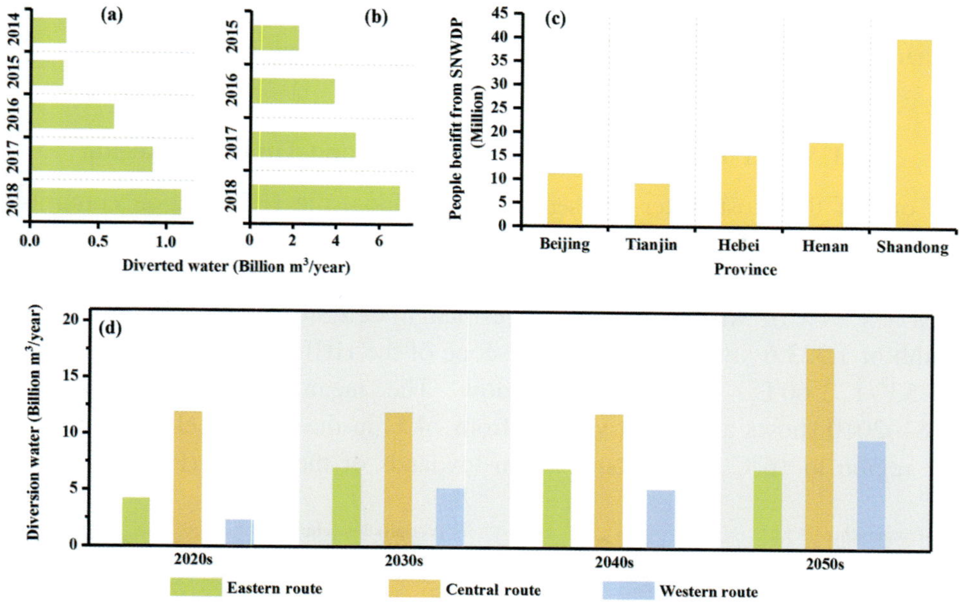

Figure 12.2 (a) Water transferred by the Eastern Route to Shandong Province since it was put into operation in 2014. (b) Water transferred by the Central Route to the HHH region since it was put into operation in 2015. (c) People benefitting from the SNWD project in different provinces since it has been operated for 4 years. (d) Designed quantities of water diverted by the SNWD project in future decades (billion m^3 yr^{-1}) (Zhang, 1999; Wei, 2000)

high as 65 billion US dollars (excluding social and ecological impacts) and its goal is to move approximately 44.8 billion m^3 of water northward each year to meet the needs of almost half a billion people by 2050.

The Eastern Route (ER) moves water from the lower Yangtze River along an existing network of rivers, lakes, reservoirs and canals, aiming at alleviating water shortages in Jiangsu, Anhui, Shandong, Hebei Provinces and Tianjin Municipality. The Middle Route (MR) transports water by gravity along the canals from Danjiangkou Reservoir on a tributary of the Yangtze, the Han River, to two provinces of Henan and Hebei and two municipalities of Beijing and Tianjin. The ER and MR include two phases. Phase I of the ER and phase I of the MR were completed in November 2013 and December 2014, respectively, and the routes are now in operation. By the end of 2018, the ER conveyed an accumulated amount of 3.1 billion m^3 of water to Shandong (Figure 12.2(a)), and the MR conveyed an accumulated amount of 17.8 billion m^3 of water (Figure 12.2(b)). Approximately 40 million people have benefitted from the SNWD project (Figure 12.2(c)). The Western Route (WR) will link the mountainous headwaters of the Yangtze River and Yellow River, by the construction of dammed reservoirs and by the digging of tunnels, aiming at alleviating water shortages in the Qinghai, Gansu, Ningxia, Shaanxi, Shanxi Provinces and in the Inner Mongolia Autonomous Region.

12.3 Water-Receiving Area of SNWD: Huang–Huai–Hai Region

As one of the most water-stressed areas of China, the HHH region comprises China's three major river basins, namely the Hai River Basin (HaiRB), Yellow River Basin (YRB) and Huai River Basin (HuaiRB) (Figure 12.1). It covers a total area of 1.34×10^6 km^2, with an east–west length of 2,344.5 km and a north–south width of 1,323.6 km. The geographic scope of the HHH region is delimited by a 95°53′–122°60′E, 32°10′–43°N window. The mean annual temperature in 1981–2010 shows a gradient varying from –4°C in the western elevated parts of the region to 16°C in the southeastern lowlands of the region (Figure 12.3(a)).

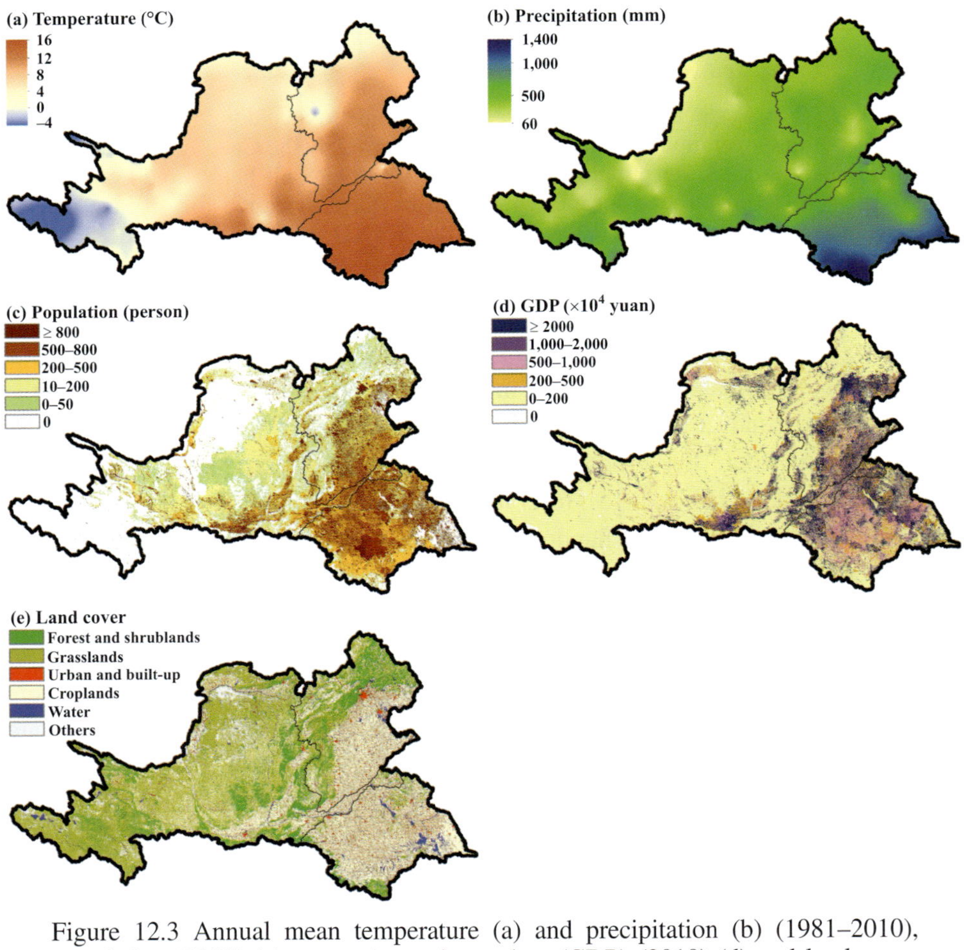

Figure 12.3 Annual mean temperature (a) and precipitation (b) (1981–2010), population (2010) (c), gross domestic product (GDP) (2010) (d) and land cover (2010) (e) in the HHH region

The mean annual precipitation in 1981–2010 was low in the northwest and high in the southeast, varying from 60 mm to 1,000 mm yr^{-1} (Figure 12.3(b)). Both future temperature and precipitation are projected to increase during the twenty-first century (Lu et al., 2013; Gu et al., 2014). The HHH region occupies an important position in China's economic and social development layout (Figure 12.3(c) and (d)). It is one of the most densely populated areas in the world and supports part of the country's largest industrial concentrations and most intensive irrigated agriculture (Figure 12.3(e)). The cultivated land, population and gross domestic product (GDP) in this region account for approximately 30, 29 and 35 per cent of the national totals in 2010, respectively (NBSC, 2011), but with a runoff contributing to 6.5 per cent of the total in China. The Huang–Huai–Hai Plain, located in the east of the HHH region, is a highly productive agricultural area and is often referred to as China's breadbasket (Shi, 2003; Kong et al., 2014). The plain is estimated to have some 1.8×10^4 km^2 of arable land, of which 56 per cent is irrigated (Tian et al., 2008), and produces 35–40 per cent and 60–80 per cent of China's wheat and corn (Liu et al., 2015), respectively.

12.4 Data and Methods

12.4.1 Data

The data used in this study are summarized in Table 12.1. The simulated runoff data for the period 1971–2099 and the simulated groundwater recharge and irrigation water withdrawal (IrrWW) data for the period 1980–2099 were obtained from the Inter-Sectoral Impact Model Intercomparison Project (ISI-MIP; Warszawski et al., 2014). These simulated data were provided at a spatial resolution of 0.5 degree, and were produced by three global gridded hydrological models (GGHMs, Table 12.2), namely, H80 (Hanasaki et al., 2008a, 2008b), LPJmL (Bondeau et al., 2007; Rost et al., 2008) and PCR-GLBWB (Wada et al., 2010, 2011). The GGHMs were forced with the bias-corrected climatic variables from the climate projections of the three global climate models (GCMs), namely, HadGEM2-ES, IPSL-CM5A-LR and GFDL-ESM2M under Representative Concentration Pathway 2.6 (RCP2.6) scenario and RCP8.5 scenario.

The 1 km gridded population and GDP data of China in 2005 were provided by the Institute of Geographic Sciences and Resources Research (IGSRR), Chinese Academy of Sciences (CAS) (Fu et al., 2014; Huang et al., 2014). The annual total population and GDP data of China during 1981–2013 were obtained from the National Bureau of Statistics of China (NBC). The quantitative projections of population and GDP for Shared Socio-economic Pathway 2 (SSP2) – a 'middle of the road' world where social, economic and technological trends do not shift

Table 12.1. Datasets used in this study

Datasets		Spatial resolution	Year	Source
Simulated runoff		$0.5° \times 0.5°$	1971–2099	Inter-Sectoral Impact Model Inter-comparison Project (ISI-MIP) (Warszawski et al., 2014)
Simulated groundwater recharge		$0.5° \times 0.5°$	1980–2099	
Simulated irrigation water withdrawal		$0.5° \times 0.5°$	1980–2099	
Population data	1-km gridded population of China	1 km × 1 km	2005	Institute of Geographic Sciences and Natural Resources Research, CAS (Fu et al., 2014)
	Historical population of China	Country	1981–2010	National Bureau of Statistics of China (NBSC)
	Projected population under SSP2	Country	2010–2100	Organization for Economic Co-operation and Development (OECD)
GDP data	1-km gridded GDP of China	1 km × 1 km	2005	Institute of Geographic Sciences and Natural Resources Research, CAS (Huang et al., 2014)
	Historical GDP of China	Country	1981–2010	National Bureau of Statistics of China (NBSC)
	Projected GDP under SSP2	Country	2010–2100	Organization for Economic Co-operation and Development (OECD)
Observed runoff		10 hydrological stations	1971–2010	Hydrological Bureau of the Ministry of Water Resources of China (HBMWRC)
Total renewable water resources		Basin	1998–2017	China's Water Resources Bulletin
Surface renewable water resources				
Intensity of irrigation water withdrawals		Province	1997–2013, 2015	Yearbook of China Water Resources
Irrigated area		Province	1997–2013, 2015	National Bureau of Statistics of China (NBSC)
Livestock water withdrawals		$0.5° \times 0.5°$	1981–2010	Institute of Geographic Sciences and Natural Resources Research, CAS (Huang et al., 2018a)
Thermoelectric water withdrawals				
Designed quantities of water diverted by the SNWD project		–	–	Zhang (1999); Wei (2000)

Table 12.2. Treatment of evaporation, runoff, groundwater recharge and irrigation for three global gridded hydrological models (GGHMs) used in this study

Model name	H80	LPJmL	PCR-GLBWB
References	Hanasaki et al. (2018)	Alam (2016)	Wada et al. (2014)
Evaporation	Bulk formula	Priestley-Taylor	Hamon
Runoff	Saturation excess, non-linear	Saturation excess	Saturation excess, Beta function
Groundwater recharge	Empirically estimated from total runoff of the surface soil layer.	Equal to the water reaching the bottom layer. Estimated by a base flow recession coefficient as a fraction of the existing groundwater store.	Estimated from the soil moisture content and hydraulic conductivity of the second soil layer (S2) and third groundwater layer (S3). Recharge equals deep percolation (S2) minus capillary rise (S3).
Irrigation water withdrawal	Computed as a function of the changed climate. Crops provided by Monfreda et al. (2008); the cropping calendar simulated by the model; irrigated areas provided by Siebert et al. (2007); irrigation efficiency; multi-cropping.	Computed as a function of the changed climate. Eleven crops and pasture; the cropping calendar simulated by the model; irrigated areas provided by Fader et al. (2010); irrigation efficiency.	Computed as a function of the changed climate. Two kinds of crop: paddy and non-paddy, provided by Portmann et al. (2010); the cropping calendar provided by Portmann et al. (2010) and Siebert and Döll (2010); irrigated areas provided by Portmann et al. (2010); flooding irrigation to paddy; irrigation efficiency; multi-cropping.

markedly from historical patterns (van Vuuren et al., 2014) – were developed for the 2010–2099 period based on the SSP Scenario Database, available at https://secure.iiasa.ac.at/web-apps/ene/SspDb.[1] The population and GDP projections were provided at country level within 5-year intervals. The projected GDP in US dollars was converted to Chinese Yuan using the official exchange rate provided by the World Bank (WB). The gridded population and GDP datasets during 1981–2099 were generated by spatially and temporally downscaled country-scale population and GDP, estimated using a simple linear downscaling method (see Gaffin et al. (2004) and Yin et al. (2017) for further details). The observed runoff data over 1971–2000 of 10 major hydrologic stations (Figure 12.1) at the main stream of the

Yellow River (YR), Hai River (HaiR) and Huai River (HuaiR) were collected from the Hydrological Year Book of the Ministry of Water Resources of China (MWRC). The total renewable water resources and surface renewable water resources during 1998–2017 of three basins were collected from China's Water Resources Bulletin of MWRC and were used to bias-adjust the simulated groundwater recharge. The intensity of irrigation water withdrawals and irrigated area were collected from the Yearbook of China Water Resources and China's National Bureau of Statistics, respectively.

12.4.2 Methods

Figure 12.4 illustrates the research flowchart of this chapter. A multiplicative adjustment method was used to bias-adjust the simulated runoff, groundwater recharge and irrigation. Based on the bias-adjusted runoff, groundwater recharge and irrigation, and the downscaled population and GDP, water supply and demand were calculated. The impact of the SNWD project on water supply risk was analysed from the recent past to the end of the twenty-first century in terms of the water supply stress index (WaSSI).

12.4.2.1 Bias-Adjustment for Runoff, Groundwater Recharge and Irrigation

As the GGHMs often exhibit considerable bias and cannot adequately reproduce long-term average volumes (Hattermann et al., 2017), we used a multiplicative adjustment method to remove bias from the model simulations (Ho et al., 2012;

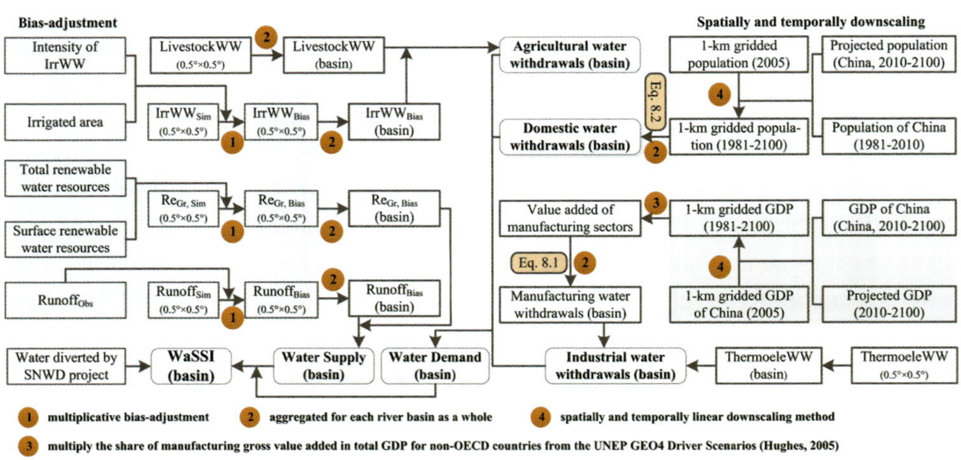

Figure 12.4 Research flowchart of this chapter. WaSSI is defined as the ratio of water demand to water supply for a specific basin

Hawkins et al., 2013). The multiplicative factor was calculated as the ratio of the observation to the model simulation. For runoff, the calibration period was from 1971 to 2000. Huangbizhuang, Lijin and Bengbu hydrological stations were selected to calculate the multiplicative bias-adjustment factors for the HaiRB, YRB and HuaiRB (Figure 12.1). Independent validation was conducted from 2001–2010 (hereafter referred to as the validation period) to assess the performance of bias-adjustment. We compared the bias-adjusted runoff of the three GGHMs with the streamflow observations at the other seven selected hydrological stations (Figures 12.1 and 12.5(a)). The bias-adjusted runoff can adequately reproduce the streamflow in the calibration period. During the validation period, the bias-adjusted runoff is similar to the observed runoff at the seven hydrological stations in the HHH region (Figure 12.5(a)). The R-squared value and relative standard deviation (RSD) between the bias-adjustment and observation are 0.99 and 14.73 per cent, respectively.

For groundwater recharge, the calibration and validation periods were 1997–2006 and 2007–2013, respectively. The statistical groundwater recharge generated from endogenous precipitation is the difference between the total renewable water resources and the surface renewable water resources (GAQUIQ and SAC, 2009). We compared the median values of bias-adjusted groundwater recharge from the three GGHMs with the statistics available for the three basins (Figure 12.5(b)). The R-squared value and RSD between the bias-adjusted and statistical groundwater recharge are 0.97 and 22.04 per cent, respectively.

The IrrWW of 31 provinces was used to adjust the bias in the simulated IrrWW in China. For a given province, the statistical IrrWW is the product of the IrrWW intensity and the irrigated area. The calibration and validation periods were 1998–2007 and 2008–2016. For the validation period, the R-squared value and RSD between the bias-adjusted IrrWW and the statistics are 0.95 and 6.81 per cent, respectively (Figure 12.5(c)).

12.4.2.2 Available Water Supply

The bias-adjusted runoff and groundwater recharge generated from endogenous precipitation were aggregated for the whole region and each basin to calculate the renewable water resources, defined as the sum of the long-term average annual runoff (including both subsurface and surface runoff) of a given river, and groundwater recharge for the given watershed generated from endogenous precipitation (FAO, 2003; UN Water, 2017). Note that to sustain the in-stream ecosystems, only some of the renewable water resources can be exploited by humans. The exploitation index refers to the annual net water withdrawal (i.e., total water withdrawal minus the return flow), including the amount transferred to other watersheds, in relation to the annual renewable water resources (FAO, 2003;

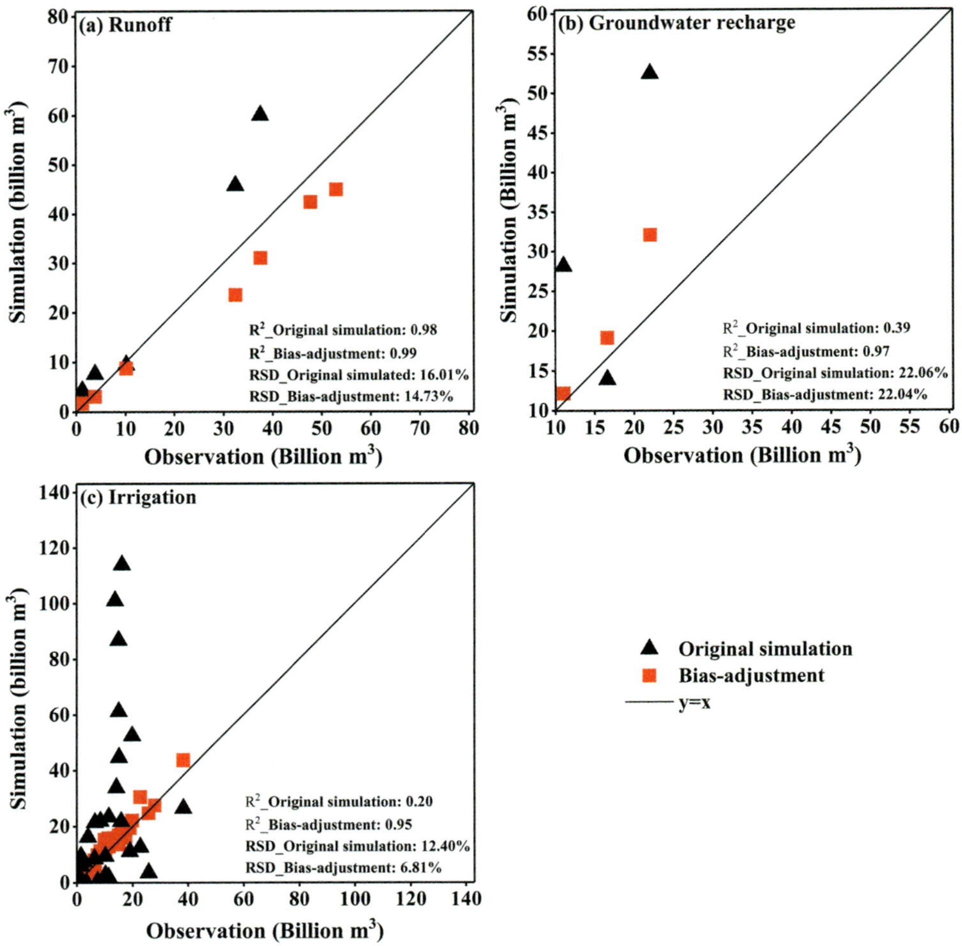

Figure 12.5 (a) Comparison of the annual mean runoff between the original and bias-adjusted simulations and the observations at seven hydrological stations in the HHH region during the validation period of 2001–2010. (b) Comparison of the annual mean groundwater recharge between the observations and the original as well as bias-adjusted simulations in the three basins during the validation period of 2008–2016. (c) Comparison of the annual mean irrigation water withdrawals between the original and bias-adjusted simulation as well as the observations in 31 provinces during the validation period of 2007–2013

MWRC, 2011). The exploitation index values of the HR, YR and HR are 126, 73 and 62 per cent, respectively (MWRC, 2011). The available water supply in each basin was estimated as the quantity of renewable water resources multiplied by the exploitation index. The available water supply for the entire region is the sum of the available water supply in the three basins. The annual runoff and groundwater recharge were calculated for each GCM–GGHM pair. There are nine model pairs (3 GCMs $\times$ 3 GGHMs) for runoff under two scenarios and groundwater recharge under RCP2.6. Meanwhile, there are six GCM–GGHM pairs for groundwater recharge under RCP8.5 because the groundwater recharge is missing in PCR-GLOBWB simulations. Based on the bias-adjusted runoff and groundwater recharge, the renewable water resources during the twenty-first century and the available water supply were estimated.

12.4.2.3 Sectoral and Total Water Demand

The total water demand was calculated as the sum of three main water-consumptive sectors: agriculture, domestic and industry. Agricultural water use includes the demands for irrigation and livestock. Industrial water use includes two main components: water withdrawals for the manufacturing sector and withdrawals for the cooling of thermoelectric plants in the electricity sector. Currently, livestock water withdrawals are very small compared to those for irrigation and thermoelectric water withdrawal data are only available from 1981–2010. The projected data related to livestock and thermoelectric water withdrawals are currently unavailable. In this study, we assumed that the water withdrawals for livestock and for the cooling of thermoelectric plants are equal to these average annual values during 1981–2009 and would remain constant. The average annual value for the entire region and each basin during 1981–2009 was calculated by aggregating the reconstructed livestock water withdrawal data and thermoelectric water withdrawal data provided by Huang et al. (2018b). To calculate the irrigation water use, the bias-adjusted IrrWW was aggregated for the entire region and each basin and the multi-model-ensemble median of IrrWW from all the available model pairs was calculated. There are nine GCM–GGHM pairs for the RCP2.6 scenario. Meanwhile, there are six GCM–GGHM pairs for the RCP8.5 scenario as the irrigation water withdrawal is missing in PCR-GLOBWB simulations.

Manufacturing water withdrawals is positively correlated with the economic metric of manufacturing gross value added (Dziegielewski et al., 2002), and was estimated by multiplying the value added of the manufacturing sector by the manufacturing water use intensity (Flörke et al., 2013; Wada et al., 2016). The water use intensity for manufacture was estimated by multiplying the manufacturing water use intensity for the base year by the change of the manufacturing water use intensity (Eq. 12.1).

$$I_{manu} = i_{manu,t0} \times (1 - S_{manu,cat})^{(t-t0)} \tag{12.1}$$

Table 12.3. Changes of the domestic water intensity (Hanasaki et al., 2013)

Per capita GDP		Domestic water intensity changes rate (L person^{-1} day^{-1} year^{-1})
Per capita GDP < 2,000 USD	$200 \leq i_{dom}$	0
	$i_{dom} < 200$	2 ($i_{dom} < 300$)
2,000 USD ≤ Per capita GDP		0

where I_{manu} is the water use intensity for manufacture (m^3 per 10,000 RMB Yuan), $I_{manu,t0}$ is the manufacturing water use intensity for the base year (m^3 per 10,000 RMB Yuan) and $S_{manu,cat}$ is the change of the manufacturing water use intensity (%). The added value of manufacturing sectors is calculated by multiplying the share of manufacturing gross value added in total GDP for non-OECD countries from the UNEP GEO4 Driver Scenarios (Hughes, 2005). In this study, the base year is 2010. The manufacturing water use intensity values in the HaiRB, YRB and HuaiRB for the base year are 70.2, 85.7 and 168.9 m^3 per 10,000 RMB Yuan, respectively. The change of the manufacturing water use intensity is 0.6 per cent (Wada et al., 2016).

Domestic water withdrawals were linked to the main driving forces of water use in the domestic sector, for example, the population (Alcamo et al., 2003), and were estimated by multiplying the population by the domestic water use intensity (Flörke et al., 2013). Following Hanasaki et al. (2013), the domestic water use intensity was calculated with Eq. (12.2).

$$I_{dom} = i_{dom,t0} + S_{dom,cat} \times (t - t0) \tag{12.2}$$

where I_{dom} is the water intensity for domestic use (L day^{-1} person^{-1} year^{-1}), $i_{dom,t0}$ is the domestic water use intensity for the base year (L day^{-1} person^{-1} year^{-1}) and $S_{dom,cat}$ is the change of the domestic water use intensity (L person^{-1} day^{-1} year^{-1}). The base year is 2010. The domestic water use intensity values in the HaiRB, YRB and HuaiRB for the base year are 112.7, 100.6 and 148.7 L day^{-1} person^{-1} year^{-1}, respectively. The change of the domestic water use intensity is shown in Table 12.3 (Hanasaki et al., 2013).

12.4.2.4 Impact of the SNWD Project on Water Supply Risk

In order to assess the impact of climate change and the SNWD on water supply risk, a modified version of the WaSSI was used (Sun et al., 2008; Eldardiry et al., 2016). We estimated the WaSSI with and without the impact of climate change and the SNWD project for each basin and the whole region in different decades of the twenty-first century. The metric was defined as the ratio of annual water demand to annual water supply for a specific watershed. It measures whether water supplies are sufficient for concurrently meeting all net withdrawal requirements within a watershed. A WaSSI greater than 1.0 suggests that the water supply cannot sustain socio-economic development and water scarcity occurs. The greater the WaSSI value, the greater the water supply stress expected.

12.5 Results

12.5.1 Projected Changes of Water Supply

Climate change is expected to accelerate water cycles and thereby increase the availability of renewable freshwater resources in the HHH region. The runoff and groundwater recharge in the HHH region and three basins under RCP2.6 and RCP8.5 from the present to the end of the twenty-first century are shown in Figure 12.6. The median values in the HHH region indicate that the temporal variance of both runoff and groundwater recharge shows a significantly increasing trend ($p < 0.05$) during the twenty-first century (Figure 12.6(a1) and (a2), Table 12.4), with the rate of 2.55 (2.06) and 0.97 (1.52) billion m^3 yr^{-1} in each decade under RCP2.6 (RCP8.5), respectively. Annual runoff and groundwater recharge are projected to be about 150 and 65 billion m^3 yr^{-1} at the end of the twenty-first century, increasing by about 14 and 22 per cent compared to the values in the historical period (1980–2009). The median values of runoff over the three basins are projected to significantly ($p < 0.05$) increase during the twenty-first century (Figure 12.6(b1–d1), Table 12.4). The rates of runoff are 0.31 (0.61), 0.77 (1.10) and 0.29 (0.54) billion m^3 yr^{-1} in each decade over HaiRB, YRB and HuaiRB under RCP2.6 (RCP8.5), respectively. The median values of groundwater recharge are projected to significantly ($p < 0.05$) increase from 2010 to 2100 over HaiRB and YRB under both RCP2.6 and RCP8.5. Over HuaiRB, the median values are projected to decrease by 0.03 billion m^3 yr^{-1} in each decade under RCP2.6, yet the median values are projected to significantly ($p < 0.05$) increase with the rate of 0.27 billion m^3 yr^{-1} in each decade under RCP8.5.

Considering the exploitation index, the available water supply is shown in Figure 12.7. The water supply will significantly ($p < 0.05$) increase during the twenty-first century under both RCP2.6 and RCP8.5. The rates of water supply are 3.00, 1.41, 0.69 and 0.20 billion m^3 yr^{-1} in each decade in the HHH region, HaiRB, YRB and HuaiRB under RCP2.6, respectively. And the rates under RCP8.5 are larger than the rates under RCP2.6. For the HHH region, the water supply is projected to be approximately 163.80 and 184.76 billion m^3 yr^{-1} at the end of the twenty-first century under RCP2.6 and RCP8.5, respectively, increasing

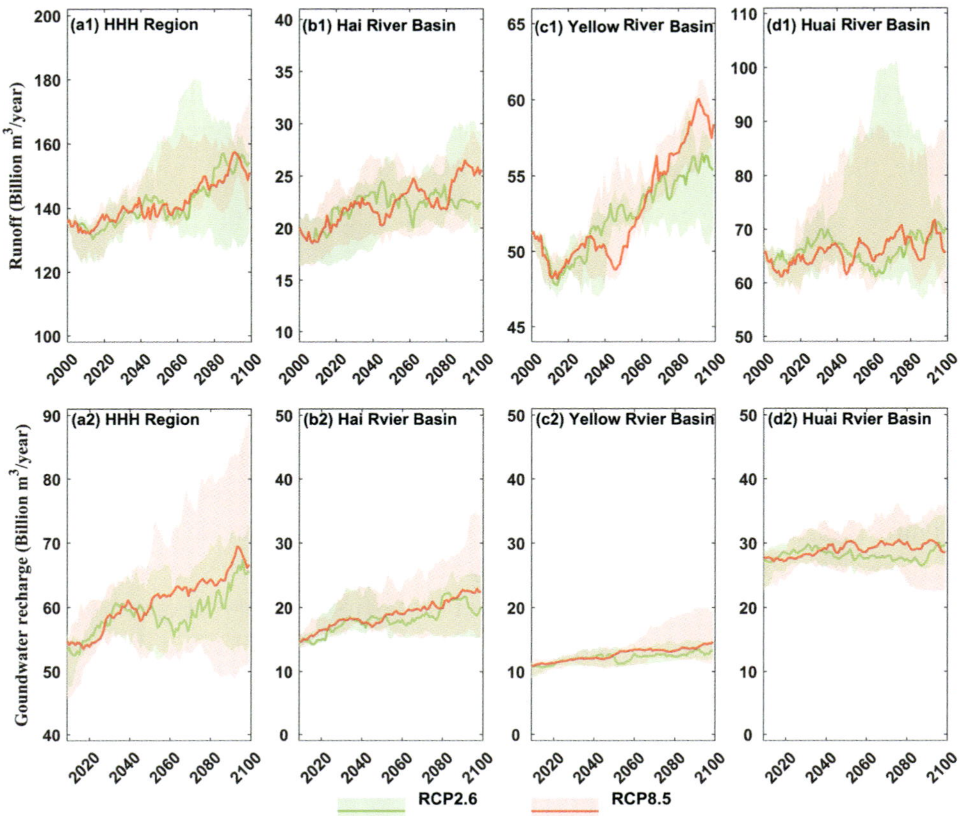

Figure 12.6 Bias-adjusted runoff (top) and groundwater recharge (bottom) in the HHH region as well as the Hai River Basin, Yellow River Basin and Huai River Basin from 2010 to 2100 under RCP2.6 and RCP8.5. The shaded band denotes the inter-quartile range and the solid line shows the median of the all GCM–GGHM pairs

by about 18.73 and 35.03 per cent from that during the historical period. Among the three basins, the water supply in HaiRB would show the largest increase, with an increase by more than 29 per cent from that during the historical period. The water supply in HuaiRB would exhibit the smallest increase, with an expected increase of less than 17 per cent from that during the historical period.

12.5.2 Projected Changes of Sectoral and Total Water Demand

In both the HHH region and the three basins, the total water demand is projected to first increase and then decrease during the twenty-first century (Figure 12.7). In the HHH region, the total water demand is projected to increase from 141 billion m^3 yr^{-1} in 2009 to close to 228 and 226 billion m^3 yr^{-1} in 2099 under RCP2.6 and

Table 12.4. Projected trends of runoff and groundwater recharge for each global gridded hydrological model (GGHM) and the median of all possible GGHM–GCM pairs

		Runoff				Groundwater recharge			
		RCP2.6		RCP8.5		RCP2.6		RCP8.5	
Basin	Median/ GGHM	Trends (billion m^3 yr^{-1} $decade^{-1}$)	R^2	Trends (billion m^3 yr^{-1} $decade^{-1}$)	R^2	Trends (billion m^3 yr^{-1} $decade^{-1}$)	R^2	Trends (billion m^3 yr^{-1} $decade^{-1}$)	R^2
HHH Region	Median	2.25*	0.78	2.06*	0.83	0.97*	0.60	1.52*	0.92
	H08	2.22*	0.61	1.00*	0.38	–0.15*	0.15	–0.46*	0.43
	LPJmL	1.84*	0.73	2.83*	0.86	2.41*	0.93	3.37*	0.97
	PCR-GLOBWB	1.74*	0.72	1.87*	0.76	1.08*	0.59	–	–
Hai River Basin	Median	0.31*	0.37	0.61*	0.75	0.66*	0.73	0.79*	0.93
	H08	0.90*	0.69	0.37*	0.20	0.12*	0.43	0.02	–0.01
	LPJmL	0.76*	0.89	0.60*	0.80	0.81*	0.59	1.65*	0.89
	PCR-GLOBWB	0.55*	0.89	0.33*	0.68	0.93*	0.76	–	–
Yellow River Basin	Median	0.77*	0.85	1.10*	0.79	0.22*	0.66	0.36*	0.93
	H08	0.59*	0.65	0.91*	0.74	0.12*	0.67	0.10*	0.33
	LPJmL	0.79*	0.86	1.24*	0.79	0.50*	0.92	1.13*	0.93
	PCR-GLOBWB	0.87*	0.87	1.12*	0.79	0.25*	0.44	–	–
Huai River Basin	Median	0.29*	0.11	0.54*	0.43	–0.03	0	0.27*	0.56
	H08	0.22	0.01	–0.42*	0.10	–0.29*	0.61	–0.49*	0.73
	LPJmL	0.29*	0.11	0.85*	0.62	0.81*	0.72	0.73*	0.77
	PCR-GLOBWB	0.22*	0.07	0.38*	0.26	0.09	0.02	–	–

Note: * represents that the corresponding increasing/decreasing trend is significant at 5 per cent level ($p < 0.05$).

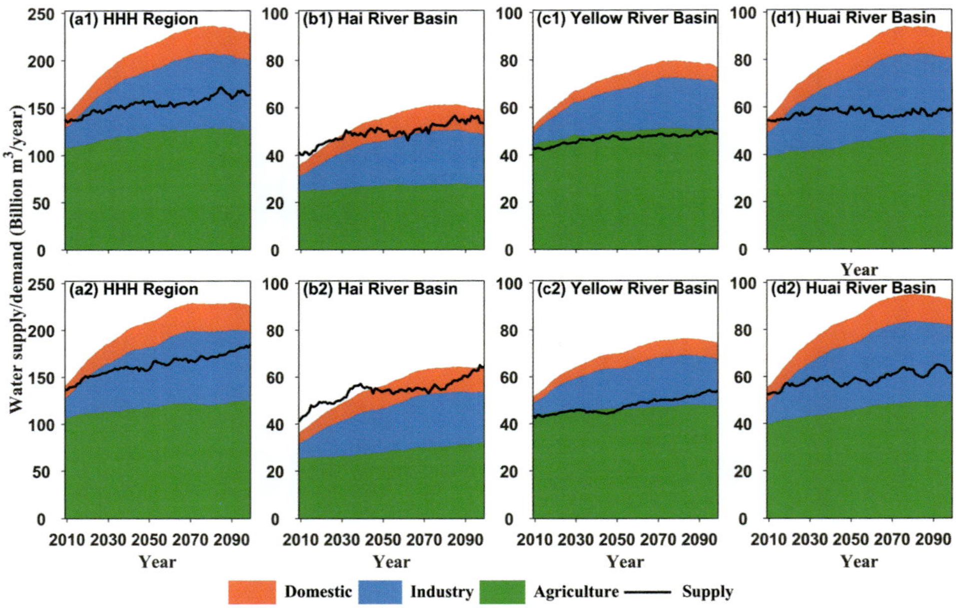

Figure 12.7 Ensemble-median of annual water supply and agricultural water demand across all GCM–GGHM pairs and the annual industrial and domestic water demand under SSP2 in the HHH region as well as the Hai River Basin, Yellow River Basin and Huai River Basin from 2009 to 2099 under RCP2.6 (top) and RCP8.5 (bottom)

RCP8.5, respectively. This increase is primarily driven by the growth in the industrial water withdrawals, accounting for approximately 32 per cent of the total in 2099. Irrigation is the dominant water use sector during the twenty-first century. Domestic water withdrawals are projected to increase before 2076, as both population and domestic water use intensities will increase, and then are projected to decrease due to a decrease in population. With the assumption that thermoelectric water withdrawals are static, manufacturing water withdrawals are projected to rapidly increase before 2080, as the value added of manufacturing sectors will increase, and then are projected to decrease slightly due to a decrease in manufacturing water use intensities. The irrigation water withdrawals, accounting for more than 55 per cent, are projected to significantly ($p < 0.5$) increase from 106.12 billion m^3 yr^{-1} in 2009 to more than about 125.84 billion m^3 yr^{-1} in 2099 (Figure 12.8). The irrigation water withdrawal under RCP8.5 is almost equal to that under RCP2.6 during 2009–2025, and then is smaller than that under RCP2.6 during 2026–2080. The total water demand and sectoral water demand are also projected to increase in each basin during the twenty-first century under both RCP2.6 and RCP8.5. The total water demand is projected to increase from about 41, 44 and 46 billion m^3 yr^{-1} in 2009 to more than about 66.5, 65.9 and

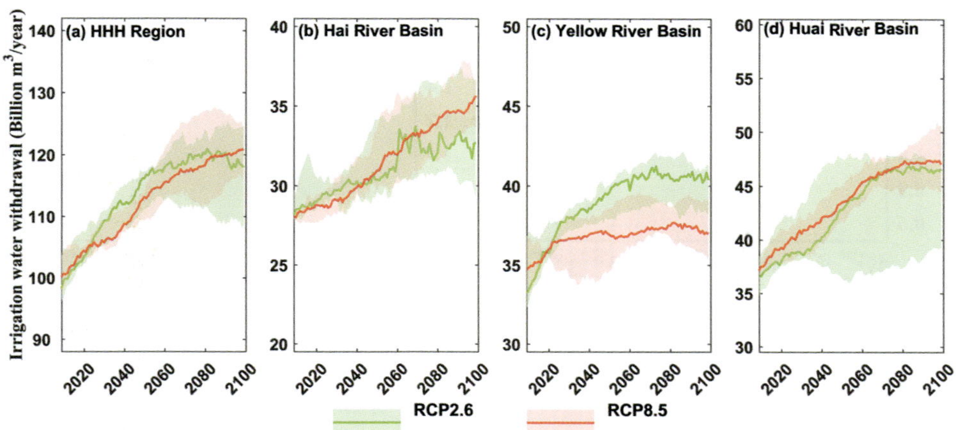

Figure 12.8 Bias-adjusted irrigation in the HHH region as well as the Hai River Basin, Yellow River Basin and Huai River Basin from 2010 to 2100 under RCP2.6 and RCP8.5. The shaded band denotes the inter-quartile range and the solid line shows the median of the all GCM–GGHM pairs

93.9 billion m^3 yr^{-1} in 2099 in HaiRB, YRB and HuaiRB, respectively. The irrigation water withdrawals are also projected to significantly ($p < 0.5$) increase in each basin, and the increased rate under RCP8.5 is expected to exceed the rate under RCP2.6 (Figure 12.8). In the HaiRB, the irrigation water withdrawal under RCP8.5 is less than that under RCP2.6 before 2043, and then is larger than that after 2043. The irrigation water withdrawal under RCP8.5 is almost as large as that under RCP2.6 during the twenty-first century in the HuaiRB. However, the irrigation water withdrawal under RCP8.5 is almost smaller than that under RCP2.6 in the YRB.

12.5.3 Impact of the SNWD Project on Water Supply Risk

Figure 12.9 shows the average annual WaSSI for the HHH region and three basins throughout the twenty-first century without and with the impact of the SNWD under RCP2.6 and RCP8.5. With the climatic and social change, the HHH region is projected to have a WaSSI greater than 1 during the twenty-first century for both climate change scenarios, which means that water demand exceeds supply. The WaSSI under RCP2.6 is projected to be larger than WaSSI under RCP8.5, likely because precipitation under RCP8.5 is projected to be higher than that under RCP2.6 in this region. In the HHH region, the WaSSI without the SNWD in each decade of the twenty-first century is projected to be greater than 1.23. Thus, to completely satisfy the water demand, at least 23 per cent of the water supply (approximately 34.94 billion m^3 yr^{-1}) has to be provided. Since the SNWD project

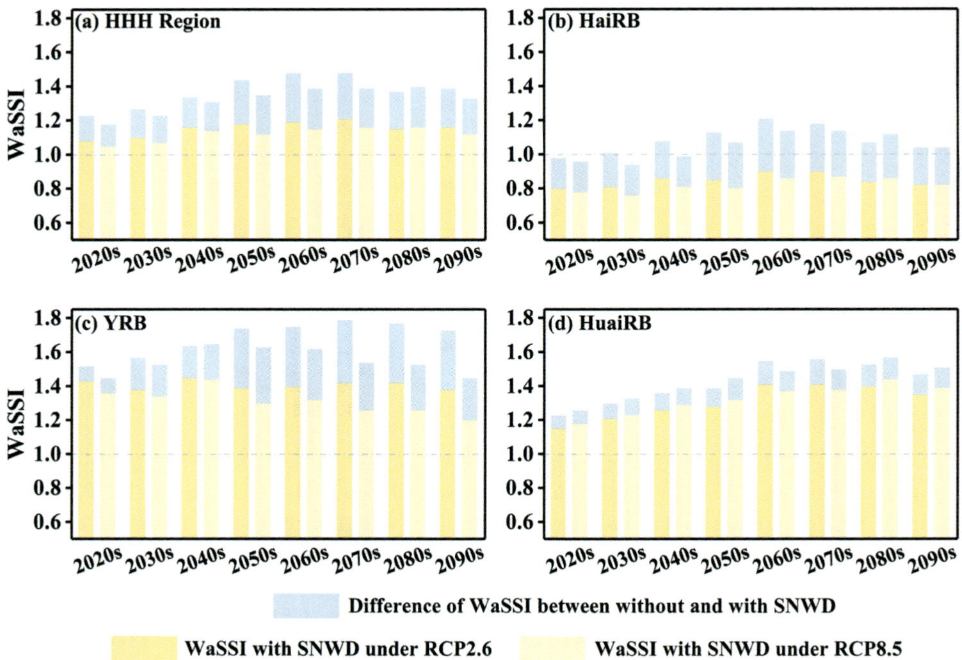

Figure 12.9 WaSSI without (whole bar) and with the impact of the SNWD (yellow bar) in the different decades of the twenty-first century in the HHH region as well as the Hai River Basin (HaiRB), Yellow River Basin (YRB) and Huai River Basin (HuaiRB). The blue bar denotes the difference of WaSSI between without and with the impact of the SNWD. Water scarcity occurs if WaSSI is greater than 1

was put into operation in 2013 for ER and 2014 for MR, millions of people have benefitted from it. According to the overall plan of the SNWD project, water diversion will increase as more conveyance systems and water plants are put into operation in the next few decades (Figure 12.2(d)). Annual water diversion is expected to reach 18.28 billion m^3 by the 2020s. As the project progresses, the amount of annual water to be diverted to the HHH region will increase to 23.96 billion m^3 in the 2030s and 34.51 billion m^3 in the 2050s (Figure 12.2(d)). With the implementation of the SNWD project, there is an opportunity to mitigate water scarcity issues faced by this region (Figure 12.9). The SNWD will reduce the WaSSI by at least 0.1, suggesting that more than 19.7 billion m^3 yr^{-1} water supply will be needed to completely mitigate water scarcity issues in different decades. Therefore, the SNWD project can significantly reduce water scarcity in the HHH region but cannot completely solve the problem.

With the impact of the SNWD project, the HaiRB is projected to have a WaSSI less than 1 under both climate change scenarios, meaning that water supply should exceed water demand during the twenty-first century. Therefore, SNWD could be a suitable approach to alleviate the problem of groundwater over-exploitation. The WaSSI under RCP2.6 is projected to be larger than WaSSI under RCP8.5 in the YRB, while the index under RCP2.6 is projected to be less than that under RCP8.5 in the HuaiRB. Despite concerns of the SNWD project's impact, the WaSSI in both YRB and HuaiRB is projected to be larger than 1 during the twenty-first century under RCP2.6 and RCP8.5, meaning that the water supply wouldn't satisfy socio-economic development needs.

12.6 Discussions and Conclusions

Traditional supply-oriented management schemes have resulted in a number of water transfer projects built to solve water supply issues in regions of water shortage (Li et al., 2014). However, inter-basin transfer may facilitate unwise and unsustainable urban and irrigation developments (Pittock et al., 2009). Water resources management is gradually turning to a demand-oriented management scheme. In China, at the national conference on water conservancy held in 2009, the 'Three Red Lines' of water resources was proposed in the No. 1 Document, including the upper limit of water resources allocation, the baseline of utilization efficiency of water resources and the upper limit of sewage discharge. According to the No. 1 Document, the upper limit of the total water use in China is 700 billion m^3 yr^{-1} to the 2030s. Irrigation is the dominant water use sector in the HHH region, accounting for more than 55 per cent of total water use. It is critical for control of the water resources amount to improve irrigation water use efficiency. To investigate the impact of improving irrigation water-use efficiency on water abundance/scarcity, we established three improving irrigation water-use efficiency scenarios: 5 per cent increase (Scenario 1), 10 per cent increase (Scenario 2) and 15 per cent increase (Scenario 3) compared to the efficiency in the historical period. Figure 12.10 shows WaSSI without and with the impact of the SNWD in the 2030s under three improving irrigation water-use efficiency scenarios in the HHH region. The WaSSI is projected to decrease with the increase of irrigation water-use efficiency under both RCP2.6 and RCP8.5. Water scarcity could not occur with the impact of the SNWD, when the increase of irrigation water-use efficiency is larger than 15 per cent compared to the efficiency in the historical period.

Projections of future runoff, groundwater recharge and irrigation under climate change contain uncertainty due to three key sources: emissions uncertainty, GCM modelling uncertainty and GGHM modelling uncertainty (Hawkins & Sutton, 2009). We used the standard deviation to quantify the uncertainty in water supply and irrigation projections. The standard deviations of runoff, groundwater recharge and irrigation from all available GCM–GGHM pairs are projected to increase over

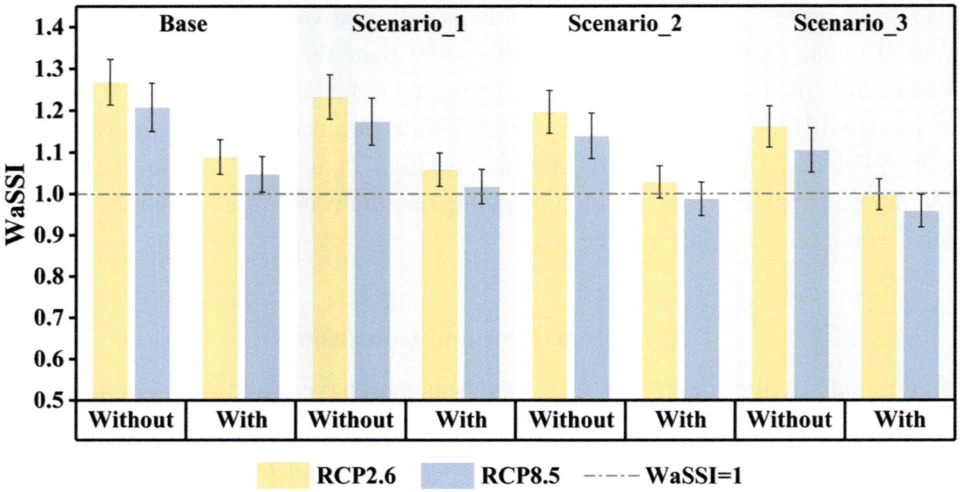

Figure 12.10 WaSSI without (yellow bar) and with the impact of the SNWD (blue bar) in the 2030s under three improving irrigation water-use efficiency scenarios in the HHH region. The bars and error bars represent the median and ±1 standard deviation of all GCM–GGHM pairs, respectively. Water scarcity occurs if WaSSI is greater than 1

time under both RCP2.6 and RCP8.5 in the HHH region (Figure 12.11(a)). The uncertainty varies with variable and emissions scenario. In the HHH region, the standard deviation of each variable in the 2090s is more than 1.5-times larger than the standard deviation in the 2030s. For runoff and groundwater recharge, the total uncertainty under RCP8.5 is larger than the uncertainty under RCP2.6. Yet, the total uncertainty of irrigation under RCP2.6 is larger than the uncertainty under RCP8.5. The uncertainty of groundwater recharge caused by GGHMs is about 5-times compared with the expected uncertainty derived from the GCMs in the 2090s (Figure 12.11(b)). Nevertheless, the uncertainty of runoff from the GCMs is more than 2.3-times the expected uncertainty from the GGHMs in the 2090s. For irrigation, the standard deviation associated with GGHMs predictions is larger than that linked with GCM under RCP2.6, whereas the standard deviation associated with GCM is larger than that linked with the GGHMs under RCP8.5. Among the three basins, the standard deviations of these three variables from all available GCM–GGHM pairs in the 2090s should be lowest in YRB under two emissions scenarios (Figure 12.11(c)). The standard deviation of groundwater recharge is projected to be highest in HaiRB, while the standard deviation of runoff and irrigation is projected to be highest in HuaiRB.

It is difficult to quantify domestic and industrial water withdrawals because these water withdrawals are influenced by many social, economic and political factors. Different SSP storylines include different qualitative scenario assumptions

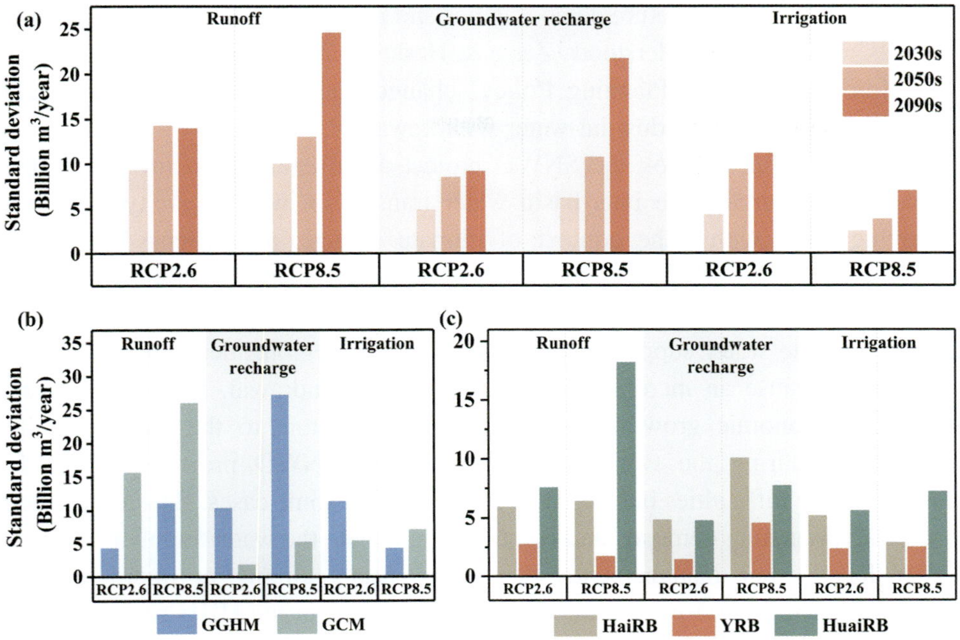

Figure 12.11 (a) The standard deviation of runoff, groundwater recharge and irrigation from all available GCM–GGHM pairs in the 2030s, 2050s and 2090s in the HHH region. (b) The model spreads caused by GGHMs and GCMs in 2090s in the HHH region. The standard deviation from the three GGHMs was firstly calculated for each GCM. The averaged GGHM standard deviations of the three GCMs were then used to assess the model spread caused by GGHMs. Using the same calculation procedure, the averaged GCM standard deviations of the three GGHMs were used to assess the model spread caused by GCMs. (c) The standard deviation of runoff, groundwater recharge and irrigation from all available GCM–GGHM pairs in the 2090s in the Hai River Basin (HaiRB), Yellow River Basin (YRB) and Huai River Basin (HuaiRB)

of future changes in demographics, human development, economy and lifestyle and technology (O'Neill et al., 2017). Even with the same set of scenario assumptions, the differences in approaches to estimating water withdrawals can result in significantly different projections (Wada et al., 2016). In China, thermoelectric power generation is responsible for roughly 10 per cent of the total freshwater withdrawal (Zhang & Anadon, 2013). Although the projection of water withdrawal for cooling the thermoelectric generation is subject to large uncertainties, the water withdrawal is expected to increase in the HHH region under different scenarios (Liao et al., 2016; Zhang et al., 2016). In this study, the assumption that the thermoelectric cooling water withdrawal is static could underestimate the water supply stress. Moreover, in October 2015, China implemented the universal two-child policy to replace the one-child policy. The new policy could affect future demographics and increase the GDP, yet its possible impacts are not well understood (Zeng & Hesketh, 2016; Huang et al., 2018b). Therefore, the 'Family Planning Policy' change makes it harder to accurately estimate domestic and industrial water withdrawals.

In this chapter, we took the SNWD project as an example to investigate the impact of the large scale inter-basin water transfer on water scarcity in water-receiving areas within the context of climate change, and discuss the future effectiveness of the SNWD within the context of improving irrigation water-use. The results show that during the twenty-first century, climate change is expected to cause available water supply increases in the HHH region, meanwhile it is also expected to cause an increase in irrigation water withdrawal. Societal changes, especially economic growth, are the major contributors to the water scarcity aggravation during the twenty-first century. The SNWD project can greatly alleviate these difficulties but might be insufficient in some cases. Besides, to keep pace with escalating demands and completely alleviate the water supply problems, better local water management, such as measures aiming at improving the efficiency of irrigation water use, must be undertaken in the HHH region.

Note

1 Website last accessed 14 August 2019.

References

Alam, M. (2016). Evaluating the benefit-cost ratio of groundwater abstraction for additional irrigation water on global scale. *Degree Project in Environmental Engineering*, Second Cycle, 30 Credits, Stockholm, Sweden.

Alcamo, J., Dŏll, P., Henrichs, T., et al. (2003). Global estimates of water withdrawals and availability under current and future 'business-as-usual' conditions. *Hydrological Sciences Journal* 48(3): 339–348.

Bondeau, A., Smith, P., Zaehle, S., et al. (2007). Modelling the role of agriculture for the 20th century global terrestrial carbon balance. *Global Change Biology* 13(3): 679–706.

CAB-SNWDP (Construction and Administration Bureau of South to North Water Diversion Project, Ministry of Water Resources of China) (2003). Introduction of south to north water diversion project's plan. *China Water Resources* B: 56–62 (in Chinese).

Cosgrove, W., & Loucks, D. (2015). Water management: Current and future challenges and research directions. *Water Resources Research* 51(6): 4823–4839.

Dziegielewski, B., Sharma, S. C., Bik, T. J., Margono, H., & Yang, X. (2002). *Analysis of Water Use Trends in the Unites States: 1950–1995*. Special Report 28. Urbana-Champaign, IL: Illinois Water Resources Center, University of Illinois.

Eldardiry, H., Habib, E., & Borrok, D. (2016). Small scale catchment analysis of water stress in wet regions of the U.S.: An example from Louisiana. *Environmental Research Letters* 11(12): 124031.

Fader, M., Rost, S., Müller, C., Bondeau, A., & Gerten, D. (2010). Virtual water content of temperate cereals and maize: Present and potential future patterns. *Journal of Hydrology* 384(3–4): 218–231.

FAO (Food and Agriculture Organization of the United Nations) (2003). *Water Report (Book 23): Review of World Water Resources by Country*. Rome: FAO.

Flörke, M., Kynast, E., Bärlund, I., Eisner, S., Wimmer, F., & Alcamo, J. (2013). Domestic and industrial water uses of the past 60 years as a mirror of socio-economic development: A global simulation study. *Global Environment Change* 23(1): 144–156.

Fu, J. Y., Jiang, D., & Huang, Y. H. (2014). 1 km grid population dataset of China (2005, 2010). *Acta Geographica Sinica* 69(Suppl.): 136–139.

Gaffin, S. R., Rosenzweig, C., Xing, X. S., & Yetman, G. (2004). Downscaling and geo-spatial gridding of socio-economic projections from the IPCC Special Report on Emissions Scenarios (SRES). *Global Environmental Change* 14(2): 105–123.

Gain, A. K., Giupponi, C., & Wada, Y. (2016). Measuring global water security towards sustainable development goals. *Environmental Research Letters* 11(12): 124015.

GAQUIQ and SAC (General Administration of Quality Supervision, Inspection and Quarantine of the People's Republic of China, and Standardization Administration of the People's Republic of China) (2009). GB/T 23598-2009, Code of practice for water resources bulletin (in Chinese).

Gu, H., Yu, Z., & Wang, J. (2014). Future extreme climates projection over Huang-Huai-Hai region of China. *Advanced Materials Research* 955: 3887–3892.

Gupta, J., & van der Zaag, P. (2008). Interbasin water transfers and integrated water resources management: Where engineering, science and politics interlock. *Physics and Chemistry of the Earth* 33(1–2): 28–40.

Haddeland, I., Heinke, J., Biemans, H., et al. (2014). Global water resources affected by human interventions and climate change. *Proceedings of the National Academy of Sciences (USA)* 111(9): 3251–3256.

Hanasaki, N., Fujimori, S., Yamamoto, T., et al. (2013). A global water scarcity assessment under shared socio-economic pathways – Part 1: Water use. *Hydrology and Earth System Sciences* 17(7): 2375–2391.

Hanasaki, N., Kanae, S., Oki, T., et al. (2008a). An integrated model for the assessment of global water resources – Part 1: Model description and input meteorological forcing. *Hydrology and Earth System Science* 12(4): 1007–1025.

Hanasaki, N., Kanae, S., Oki, T., et al. (2008b). An integrated model for the assessment of global water resources – Part 2: Applications and assessments. *Hydrology and Earth System Science* 12(4): 1027–1037.

Hanasaki, N., Yoshikawa, S., Pokhrel, Y., & Kanae, S. (2018). A global hydrological simulation to specify the sources of water used by humans. *Hydrology and Earth System Science* 22(1): 789–817.

Hattermann, F. F., Krysanova, V., Gosling, S., et al. (2017). Cross-scale intercomparison of climate change impacts simulated by regional and global hydrological models in eleven large river basins. *Climatic Change* 141: 561–576.

Hawkins, E., & Sutton, R. (2009). The potential to narrow uncertainty in regional climate predictions. *Bulletin of the American Meteorological Society* 90(8): 1095–1107.

Hawkins, E., Osborne, T. M., Ho, C. K., & Challinor, A. J. (2013). Calibration and bias correction of climate projections for crop modelling: An idealized case study over Europe. *Agricultural and Forest Meteorology* 170: 19–31.

High-level Panel on Water (2018). *Making Every Drop Count: An Agenda for Water Action: High-Level Panel on Water Outcome Document*. New York: High-Level

Panel on Water, United Nations Division for Sustainable Development (UN DESA), 14 March 2018.

Ho, C., Stephenson, S., Collins, M., Freeo, C., & Brown, S. (2012). Calibration strategies: A source of additional uncertainty in climate change projections. *Bulletin of the American Meteorological Society* 93(1): 21–26.

Hoekstra, A. (2014). Water scarcity challenges to business. *Nature Climate Change* 4: 318–320.

Huang, J., Qin, D., Jiang, T., et al. (2018a). Effect of fertility policy changes on the population structure and economy of China: From the perspective of the Shared Socioeconomic Pathways. *Earth's Future* 7(3): 250–265.

Huang, Y. H., Jiang, D., & Fu, J. Y. (2014). 1 km grid GDP data of China (2005, 2010). *Acta Geographica Sinica* 69(Suppl.): 140–143.

Huang, Z., Hejazi, M., Li, X., et al. (2018b). Reconstruction of global gridded monthly sectoral water withdrawals for 1971–2010 and analysis of their spatiotemporal patterns. *Hydrology and Earth System Sciences* 22(4): 2117–2133.

Hughes, B. B. (2005). *UNEP GEO4 Diver Scenarios (fifth draft)*. Denver, CO: Josef Korbel School of International Studies, University of Denver.

Intergovernmental Panel on Climate Change (IPCC) (2014). *Summary for policymakers Climate Change 2014: Impacts, Adaptation, and Vulnerability: A. Global and Sectoral Aspects, Contribution of Working Group II to the Fifth Assessment Report of the Intergovernmental Panel on Climate Change*, ed. C. B. Field, V. R. Barros, D. J. Dokken, et al. Cambridge: Cambridge University Press.

Kong, X., Lal, R., Li, B., et al. (2014). Fertilizer intensification and its impacts in China's HHH plains. In D. L. Sparks (ed.), *Advances in Agronomy* (pp. 135–169). Oxford: Elsevier Science & Technology.

Kummu, M., Ward, P. J., de Moel, H., & Varis, O. (2010). Is physical water scarcity a new phenomenon? Global assessment of water shortage over the last two millennia. *Environmental Research Letters* 5: 034006.

Li, W., Sankarasubramanian, A., Ranjithan, R. A., & Brill, E. D. (2014). Improved regional water management utilizing climate forecasts: An interbasin transfer model with a risk management framework. *Water Resources Research* 50(8): 6810–6827.

Liao, X., Hall, J., & Eyre, N. (2016). Water use in China's thermoelectric power sector. *Global Environmental Change* 41: 142–152.

Liu, C., & Zhang, H. (2002). South-to-north water transfer schemes for China. *Water Resources Development* 18(3): 453–471.

Liu, Y., Long, H., Li, T., & Tu, S. (2015). Land use transitions and their effects on water environment in Huang-Huai-Hai plain, China. *Land Use Policy* 47: 293–301.

Lu, G., Xiao, H., Wu, Z., Zhang, S., & Li, Y. (2013). Assessing the impacts of future climate change on hydrology in Huang-Huai-Hai region in China using the PRECIS and VIC models. *Journal of Hydrologic Engineering* 18(9): 1077–1087.

Ma, J. (2004). *China's Water Crisis*. Norwalk, CT: EastBridge.

Milly, P., Betancourt, J., Falkenmark, M., et al. (2008). Stationarity is dead: Whither water management? *Science* 319(5863): 573–574.

Monfreda, C., Ramankutty, N., & Foley, J. A. (2008). Farming the planet, part 2: The geographic distribution of crop areas and yields in the year 2000. *Global Biogeochemical Cycles* 22(1): GB1022.

MWR (Ministry of Water Resources) (2002). *South–North Water Transfer Project Masterplan (Summary)*. Beijing: Ministry of Water Resources (in Chinese).

MWRC (Ministry of Water Resources of the People's Republic of China) (2011). *National Integrated Water Resources Planning of China (2000–2030)*. Beijing: Ministry of Water Resources (in Chinese).

NBSC (National Bureau of Statistics of China) (2011). *China Statistical Yearbook 2010*. Beijing: China Statistics Press (in Chinese).

NDRC (National Development & Reform Commission) (2007). *China's National Climate Change Programme* [English version]. Beijing: National Development & Reform Commission.

O'Neill, B., Kriegler, E., Ebi, K., et al. (2017). The roads ahead: Narratives for shared socioeconomic pathways describing world futures in the 21st century. *Global Environmental Change* 42: 169–180.

Office of the South-to-North Water Diversion Project Construction Committee, State Council (PRC) (2016). The south-to-north water diversion project. *Engineering* 2 (3): 265–267.

Pittock, J., Meng, J., & Chapagain, A. (2009). *Interbasin Water Transfers and Water Scarcity in a Changing World – A Solution or a Pipedream?* Germany: World Wide Fund (WWF).

Portmann, F. T., Siebert, S., & Döll, P. (2010). MIRCA2000 – Global monthly irrigated and rainfed crop areas around the year 2000: A new high-resolution data set for agricultural and hydrological modeling. *Global Biogeochemical Cycles* 24(1): GB1011.

Rajagopalan, B., Nowak, K., Prairie, J., et al. (2009). Water supply risk on the Colorado River: Can management mitigate? *Water Resources Research* 45(8): W08201.

Rost, S., Gerten, D., Bondeau, A., Lucht, W., Rohwer, J., & Schaphoff, S. (2008). Agricultural green and blue water consumption and its influence on the global water system. *Water Resources Research* 44(9): W09405.

Schewe, J., Heinke, J., Gerten, D., et al. (2014). Multimodel assessment of water scarcity under climate change. *Proceedings of the National Academy of Sciences (USA)* 111 (9): 3245–3250.

Shi, Y. C. (2003). Comprehensive reclamation of salt-affected soils in China's Huang-Huai-Hai plain. In S. S. Goyal, S. K. Sharma, & D. W. Rains (eds.), *Crop Production in Saline Environments: Global and Integrative Perspectives* (pp. 163–179). New York: Food Products Press.

Shiklomanov, I. (2000). Appraisal and assessment of world water resources. *Water International* 25(1): 11–32.

Shumilova, O., Tockner, K., Thieme, M., Koska, A., & Zarfl, C. (2018). Global water transfer megaprojects: A potential solution for the water-food-energy nexus? *Frontiers in Environmental Science*, 6: 150.

Siebert, S., & Döll, P. (2010). Quantifying blue and green virtual water contents in global crop production as well as potential production losses without irrigation. *Journal of Hydrology* 384(3–4): 198–217.

Siebert, S., Döll, P., Feick, S., Hoogeveen, J., & Frenken, K. (2007). *Global Map of Irrigation Areas Version 4.0.1*. Frankfurt am Main: Institute of Physical Geography, University of Frankfurt.

Sun, G., McNulty, S. G., Myers, J. A. M., & Cohen, E. C. (2008). Impacts of multiple stresses on water demand and supply across the southeastern United States. *Journal of the American Water Resource Association* 44(6): 1441–1457.

Tian, Z., Shi, J., Gao, Z., & Tubiello, F. N. (2008). Assessing the impact of future climate change on wheat production in Huang-Huai-Hai Plain in China based on GIS and crop model. *Remote Sensing and Modeling of Ecosystems for Sustainability*, 7083.

UN Water (United Nations Water). (2017). Level of water stress: Freshwater withdrawal in percentage of available freshwater resources (Version 19 January 2017). Integrated Monitoring Guide for SDG6: Step-by-step monitoring methodology for indicator

6.4.2. Available from www.unwater.org/publications (Last accessed 17 August 2019).

Verma, S., Kampman, D., der Zaag, P., & Hoekstra, A. (2009). Going against the flow: A critical analysis of inter-state virtual water trade in the context of India's National River Linking Program. *Physics and Chemistry of the Earth* 34(4–5): 261–269.

van Vuuren, D., Kriegler, E., O'Neill, B., et al. (2014). A new framework for climate change research: Scenario matrix architecture. *Climatic Change* 122(3): 373–386.

Wada, Y., Flörkem, M., Hanasaki, N., et al. (2016). Modeling global water use for the 21st century: The Water Futures and Solutions (WFaS) initiative and its approaches. *Geoscientific Model Development*, 9(1): 175–222.

Wada, Y., van Beek, L., van Kempen, C., Reckman, J., Vasak, S., & Bierkens, M. (2010). Global depletion of groundwater resources. *Geophysical Research Letters* 37: L20402.

Wada, Y., van Beek, L., Viviroli, D., Duee, H. H., Weingartner, R., & Bierkens, M. (2011). Global monthly water stress: 2 Water demand and severity of water stress. *Water Resources Research*, 47(7): W07518.

Wada, Y., Wisser, D., & Bierkens, M. F. P. (2014). Global modeling of withdrawal, allocation and consumptive use of surface water and groundwater resources. *Earth System Dynamics* 5(1): 15–40.

Warszawski, L., Frieler, K., Huber, V., Piontek, F., Serdeczny, O., & Schewe, J. (2014). The inter-sectoral impact model intercomparison projection (ISI-MIP): Project framework. *Proceedings of the National Academy of Sciences (USA)* 111(9): 3228–3232.

Wei, C. (2000). *South to North Water Transfer Project in China*. Beijing: China Agriculture Press (in Chinese).

WWAP (World Water Assessment Programme) (2009). *The United Nations World Water Development Report 3: Water in a Changing World*. Paris: UNESCO Publishing.

Xia, J. (2012). Climate change impact on water security and adaptive management in China: Introduction. *Water International* 37(5): 509–511.

Xia, J., Qiu, B., & Li, Y. Y. (2012). Water resources vulnerability and adaptive management in the Huang, Huai and Hai river basins of China. *Water International* 37(5): 523–536.

Yin, Y., Tang, Q., Liu, X., & Zhang, X. (2017). Water scarcity under various socio-economic pathways and its potential effects on food production in the Yellow River basin. *Hydrology and Earth System Sciences* 21(2): 791–804.

Zeng, Y., & Hesketh, T. (2016). The effects of China's universal two-child policy. *Lancet* 388(10054): 1930–1938.

Zhang, C., & Anadon, L. D. (2013). Life cycle water use of energy production and its environmental impacts in China. *Environmental Science and Technology* 47(24): 14459–14467.

Zhang, C., Zhong, L., Fu, X., Wang, J., & Wu, Z. (2016). Revealing water stress by the thermal power industry in China based on a high spatial resolution water withdrawal and consumption inventory. *Environmental Science and Technology* 50(4): 1642–1652.

Zhang, Q., Xu, Z., Shen, Z., Li, S., & Wang, S. (2009). The Han River watershed management initiative for the South-to-North Water Transfer project (middle route) of China. *Environmental Monitoring and Assessment* 148: 369–377.

Zhang, X. (1999). *South to North Water Transfer Project: Supporting Project for China's Sustainable Development*. Beijing: China Water & Power Press (in Chinese).

13

Broadening and Deepening the Rainfall-Induced Landslide Detection

Practices and Perspectives at a Global Scale

GUOQIANG JIA, QIUHONG TANG, STEFANO LUIGI GARIANO, MASSIMO MELILLO, XIMENG XU, GUOYONG LENG AND XU LI

13.1 Introduction

Landslides cause significant damage to societies and the environment globally (e.g., Guzzetti et al., 1999; Dai et al., 2002; Kirschbaum et al., 2010; Froude & Petley, 2018; Haque et al., 2019). According to the available data, the United Nations Office for Disaster Risk Reduction (UNDRR, 2019) reported a worldwide incidence of landslide in 104 countries, causing 39,922 deaths (i.e., 1,559 per year) during 1997–2017. However, this figure is not complete due to unreported or misrelated deaths (Froude & Petley, 2018). A newly updated Global Fatal Landslide Database (GFLD; Petley, 2019) recorded 64,218 deaths (i.e., 4,587 per year) in 2004–2017. Haque et al. (2019) addressed an increasing trend of deadly landslide occurrence during 1995–2014 in 128 countries, and showed a larger number of casualties: averaging 8,877 people killed per year by primary and secondary landslide hazards. With the increasing availability of global datasets, scientists are now trying to quantify landslide hazards and risks at the global scale. Pioneering research work can be dated back to the first worldwide landslide dataset compiled by the International Association of Engineering Geology (IAEG) Commission on Landslides in 1971 (UNESCO, 1973). Since then, important efforts made by the research communities have greatly contributed to global assessments of landslide hazards (e.g., Caine, 1980; Sassa, 2004; Hong et al., 2006; Guzzetti et al., 2008; Kirschbaum et al., 2010; Gariano & Guzzetti, 2016; Kirschbaum & Stanley, 2018; Benz & Blum, 2019; Evans & Delaney, 2019). Specifically, rainfall-induced landslides are the main type of landslide disasters

This research is supported by the National Natural Science Foundation of China (41730645), the Strategic Priority Research Program of the Chinese Academy of Sciences (XDA20060402), and Newton Advanced Fellowship. We are grateful to Froude and Petley (https://blogs.agu.org/landslideblog/) and the National Aeronautics and Space Administration landslide team (https://pmm.nasa.gov/landslides/index.html) for their efforts on the free available global landslide datasets, and groups and individuals for other available environmental datasets. We thank colleagues from the Research Institute for Hydrogeological Protection (IRPI), National Research Council (CNR), Italy for their kind support.

since most landslides are initiated by short-term downpours or long-term drizzles (e.g., Hong et al., 2006).

The reasons for a global assessment of landslide hazards and risks are manifold. First, a global assessment is a way to illustrate the spatial patterns of possible landslide occurrence, given the fact that patterns of landslide occurrence vary with diverse geological and environmental factors (Guzzetti et al., 2007; Ma et al., 2015; Peruccacci et al., 2017; Palladino et al., 2018). Second, a large-scale (global) assessment provides opportunities for different regions to compare and combine their efforts as a uniform reference, and can be used to identify better strategies for disaster mitigation and reduction (Guzzetti et al., 2008; Van Den Eeckhaut et al., 2012; Broeckx et al., 2018). Third, a global assessment is a primary proxy for data-scarce regions and it is easier to focus on areas of interest when detailed regional information is available. By comparison, a local or regional evaluation is always geologically limited and cannot be applied to other areas (e.g., Guzzetti et al., 1999; Harp et al., 2009).

Attempts have been made to detect rainfall-induced landslides by establishing and explaining the relationship among landslide occurrence and contributing factors, which are grouped as the preparatory environmental 'cause' and dynamic meteorological 'trigger' (Dai et al., 2002; Bogaard & Greco, 2018). In this context, Caine (1980) proposed the first global empirical model (rainfall mean intensity-duration threshold) for landslide occurrence by linking triggering rainfall conditions and landslide occurrence. Then, Nadim et al. (2006) first mapped the global landslide susceptibility by relating influencing environmental factors and landslide occurrence. These studies were followed by different rainfall thresholds (reviewed in Guzzetti et al., 2008; Segoni et al., 2018) and susceptibility maps (e.g., Hong et al., 2007b; Farahmand & Aghakouchak, 2013; Reichenbach et al., 2018) based on different landslide types, methods and datasets. Using a framework that combines susceptibility and rainfall thresholds of landslide occurrence for comprehensive landslide detection (Hong et al., 2007a; Kirschbaum et al., 2009), Kirschbaum and Stanley (2018) launched a real-time prediction system by combining rainfall thresholds derived from satellite-based rainfall data and susceptibility obtained from high-resolution environmental factors.

Existing versions of global modelling provide insights into the relationship between contributing factors and landslide occurrence to some extent. Rainfall thresholds and susceptibility have been two useful tools for global landslide detection. However, it is still challenging to generalize the physical process behind slope instability at the global scale. First, susceptibility analysis provides no temporal detection of landslide occurrence. Then, rainfall thresholds for landslide occurrence are usually determined empirically through a power-law equation (Guzzetti et al.,

2008) or quantile values of antecedent rainfall index (Kirschbaum & Stanley, 2018) and uniquely at a regional or global scale without considering their spatial differences (Guzzetti et al., 2007). Many studies reveal that environmental factors impact rainfall thresholds by influencing rainfall and landslide patterns both spatially and temporally (e.g., Caine, 1980; Guzzetti et al., 2008; Peruccacci et al., 2017; Palladino et al., 2018), and these impacts are still not depicted well, though some studies have proposed a two-threshold framework of susceptibility and rainfall index (Kirschbaum & Stanley, 2018; Monsieurs et al., 2018).

In this chapter, we first review the main progress made over the past 40 years in the detection practices of global rainfall-induced landslide (Section 13.2). We build knowledge from landslide datasets compilation, susceptibility analysis and triggering threshold identification. In Sections 13.3 and 13.4, we suggest an approach to fill the literature gap by proposing a distributed global rainfall threshold model for rainfall-induced landslide detection by including the effect of environmental factors on rainfall thresholds. Finally, we discuss our results and the implications for global rainfall-induced landslide detection in Section 13.5.

13.2 Practices for Global Rainfall-Induced Landslide Detection

13.2.1 Landslide Datasets Compilation

Getting a dataset that provides reliable and accurate information on a large number of landslide events is necessary for a rigorous detection of landslide occurrence in large areas (e.g., Intrieri et al., 2013). Spatial maps of occurrence sites highlight clusters of landslides, from which areas of high incidence or hotspot areas can be detected (Kirschbaum et al., 2015; Froude & Petley, 2018). Landslide datasets are also vital to modelling performance evaluation (Kirschbaum et al., 2009; Kirschbaum & Stanley, 2018; Jia et al., 2020).

The first effort to gather global landslide events data can be dated back to the 1970s, when the IAEG documented 28 severe events, which mainly occurred in Japan, Colombia, Peru, Canada and Europe, in 1971 (UNESCO, 1973). Notably, 240 people died in a highly lethal event that occurred in Chungar, Peru on 18th March 1971. Caine (1980) compiled the first rainfall-induced landslide dataset from published literature, containing 73 events with their corresponding rainfall conditions during 1957–1979. Since then, small datasets have been proposed, however the limitations of their spatial domains do not allow for global assessment. Guzzetti et al. (2008) updated Caine's work by establishing a large rainfall-induced landslide dataset with 2,626 records from 1917 to 2005. The dataset included detailed fields such as occurrence location and accuracy, landslide types, climate and triggering rainfall information. The efforts toward a more accurate and larger dataset began with studies by Petley et al. (2005) and Kirschbaum et al., (2010). The authors improved the completeness and representativeness of the dataset and introduced the estimation of location accuracy. Consequently, the Global Landslide Catalogue (GLC) and Global Fatal Landslide Database (GFLD) were created (in Figure 13.1; Kirschbaum et al., 2015; Petley, 2019). Following similar compiling strategies, another dataset of deadly landslides was launched by Haque et al. (2019), which considered a longer time period (20 years). They used strict cross-checks with available international and country-specific sources.

The main information for compilations of a global landslide include: (1) the features or types of landslides; (2) the types of information source (e.g., English newspapers and media, government and organization reports, academic papers and hazard databases); (3) the location and date of occurrence: the location is always available as a description of administrative unit instead of latitude and longitude. Thus, some map services, such as Google Earth and ArcGIS, are used for estimation and cross-check. Date information is mainly obtained from the original sources; (4) estimation of location precision: spatial accuracy is derived from location description of landslide event reports or polygon areas of administrative units are used; (5) the triggers and (6) casualties and economic loss (Kirschbaum et al., 2010, 2015; Petley 2012; Haque et al., 2016, 2019; Froude & Petley, 2018). Detailed information is displayed in Table 13.1.

13.2.2 Landslide Susceptibility Analysis

Landslide susceptibility is considered as a spatial probability index of landslide occurrence, depicting where there is the likelihood of a landslide occurrence in an area on the basis of local terrain conditions (e.g., Guzzetti et al., 2005; Reichenbach et al., 2018). This index assumes that the environmental causes are static over a long period. Early global landslide susceptibility maps were created empirically, such as in Nadim et al. (2006) and Hong et al. (2007b). They assigned susceptibility values for each contributing factor and used an empirical relationship to estimate their global effects on susceptibility. The works of Nadim et al. (2006) and Hong et al. (2007b) were updated with new and high-resolution datasets by Giuliani and Peduzzi (2011; https://preview.grid.unep.ch/index.php) and by Stanley and Kirschbaum (2017; https://pmm.nasa.gov/applications/global-landslide-model), respectively.[1] Furthermore, they benefitted from the application of statistical method and landslide information. Farahmand and AghaKouchak (2013) introduced an inventory-based landslide model to link landslide and non-landslide events and their contributing factors by using a support vector machine (SVM)

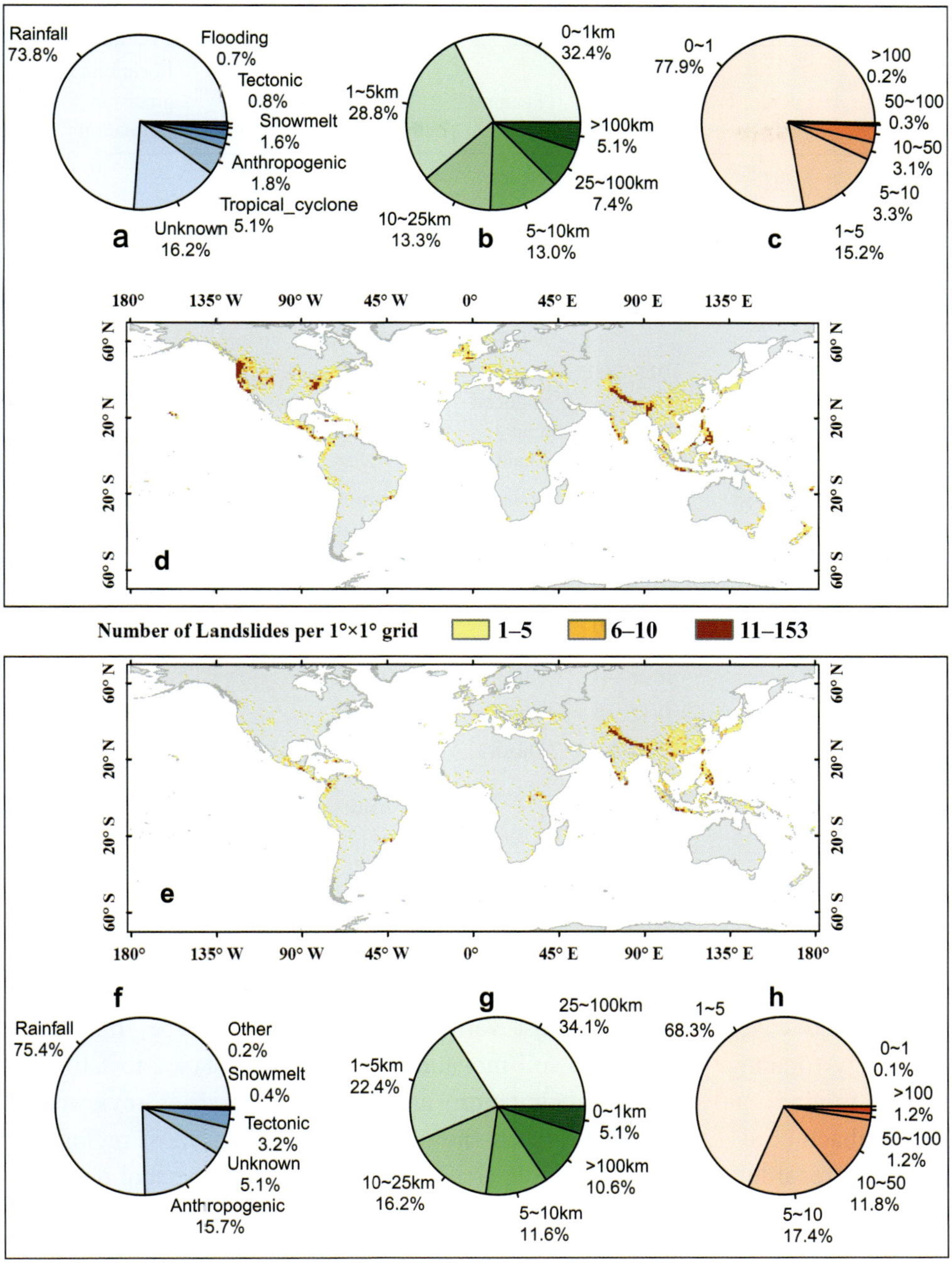

Figure 13.1 Spatial distribution and statistics of reported landslide occurrence in the Global Landslide Catalogue (GLC, a–d; 10,988 records in 2007–2017) and Global Fatal Landslide Database (GFLD, e–h; 5,490 records in 2004–2017). Spatial density (number of landslides in each $1^\circ \times 1^\circ$ grid; d, e), percentage of different triggers (a, f), location accuracies (b, g) and fatalities (c, h) are displayed

Table 13.1. Summary of the existing global landslide datasets

References	Records (#)	Data period	Feature/ type	Fields	Location precision estimation
UNESCO (1973)	28	1971	Severe	Occurrence location, date, terrain, characteristics, cause, damage	–
Caine (1980)	73	1957–1979	Rainfall-induced	Rainfall, duration, intensity	–
Guzzetti et al. (2008)	2,626	1917–2005	Rainfall-induced and shallow	Location, date, type, rainfall duration and intensity, climate	Geographical information (site, village or town, km^2)
Kirschbaum et al. (2015)	10,988	2007–2017	Rainfall-induced	Location, date, trigger, type, size, location accuracy, fatality	Information available in landslide reports (km)
Froude and Petley (2018)	5,490	2004–2017	Fatal	Location, date, trigger, location accuracy, fatality	Area of administrative units (villages or states, km^2)
Haque et al. (2019)	3,876	1995–2014	Deadly	Location, date, trigger, location accuracy, casualties, economic loss	Rough descriptions in reports or publications (km)

discriminator. Lin et al. (2017) proposed a logistic regression model with similar predictive performance compared to existing regional and global susceptibility maps. Inspection of the environmental factors, morphology (slope, elevation and relief), land cover (including forest variation), geology (soil type, lithology), hydrology (drainage density and soil moisture), seismicity (distance to fault and ground motion) and climate (precipitation) are commonly used, as reviewed in Reichenbach et al. (2018). These maps show similar outputs in most regions, in particular regarding highly susceptible areas, and are helpful for spatial detection of possible landslide occurrence (detailed in Table 13.2).

13.2.3 Triggering Threshold Identification

While susceptibility can be viewed as the spatial likelihood of landslide occurrence, the introduction of time-variant rainfall data adds a temporal

Table 13.2. Summary of existing global susceptibility maps

References	Environmental factors	Methods	Landslide information
Nadim et al. (2006)	Slope, lithology, soil moisture, precipitation, seismicity	Empirical classification and equation	~300 landslides for validation
Hong et al. (2007b)	Slope, soil type, soil texture, elevation, land cover, drainage density	Empirical classification and linear combination	No
Stanley and Kirschbaum (2017)	Slope, distance to fault, presence of road, forest loss, geology	Heuristic fuzzy approach	9 datasets for validation
Farahmand and AghaKouchak (2013)	Precipitation, slope, land cover	Support Vector Machines (SVM)	581 landslides (the total) for training and validation
Lin et al. (2017)	Relief, precipitation, lithology, ground motion, soil moisture	Logistic regression	2,005 landslides (the total) for training and validation

dimension to the prediction of landslide modelling. Rainfall thresholds have been a simple and practical tool for landslide modelling since the first global rainfall mean intensity-duration (*ID*) threshold for shallow landslides and debris flows was proposed by Caine (1980), followed by *ID* thresholds for debris flows (Innes 1983), soil slip (Clarizia et al., 1996) and shallow landslides (Crosta & Frattini 2000). Guzzetti et al. (2007) reviewed the classification, fitting method and uncertainties of existing rainfall thresholds at various scales, introduced global *ID* thresholds of different percentiles by using a Bayesian statistical approach and proposed normalized *ID* thresholds that are independent of the climatic effect (Guzzetti et al., 2008). Finally, the new global *ID* threshold may be considered as a minimum threshold for all the existing thresholds, so reducing the number of missing alarms. These thresholds are computed by using gauged rainfall data, whereas satellite-based rainfall thresholds have been proposed to test the potential application of satellite-based rainfall data for real-time landslide detection (Hong et al., 2006; Jia et al., 2020).

Rainfall thresholds have been combined with susceptibility to conduct a comprehensive landslide prediction. Hong et al. (2007a) used the global *ID* threshold and landslide susceptibility as two criteria of possible landslide detection. Kirschbaum and Stanley (2018) introduced rainfall thresholds defined by a seven-day antecedent rainfall index obtained from a satellite-based rainfall data and applied together with a high-resolution susceptibility map in a real-time landslide prediction system.

Along with landslide dataset compilation and methods of landslide detection, the evaluation of model performance is crucial. Here, we merely addressed the achievements and limitations of landslide detection practices as a preliminary analysis and support to our distributed global rainfall thresholds for landslide occurrence.

13.3 Data and Methods

13.3.1 Environmental Factors and Landslide Datasets

To establish global distributed rainfall thresholds, the identification of robust links between triggering rainfall and environmental factors is required. It is a complex and challenging task including various contributing factors such as morphology, hydrology, geology, land cover and climate (e.g., Dai et al., 2002; Hong et al., 2007b; Reichenbach et al., 2018). The first step is the preparation of reliable input datasets.

Owing to recent advances in remote sensing, newly updated and reliable global datasets with high resolution are available. Here, we used numeric variables instead of categorical variables that are used in susceptibility analysis, to explore the quantitative relationship between the triggering rainfall thresholds and preparatory environmental factors. For land cover data, the percentage of covered area for each category in each pixel is used (0~100 per cent; maps of dominant class and grassland area percentage in each pixel are shown in Figure 13.2(c) and (d), respectively), while the percentage of silt and clay in subsoil (30–100 cm) is used as a proxy of different soil class (noted as 0~32,050, soil unit identifier, representing different soil classes; maps of soil class and clay percentage in each pixel are shown in Figure 13.2(e) and (g), respectively). Datasets information is summarized in Table 13.3 and the spatial distribution of the investigated variables is shown in Figure 13.2. Two precipitation datasets are included, of which mean annual precipitation (*MAP*) in 1970–2000 is used as a preparatory environmental variable related to climate conditions from the WorldClim2 datasets (Fick & Hijmans, 2017) and data series of half-hourly rainfall are used to calculate triggering rainfall for reported landslide events from the Global Precipitation Measurement (GPM) rainfall datasets. According to the available datasets, the modelling scale in our analysis is set as 0.00833° (~1 km), and the study area ranges from 62°N to 55.7°S and 138°W to 180°E, which has been determined according to global landmasses and data availability. River density was resampled at this resolution.

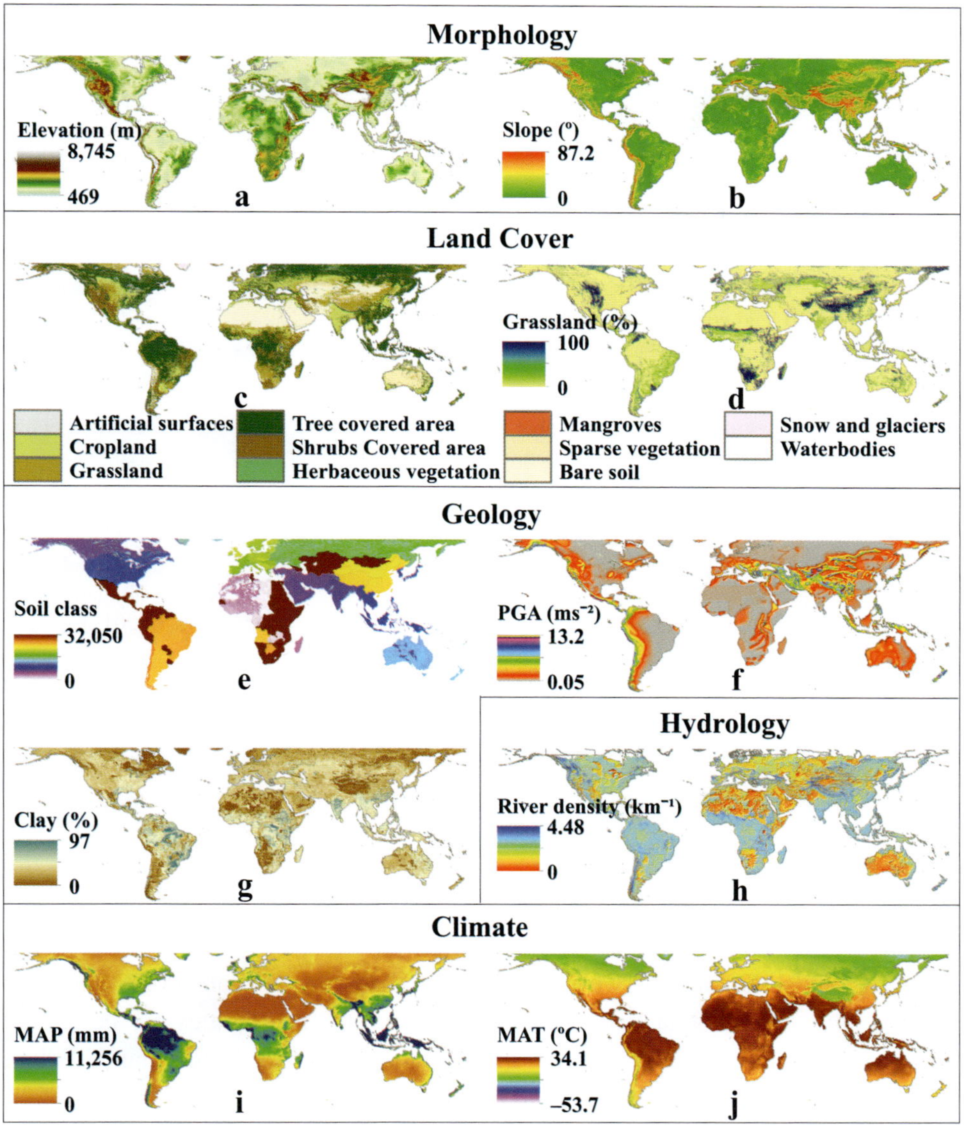

Figure 13.2 Global distribution of environmental factors for landslide detection used in this chapter. (a) elevation; (b) slope; (c) land cover; (d) percentage of grassland; (e) soil class; (f) peak ground acceleration (PGA); (g) percentage of clay; (h) river density; (i) mean annual precipitation (*MAP*) and (j) temperature (*MAT*). For land cover data, the percentage of each land cover type is used for analysis; for the soil dataset, soil classes are noted as 0~32,050, based on Harmonized world soil database V1.2 classification. The percentages of silt, clay and sand in subsoil vary with classes, and thus are used as proxies for different soil classes; for the climate dataset, *MAP* and *MAT* derived from weather station data in 1970–2000 are used

Table 13.3. Landslide and environmental factor datasets and variables used in this chapter

Factors	Variables	Datasets	Resolution (km)	References
–	Latitude, longitude	–	–	–
Morphology	Elevation, slope	Global Multi-resolution Terrain Elevation Data 2010	~1	Danielson and Gesch (2011)
Land cover	Area percentage of 10 land cover types	Global Land Cover-SHARE (GLC-share)	~1	Latham et al. (2014)
Hydrology	River density	Global River Network	~12.5	Schneider et al. (2017)
Geology	Peak ground acceleration	Global Seismic Hazard Assessment Programme	~1	Giardini et al. (2003)
	Percentage of two soil structure types	Harmonized world soil database V1.2	~1	FAO et al. (2012)
Climate	Mean annual precipitation, temperature	WorldClim2	~1	Fick and Hijmans (2017)
Precipitation	Half-hourly rainfall	Global Precipitation Measurement IMERG	~10	Huffman et al. (2019)
Landslide	Record	Global Landslide Catalogue	–	Kirschbaum et al. (2015)
		Global Fatal Landslide Database	–	Froude and Petley (2018)

As for landslide datasets, we merged the available state-of-art global landslide datasets and excluded landslide records with high uncertainties. A combined dataset of GLC and GFLD with high-location-accuracy (less than 10 km, corresponding to the spatial resolution of GPM IMERG rainfall data) and rainfall-induced records was obtained. Firstly, 5,905 records in the GLC which occurred in the period 2007–2017 was extracted. Then, 1,211 records in the GFLD were extracted and compared with the GLC records. Only those occurring more than 1 day before or after and more than 10 km away from all GLC records were selected. Finally, we intersected the point data of landslide records with all the environmental variable raster maps and excluded points with no values. The total number of landslide records in the final dataset is 6,128, which was split as training (4,169 records, from 2007 to 2014) and test datasets (1,959 records, from 2015 to 2017).

13.3.2 Triggering Rainfall Identification for Landslide Events

After extracting the environmental variables (raster maps) at the landslide points, we assigned the preparatory causes for each landslide event. In our model, the triggering rainfall events are required as well. Some empirical or practical methods have been used to reconstruct triggering rainfall events, such as field survey or monitoring in landslide datasets compiled with rainfall conditions (Caine, 1980; Guzzetti et al., 2008) and conceptual critical rainfall events for landslide occurrence (Aleotti, 2004). We employed an automatic and ensemble algorithm proposed by Melillo et al. (2015, 2016, 2018). Dry period with no rain is defined as a division of successive individual rainfall events. The triggering rainfall events, including cumulated event rainfall, E, and duration, D (hereinafter denoted as reconstructed E and D), are extracted by searching for the latest rainfall events related to occurrence time of landslide records. A modified version of the algorithm for global satellite-based rainfall data are detailed in Jia et al. (2020). We updated the distributed dry period parameters by using the CRU TS4.03 dataset (University of East Anglia Climatic Research Unit et al., 2020).

13.3.3 Modelling for Global Distributed Rainfall Thresholds

An assumption is that rainfall thresholds for landslide occurrence vary with geo-environmental factors. Modelling pixels in our study can be classified as two types: the one with reported landslide occurrence and the other with no reported landslide occurrence. Thresholds for modelling pixels where landslide occurrence is reported can be defined by the triggering rainfall events related to reported landslides. For example, a reconstructed triggering rainfall event of one landslide record is $D = 98$ h and $E = 102$ mm, which assumed that the triggering conditions for the modelling pixel where the landslide occurs is 102 mm when the duration is 98 hours. We assume that rainfall thresholds defined by cumulated event rainfall is a function of rainfall duration and environmental variables and could be simulated and useful for threshold prediction in modelling pixels where no landslide is reported. Specifically, we proposed the distributed threshold as follows:

$$E = f(D, env_1, env_2, ...) \tag{13.1}$$

where $env_i (i = 1, 2, \ldots)$ represents the environmental variables corresponding to landslide occurrence (in Table 13.3). Previous research revealed a linear relationship of cumulated event rainfall and rainfall duration (ED rainfall threshold) in logarithmic coordinates (e.g., Guzzetti et al., 2008; Jia et al., 2020). Accordingly, we fit Eq. (13.1) with a multiple linear regression model (MLR) by using logarithmic metrics of E and D, i.e.,

$$\log E = f(\log D, env_1, env_2, \ldots) = \alpha \log D + \beta_1 env_1 + \beta_2 env_2 + \cdots \quad (13.2)$$

We trained and tested our models using the R open-source software for advanced statistical computing and graphics, release 3.3.3 (www.r-project.org).[2] For the training, the relevance of geo-environmental variables and E was tested by considering results of the student's test and fitted by using a stepwise method for variable selection. To conduct a performance evaluation, we analysed and assessed mean absolute error (MAE) and coefficient of determination (R^2), for training and testing datasets by comparing values of the reconstructed E (in Section 13.3.2) and MLR modelled E, as shows in Eq. (13.3). Finally, the validated model was used for estimation of global distributed rainfall thresholds for sub-daily and multi-daily rainfall durations.

$$MAE = \frac{1}{n}\sum_{i=1}^{n}|E_i^m - E_i^r|$$
$$R^2 = 1 - \frac{\sum_{i=1}^{n}(E_i^m - E_i^r)^2}{\sum_{i=1}^{n}(E_i^r - \bar{E}^r)^2}, \quad (13.3)$$

where E_i^m and $E_i^r (i = 1, 2, \ldots, n)$ represent the MLR modelled and reconstructed rainfall, E (mm), respectively; n is the total number of records, and $\bar{E}^r$ is the mean value of all the E_i^r.

13.4 Results

13.4.1 Characteristics of Triggering Rainfall Events

Figure 13.3 scatters the reconstructed E and D for the training and testing landslide datasets. The rainfall duration ranges from 0.5 hours to 1,786 hours (~74 days). We tested this output by fitting an empirical rainfall ED threshold based on training data, and a 5 per cent ED frequentist threshold was used (methods in Brunetti et al., 2010 and Peruccacci et al., 2012; algorithm and tool in Melillo et al., 2018). The threshold is found to perform well for testing data, and similar to thresholds proposed by Guzzetti et al. (2008; $E = 2.20D^{0.56}$). Furthermore, we compared the characteristics of triggering rainfall for training and testing data. Empirical cumulative distribution functions ($ECDF$) of reconstructed E and D are almost the same for the two datasets, indicating triggering rainfall conditions for landslide occurrence are relatively stable in the modelling period.

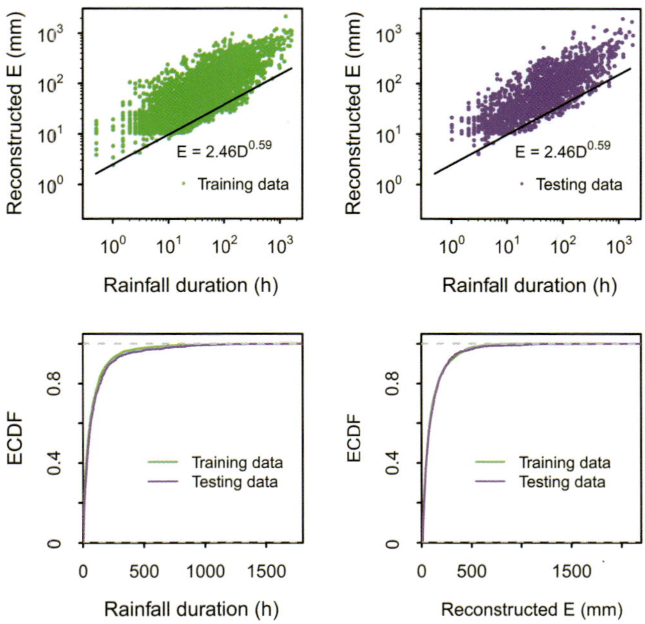

Figure 13.3 Power–law relationship and probability distribution (empirical cumulative distribution function, *ECDF*) of reconstructed cumulate event rainfall (reconstructed *E*) and rainfall duration (*D*) conditions that have triggered landslide occurrence in training and testing datasets

13.4.2 Multiple Linear Regression Model for Global Distributed Rainfall Thresholds

A preliminary analysis indicates that not all the variables are relevant to the initiation of landslide events. For the multiple linear regression method, the significance of student distribution test was evaluated (*p* values in Table 13.4). Then, we used stepwise methods for a better explanatory result and 13 variables were included in the equation (in Table 13.4). The results indicate that rainfall duration contributes greatly to triggering cumulated event rainfall, in line with the common practice of rainfall threshold application for landslide modelling (e.g., Caine, 1980; Guzzetti et al., 2008). Effects of other factors are comparatively small, while coefficients of slope, elevation, river density, percentage of silt, grassland and herbaceous vegetation are negative, suggesting that high values of these variables threaten the slope stability.

Figure 13.4 shows comparisons between reconstructed and MLR modelled *E* (logarithmic metrics, log(mm)) in training, testing and training plus testing datasets. Most values range 1–3 (10 mm–1,000 mm). Values of *MAE* are relatively small, about 0.21 log(mm). Values of R^2 are more than 0.66,

Table 13.4. Summary of multiple linear regression for distributed rainfall thresholds

Variables	Range	Preliminary analysis			Stepwise methods		
		Coefficient	Standard error	Pearson's Coefficient	Coefficient	Standard error	p
D (h)	0~1786	0.55005	0.0071	$<10^{-4}$	0.55035	0.0071	$<10^{-4}$
Latitude (°)	–55.7~62.0	0.00095	0.0003	0.0005	0.00080	0.0003	0.0022
Longitude (°)	–180~180	0.00043	0.0001	$<10^{-4}$	0.00041	0.0001	$<10^{-4}$
Mean annual temperature (°C)	–53.7~34.1	0.00818	0.0012	$<10^{-4}$	0.00739	0.0011	$<10^{-4}$
MAP (mm)	0~11,256	0.00004	0	$<10^{-4}$	0.00005	0	$<10^{-4}$
Slope (°)	0~87.2	–0.00097	0.0002	$<10^{-4}$	–0.00086	0.0002	0.0002
Elevation (m)	469~8745	–0.00001	0	0.1173	–0.00001	0	0.0652
River density (km^{-1})	0~4.48	–0.04915	0.0183	0.0073	–0.04604	0.0181	0.0112
% of silt (%)	0~100	–0.00096	0.0006	0.0910	–0.00094	0.0006	0.0939
% of clay (%)	0~100	0.00094	0.0004	0.0364	0.00109	0.0004	0.0136
% of grassland (%)	0~100	–0.00094	0.0003	0.0032	–0.00084	0.0002	0.0006
% of herbaceous vegetation (%)	0~100	–0.00080	0.0003	0.0061	–0.00068	0.0002	0.0007
% of mangroves (%)	0~100	0.00614	0.0035	0.0773	0.00631	0.0035	0.0688
% of bare soil (%)	0~100	–0.00086	0.0005	0.0933	–	–	–
% of shrubs covered area (%)	0~100	–0.00077	0.0006	0.2149	–	–	–
% of artificial surfaces (%)	0~100	–0.00028	0.0002	0.2538	–	–	–
% of cropland (%)	0~100	–0.00023	0.0003	0.3658	–	–	–
% of sparse vegetation (%)	0~100	0.00045	0.0007	0.4925	–	–	–
% of tree covered area (%)	0~100	0.00008	0.0002	0.7483	–	–	–
% of snow and glaciers (%)	0~100	–0.00085	0.0027	0.7521	–	–	–
Peak ground acceleration (ms^{-2})	0.05~13.2	–0.00026	0.0015	0.8658	–	–	–
(Intercept)	–	0.81351	0.0384	$<10^{-4}$	0.80085	0.0322	$<10^{-4}$

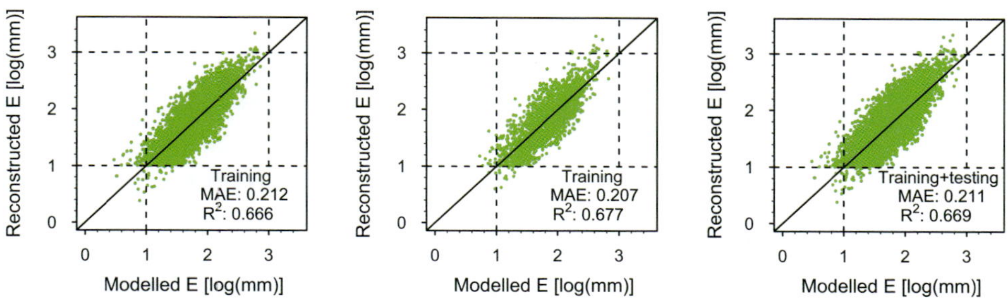

Figure 13.4 Reconstructed triggering rainfall versus modelled triggering rainfall that have caused landslide occurrence in training, testing and training plus testing datasets in logarithmic coordinates

equivalent to Pearson's correlation coefficient over 0.81, suggesting a good performance of the MLR model.

13.4.3 Model Application

We extended the application of the established MLR model around the global land, to predict the cumulated event rainfall as a threshold for landslide occurrence of different rainfall durations in each modelling pixel. Rainfall durations of 0.5, 1, 3, 6, 12 hours, 1, 3, 7, 15 and 30 days were considered (in Figure 13.5). Cumulated event rainfall increases with longer rainfall duration. In sum, high thresholds are detected in the south, southeast and west of Asia, east of America, north of South America, most of north and middle Africa and in the inner part of Oceania. These results coincide with heavy rainfall in the former three regions (in Figure 13.6(a)) and low landslide occurrence in the last two regions (in Figure 13.1(d) and (e)). Low estimated thresholds are observed in the west of North America, west alpine areas of South America, Alps in Europe, south and partly east of Africa, Persian Plateau and Tibetan Plateau (Himalaya) in Asia, New Zealand and partly east of Australia in Oceania, most of which are highly affected by landslides (in Figure 13.1(d) and (e)). Kirschbaum and Stanley (2018) proposed global distributed rainfall thresholds by using a 7-day antecedant rainfall index. The spatial distribution of these global thresholds is consistent with our results. In our results, the thresholds of north Africa are higher and those in Europe are lower than the 7-day antecedant rainfall thresholds in Kirschbaum and Stanley (2018). The reason lies in that their rainfall index is an outcome of 7-day rainfall, which is more related to the spatial distribution of precipitation. Conversely, the rainfall thresholds here defined indicate the critical rainfall conditions of landslide occurrence. Thus, there are high thresholds in north Africa, where landslides are

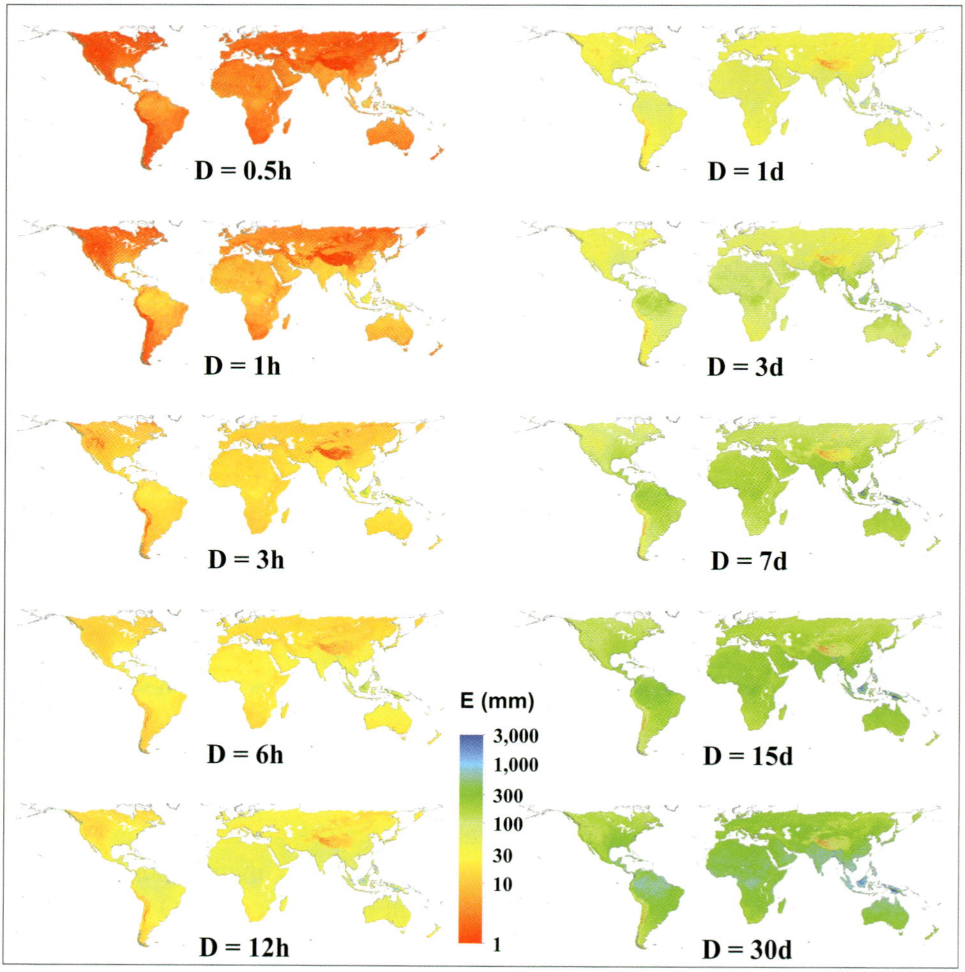

Figure 13.5 Global distributed rainfall thresholds of cumulated event rainfall (*E*) for sub-daily (0.5, 1, 3, 6 and 12 hour) and multi-daily (1, 3, 7, 15 and 30 day) rainfall durations based on a multiple linear regression (MLR) model

not likely to occur, and low thresholds in Europe, which is characterized by high incidence of landslides.

Since the spatial pattern of possible landslide occurrence does not merely depend on the rainfall thresholds at a given duration but also on the rainfall amount or intensity in a given region, we used the ratio of mean monthly precipitation over the distributed rainfall thresholds of 30 days pixel to pixel as a normalized rainfall index for landslide occurrence. This index represents the probability of landslide occurrence (in Figure 13.6(b)). The mean monthly precipitation data are obtained from monthly IMERG data from 2001 to 2018 (in Figure 13.6(a)). For regions where the normalized index is larger than 1.0, the mean monthly precipitation is

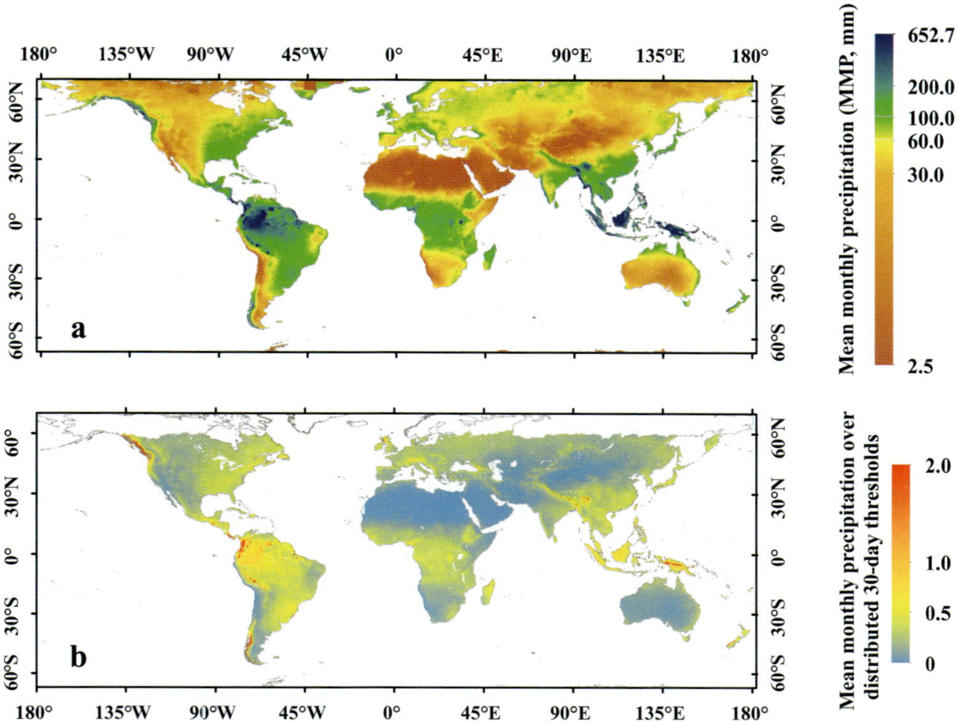

Figure 13.6 Spatial distribution of mean monthly precipitation (a) of GPM IMERG monthly data from 2001 to 2018 and normalized rainfall index for landslide occurrence (b; mean monthly precipitation divided by 30-day distributed thresholds pixel to pixel)

more than the 30-day rainfall thresholds, which indicates a high probability of landslide occurrence in these areas; averagely landslides are likely to occur during half a year (rainfall amount over mean monthly precipitation). For regions where the normalized index is about zero, cumulated rainfall tends to be considerably less than the rainfall thresholds, which implies a low probability of landslide occurrence. We intersected the normalized index map with landslide points in the combined datasets (in Section 13.3.1). About 20 per cent of landslides are reported in regions where the normalized index is larger than 0.5 and about 80 per cent of landslides are reported in regions where the normalized index is larger than 0.2. Thus, a certain threshold of the normalized index can be used as a representation of high and low probability of landslide occurrence.

The following is a practical application process of our model for (near) real-time nowcasting. When the latest rainfall data is available, we can search for backward time periods of 0.5, 1, 3, 6 hours and so on, to calculate the cumulated rainfall for each pixel and then compare the cumulated rainfall maps with our threshold maps related to different rainfall durations. Thus, the pixels whose thresholds are exceeded are defined as threatened areas. The combined mapping of threatened areas for different durations could be used as the final nowcasting.

13.5 Discussions and Conclusion

In the context of population expansion and climate change, landslide modelling and detection are relevant to better understand and predict landslide events and reduce landslide risk. In this chapter, we presented global distributed rainfall thresholds for modelling pixels of high resolution (~1 km^2), based on a relationship between triggering rainfall conditions and environmental factors in pixels where landslide events were recorded. Our models assume that rainfall thresholds are impacted by environmental factors contributing to landslide occurrence and are statistically related with geo-environmental factors. The results are consistent with existing global or regional analysis (e.g., Guzzetti et al., 2007; Peruccacci et al., 2017; Segoni et al., 2018). Moreover, it is true that the theoretical thresholds are below the reconstructed rainfall conditions for each pixel. Practically, we used the reconstructed rainfall conditions as thresholds in pixels for model fitting and validation. Only when more landslide records are available at certain pixels can the thresholds be updated. Improvements are required for this assumption to hold in the context of our research, which would require the availability of more landslide records or changing the study focus on a regional application. This work serves as an attempt to highlight existing empirical or unique thresholds for landslide occurrence, by proposing global distributed rainfall thresholds of different rainfall durations (sub-daily and multi-daily) and potential applications for normalized rainfall index calculation and threatened areas nowcasting.

Further exploration is still required for an operational application. A first potential improvement could be made about the fitting methods. While multiple linear regression models provide a convenient way to link cause and effect, they may obscure a more complex relationship that may exist between landslide occurrence and geo-environmental conditions. One indication about a potential mismatch between the proposed model and the true relationship is suggested by the presence of the intercept (in Table 13.4), behind which what is the underlying mechanism and what are the influencing factors related to the intercept's contribution. As a potential improved modelling framework, non-linear methods, such as support vector regression (SVR) and the artificial neural network (ANN), can accommodate more features of input data. They have been widely used in non-linear regression and statistical analysis. These techniques have also been used in landslide spatial modelling and exhibit good performance (e.g., Ermini et al., 2005; Chen et al., 2017; Huang & Zhao, 2018). Second, the availability of more accurate landslide information could be used for further validation and performance evaluation of the models. In this study, we have benefitted from GFLD and GLC datasets, which resulted from considerable efforts from the research communities, which is continuing to improve the accuracy and coverage of the data, as illustrated by the Cooperative Open Online Landslide Repository (COOLR), a mission recently launched by the NASA landslide team (Juang et al., 2019). Third, one can apply a distributed threshold model on a regional or national scale, through the use of detailed and accurate landslide information for specific regions. Finally, the selection of geo-environmental factors requires robust criteria. We acknowledge the difficulty in operating a complex modelling system on a global scale. Regional or national analyses (e.g., Broeckx et al., 2018) suggest potential new avenues of research, such as the introduction of non-parametric tests and measures of goodness-of-fit. Additionally, forecasting or prediction of potential landslide occurrence in future scenarios could be based on historical simulated rainfall for model fitting, calibration and validation (e.g., Alvioli et al., 2018).

This study serves as a starting point for a better understanding of global rainfall-induced landslide prediction by proposing a modelling framework for distributed rainfall thresholds that considers the interaction between triggering rainfall events and environmental 'cause' for landslide occurrence.

Notes

1 Both websites last accessed 26 December 2019.
2 Website last accessed 29 March 2020.

References

Aleotti, P. (2004). A warning system for rainfall-induced shallow failures. *Engineering Geology* 73(3–4): 247–265.

Alvioli, M., Melillo, M., Guzzetti, F., et al. (2018). Implications of climate change on landslide hazard in Central Italy. *Science of the Total Environment* 630: 1528–1543.

Benz, S. A., & Blum, P. (2019). Global detection of rainfall-triggered landslide clusters. *Natural Hazards and Earth System Sciences* 19(7): 1433–1444.

Bogaard, T., & Greco, R. (2018). Invited perspectives: Hydrological perspectives on precipitation intensity-duration thresholds for landslide initiation: Proposing hydro-meteorological thresholds. *Natural Hazards and Earth System Sciences* 18(1): 31–39.

Broeckx, J., Vanmaercke, M., Duchateau, R., & Poesen, J. (2018). A data-based landslide susceptibility map of Africa. *Earth-Science Reviews* 185: 102–121.

Brunetti, M., Peruccacci, S., Rossi, M., Luciani, S., Valigi, D., & Guzzetti, F. (2010). Rainfall thresholds for the possible occurrence of landslides in Italy. *Natural Hazards and Earth System Sciences* 10(3): 447–458.

Caine, N. (1980). The rainfall intensity-duration control of shallow landslides and debris flows. *Geografiska Annaler Series A* 62(1–2): 23–27.

Chen, W., Pourghasemi, H. R., Kornejady, A., & Zhang, N. (2017). Landslide spatial modeling: Introducing new ensembles of ANN, MaxEnt, and SVM machine learning techniques. *Geoderma* 305(3): 314–327.
Clarizia, M., Gullà, G., & Sorbino, G. (1996). Sui meccanismi di innesco dei soil slip. In *International Conference on Prevention of Hydrogeological Hazards: The Role of Scientific Research*, Alba, Italy, pp. 585–597. (in Italian)
Crosta, G. B., & Frattini, P. (2000). Rainfall thresholds for triggering soil slips and debris flows. In *Mediterranean Storms 2000 – Proceedings of the Second EGS Plinius Conference*, Siena, Italy, pp. 463–487.
Dai, F. C., Lee, C. F., & Ngai, Y. Y. (2002). Landslide risk assessment and management: An overview. *Engineering Geology* 64(1): 65–87.
Danielson, J. J., & Gesch, D. (2011). Global Multi-resolution Terrain Elevation Data 2010 (GMTED2010). US Geological Survey. Available from http://eros.usgs.gov/#/Find_Data/Products_and_Data_Available/GMTED2010 (Last accessed 29 March 2020).
Ermini, L., Catani, F., & Casagli, N. (2005). Artificial neural networks applied to landslide susceptibility assessment. *Geomorphology* 66(1–4): 327–343.
Evans, S. G., & Delaney, K. B. (2019). Taking the pulse of global landslide occurrence 2010–2018. *Geophysical Research Abstracts* 21: EGU2019–11815.
FAO, IIASA, ISRIC, ISS-CAS, & JRC (2012). *Harmonized World Soil Database V 1.2*. Rome: FAO.
Farahmand, A., & Aghakouchak, A. (2013). A satellite-based global landslide model. *Natural Hazards and Earth System Science* 13(5): 1259–1267.
Fick, S. E., & Hijmans, R. J. (2017). WorldClim 2: New 1-km spatial resolution climate surfaces for global land areas. *International Journal of Climatology* 37(12): 4302–4315.
Froude, M. J., & Petley, D. N. (2018). Global fatal landslide occurrence from 2004 to 2016. *Natural Hazards and Earth System Sciences* 18(8): 2161–2181.
Gariano, S. L., & Guzzetti, F. (2016). Landslides in a changing climate. *Earth Science Reviews* 162: 227–252.
Giardini, D., Gruenthal, G., Shedlock, K., & Zhang, P. (2003). The GSHAP global seismic hazard map. In W. H. K. Lee, H. Kanamori, P. C. Jennings, & C. Kisslinger (eds.), *International Handbook of Earthquake and Engineering Seismology* (Vol. 81, pp. 1233–1239). Amsterdam: Academic Press.
Giuliani, G., & Peduzzi, P. (2011). The PREVIEW global risk data platform: A geoportal to serve and share global data on risk to natural hazards. *Natural Hazards and Earth System Science* 11(1): 53–66.
Guzzetti, F., Carrara, A., Cardinali, M., & Reichenbach, P. (1999). Landslide hazard evaluation: A review of current techniques and their application in a multi-scale study, *Central Italy. Geomorphology* 31(1–4): 181–216.
Guzzetti, F., Peruccacci, S., Rossi, M., & Stark, C. P. (2007). Rainfall thresholds for the initiation of landslides in central and southern Europe. *Meteorology and Atmospheric Physics* 98(3–4): 239–267.
Guzzetti, F., Peruccacci, S., Rossi, M., & Stark, C. P. (2008). The rainfall intensity-duration control of shallow landslides and debris flows: An update. *Landslides* 5(1): 3–17.
Guzzetti, F., Reichenbach, P., Cardinali, M., Galli, M., & Ardizzone, F. (2005). Probabilistic landslide hazard assessment at the basin scale. *Geomorphology* 72 (1–4): 272–299.
Haque, U., Blum, P., da Silva, P. F., et al. (2016). Fatal landslides in Europe. *Landslides* 13 (5): 1545–1554.

Haque, U., da Silva, P. F., Devoli, G., et al. (2019). The human cost of global warming: Deadly landslides and their triggers (1995–2014). *Science of the Total Environment* 682: 673–684.

Harp, E. L., Reid, M. E., McKenna, J. P., & Michael, J. A. (2009). Mapping of hazard from rainfall-triggered landslides in developing countries: Examples from Honduras and Micronesia. *Engineering Geology* 104(3–4): 295–311.

Hong, Y., Adler, R. F., & Huffman, G. (2007a). An experimental global prediction system for rainfall-triggered landslides using satellite remote sensing and geospatial datasets. *IEEE Transactions on Geoscience and Remote Sensing* 45(6): 1671–1680.

Hong, Y., Adler, R., & Huffman, G. (2007b). Use of satellite remote sensing data in the mapping of global landslide susceptibility. *Natural Hazards* 43(2): 245–256.

Hong, Y., Alder, R., & Huffman, G. (2006). Evaluation of the potential of NASA multi-satellite precipitation analysis in global landslide hazard assessment. *Geophysical Research Letters* 33(22): 1–5.

Huang, Y., & Zhao, L. (2018). Review on landslide susceptibility mapping using support vector machines. *Catena* 165(3): 520–529.

Huffman, G. J., Bolvin, D. T., Braithwaite, D., et al. (2019). *Algorithm Theoretical Basis Document (ATBD) Version 06: NASA Global Precipitation Measurement (GPM) Integrated Multi-satellitE Retrievals for GPM (IMERG).* Greenbelt, MD: National Aeronautics and Space Administration (NASA). Available from https://pmm.nasa.gov/sites/default/files/document_files/IMERG_ATBD_V06.pdf (Last accessed 29 March 2020).

Innes, J. L. (1983). Debris flows. *Progress in Physical Geography* 7(4): 469–501.

Intrieri, E., Gigli, G., Casagli, N., & Nadim, F. (2013). Early warning system: Toolbox and general concepts. *Natural Hazards and Earth System Science* 13(1): 85–90.

Jia, G., Tang, Q., & Xu, X. (2020). Evaluating the performances of satellite-based rainfall data for global rainfall-induced landslide warnings. *Landslides* 17(2): 283–299.

Juang, C. S., Stanley, T. A., & Kirschbaum, D. B. (2019). Using citizen science to expand the global map of landslides: Introducing the Cooperative Open Online Landslide Repository (COOLR). *PLoS One* 14(7): 1–28.

Kirschbaum, D. B., Adler, R., Hong, Y., Hill, S., & Lerner-Lam, A. (2010). A global landslide catalog for hazard applications: Method, results, and limitations. *Natural Hazards* 52(3): 561–575.

Kirschbaum, D. B., Adler, R., Hong, Y., & Lerner-Lam, A. (2009). Evaluation of a preliminary satellite-based landslide hazard algorithm using global landslide inventories. *Natural Hazards and Earth System Science* 9(3): 673–686.

Kirschbaum, D. B., & Stanley, T. (2018). Satellite-based assessment of rainfall-triggered landslide hazard for situational awareness. *Earth's Future* 6(3): 505–523.

Kirschbaum, D. B., Stanley, T., & Zhou, Y. (2015). Spatial and temporal analysis of a global landslide catalog. *Geomorphology* 249: 4–15.

Latham, J., Cumani, R., Rosati, I., & Bloise, M. (2014). *Global Land Cover SHARE (GLC-SHARE).* Rome: Food and Agriculture Organization of the United Nations.

Lin, L., Lin, Q., & Wang, Y. (2017). Landslide susceptibility mapping on a global scale using the method of logistic regression. *Natural Hazards and Earth System Sciences* 17(8): 1411–1424.

Ma, T., Li, C., Lu, Z., & Bao, Q. (2015). Rainfall intensity-duration thresholds for the initiation of landslides in Zhejiang Province, China. *Geomorphology* 245: 193–206.

Melillo, M., Brunetti, M. T., Peruccacci, S., Gariano, S. L., & Guzzetti, F. (2015). An algorithm for the objective reconstruction of rainfall events responsible for landslides. *Landslides* 12(2): 311–320.

Melillo, M., Brunetti, M. T., Peruccacci, S., Gariano, S. L., & Guzzetti, F. (2016). Rainfall thresholds for the possible landslide occurrence in Sicily (Southern Italy) based on the automatic reconstruction of rainfall events. *Landslides* 13(1): 165–172.

Melillo, M., Brunetti, M. T., Peruccacci, S., Gariano, S. L., Roccati, A., & Guzzetti, F. (2018). A tool for the automatic calculation of rainfall thresholds for landslide occurrence. *Environmental Modelling and Software* 105: 230–243.

Monsieurs, E., Dewitte, O., & Demoulin, A. (2018). A susceptibility-based rainfall threshold approach for landslide occurrence. *Natural Hazards and Earth System Sciences* 19(4): 775–789.

Nadim, F., Kjekstad, O., Peduzzi, P., Herold, C., & Jaedicke, C. (2006). Global landslide and avalanche hotspots. *Landslides* 3(2): 159–173.

Palladino, M. R., Viero, A., Turconi, L., et al. (2018). Rainfall thresholds for the activation of shallow landslides in the Italian Alps: The role of environmental conditioning factors. *Geomorphology* 303: 53–67.

Peruccacci, S., Brunetti, M. T., Gariano, S. L., Melillo, M., Rossi, M., & Guzzetti, F. (2017). Rainfall thresholds for possible landslide occurrence in Italy. *Geomorphology* 290(4): 39–57.

Peruccacci, S., Brunetti, M. T., Luciani, S., Vennari, C., & Guzzetti, F. (2012). Lithological and seasonal control on rainfall thresholds for the possible initiation of landslides in central Italy. *Geomorphology* 139–140: 79–90.

Petley, D. (2012). Global patterns of loss of life from landslides. *Geology* 40(10): 927–930.

Petley, D. (2019). *Global Fatal Landslide Database (GFLD)* Version 2, pp. 1–24. University of Sheffield. Online publication.

Petley, D. N., Dunning, S. A., & Rosser, N. J. (2005). The analysis of global landslide risk through the creation of a database of worldwide landslide fatalities. In O. Hungr, R. Fell, R. Couture, & E. Eberhardt (eds.), *Landslide Risk Management* (pp. 367–374). Amsterdam: Academic Press.

Reichenbach, P., Rossi, M., Malamud, B. D., Mihir, M., & Guzzetti, F. (2018). A review of statistically-based landslide susceptibility models. *Earth-Science Reviews* 180(3): 60–91.

Sassa, K. (2004). The International Consortium on Landslides. *Landslides* 1(1): 91–94.

Schneider, A., Jost, A., Coulon, C., Silvestre, M., Théry, S., & Ducharne, A. (2017). Global-scale river network extraction based on high-resolution topography and constrained by lithology, climate, slope, and observed drainage density. *Geophysical Research Letters* 44(6): 2773–2781.

Segoni, S., Piciullo, L., & Gariano, S. L. (2018). A review of the recent literature on rainfall thresholds for landslide occurrence. *Landslides* 15(8): 1483–1501.

Stanley, T., & Kirschbaum, D. B. (2017). A heuristic approach to global landslide susceptibility mapping. *Natural Hazards* 87(1): 145–164.

UNDRR. (2019). *Global Assessment Report on Disaster Risk Reduction*, Geneva: UNDRR.

UNESCO. (1973). *Annual Summary of Information on Natural Disasters No.6:1971*. Paris: UNESCO.

University of East Anglia Climatic Research Unit, Harris, I. C., & Jones, P. D. (2020). CRU TS4.03: Climatic Research Unit (CRU) Time-Series (TS) version 4.03 of High-Resolution Gridded Data of Month-by-Month Variation in Climate (Jan. 1901–Dec. 2018). Centre for Environmental Data Analysis, 22 January. Online publication.

Van Den Eeckhaut, M., Hervás, J., Jaedicke, C., Malet, J. P., Montanarella, L., & Nadim, F. (2012). Statistical modelling of Europe-wide landslide susceptibility using limited landslide inventory data. *Landslides* 9(3): 357–369.

14

Estimating Aquifer Depth in Arid and Semi-arid Watersheds using Statistical Modelling of Spectral MODIS Products

SEYED RASHID FALLAH SHAMSI, PARISA ANSARI, MASOUD MASOUDI AND HAMID REZA POURGHASEMI

14.1 Introduction

During the recent decades, a major part of the world has suffered from huge consequences associated with climate change. Climate change has caused harsh fluctuation in annual precipitations and local hydrological cycles. A considerable decline in annual precipitation and changes in hydrological regime of rivers have caused surface water resource scarcity. During recent years, dependency on underground water resources is turning critical for agricultural and industrial uses. Low precipitations along with a changing annual distribution on the one hand and rising demands for fresh water on the other hand encourage communities to develop programmes to boost harvesting underground water reserves. Arid and semi-arid environments are facing more severe challenges, such as droughts, flash floods, etc., which cause more dependency of local communities on underground water resources.

Iran depends on its ground water reserves given that the annual level of precipitation is about 250 millimetres. Underground water, stored in the aquifers, is the main source of fresh water in this region. Therefore, quantity and quality measures of aquifers are of concern for the governors and the local communities that require fresh water for their activities. Currently, Piezometer and Observation Wells are being used to monitor ground water accessibility and quality by Iran Water & Waste Water Company (IWWWC) at national and provincial levels. IWWWC policymakers are also planning and encouraging local private sectors to harvest ground water by using diesel-electrical pumping wells, in parallel to public dams' facilities for collecting water throughout its watersheds. Currently, to monitor underground water reserves and its accessibility, IWWWC checks the

The authors would like to thank the Vice-Chancellor of Research Office, Shiraz University for the financial support of the research through a grant for graduate studies. The Water and Waste Water Company of Fars province is also acknowledged for providing the aquifer data of the piezometer and the observation wells of the plains from the province archive.

piezometers and the observation wells on a monthly basis. The wells are distributed throughout the plains of the province, important for agricultural activities. The geographical position of the wells is registered by precise handsets of the Global Positioning Systems (GPS). Water samples are collected from the wells and the aquifer depth and underground flow pressure are measured by local contractor companies. The collected data is lodged into a local database of IWWWC as a part of a national database run by the Ministry of Energy of the Iranian Government.

In the southern part of Iran, Fars province and its vast watersheds are also located in arid and semi-arid regions. In the province, Sarvestan and Lamerd townships are important regions both according to agricultural activities and annual demand for fresh water by the local communities. To monitor the aquifers through periodical mensuration, 97 and 46 piezometer wells have been installed and registered in Sarvestan and Lamerd plains, respectively. The field-base mensuration is of interest for the managing companies because of its precision, sampling opportunity and to check the wells instruments and installations, etc. However, direct mensuration and field check of the wells are costly and time-consuming because of the spatial distribution, in addition to the harsh weather of an arid environment during field works throughout a year. The measures are generally given for the geo-located wells, without providing any information on the aquifers in the surroundings. As the wells are not installed for all watersheds and plains, there is no information on these aquifers, even on the depth of the watershed level. Therefore, many studies were conducted to develop methods for estimating aquifer parameters of the plains and watersheds in the absence of the piezometers and observation wells.

Within a fast-growing space technology context, the applications of remotely-sensed data and remote sensing techniques are of interest in various fields of research. This supports mapping activities by covering large areas repeatedly. Remote sensing techniques are more precise and faster, suitably designed especially for natural resources and environmental studies. Besides, they significantly reduce the necessity of intensive field works and consequently reduce their costs. Multi-temporal and spectral characteristics of remotely-sensed data and progressive high-resolution products offer promising applications in remote sensing in the context of earth sciences. The high sensitivity of shortwave infrared and thermal spectrum to water bodies and cool surfaces, respectively, make spectral passive remote sensing suitable for hydrological studies.

Due to the high potential of radiometric high resolution day and night images, remote sensing technique methods have been widely applied in hydrological research and aquifer studies (Kustas & Norman, 1996). Previous research indicates that there is a strong correlation between soil moisture and heat capacity at the earth surface and the aquifer depth (Heilman & Moore, 1982) due to capillary action through soil porosity and tubes between the aquifer and the surface. Soil moisture significantly affects the surface temperature by cooling it down due to an evaporation process and high heat of vaporization of water. From its inception, spectral-thermal remote sensing has been used to process thermal images and estimate the surface soil moisture to indirectly detect annual fluctuations of aquifer depth (Kappelmeyer, 1957; Chase, 1969; Heilman & Moore, 1982; Bastiaanssen, 2003). Both the National Aeronautics and Space Agency (NASA) of the United States and the European Space Agency (ESA) have allocated research funds such as GRACE (US NASA and ESA Joint-Project 'Gravity Recovery and Climate Experiment 2002–2017') on the application of remote sensing techniques in aquifer studies (Rodell & Famiglietti, 2002). In these early projects, most results were not satisfactory because of the sensor limitations, low data quality and the fact that image processing techniques were not fully matured.

To explore aquifer depth in arid environments, land surface temperature differences in the day and night have been measured through thermal infrared spectrum analysis (Urqueta et al., 2018). Integrating optical and thermal remote sensing can increase the accuracy of the aquifer depth estimation. However, it is particularly difficult to merge near-infrared and thermal spectrum analysis to explore soil characteristics and moisture, respectively. Multi-sensor thermal remote sensing which are of interest include MODIS and NOAA-AVHRR thermal images. They can be used to study soil moisture and daily changes of land surface temperature (LST) (Fang et al., 2018). By developing new remotely-sensed data such as radar images, microwave imaging (Sutanudjaja et al., 2013) and ancillary data such as the digital elevation model (DEM), fusion and merging operations have been considered as complementary methods of estimation in aquifer studies (Caselles et al., 2014). Indirectly estimated depth of aquifers, using Vegetation Indices (VIs) products such as the Normalized Difference Vegetation Index (NDVI) and LST have been investigated by researchers through methods that compute probable correlations between vegetation cover on the surface and aquifer depth below the surface (Tiyip et al., 2005). Thermal Infra-Red (TIR) and LST products have been investigated to detect shallow aquifers from the early days of remote sensing (Chase, 1969). Considering earth surface heat as the only source of infrared radiation makes LST products a suitable indicator of a cooler surface with a shallow aquifer (Quiel, 1975). Attempts have been made to integrate potential evapotranspiration (ETP) time-series to remotely-sensed data to increase the accuracy of aquifer depth estimation from satellite images. There is also promising research that indicates a positive and significant corelation between them (Ndou et al., 2018). In recent years, mathematical modelling has been used as aquifer depth indirect estimation techniques. It includes differential models and statistical models based on remotely-sensed images, ETP data, VIs, soil moisture and LST, simultanously (Alkhaier et al., 2009).

A wide range of techniques and methods were developed in remote sensing applications for information extraction from the image data. The MODIS data spectral-radiometric characteristics and a variety of its products can be used in a large domain of earth surface applications. MODIS data can be used in statistical models, such as the geostatistical analysis for earth surface studies (Fallah Shamsi et al., 2012; Zare et al., 2019). However, using ancillary data, statistical and mathematical modelling are complementary to image classification and mapping techniques. Applications vary according to the complexity of the study area, the image data, image products, remote sensing indices, robustness of the method, ancillary data and the characteristics of the aquifer. The hypotheses of the current research are that MODIS image products, especially day- and night-time images, are suitable for the statistical modelling of the aquifer depth. The main objective is to build models to estimate the aquifer depth at watershed level, in the absence of piezometer and observation wells.

This chapter will present a simple and low-cost methodology, designed for aquifer depth estimation in an arid environment using spectral satellite images. We consider different scenarios that include several models and validation processes applied to the investigated data and image products. For the sake of conciseness, this chapter presents the main results and opens the door to future research.

14.2 Materials and Methods

14.2.1 Study Area

In this research, the study area includes two famous and important plains – Sarvestan and Lamerd – located in Fars province, southern Iran. Figure 14.1 indicates the locations of the plains in Iran and in the Middle East. To simplify research conditions, we selected two watersheds with thin vegetation cover to avoid any spectral interference in the modelling process.

14.2.1.1 Sarvestan Plain

The plain is located at 52°, 14' E and 29°, 27' N, near to the township of Sarvestan. Its altitude is 1,500 metres above sea level. Annual average precipitation and average temperature are 320 millimetres and 17.2°C, respectively. The climate is arid and annual evapotranspiration is about 2,190 millimetres annually, which is nine-times higher than the precipitation received. Maharloo Lake, one of the most famous salty lakes of Iran, is located in this basin. In the

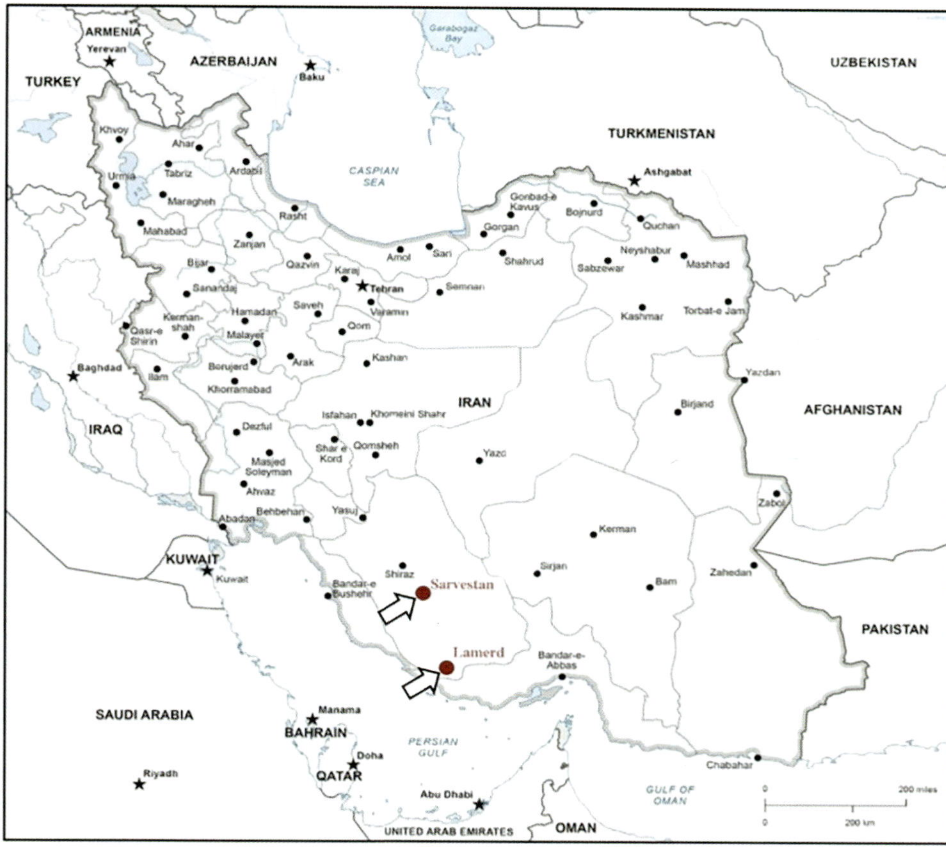

Figure 14.1 The locations of the study areas in Iran and the Middle East

region, a deep aquifer has extended up, close to the earth's surface. The soil is generally saline, composed of clay and silt in the form of thick sedimentation layers. The major plants are *Tamarix* spp., *Salicornia* Spp. and a thin cover of halophyte vegetation. In the plain, water harvesting supports the local farmers for cereal cultivation, such as wheat and barley and salinity-resistant orchard plantations, including pistachio and olive trees.

14.2.1.2 Lamerd Plain

The plain is located at 54°, 52' E and 28°, 27' N, near to the township of Lamerd. Its altitude is 450 metres above sea level. Annual average precipitation and average temperature are 250 millimetres and 22°C, respectively. The climate is arid and annual evapotranspiration is about 4,000 millimetres annually, which is 16-times higher than the precipitation received. In the region, the aquifer depth is also close to the earth's surface. However, the aquifer is generally a thin layer itself. Soils are composed of evaporites sedimentation layers. The major plants are a thin cover of xerophyte-cryptophytic vegetation from *Papilionaceae* Spp., *Labiatae* Spp., *Gramineae* Spp., *Compositae* Spp. and *Chenopodiaceae* Spp. Collecting surface runoff water and harvesting ground water, in the plain, help the local farmers for cereal cultivation (such as winter wheat) and orchard plantations (such as heat-tolerant date palm).

14.2.2 Research Data

14.2.2.1 Field Data

In Iran, aquifer depth measures are recorded on a monthly basis in piezometer and observation well facilities distributed across the major plains. As shown in Figure 14.2, a piezometer is a well, used to estimate water pressure in an aquifer by measuring pressure of groundwater at its location. Besides, the network of observation wells helps monitor aquifer level over time. The Water and Waste Water Company (WWWC) of Fars province has developed an intensive piezometer and observation well network for all plains within the province. A contractor company also records monthly observations of the wells using facilities installed on the wells and store the collected data directly in the provincial database and the national archive of the Ministry of Energy.

In this research, the records of the wells have been extracted from the Province database and monthly archives of WWWC of Fars province, including 95 and 46 wells, collected in early June 2017, for both Sarvestan and Lamerd plains, respectively. The minimum depths are about 0.8 and 2.5 metres for Sarvestan and Lamerd watersheds, respectively. The mean depth of the Sarvestan aquifer is around 32 metres, twice the depth of the one in the Lamerd plain, at around 16 metres depth from the surface. The maximum depths are 160 and 61 metres for Sarvestan and Lamerd plains, respectively. In the Maharloo Lake bank, we checked the aquifers

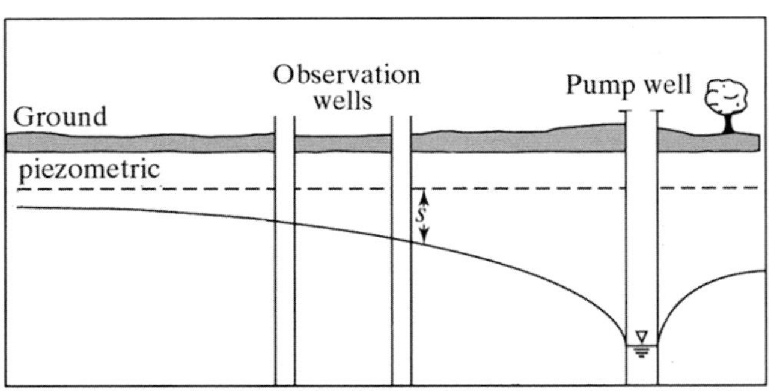

Figure 14.2 Piezometric level and the observation wells in an aquifer

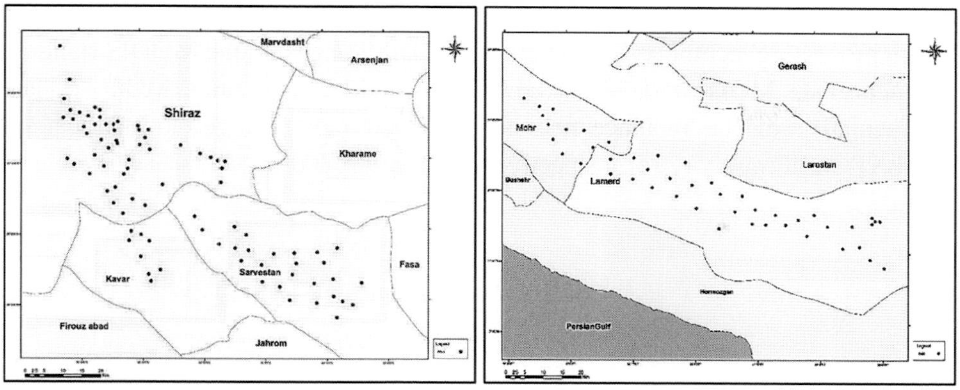

Figure 14.3 Distribution of the piezometric and the observation wells in Sarvestan plain (left) and Lamerd plain (right)

with a depth of less than 1 metre by drilling the surface using a hand auger. Figure 14.3 indicates the distribution of the piezometric wells in the plains.

14.2.2.2 Remotely-Sensed Data

In the recent years, large datasets have been generated from repeated observations of the earth surface through remotely-sensed imaging data. High-frequency (hourly) data can be used to identify changes in short-lag frequencies. Some of these archives are free of charge for the users, for example, Moderate Resolution Imaging Spectro-radiometer (*MODIS*) images. There are huge possibilities for producing new products from the initial satellite images.

MODIS is a multi-spectral sensor on the Terra platform – sun synchronous polar orbit satellite, designed and developed jointly by Canada, the United States and Japan in 2000. On its track, the sensor covers a 2,330-kilometre swath width for each passage and records 12-bit spectral Reflectance/Radiance images of the earth surface within a 0.4–14.4 μm electromagnetic range (visible to thermal domain), divided into 36 spectral bands. However, besides its high-performing spectral and radiometric capabilities, the spatial resolution of the sensors is quite low (250, 500 and 1,000 metres).

MODIS images cover a repetitive track in an average spatial resolution and temporal resolution of a couple of days (Zhang et al., 2006). High radiometric sensitivity of the sensor to spectral behaviour of earth surface objects makes it suitable for hydrological studies (Jensen, 2000). Currently, the Iranian Space Agency (ISA) and Centre for Oceanic and Atmospheric Research, College of Agriculture, Shiraz University have their own receiver antenna facilities for image data acquisition, directly from Terra/Aqua platforms.

According to the technical advice of the developers, bands 3–7, 20–23 and 31–32 are suitable for solid earth surface applications. However, recent methods can now produce a wide variety of image products, namely MOD xx, by applying image processing techniques on the initial MODIS images. In the MODIS standard products list, MOD 01-L1A, MOD 02-L1B, MOD 03 and MOD 11 are representative of the radiance counts, calibrated radiance, geo-located dataset and LST and Emissivity (ε), respectively (Qu et al., 2006).

14.2.3 The Method

14.2.3.1 Image Products of MODIS Data

MOD02 and MOD11 were acquired from the archives for both of the regions on 25 May 2017 and 2019. This is the exact date of checking the wells for data collection. The standard level of MOD02B1 includes bands of one kilometre spatial resolution day- and night-time images which have been used as the input for the products, including day- and night-time radiance (Rad.), reflectance (Rlf.), emissivity (ε), brightness temperature (T_B) and LST. However, a MODIS image date set of the regions was also acquired on 24 November 2019 to be used in the validation process of the results. A list of MODIS band number, bandwidth and the general application they were developed for are presented in Table 14.1.

In this research, to have radiance images, a calibrated and processed MODIS L1B dataset has also been acquired from the archives for the date. MODIS radiance products are radiometrically calibrated using an advanced algorithm and geographically geo-located. The reflectance bands, over the spectral range 0.4–2.3 μm and the emissive infrared bands 3.6–14.4 μm, are also used to diversify model input variables in this research. Both HDF-format radiance and reflectance images are scaled to 16-bit data arrangement. The products are provided by MODIS project, based on three categories of spatial resolution in the Level 1B standard level. As shown in Table 14.1, spatial resolution for b_1 and b_2 is 250 metres, while it is 500 metres for b_3 to b_7. Spatial resolution for all reflective b_8 to b_{19} and b_{26}, b_{20} to b_{25} and b_{27} to b_{36} is 1,000 metres. The reflection bands are registered just for day-time (form b_1 to b_{19}). All emissive bands plus reflective b_{26} are being scanned for both day- and night-time. The L1B dataset also presents a geo-location ancillary dataset, accompanying one-kilometre images. One of the helpful products is the brightness temperature ($T_{B\lambda}$) counts, comes from MODIS02 initial images using transformation algorithms (Jensen, 2000). It is a measure of the radiance, expressed in term of the temperature of an equivalent black body. It is usually used as a representative for water vapour and surface temperature in spectral remote sensing. One of the most famous transformation techniques to calculate $T_{B\lambda}$ from thermal infra-red domain is the Plank relation (Sobrino et al., 2004). Its core assumption is based on the

Table 14.1. MODIS spectral bands and the general applications developed for earth studies (Ahmad et al., 2002)

No.	General Use	Bandwidth	µm	Band (b)
Reflective Solar Bands (250 m spatial resolution)				
1	aerosol, cloud,	0.620–0.670	0.659	1
2	land	0.841–0.876	0.865	2
Reflective Solar Bands (500 m spatial resolution)				
3	aerosol & cloud	0.459–0.479	0.47	3
4	optical thickness	0.545–0.565	0.555	4
5	cloud phase, cloud	1.230–1.250	1.24	5
6	effective radius, cloud	1.628–1.652	1.64	6
7	mask, snow, land	2.105–2.155	2.13	7
Reflective Solar Bands (1 km spatial resolution)				
8	Ocean colour	0.405–0.420	0.412	8
9	chlorophyll	0.438–0.448	0.443	9
10	phytoplankton	0.483–0.493	0.488	10
11	biogeochemistry	0.526–0.536	0.531	11
12	sediments	0.546–0.556	0.551	12
13	atmosphere	0.662–0.672	0.667	13
14	fluorescence	0.673–0.683	0.678	14
15	atmosphere	0.743–0.753	0.748	15
16	aerosol	0.862–0.877	0.869	16
17	atmospheric total	0.890–0.920	0.905	17
18	precipitable water	0.931–0.941	0.936	18
19	vapour, clouds	0.915–0.965	0.94	19
20	cirrus cloud	1.360–1.390	1.375	26
MWIR Thermal Emissive Bands (1 km spatial resolution)				
21	cloud & surface,	3.660–3.840	3.75	20
	temperature, fire &	3.929–3.989	3.96	21
	volcano, sea surface temperature	3.929–3.989	3.96	22
		4.020–4.080	4.05	23
22	atmospheric	4.433–4.498	4.47	24
	temperature profile	4.482–4.549	4.52	25
LWIR Thermal Emissive Bands (1 km spatial resolution)				
23	tropospheric water	6.535–6.895	6.72	27
	vapor	7.175–7.475	7.33	28
24	cloud particle radius	8.400–8.700	8.55	29
	total column ozone	9.580–9.880	9.73	30
25	cloud, surface	10.780–11.280	11.03	31
	temperature, fire	11.770–12.270	12.02	32
26	cloud top height,	13.185–13.485	13.34	33
	temperature,	13.485–13.785	13.64	34
	pressure, temperature profile	13.785–14.085	13.94	35
		14.085–14.385	14.24	36

theoretical black body object. However, as physics says, the objects are not of standard physical black object in our surroundings. Irradiance Power ($IP_{b\lambda}$) is also another product which is widely used for surface and energy balance studies. It is defined as a proportion of an object's radiance to its black object radiance, generally estimated using indirect methods (Sobrino et al., 2004). It deeply depends on surface emissivity ($\varepsilon_{b\lambda}$), which is the effectiveness of surface material emitting energy as thermal radiation (Price, 1983). As $IP_{b\lambda}$ is an important factor for calculating LST count, satellite radiative thermal images are usually used in research (Jiménez-Muñoz & Sobrino, 2003). In this research, radiative temperature was calculated using MODIS b_{31} and b_{32}, ranging between 11 μm and 12 μm electromagnetic wavelengths, as shown in Table 14.1. The method, based on Normalized Differenced Vegetation Index (NDVI), was used for estimating $IP_{b\lambda}$ (Valor & Caselles, 1996; Sobrino & Raissouni, 2000). As a vegetation index, NDVI is developed to be efficiently sensitive to earth surface cover. By ranging values from −1 to 1, it represents land covers from dense vegetation to bare soils, similar to the condition of the study areas. $\varepsilon_{b\lambda}$ – and consequently $IP_{b\lambda}$ – was also extracted from the MOD11 dataset using the MODIS Conversion Toolkit (MCTK) (White, 2014) in remote sensing software for comparing the results. LST is also one of the most important variables considered in this research, as a representative for physical energy processes on the earth surface (Karnieli et al., 2010). Therefore, LST is used frequently in hydrological and evapotranspiration studies (Bastiaanssen, 2003; Hansen et al., 2010). Because of its temporal change, a set of accurate LST measures have to be entered into the modelling process (Li et al., 2013). Nowadays, LST products derived from radiation models are still of interest as the early era of spectral remote sensing. Thermal Infra-Red (TIR) images, such as those provided by MODIS products, are used for estimating LST measures indirectly. However, this is a low accuracy method and depends on atmospheric parameters in addition to land surface temperature and irradiance power (Li & Becker, 1993). Hence, a separate window method has been considered for transforming MODIS data to the LST product (Zhang et al., 2006). To produce a more accurate LST count from MODIS images, the model was calibrated using an improved set of coefficients, based on the data collected by field weather stations. Then the calibrated LST product (LST_c), particularly developed for Iran, has been entered in the procedure (Modiri & Modiri, 2016). Besides using the locally calibrated models to produce LST products, a separate LST product has been extracted from MCTK extension for comparing the results. In this research, a set of novel products (indices) have been developed serving as independent variables for statistical modelling process. The indices were developed based on day- and night-time difference and ratio expressions, applied on standard MODIS initial

Table 14.2. The products and the novel indices, developed for day- and night-time MODIS data

No.	Prod. & Indices	Description
1	ε_{Db28}	Emissivity (ε) products for Band 28 (Day-)
2	ε_{Nb28}	Emissivity (ε) products for Band 28 (Night-)
3	ε_{Db31}	Emissivity (ε) products for Band 31 (Day-)
4	ε_{Nb31}	Emissivity (ε) products for Band 31 (Night-)
5	ε_{Db32}	Emissivity (ε) products for Band 32 (Day-)
6	ε_{Nb32}	Emissivity (ε) products for Band 32 (Night-)
7	T_{BDb28}	Brightness Temperature for Band 28 (Day-)
8	T_{BNb28}	Brightness Temperature for Band 28 (Night-)
9	T_{BDb31}	Brightness Temperature for Band 31 (Day-)
10	T_{BNb31}	Brightness Temperature for Band 31 (Night-)
11	T_{BDb32}	Brightness Temperature for Band 32 (Day-)
12	T_{BNb32}	Brightness Temperature for Band 32 (Night-)
13	LST_{DMOD11}	Land Surface Temperature for MOD11 (Day-)
14	LST_{NMOD11}	Land Surface Temperature for MOD11 (Night-)
15	LST_C	Calibrated Land Surface Temperature for MOD 11 of Iran (Day-)
16	$(\varepsilon_{Db28} - \varepsilon_{Nb28})$	Day- & Night- Emissivity (ε) Difference for Band 28
17	$(T_{BDb28} - T_{BNb28})$	Day- & Night- Brightness Temperature Difference for Band 28
18	$(\varepsilon_{Db31} - \varepsilon_{Nb31})$	Day- & Night- Emissivity (ε) Difference for Band 31
19	$(T_{BDb31} - T_{BNb31})$	Day- & Night- Brightness Temperature Difference for Band 31
20	$(\varepsilon_{Db32} - \varepsilon_{Nb32})$	Day- & Night- Emissivity (ε) Difference for Band 32
21	$(T_{BDb31} - T_{BDb32})$	Day- Brightness Temperature Difference for Bands 31 and 32
22	$(\varepsilon_{Db31} - \varepsilon_{Db32})$	Day- Emissivity (ε) Difference for Bands 31 and 32
23	$(T_{BDb32} - T_{BNb32})$	Day- & Night- Brightness Temperature Difference for Band 32
24	$(\varepsilon_{Nb31} - \varepsilon_{Nb32})$	Night- Emissivity (ε) Difference for Bands 31 and 32
25	$(T_{BNb31} - T_{BNb32})$	Night- Brightness Temperature for Bands 31 and 32
26	$(\varepsilon_{Db31} - \varepsilon_{Nb32})$	Day- & Night- Emissivity (ε) Difference for Bands 31 and 32
27	IP_{b31}	Irradiance Power product for Band 31
28	IP_{b32}	Irradiance Power product for Band 32
29	$(\varepsilon_{Db31} / \varepsilon_{Db32})$	Day- Emissivity (ε) Ratio for Bands 31 and 32
30	$(LST_D - LST_N)$	Day- & Night- Land Surface Temperature (LST) Ratio
31	R_{Db31}/R_{Db31}	Day- Reflectance Ratio for Band 31
32	$(R_{Db31} - R_{Db32})/2$	Mean of Day- Reflectance for Bands 31 and 32
33	$(Rad_{Db31} - Rad_{Db32})$	Day- Radiance Difference for Bands 31 and 32
34	$(Rad_{Nb31} - Rad_{Nb32})$	Night- Radiance Difference for Bands 31 and 32

images and products, such as Rad., $T_{B\lambda}$, $\varepsilon_{B\lambda}$ and LST, as shown in Table 14.2. The main objective is to integrate the capability of day- and night-time initial products in one final product, using a procedure currently accepted in remote sensing techniques.

14.2.3.2 Field Data Exploration

The aquifer depth records from 2017 to 2019 have been extracted from Province archives, which count about 95 and 46 wells in the Sarvestan and Lamerd plains, respectively. Ancillary field measures, such as auger-sampling records, were also stored in a separate database for further analysis. First, we used ArcGIS software to transform the GPS coordinates of the Wells into geographic layers. Then, we used SPSS software to explore the depth data and estimated descriptive parameters associated with the records.

14.2.3.3 Statistical Modelling

The Pearson Correlation Analysis First, we carried out a correlation analysis by computing the Pearson Correlation Coefficient (PCC) on 34 MODIS image products. Table 14.3 shows the covariance of the variables estimated among each pair of variables with PCC and divided by the product of their standard deviations. The PCC, calculated for the products, serves as a criterion for selecting the variables that will be used in the depth modelling process. Besides, the aquifer depth measures have been classified into clusters, using a 5-metre equal-interval classification method, to study the effects of depth classification on the probable within-class correlations analysis.

Simple Linear Regression (SLR) Simple Linear Regression (SLR) is the most common form of regression analysis, in which a linear function is fitted over the data according to mathematical criteria. In this research, we first consider a simple linear regression (SLR) model with a single explanatory variable that assumes a linear function between MODIS products and aquifer depth as independent and dependent variables, respectively. In addition to the square residuals of models, validation criteria, including Relative Error (RE), Coefficient of Efficiency (CE) and Root Mean Square Error (RMSE), have been calculated using a separate control dataset of observation records from Sarvestan and Lamerd plains.

First, the measures of the independent variables were extracted on the exact location of the 95 wells of Sarvestan plain and the 46 wells of Lamerd plain, using the GIS layer of the location of the wells. Then, we analysed the correlation among each pair of variables, ranging all aquifer depths in the regions. The results of the analysis using several methods to identify correlation failed to show any evidence of an association between the dependent and the independent variables. Figures 14.4 and 14.5 show the very low correlation observed between the MODIS products and aquifer depth for the Sarvestan and Lamerd plains, respectively.

Then, as a solution, an equal-interval clustering operation has been applied on the aquifer depth records of the plains. Then the correlation analysis was repeated within the clusters for the variables. According to the significant rise in the PCC, in this step, the SLR model development was carried out using a lower number of wells, after clustering operation. Figures 14.6 and 14.7 indicate SLR trend lines

Table 14.3. The Pearson correlation coefficients for the products and the novel indices for the 0–5 metre depth class of the study area

No.	Prod. & Indices	Sarvestan (0–5 m)	Lamerd (2–5 m)
1	ε_{Db28}	0.759**	0.833*
2	ε_{Nb28}	−0.075	0.623
3	ε_{Db31}	0.504	−0.453
4	ε_{Nb31}	−0.059	−0.132
5	ε_{Db32}	0.425	−0.475
6	ε_{Nb32}	−0.016	−0.118
7	T_{BDb28}	0.760**	0.833*
8	T_{BNb28}	0.281	−0.614
9	T_{BDb31}	0.491	−0.454
10	T_{BNb31}	−0.063	−0.138
11	T_{BDb32}	0.516	−0.470
12	T_{BNb32}	−0.061	−0.125
13	LST_{DMOD11}	0.372	−0.448
14	LST_{NMOD11}	−0.017	−0.214
15	LST_C	0.481	0.304
16	$(\varepsilon_{Db28} - \varepsilon_{Nb28})$	0.728**	0.962**
17	$(T_{BDb28} - T_{BNb28})$	0.745**	0.951**
18	$(\varepsilon_{Db31} - \varepsilon_{Nb31})$	0	−0.338
19	$(T_{BDb31} - T_{BNb31})$	0	0
20	$(\varepsilon_{Db32} - \varepsilon_{Nb32})$	0	−0.354
21	$(T_{BDb31} - T_{BDb32})$	−0.362	0.211
22	$(\varepsilon_{Db31} - \varepsilon_{Db32})$	0.364	−0.307
23	$(T_{BDb32} - T_{BNb32})$	0	0
24	$(\varepsilon_{Nb31} - \varepsilon_{Nb32})$	−0.336	−0.281
25	$(T_{BNb31} - T_{BNb32})$	−0.494	−0.225
26	$(\varepsilon_{Db31} - \varepsilon_{Nb32})$	−0.233	0
27	IP_{b31}	−0.241	−0.198
28	IP_{b32}	−0.207	−0.198
29	$(\varepsilon_{Db31} / \varepsilon_{Db32})$	0.226	−0.201
30	$(LST_D - LST_N)$	0.319	−0.253
31	R_{Db31} / R_{Db31}	0	0
32	$(R_{Db31} - R_{Db32})/2$	0	0
33	$(Ra_{Db31} - Ra_{Db32})$	0	0
34	$(Ra_{Nb31} - Ra_{Nb32})$	0	0

* Significant for $P < 0.05$; ** Significant for $P < 0.01$.

and functions, estimated for aquifer depth measures of 0–5 metres and the products for Sarvestan and Lamerd plains, respectively. Tables 14.4 and 14.5 present the best of the SLR aquifer depth models (0–5 metres) for the plains.

Multiple Regression Models Regression analysis is primarily used for predicting, inferring and revealing unknown relationships between the independent and

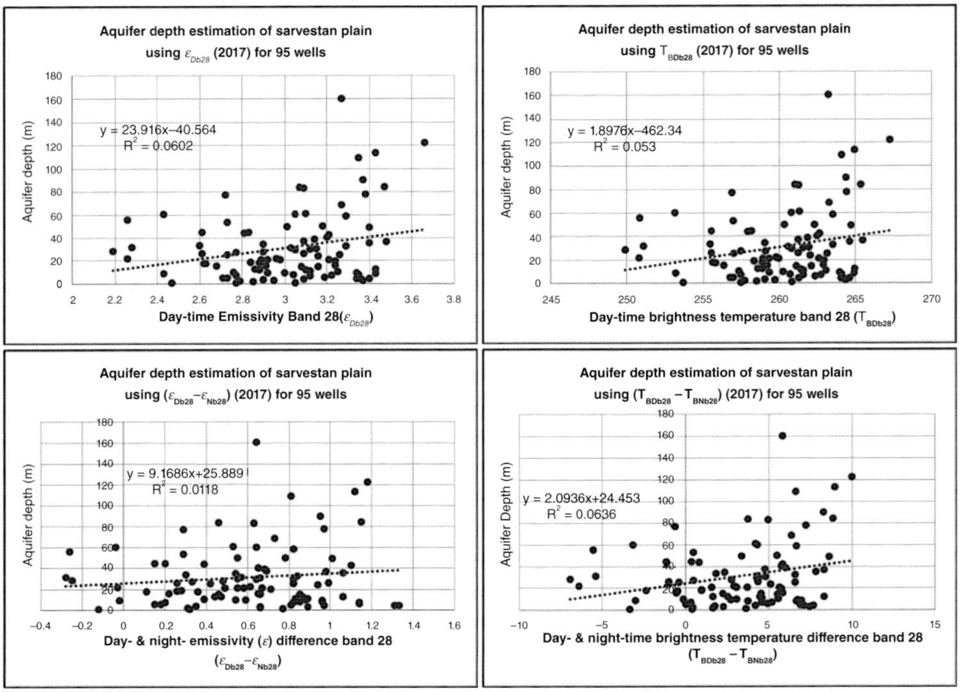

Figure 14.4 Aquifers depth estimation models for 95 wells in Sarvestan plain (0–170 metres)

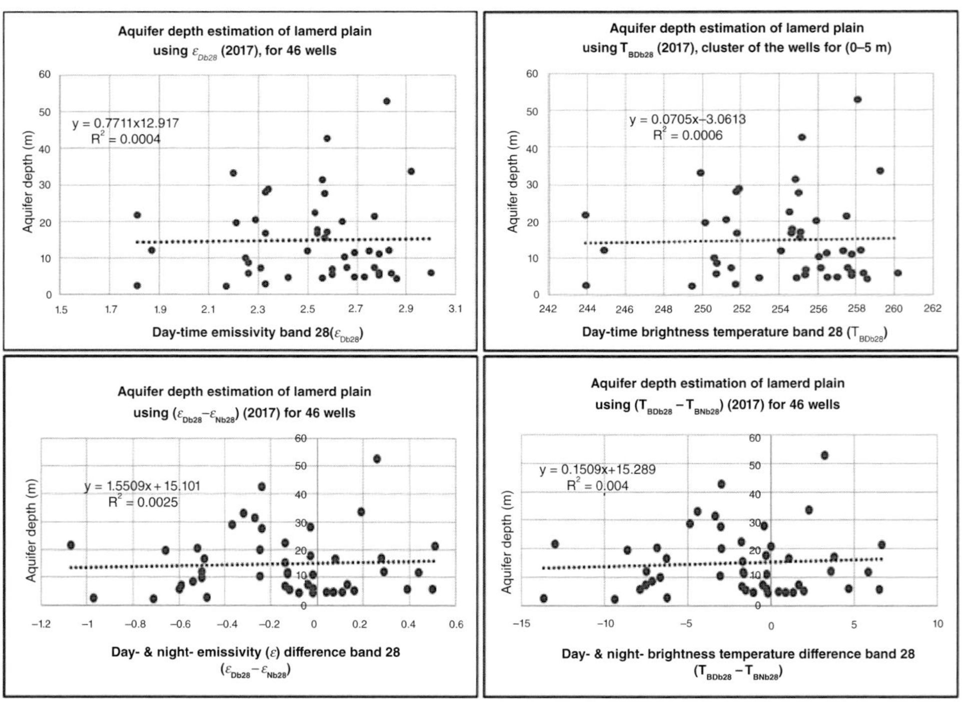

Figure 14.5 Aquifers depth estimation models for about 46 wells in Lamerd plain (0–60 metres)

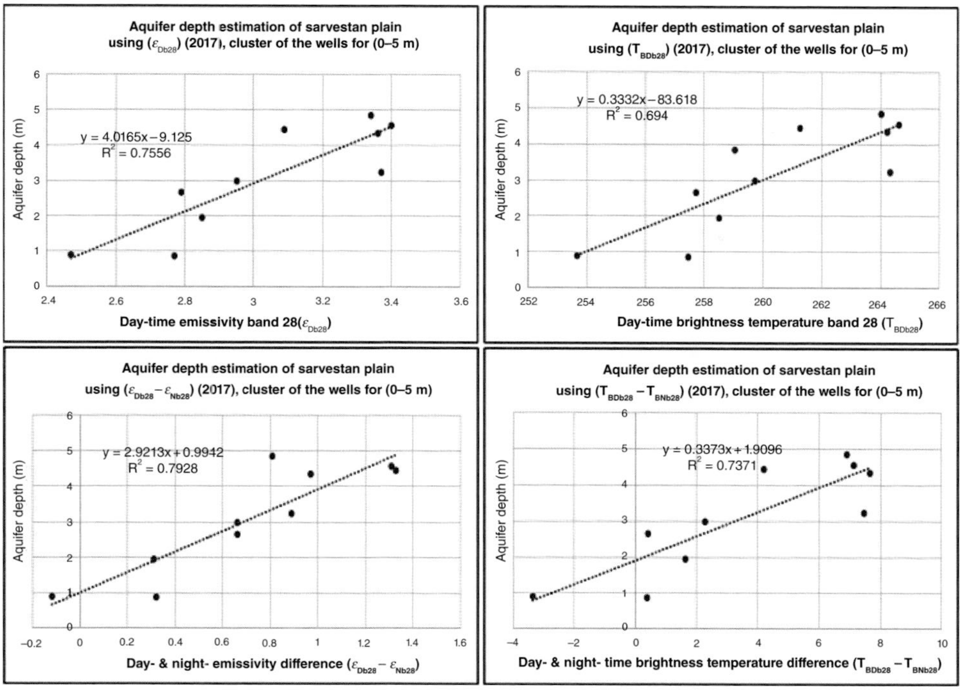

Figure 14.6 Simple linear models for the aquifer depth estimation (0–5 metres), Sarvestan plain

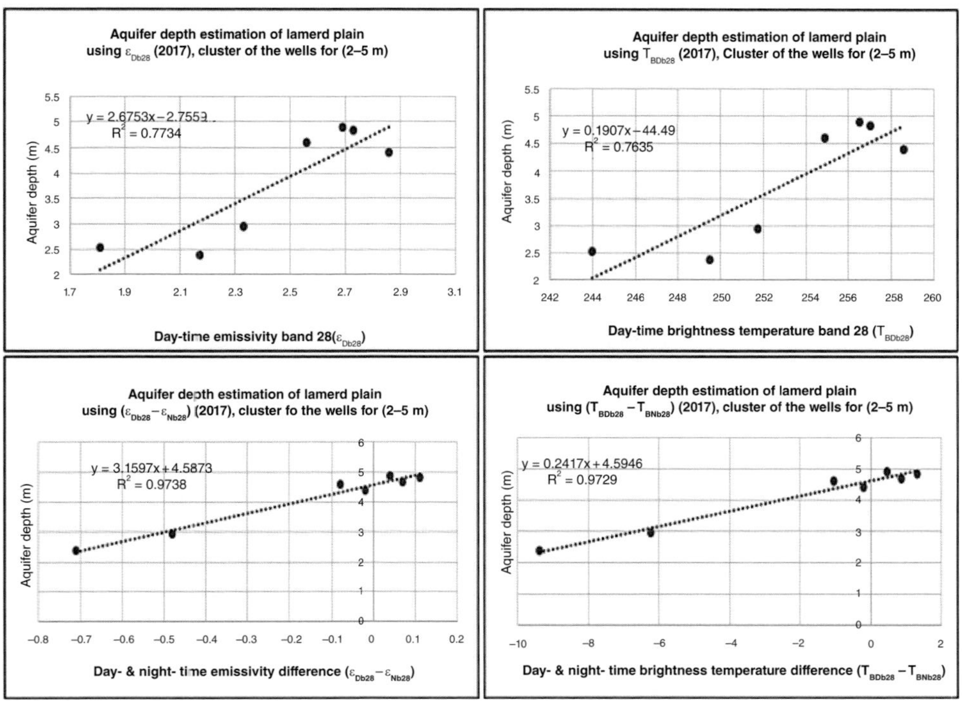

Figure 14.7 Simple linear models for the aquifer depth estimation (0–5 metres), Lamerd plain

Table 14.4. Aquifer depth linear estimation models for Sarvestan plain

Depth (m)	No.	The models	R	R^2
0–5	**1**	$\mathbf{AD_e = -82.3 + 0.33\ T_{BDb28}}$	**0.760**	**0.575**
	2	$AD_e = -8.5 + 0.38\ \varepsilon_{Db28}$	0.759	0.575
	3	$AD_e = 1.93 + 0.32\ (T_{BDb28} - T_{BNb28})$	0.745	0.556
	4	$AD_e = 1.25 + 2.53(\varepsilon_{Db28} - \varepsilon_{Nb28})$	0.728	0.530

Table 14.5. Aquifer depth linear estimation models for Lamerd plain

Depth (m)	No.	The models	R	R^2
2–5	**1**	$\mathbf{AD_e = -44.13 + 0.19\ T_{BDb28}}$	**0.833**	**0.694**
	2	$AD_e = -2.56 + 2.64\ \varepsilon_{Db28}$	0.833	0.694
	3	$AD_e = 4.55 + 0.18\ (T_{BDb28} - T_{BNb28})$	0.951	0.904
	4	$AD_e = 4.56 + 2.54\ (\varepsilon_{Db28} - \varepsilon_{Nb28})$	0.962	0.925

dependent variables of the dataset. Within a multiple linear regression model (MLR) framework, we used a stepwise procedure to select the variables so that the predictive power of the model is maximized. At each iteration, a variable taken from the set of the products is either added or removed from the set of explanatory variables. The results of the selected models, including the square residuals, are presented in Tables 14.6 and 14.7 for the Sarvestan and Lamerd plains, respectively.

Model Validation We validate the model by analysing the goodness-of-fit, which provides information on the predictive performance of the models. We test how well the models predict the aquifer depth of the plains based on a new set of remotely-sensed input data. Tables 14.4–14.7 show the independent dataset used for Sarvestan and Lamerd MODIS products used to predict the aquifer depth of the plains in 2019. The aquifer depths were already measured by the WWWC of Fars province for the wells in 2019. Based on the predicted and observed aquifer depth measures of the plains in 2019, we used a regression model to explore and to estimate the variation of the observed measures of aquifer depth. Figure 14.8 indicates the model estimating the aquifer depth in Sarvestan for 2017. In addition, we compare the validity of the model among different regions, by predicting the aquifer depth for Sarvestan plain in 2019 using models fitted on data from Lamerd in 2017 (Figure 14.8).

We used numerical methods to validate the models. The results are as follows: average RE, RMSE and CE: 12 per cent, 0.55 depth and 52 per cent efficiency, respectively. We explore several ways to select the most suitable model, based on

Table 14.6. SLMR models for aquifer depth estimation (AD_e), Sarvestan plain

Depth (m)	No.	The models	R	R^2
0–5	**1**	$\mathbf{AD_e = -83.55 + 0.336\ T_{BDb28} - 2.193\ (T_{BDb31} - T_{BDb32})}$	**0.858**	**0.736**
	2	$AD_e = -61.74 + 2.312\ (\varepsilon_{Db28} - \varepsilon_{Nb28}) - 2.001 T_{BDb32})$	0.837	0.703
5–10	1	$AD_e = -230.185 + 0.809\ LST_N$	0.594	0.309
	2	$AD_e = -39.850 + 5.552\ \varepsilon_{Nb31}$	0.525	0.276
	3	$AD_e = -201.058 + 0.731\ T_{BNb31}$	0.524	0.275
	4	$AD_e = -199.693 + 0.709\ T_{BNb32}$	0.521	0.272
	5	$AD_e = -41.171 + 6.049\ \varepsilon_{Nb32}$	0.507	0.257
10–20	1	$AD_e = -16.178 + 12.712\ \varepsilon_{Nb28}$	0.556	0.309
20–25	1	$AD_e = 47.736 - 11.030\ \varepsilon_{Nb28}$	0.902	0.814
25–30	1	$AD_e = -2{,}903.797 + 5{,}952.420\ IP_{32} - 2{,}935.541\ IP_{31} + 1.864\ \varepsilon_{Db32}$	0.923	0.851
30–45	1	$AD_e = -5352.503 - 5382.807\varepsilon_{Db31} + 10{,}962.008\ \varepsilon_{Db32}$	0.724	0.524
45–50	1	$AD_e = 35.907\ T_{BDb20} - 5.462\ (\varepsilon_{Db28} - \varepsilon_{Nb28})$	0.993	0.986
50–55	1	$AD_e = 35.907\ T_{BDb20} - 5.462\ (\varepsilon_{Db28} - \varepsilon_{Nb28})$	0.993	0.986
55–60	1	$AD_e = -137.946 - 7.555\ (T_{BDb31} - T_{BDb32}) + 0.837\ LST_D - 0.308\ T_{BDb28}$	0.971	0.944
	2	$AD_e = -943.737 - 140.622\ (\varepsilon_{Db31} - \varepsilon_{Db32}) + 3.551\ LST_{CD}$	0.915	0.826

Table 14.7. SLMR for aquifer depth estimation (AD_e), Lamerd plain

Depth (m)	No.	The models	R	R^2
0–5	**1**	$\mathbf{AD_e = -44.127 + 0.197\ T_{BDb28}}$	**0.833**	**0.694**
	2	$AD_e = 4.550 + 0.184\ (T_{BDb28} - T_{BNb28})$	0.951	0.904
	3	$AD_e = 4.557 + 2.543\ (\varepsilon_{Db28} - \varepsilon_{Nb28})$	0.962	0.925
	4	$AD_e = 4.554 + 16.356\ (\varepsilon_{Db28} - \varepsilon_{Nb28}) - 1.012\ (T_{BDb28} - T_{BNb28})$	0.989	0.979

the highest R^2, R and CE and lower RE and RMSE. As an additional criterion, we consider the simplicity of the model, which is solely determined on the number of variables in the model since all investigated models are from the same class (i.e. linear models).

14.3 Results

The results failed to demonstrate a significant correlation between the variable of interest when all observations – when we consider all aquifer depth observations of all plains – are used in the models. Early findings are more conclusive with regard

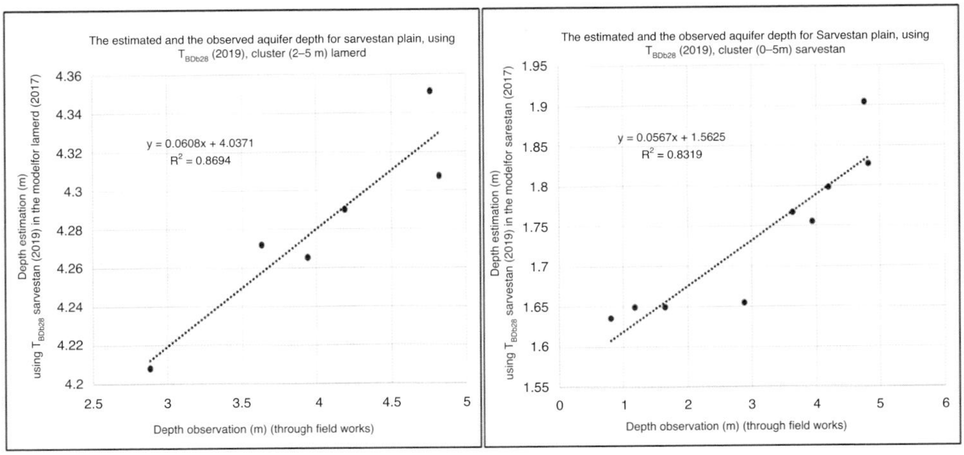

Figure 14.8 Predicting Sarvestan plain's aquifer depth measures using the parameters estimated with models developed for Lamerd plain (depth 0–5 metres)

to the wells, which show a clustering pattern that can be observed within equal-interval depths of 5 metres.

Table 14.3 shows that ε_{Db28} and T_{BDb28} are the most correlated variables associated with aquifer depth (0–5 metres), for both the Sarvestan and Lamerd plains. The PCC of the products of Lamerd plain is also higher than that observed for the Sarvestan plain, which represents a 1 per cent difference, with figures of about 0.07 for the Sarvestan plain and 0.08 for the Lamerd plain. Among the indices, the day- and night-time emissivity difference of Band28 ($\varepsilon_{Db28} - \varepsilon_{Nb28}$) and day- and night-time brightness temperature difference of Band28 ($T_{BDb28} - T_{BNb28}$) are the most correlated indices associated with aquifer depth, for both Sarvestan and Lamerd plains.

The PCC of the indices in the Lamerd plain (0.24) is higher than in the Sarvestan plain (0.23). The PCC for ε_{Db28}, T_{BDb28}, ($\varepsilon_{Db28} - \varepsilon_{Nb28}$) and ($T_{BDb28} - T_{BNb28}$) are statistically significant at p values of 1 and 5 per cent. However, the PCC is also high but significant for ε_{Db31}, ε_{Db32}, T_{BDb31}, T_{BDb32}, LST_{DMOD11}, LST_C, ($\varepsilon_{Nb31} - \varepsilon_{Nb32}$), ($T_{BNb31} - T_{BNb32}$) and ($LST_D - LST_N$).

Table 14.4 indicates that the models based on T_{BDb28} and ($\varepsilon_{Db28} - \varepsilon_{Nb28}$) predict the aquifer depth (0–5 metres) for the Sarvestan plain with a square residual of 0.76 and 0.79 per cent, respectively. Table 14.5 shows that the square residuals from the models using the Lamerd plain data are higher than 0.97 per cent for both ($\varepsilon_{Db28} - \varepsilon_{Nb28}$) and ($T_{BDb28} - T_{BNb28}$), respectively.

For the Sarvestan plain, Table 14.4 indicates that the regression coefficients (R) of models 1–4 are decreasing, using day- and night-time difference indices for emissivity and brightness temperature. The coefficients are considerably higher for

Lamerd plain, as presented in Table 14.5. It seems that the contribution of the indices is higher in the SLR models carried out with the Lamerd plain data.

Table 14.6 shows the results of the regression coefficients (R) increased with predicting variables of ε_{Db31}, ε_{Db32}, LST_D, LST_{CD}, $(\varepsilon_{Db31} - \varepsilon_{Db32})$ and $(T_{BDb31} - T_{BDb32})$, even for the deeper aquifer categories (45–60 metres) in the Sarvestan plain. However, there is an exception for aquifer depth (5–10 metres), which shows weaker results even in comparison with the SLR models. Based on the regression coefficients (R), the SMLR models show better results for Lamerd plain, when more predictive variables are used, as shown in Table 14.7. In Lamerd plain, the SMLR modelling was carried out with regard to the aquifer depth (0–5 metres) only, as the plain was used as a test region for the validation analysis of Sarvestan plain estimation models.

The model fitted with data from Sarvestan plain in 2017 shows good performance in predicting the aquifer depth of the plain in 2019. Figure 14.8 shows that the model estimates more than 0.83 per cent of the variation of the aquifer depth. In addition, the models show good inter-region predictive performance, with promising results in the prediction of the aquifer depth of Sarvestan plain in 2019 based on the parameter values estimated from a model fitted with Lamerd plain data in 2017. The model explains more than 0.86 per cent of the variation of the aquifer depth, after 2 years of the model establishment. Despite the large part of the variation explained by the models in validation process, the results of the numerical validation criteria were comparatively low, with 12 per cent, 0.55 metre depth and 52 per cent efficiency for the average RE, RMSE and CE, respectively.

14.4 Discussion

The results confirm that emissivity products and their derivatives such as $T_{B\lambda}$ and LST play a major role and their association with the aquifer depth is statistically significant. Band 28 shows good predictive power, which might be due to its sensitivity to vapour detection, seemingly originated from shallow parts of the aquifers through capillary action. High evaporation from the soil surface in an arid environment may accelerate capillary action and increase the capillary fringe over the aquifer. By increasing water vapor above the earth surface, its influence on the b_{28} behaviour and response extend to deeper aquifers.

Day- and night-time MODIS products are also important factors that are significantly associated with the variation of the aquifer depth in arid and semi-arid environments (Urqueta et al., 2018). In desert regions, the high variation of the temperature between the day and the night may explain variations in deeper aquifers, which has been illustrated in this present research case study. The

Sarvestan plain is located in the Lake banks, which are generally covered by irregular salt crusts caused by capillary action (Zare et al., 2019). The evaporites layer also affects the land surface thermal behaviour and may affect the modelling results in these regions (Fallah Shamsi et al., 2017).

The results suggest that emissivity, brightness temperature and land surface temperature are important factors that contribute to explain the variations of aquifer depth. ε_{Db28}, and its derivatives, such as T_{BDb28} and LST, are affected indirectly by surface materials, background effects of soil types and by cooling effects of the capillary action of the aquifer moisture beneath the surface, respectively. The geological structure and materials are also important characteristics of the subsurface of the plain, which may affect the soil properties of the hydrogeology by capillarity action.

Despite the significant correlation between emissivity-based products and aquifer depth measures observed in the study areas, the results of the models cannot be well generalized. Various sources of error affect the reliability of the model, such as errors on the surface spectral behaviour in TIR, reflective and emissive electromagnetic ranges.

Land management, annual agricultural activities, frequently ploughing and irrigation methods affect the earth's surface materials, texture and its spectral behaviour, especially for old plains with a long history of agricultural practices.

The emissivity is influenced by soil types (Fallah Shamsi et al., 2012) and consequently soil types affect the estimation of aquifer depth within a regression framework. The analysis of data becomes more complex when the investigated wells are in the same soil type but show various depths or when wells are of similar lengths but observed in different soil types. Using the day- and night- time difference indices reduces the variation slightly. It provides better results than those using initial image bands of the MODIS dataset, as illustrated in this research.

For the regions in the Lake banks, the estimated results are weaker and the model validation criteria are considerably low, while better results are expected for shallow aquifers. It seems that water vapour dump, originating from the lake surface may affect b_{28}, b_{31}, b_{32} and emissivity products due to their high sensitivity to cloud and water vapor.

Precedent studies indicate that the aquifer upper level is bounded within an interval, represented by a mean depth plus a stochastic term. This stochastic additive term, associated with the depth variation, affects the modelling results, especially when simple methods such as the SLR or SMLR are used to estimate the depth. Therefore, additional variability associated with the stochastic behaviour of aquifer depth, rooted in aquifer structure, soil types, sediments and bed rock formation were not accounted in the suggested models. This might explain the high correlation observed among the investigated variables while exhibiting a relatively low predictive power of the models.

The results show a good ability of the models to predict aquifer depth of the Sarvestan plain in 2019 using the data from the same plain in 2017. The cross-plains estimations – using past data from one plain to predict future observations in another plain – are similarly satisfactory. Our research does not consider temporal variations, and excludes potential seasonal effects (i.e., radiation, temperature, vapour content and atmospheric condition in general). We exclusively considered short time intervals, June 2017 and 2019 to ensure similar conditions. However, local atmospheric conditions, such as the cooling effect of surface winds, were neglected (Alkhaier et al., 2009; Brenning et al., 2012).

14.5 Conclusion

This current research has proposed a method of estimating aquifer depth. Our approach aimed at providing a simple modelling framework, requiring little field data along with reduced costs associated with remote sensing images that do not need sophisticated data processing equipment and facilities.

Despite the limitations of the methods, our results bring an engineering solution to tackle water scarcity, a key strategic problem in arid environments. However, further research based on more sophisticated statistical models could bring light into the mechanisms behind the processes associated with aquifers of deeper depths. A set of novel indices were to summarize day- and night-time behaviours of spectral remotely-sensed data into simple values, which contributes to generate new products used in daily remote sensing techniques applied to natural resource and environmental studies. In practice, more efficient models of this type simply could be promoted to aquifer depth mapping instrument. This could be used to identify the locations that are more suitable to contain wells. The results of the models are promising but, with their current forms, there are not applicable yet. New remotely-sensed data along with more sophisticated models should improve the performance of the models. This current research has offered a new research perspective in the investigation using data and products from passive spectral remote sensing techniques that could be expanded beyond the field of aquifer studies.

References

Ahmad, S. P., Salomonson, V. V., Barnes, W. L., Xiong, X., Leptoukh, G. G., & Serafino, G. N. (2002). MODIS radiances and reflectance for earth system science studies and environmental applications. *Proceedings of the 18th International Conference on*

Interactive Information and Processing Systems (IIPS) for Meteorology, Oceanography, and Hydrology, 13–17 January 2002, Orlando, Florida, pp. 188–192.

Alkhaier, F., Schotting, R. J., & Su, Z. (2009). A qualitative description of shallow groundwater effect on surface temperature of bare soil. *Hydrology and Earth System Sciences* 13(9): 1749–1756.

Bastiaanssen, W. G. M. (2003). Retrieving soil moisture storage in the unsaturated zone using satellite imagery and bi-annual phreatic surface fluctuation. *Irrigation and Drainage Systems* 17(3): 141–161.

Brenning, A., Peña, M., Long, S., & Soliman, A., (2012). Thermal remote sensing of ice-debris landforms using ASTER: An example from the Chilean Andes. *Cryosphere* 6 (2): 367–382.

Caselles, E., Pitarch, C., & Caselles, V. (2014). Estimation of the water table depth of the Calarasi district Island (Romania) at the Danube River using ASTER/DEM data. *European Journal of Remote Sensing* 47(1): 169–180.

Chase, M. E. (1969). Airborne remote sensing for groundwater studies in prairie environment. *Canadian Journal of Earth Sciences* 6(4): 737–741.

Fallah Shamsi, S. R., Ebrahimi, Z., Ekhtesasi, M., & Kompani-Zare, M. (2017). Segmentation of playa geomorphological facies in desert regions, using integrated remote sensing and statistical modeling of soil properties. *Journal of Remote Sensing Technology* 5(1): 1–9.

Fallah Shamsi, S. R., Zare, S., & Abtahi, S. (2012) Soil salinity characteristics using moderate resolution imaging spectro-radiometer (MODIS) images and statistical analysis, *Archives of Agronomy and Soil Science* 10(4): 471–489.

Fang, B., Lakshmi, V., Bindlish, R., & Jackson, T. J. (2018). Downscaling of SMAP soil moisture using land surface temperature and vegetation data. *Vadose Zone Journal* 17(1): 1–15.

Hansen, J., Ruedy, R., Sato, M., & Lo, K. (2010). Global surface temperature change. *Reviews of Geophysics*, 48(4): RG4004.

Heilman, J. L., & Moore, D. G. (1982). Evaluating near-surface soil moisture using heat capacity mapping mission data. *Remote Sensing of Environment* 12(2): 117–121.

Jensen, J. R. (2000). *Remote Sensing of the Environment: An Earth Resource Perspective* (2nd ed., p. 544). Upper Saddle River, NJ: Prentice-Hall, Inc.

Jiménez-Muñoz, J. C., & Sobrino, J. A. (2003). A generalized single-channel method for retrieving land surface temperature from remote sensing data. *Journal of Geophysical Research* 108(D22): 4688–4695.

Kappelmeyer, O. (1957). The use of near surface temperature measurements for discovering anomalies due to causes at depths. *Geophysical Prospecting* 5(3): 239–258.

Karnieli, A., Agam, N., Pinker, R. T., Anderson, M., Imhoff, M. L., & Gutman, G. G. (2010). Use of NDVI and land surface temperature for drought assessment: Merits and limitations. *Journal of Climate* 23(3): 618–633.

Kustas, W. P., & Norman, J. M. (1996). Use of remote sensing for evapotranspiration monitoring over land surfaces. *Hydrological Sciences Journal* 41(4): 495–516.

Li, Z. L., & Becker, F. (1993). Feasibility of land surface temperature and emissivity determination from AVHRR data. *Remote Sensing of Environment* 43(1): 67–85.

Li, Z. L., Wu, H., Wang, N., et al. (2013). Land surface emissivity retrieval from satellite data. *International Journal of Remote Sensing* 34(9–10): 3084–3127.

Modiri, S., & Modiri, M. (2016). Calibration of separate window model factors to calculate land surface temperature using MODIS images. *European Online Journal of Natural and Social Sciences* 5(2): 546–558.

Ndou, N. N., Palamuleni, L. G., & Ramoelo, A. (2018). Modelling depth to groundwater level using SEBAL-based dry season potential evapotranspiration in the upper Molopo river catchment, South Africa. *The Egyptian Journal of Remote Sensing and Space Science* 21(3): 237–248.

Price, J. C. (1983). Estimating surface temperatures from satellite thermal infrared data: A simple formulation for the atmospheric effect. *Remote Sensing of Environment* 13 (4): 353–361.

Qu, J. J., Gao W., Kafatos, M., Murphy, R. E., & Salomonson, V. V. (2006). *Earth Science Satellite Remote Sensing, Vol. 1: Science and Instruments*. New York: Springer.

Quiel, F. (1975). Thermal/IR in geology. *Photogrammetry Engineering and Remote Sensing* 41(3): 341–346.

Rodell, M., & Famiglietti, J. S. (2002). The potential for satellite-based monitoring of groundwater storage changes using GRACE: The High Plains aquifer, Central US. *Journal of Hydrology* 263(1–4): 245–256.

Sobrino, J. A., & Raissouni, N. (2000). Toward remote sensing methods for land cover dynamic monitoring, application to Morocco. *International Journal of Remote Sensing* 21(2): 353–366.

Sobrino, J. A., Soria, G., & Prata, A. J. (2004). Surface temperature retrieval from Along Track Scanning Radiometer 2 data: Algorithms and validation. *Journal of Geophysical Research Atmospheres* 109(11): D11101.

Sutanudjaja, E., De Jong, S., van Geer, F., & Bierkens, M. (2013). Using ERS space-borne micro-wave soil moisture observations to predict groundwater head in space and time. *Remote Sensing of Environment* 138: 172–188.

Tiyip, T., Cui, J. Y., & Ding, J. L. (2005). Study on the means of groundwater distribution beneath the oasis-desert ecotone in an arid area by using thermal infrared data. *Arid Land Geography* 28(2): 252–257.

Urqueta, H., Jodar, J., Herrera, C., et al. (2018). Land surface temperature as an indicator of the unsaturated zone thickness: A remote sensing approach in the Atacama Desert. *Science of the Total Environment* 612: 1234–1248.

Valor, E., & Caselles, V. (1996). Mapping land surface emissivity from NDVI: Application to European, African and South American areas. *Remote Sensing of Environment* 57: 167–184.

White, D. A. (2014). The MODIS conversion toolkit (MCTK) user's guide, ITT Visual Information Solutions. Available from http://nsidc.org/data/modis/tools.html (Last accessed 3 September 2021).

Zare, S., Fallah Shamsi, S. R., & Abtahi, S. A. (2019). Weakly-coupled geo-statistical mapping of soil salinity to stepwise multiple linear regression of MODIS spectral image products. *Journal of African Earth Sciences* 152: 101–114.

Zhang, Z.-M., He, G., Xiao, R.-B., Wang, W., & Ouyang, Z. (2006). Land surface temperature retrieval of Beijing city using MODIS and TM data. *Proceedings of the IEEE International Geoscience and Remote Sensing Symposium, IGARSS*, 31 July–4 August, Denver, CO, pp. 1094–1096.

Part III

Sustainable Water Management under Future Uncertainty

15

Managing Urban Flood Risk and Building Resilience in a Changing Climate

YUELING WANG, QIUHONG TANG AND NIGEL WRIGHT

15.1 Introduction

Urban flooding is a specific type of flooding, which is prevalent in urban areas and caused by an exceedance of the capacity of both natural and artificial drainage. The widespread impacts in urban areas of impermeable surfaces, inadequate drainage capacity and high building densities during extreme precipitation induce excessive runoff discharge (Gaitan et al., 2016; Yao et al., 2016). Urban flooding can be driven by different sources, such as pluvial flooding, fluvial flooding and coastal flooding (Saul et al., 2011). These types of urban flooding are related to a range of events, most commonly short duration–high intensity precipitation, prolonged moderate rainfall, rapid snowmelt, river overbank flow, backward flow from the drainage network, tidal surge or waves and coastal tsunamis (e.g., Sörensen & Mobini, 2017; Sadler et al., 2018; Griffiths et al., 2019; Rubinato et al., 2019; Wang et al., 2019). These events can cause considerable impacts on local residents, properties, assets, infrastructure systems, the economy and the environment (Lu & Ran 2011; Zheng et al., 2013).

The potential damage of urban flooding is diverse, including direct impact, such as casualties and damaged infrastructure; indirect impact, such as increased traffic congestion; secondary effect on people who rely on services or products from the affected areas; and intangible impacts on environmental quality (Petersen, 2001). According to data from the United Nations in 2018, 55 per cent of the world's population lives in urban areas (UN, 2018). Assets and properties are increasingly concentrated in urban areas worldwide. The global direct economic loss caused by flood events between 1980 and 2018 exceeds $1.03 trillion, and the worldwide fatalities due to catastrophic floods and other relevant floods are 115,542 and 223,482, respectively (NatCatSERVICE, 2019 values). Based on the global flood

This research is supported by a Royal Society Newton Advanced Fellowship and National Natural Science Foundation of China (41730645, 41890823, 51609231).

risk projections without mitigation, Winsemius et al. (2016) showed that global total damage may increase by a factor of up to 20 by the end of the century.

The best way to investigate the main drivers of urban flooding is to better understand the mechanisms and characteristics behind the observed phenomenon. For example, pluvial flooding, caused by normal rainfall, occurs frequently but does not pose a serious threat to human life or infrastructure; it may only result in traffic delays, strains on the drainage system and property damage (Moftakhari et al., 2018). In contrast, pluvial flooding due to extreme rainstorms may inundate large parts of a city for several days, induce huge property damage and pose serious health problems (e.g. Vahedifard et al., 2016; Watson et al., 2018). An urban flooding disaster ultimately results from an imbalance between discharge and drainage capacity. In the new era dominated by human beings – the Anthropocene – anthropogenic factors have become an important driving force in changing flood risk (Tang, 2020). Human influence on floods includes a range of factors, including human-induced climate change, making hydrologic extremes more frequent and severe; land-use and land-cover change, such as deforestation and urbanization modifying the hydrologic response to climate extremes; and artificial flood control measures, changing vulnerability to floods. On the one hand, rapid changes in the global water system may increase the threats of hydrological extremes. On the other hand, population growth and socio-economic development have led to an increase in global exposure to flood hazard, all of which means that the vulnerability of social-ecological systems needs to be reduced by flood risk mitigation measures (Tang, 2020). There is an urgent need to take human behaviour and socio-economic feedback into consideration for building urban resilience to floods.

One of the major drivers behind flood hazards is climate change, causing significant changes in rainfall/precipitation patterns and sea level (IPCC, 2014). Climate change predictions show an increasing trend of extreme precipitation. However, this trend varies at both temporal and spatial scales. The Special Report of the Intergovernmental Panel on Climate Change (IPCC) notes that ‘It is likely that the frequency of heavy precipitation or the proportion of total rainfall from heavy falls will increase in the twenty-first century over many areas of the globe’ and ‘Overall there is low confidence in projections of changes in fluvial floods. Confidence is low due to limited evidence and because the causes of regional changes are complex’ (IPCC, 2012, p. 13). The Synthesis Report of the IPCC announced that ‘In urban areas, climate change is projected to increase risks for people, assets, economies and ecosystems, including risks from heat stress, storms and extreme precipitation, inland and coastal flooding, landslides, air pollution, drought, water scarcity, sea level rise and storm surges (very high confidence)’ (IPCC, 2014, p. 69).

With the deterioration of climate conditions, rapid and unplanned urbanization could generate high vulnerability in hazardous areas, especially in developing and emerging economies. The proportion of the world's urban population is expected to increase to 68 per cent by 2050, with nearly 90 per cent of this increase taking place in Asia and Africa (UN, 2018). The 2018 Revision of World Urbanization Prospects produced by the Population Division of the UN Department of Economic and Social Affairs (UN, 2018) notes that the future growth of the world's urban population is expected to be highly concentrated in a few countries. It also states that, between 2018 and 2050, India, China and Nigeria together will account for 35 per cent of this growth. 'By 2050, it is projected that India will have added 416 million urban dwellers, China 255 million and Nigeria 189 million' (UN, 2019, p. 1). These risks will be amplified for exposed areas within countries that lack the essential infrastructure and services. These complicated situations raise the significance of and necessity for flood risk management in order to tackle the impact of climate change and increase the resilience of cities.

Flood risk management is a decision-making process under uncertainty, considering natural and anthropogenic factors. This process can be summarized in the following steps: data and information gathering, modelling and risk assessment, decision-making, implementation, reviewing and modification to adapt to changing circumstances and societal preferences (Hall et al., 2003; Schanze, 2006). Traditionally, extensive flood-control infrastructures have been built to protect cities from flood hazard, although they are now accepted to have an adverse impact on the natural environment and to increase long-term flood risk (Burby et al., 2000; Smits et al., 2006; Sörensen et al., 2016) without sufficient consideration of vulnerability (Merz et al., 2010). In comparison, modern flood risk management aims to develop an integrated approach for sustainable urban water management and sustainable urban development. This new approach is shifting the focus from flood control to flood adaptation to increase urban resilience to flood hazard. In this context, resilience represents the urban capacity of preparing for, absorbing, adapting to and recovering from flood disasters (Sendzimir et al., 2007; Liao, 2012; Chelleri et al., 2015). Numerical modelling is the best way to represent the future climate scenarios and estimate the impact of climate change on urban flooding.

Modelling is now an indispensable part of flood risk assessment and management. It enables the investigation of the interaction between climate change, urban flooding process and socio-economic response (Beven, 2012; Nkwunonwo et al., 2020). An integrated modelling framework is required to support comprehensive consideration of flood risk in order to accomplish sustainable urban development strategies over the next 50 years. It offers suitable flooding mitigation and adaptation strategies that may quantify the impacts of climate change on urban flooding hazard and identify areas at risk while accounting for the uncertainty of climate projections (e.g. Ward et al., 2013; Winsemius et al., 2016). The information produced by the models can be used to: (1) address long-term risks associated with several climate change scenarios of extreme precipitation and sea-level rise; (2) predict the induced changes in urban flooding processes and (3) assess the probable consequence of flood hazard on the local population, properties, assets and infrastructures due to the socio-economic development and population growth.

Urban flood hazards, in terms of pluvial, fluvial and coastal flooding, are projected to be intensified under climate change. Furthermore, the socio-economic development and population density in urban areas may dramatically increase flood vulnerability and raise an urgent requirement for integrated flood risk management to mitigate and adapt to flood hazards whilst achieving the Sustainable Development Goals (Tang, 2020). Therefore, this chapter delivers an up-to-date literature review on urban flooding under climate change and discusses the main drivers of urban flooding. It then investigates progress and advances in urban flood risk management in developing, emerging, and developed countries and sets out an integrated modelling framework to support flood risk management and decision-making within the concept of urban flood resilience.

15.2 Natural and Anthropogenic Drivers

Risk can be conceptualized as the combination of hazard and vulnerability (UNISDR, 2004). Hazard is defined by the United Nations Inter-Agency Secretariat of the International Strategy for Disaster Reduction (UNISDR) as 'a potentially damaging physical event, phenomenon or human activity that may cause the loss of life or injury, property damage, social and economic disruption or environmental degradation' (UNISDR, 2004, p. 4). The UNISDR define vulnerability as 'the conditions determined by physical, social, economic, and environmental factors or processes, which increase the susceptibility of a community to the impact of hazards' (UNISDR, 2004, p. 7). Hazard is the cause and vulnerability is the likely effect. Vulnerability here includes both exposures, i.e., the physical aspects of vulnerability, and non-physical aspects of vulnerability. Exposure refers to people, property, systems, or other elements in flood hazard zones that are thereby subject to potential losses. Non-physical aspects of vulnerability refer to the characteristics of a system or asset that make it susceptible to the damaging effects of flood. Urban flood hazard and vulnerability are clearly interrelated and are both affected by many natural and anthropogenic factors. Here an impact factor that leads to a better or worse condition of hazard and vulnerability is considered as a driver. It can be related to climate change, rapid urbanization, flood defence infrastructure, economic growth and public awareness. Recent research in urban flood risk has focused on two main drivers, climate change and urbanization (e.g., Miller & Hutchins, 2017; Berndtsson et al., 2019; Cobbinah et al., 2019). The latter is closely associated with socio-economic development.

Below, we discuss the main drivers of urban flood hazard, which include pluvial flooding, fluvial flooding and coastal flooding.

15.2.1 Pluvial Flooding

Pluvial or rainfall flooding in urban areas is a natural phenomenon where the surface water cannot be timely discharged into a river channel or sewerage system due to the length of time or intensity of a rainstorm (Kolsky, 1999; ten Veldhuis et al., 2010). Precipitation is the primary factor causing pluvial flooding disaster in urban areas and is beyond immediate control. The Synthesis Report of IPCC reported that the number of heavy precipitation events have been increasing in some regions since 1950, and predicts that they will become more intense and frequent over most of the mid-latitude land masses (IPCC, 2014). The projected changes in extreme precipitation due to climate change will increase flood risk for people, assets, economies and ecosystems (IPCC, 2014).

In urban areas, most of the underlying surface is impermeable. The infiltration capacity is therefore restricted, and in most cases depends on the designed sewer system and the few remaining natural spaces (e.g., Ungaro et al., 2014). Given that climate change is expected to increase the frequency of heavy precipitation in the twenty-first coming decades in many regions (IPCC, 2014), the design standard of drainage systems must be improved. The increase of impermeable surfaces due to urbanization also reduces surface roughness and thereby increases flood risk (Scholz, 2013). Hence, high imperviousness of the underlying surface makes urban areas significantly more sensitive to changes in precipitation caused by climate change (Scholz, 2013).

15.2.2 Fluvial Flooding

Cities are mainly located on rivers due to easy access to water supply, transportation and land suitable for urbanization. Extreme precipitation, snowmelt or dam failure in upstream areas leads to increased flow in the rivers in urban areas and the consequent overbank flow which inundates low-lying urban areas. The urban sewer system cannot drain a large amount of high-speed flow in a short time. The induced severe streamflow poses a threat to the local residents, assets, infrastructure system and environment. The main drivers of fluvial flooding risk are related to climatic and anthropogenic factors including upstream extreme precipitation, decreasing permeability, high building density, an inadequate sewer system and socio-economic growth (Berndtsson et al., 2019).

Increases in precipitation with climate change are not confined to urban areas. The Synthesis Report of IPCC noted 'there are likely more land regions where the number of heavy precipitation events has increased than where it has decreased' (IPCC, 2014, p. 8). Robust evidence suggests that the proportion of global population affected by major river floods will increase in the twenty-first century as global warming intensifies (IPCC, 2014). Furthermore, the global drivers of large-scale river flood risk have been projected by Winsemius et al. (2016). Their study shows that climate change is the dominant driver of an expected increasing flood risk. However, the results show regional disparities for different income regions. In southeast Asia, the effect of climate change is dwarfed by the rapid socio-economic growth.

15.2.3 Coastal Flooding

Coastal flooding is a common natural hazard in coastal areas where a surge wave with devastating force overtops or destroys coastal defences and inundates the coastal urban area. It is usually accompanied by extreme weather events such as heavy rain and strong winds. In these circumstances, the urban drainage system will be overloaded or even damaged. Coastal floods are always characterized by high water levels, destruction and brevity of warning. Predictions are usually one or two days, or even only hours ahead, and therefore may fail to inform the population at risk in a timely and efficient manner of future events. Coastal flooding therefore poses a great threat to local people and properties. Whilst coastal flooding is less frequent, the potential impact is greatest.

Sea level rise is the main climatic driver influencing coastal flooding hazard. The global mean sea level rose by 0.19 metre over the period 1901 to 2010 (IPCC, 2014). Glacier mass loss and ocean thermal expansion due to climate warming has contributed about 75 per cent of the observed global mean sea level rise since the early 1970s (IPCC, 2014). The global mean sea level will continue rising in the twenty-first century under all Representative Concentration Pathways (RCPs), at a faster rate than the 2.0 mm yr^{-1} observed during 1971–2010 (IPCC, 2014). This trend will pose a great threat to the residents, assets, infrastructure, economy and ecosystems in coastal urban areas, especially for those lacking a resilient infrastructure or located in exposed areas.

15.2.4 Summary

The extreme precipitation and rising sea-level due to climate change directly increase the probability of occurrence of urban flooding hazard, especially for the areas with unresponsive urban planning regimes (e.g., Winsemius et al., 2016; Cobbinah et al., 2017, 2019). The increasing impermeable surface and building density as well as the insufficient drainage capacity in urban areas also increase the probability of occurrence of flooding hazard (e.g., Bruwier et al., 2020). Furthermore, the concentration of population, assets and values in urban areas increases the magnitude of vulnerability (e.g., Tanoue et al., 2016). However, appropriate urban planning may effectively mitigate the magnitude of flooding damage due to climate change (e.g., Wang et al., 2019), by means of reducing greenhouse gas emission, increasing a more natural underlying surface and drainage capacity and providing natural flooding pathways. These drivers are closely interrelated, interactive and complicated due to the various flooding sources and hydrologic conditions. To achieve the UN sustainable development goals for a city, it is indispensable to consider climate change impact in urban planning to mitigate and adapt to the increasing flood risk.

15.3 Flood Risk Management

As can be seen from the discussion in Section 15.2, climate change and urbanization are increasing flood risk in urban areas. In recent decades, the response to flooding has moved away from an approach based merely on defence or hazard reduction. The approach is now focused on both hazard and vulnerability which recognizes that it is not always possible to significantly reduce the hazard and, in this situation, it is important to reduce vulnerability in order to reduce the overall risk. This approach also allows for the selection of the most cost-efficient response which could be either to reduce hazard or reduce vulnerability depending on the specific environmental, geographic or socio-economic factors in a particular location.

Accordingly, flood risk management has moved from flood defence engineering to flood mitigation. These changes assume that flood risk management is not a 'one-off' activity but a continuous process of adaptive management (Hall et al., 2003; Hutter & Schanze, 2008; Saul et al., 2011). For disaster risk reduction, the United Nations approved three notable agreements in 2015, that is, the Sendai Framework for Disaster Risk Reduction 2015–2030 (Sendai Framework), the Sustainable Development Goals (SDGs) and the Paris Agreement on Climate Change (Aitsi-Selmi et al., 2016). Based on these frameworks, most countries have implemented various urban sustainable development plans to deal with the problems encountered in urban development, especially for addressing increasing urban flooding risk.

A further consequence of the move to system-wide flood risk management is that the approach can consider a wide range of responses to flooding, including 'hard' and 'soft' approaches. The former include defence walls, storage basins, soakaways, widespread storage at source in urban areas and property level protection, whilst the latter include evacuation plans, land use planning, insurance and upstream tree planting. Each of these, along with other relevant methods, can apply to flood risk management, and a system-wide risk-based approach should consider and evaluate all possible responses.

Further to the development of flood risk management as a continuous process of adaptive management, the concept of flood resilience (Gersonius, 2012) has also been developed in the framework of urban water management (e.g., Andoh & Iwugo, 2002; Brown et al., 2009; Sayers et al., 2013; Meerow et al., 2016; Bertilsson et al., 2019; Cariolet et al., 2019). As defined in Sayers et al. (2013) resilience is 'The ability of an individual, community, city or nation to resist, absorb or recover from a shock (such as an extreme flood), and/or successfully adapt to adversity or a change in conditions (such as climate change or an economic downturn) in a timely and efficient manner' (p. 24). The urban system is considered as a complex and adaptive integrated system involving multiple social, economic, ecological, political, technical factors at various temporal and spatial scales. In essence, flood resilience is based on an acceptance that, in this case, an urban system cannot be fully restored to its antecedent state after an event, but that it will adapt and return to a functioning state over a period after the event. This entails the adoption of responses to flood risk that can be adapted over time and places a greater emphasis on adaptability under uncertain future conditions rather than resistance to impacts and the maintenance of the original design (Gersonious et al., 2016). Hence, flood resilience is complex and dynamic and difficult to quantitatively model and estimate. Since it is closely related to vulnerability, food resilience has often been associated with vulnerability in earlier studies (e.g., Koks et al., 2015). Nowadays, the assessment approach of urban resilience has been studied and categorized into qualitative and quantitative analysed methods (e.g., Mugume et al., 2015; Miguez & Veról, 2017; Bertilsson et al., 2019).

The coordinated planning, design and operation of closely coupled urban water systems is necessary to achieve urban flood resilience in an uncertain future. Building urban flood resilience will involve dealing with future uncertainties associated with extreme weather events driven by global climate change, anthropogenically altered hydrological extremes and rapid urbanization (Fenner et al., 2019). To tackle the challenges, recent research progress has been made in various aspects of urban flood resilience, including the blue–green path (O'Donnell et al., 2019), adaptation pathways for drainage infrastructure planning, interoperability (Vercruysse et al., 2019), and urban metabolism modelling (Behzadian & Kapelan, 2015). These developments will help find integrated solutions to reduce urban flood risk.

The development of flood risk management and resilience depends on the economic development of the investigated country. To compare different systems and contexts, Sections 15.3.1–15.3.4 will consider flood risk management and resilience in developing, emerging and developed economies, taking Africa, China, the United States and the United Kingdom as examples. Developed countries tend to have a history of investment in flood defence and a more recent development of resilience at a national scale. In contrast, the least developed countries tend to have little investment at national scale although, on a local scale, communities may have taken their own resilience measures based on a long history of living with flooding. Flood risk is often highest in rapidly developing economies, as economic development often outpaces investment, leading to significant risk in newly developed urban areas.

15.3.1 In Developing Economies: Africa

Over the past several decades, cities in sub-Saharan Africa have experienced increasing threats from flooding, impacting the local citizens, infrastructure, assets, economy and policies (Tiepolo, 2014). A large increase of flood frequency in eastern Africa has been reported (Hirabayashi et al., 2013), based on the projections under high-concentration scenarios (van Vuuren et al., 2011). Furthermore, in developing countries, the rapid urbanization also increases the anthropogenic pressure in urban areas, leading to increased vulnerability to flooding. In addition, the complex political contexts that affect many governments in the African continent can cause inadequate governance systems and economic challenges (Filho et al., 2018). A major challenge for African governments is the intensive migration from rural to urban areas which increases the need for urban adaptation strategies for flooding. In-depth local research on adaptation strategies adopted in African cities is limited (Cabral et al., 2017; Nkhonjera, 2017; Filho et al., 2018), so it is difficult to evaluate and improve existing strategies.

With the assistance of international organizations, sub-Saharan African governments have launched large-scale initiatives of physical infrastructure and national response plans for mitigation and adaptation strategies to enhance urban resilience to flooding disaster (Diagne et al., 2012; Pelling & Wisner, 2012; Bottazzi et al., 2019). However, as a top-down approach, capacity for long-term maintenance of large infrastructure, cooperation of stakeholders at all levels and poverty that affect many individuals are challenging issues in the current attempts to implement strategies to combat floods. These approaches need to be considered jointly with bottom-up strategies that adequately include the local population in the political decision-making and adaptive management, in order to take full advantage of the national approaches (Bottazzi et al., 2019). As well as the focus on physical infrastructure, soft measures are also encouraged in integrated flood risk management initiatives, for example, improving urban greening, urban planning and environmental awareness (e.g., Maheu, 2012; O'Donnell et al., 2018). Although the innovative approaches in sub-Saharan Africa are currently limited, they provide a pathway for integrating with existing measures and improving the resilience of the urban poor to flood risk.

15.3.2 In an Emerging Economy: China

In recent decades, with climate warming and rapid urbanization, urban flooding has become the most catastrophic and pervasive disaster in China, causing huge economic losses and large numbers of casualties (Lu & Ran, 2011; Ministry of Water Resources of the People's Republic of China, 2013; Zheng et al., 2013). In December 2013, President Xi Jinping promoted the concept of the 'sponge city' and announced that rainwater retention and natural strategies should be considered as the first priority when upgrading existing urban drainage systems. The 'sponge city' is a recent concept for urban flood risk management. It means that the city, like a sponge, should have the capability to absorb, store, infiltrate, purify rainwater and utilize it when needed, in order to improve the function of the urban ecosystem and reduce the occurrence of urban flood disaster. In this way a city will exhibit good resilience in adapting to environmental changes and mitigating urban flooding hazard. In the framework of sponge city, China is experiencing a transition toward sustainable and high-quality development in an exploratory stage from 2014 to 2020. However, this exploration process brings both opportunities and challenges.

A first step consists of understanding the mechanisms behind urban flooding hazard, risk and management under a changing environment (e.g., climate change and urbanization). This includes urban hydrology and hydrodynamics (e.g., Fletcher et al., 2013; Müller & Haberlandt, 2018; Macchione et al., 2019), the hydrological impact of urban underlying surface (e.g., Schwartz & Smith, 2016; Löschner & Nordbeck, 2019; Wang et al., 2020), models for a flood risk assessment (e.g., Wang et al., 2015; Cai et al., 2019; Costabile et al., 2020), management strategy (e.g., Jenkins et al., 2017; Waghwala & Agnihotri, 2019; Mai et al., 2020), and national policy and regulations (e.g., Hegger et al., 2020; Metz et al., 2020). Traditional physical engineering measures should be combined with nature-based measures in a resilient hybrid system. However, due to the knowledge gap and lack of experience, China decided to explore the practicalities of the sponge city approach in 30 'pilot cities' in an approach of 'learning by doing'. The national government of the PRC is investing more than ¥12 billion RMB ($1.7bn) in 30 pilot cities over a three-year period, and is also encouraging private participation in a 'Public–Private Partnership' (PPP) financing scheme (e.g., Ke et al., 2010; Wang et al., 2017). Meanwhile the government strongly supports experimental studies and scientific research into the 'sponge city' approach (e.g. Chan et al., 2018; Zhang et al., 2019; Qiao et al., 2020; Li et al., 2020). A progressive goal of the 'sponge city' initiative is to minimize the impact of urban development and construction on the ecological environment, in order to absorb and utilize 70 per cent of rainwater. It is planned to achieve this goal in more than 20 per cent of urban areas by 2020 and in more than 80 per cent of urban areas by 2030. The phased accomplishment of the 'sponge city' initiative is expected by 2030.

15.3.3 In a Developed Economy: The United States

According to the National Weather Service (NWS), 32 flooding events causing over a billion dollars of damage occurred in the United States between 1980 and 2019, generated a total $146.5 billion of damage and 555 deaths (NOAA, 2020). The 16 events between 2011 to 2019 is the same number as those in the whole period from 1980 to 2010 (NOAA, 2020). This suggests that the frequency and severity of flood disasters have increased in this country in the last decade (NOAA, 2020). In the same period, urbanization and socio-economic development along the United States coastline and floodplains have increased urban vulnerability to flood disaster. Individuals and communities of both inland and coastal regions are facing an increasing threat due to climate-change induced extreme precipitation and sea-level rise. In this context, the government needs to take action for flood mitigation to reduce flood risk to human life and property.

Strategies for flood mitigation can be divided into three categories: structural, non-structural and semi-structural measures. The physical structures for flood mitigation are assessed by the American Society of Civil Engineers (ASCE) every four years. The latest ASCE report showed the American infrastructure getting an overall rating of 'D' (Cigler, 2017). Besides the cost of maintenance, the failure/collapse of the engineered structural mitigation needs continuous attention. Crucial tools for non-structural mitigation are tax credits, rebates, grants and cost-sharing arrangements, while insurance is a major economic means (Cigler, 2017). The US government established the National Flood Insurance Programme (NFIP) in 1968 for flood mitigation (Dixon et al., 2006). The NFIP, based on a voluntary principle, involves federal and state governments, private insurance companies and local communities, aiming to enhance flood risk management in flood prone areas. However, due to the increasing flood risk to community, the Federal Emergency Management Agency (FEMA) implemented the Community Rating System (CRS) in 1990 to further reduce flood damage and exposure. The CRS offers discounts on flood insurance premiums to communities that participate in the flood risk management initiatives (Kunreuther & Roth, 1998; Noonan & Sadiq, 2018). Based on the link between community-level and household-level mitigation, CRS designates four categories of activities as public information, mapping and regulation, flood damage reduction and warning and response (Cigler, 2017). FEMA also emphasizes the significance of natural habitats and ecosystems as natural defence (FEMA, 1994). Semi-structural approaches, such as wetland, green infrastructure and floodplain restoration, have been widely studied by academic researchers (e.g., Costanza et al., 2008) and increasingly implemented by communities and homeowners for a nature-based flood risk management. The experience in the United States demonstrates that systematically learning, practicing and sharing mitigation successes is a good pathway to improve the strategies of flood risk management.

15.3.4 In a Developed Economy: The United Kingdom

Flooding is the most common natural hazard in the United Kingdom threatening local residents and properties. Over two million properties are at risk in the United Kingdom with average annual damage of £1.4 billion (Foresight, 2004). Nevertheless, the annual losses are projected to increase under all the emission scenarios of climate change by the 2080s and, according to the high emission scenario, the increase is expected to be around £27 billion in the 2080s (Foresight, 2004). Flood defence has been the traditional measure to protect not only people and private properties, but also public infrastructure and assets. The annual cost of maintaining flood defence is estimated to increase to over £1 billion by 2035 (Bennet & Hartwell-Naquib, 2014). However, non-structural measures, such as private precautions, land-use planning and insurance arrangements, have become an indispensable part of flood risk management in the United Kingdom (Van den Hurk et al., 2014). Due to the climate change and urbanization, adaptive approaches are urgently needed to account for uncertainty and improve the resilience to the changing environment.

In the United Kingdom, water management is operated by a top-down and exclusionary approach and supported by agencies and water companies (Fritsch & Benson, 2013), and modified by implementing the Water Framework Directive (WFD) into a process of multi-actor institutions and participatory at the regional scale (Bensona et al., 2014). Flood risk management in the United Kingdom is based on flood risk assessment and quantified according to the probability of flooding occurrence estimated by an annual exceedance probability and the severity of flooding consequences on human health, infrastructure and socio-economic and environmental aspects. A central government department, the

Department for Environment Food and Rural Affairs (DEFRA), is responsible for flood policy, which is then operationalized by non-departmental government bodies: the Environment Agency (EA) in England, the Northern Ireland Environment Agency (NIEA) in Northern Ireland, Natural Resources Wales in Wales and the Scottish Environment Protection Agency (SEPA) in Scotland. However, urban flood risk management in England is fragmented (Dawson et al., 2008). It is operated by the EA, the Lead Local Flood Authorities (LLFAs), water companies, Internal Drainage Boards (IDB), private insurance organizations and property owners (Bubeck et al., 2015; Rubinato et al., 2019). The EA is responsible for in-depth flood risk assessment and planning of flood risk management at national and regional scales in England (Miller & Hutchins, 2017). In the framework of the Flood and Water Management Act 2010, LLFAs are responsible for developing, maintaining, applying and monitoring the strategies of flood risk management for local level controlling (UK Government, 2010). Although local flood risk management faces specific challenges such as resource constraints, public engagement and expectations, scientific technique and knowledge, and also differences on objectives, priorities and regulatory environments (DEFRA, 2017; Rubinato et al., 2019), these non-structural measures play a significant role of flood risk management in the United Kingdom.

15.3.5 Summary

Climate change is projected to increase urban flood risk to people, assets, economies and the ecosystem (IPCC, 2014) while the consequences vary across regions due to different income levels (Winsemius et al., 2016). The anthropogenic factors during the unplanned urbanization may exacerbate the climatic impact on urban flood risk. Therefore, most countries have developed and adopted appropriate strategies of flood risk management in urban planning and development, to achieve the goals of sustainable cities and communities. To tackle the increasing urban flood problem, the traditional structural flood defence strategies are being complemented by non-structural and semi-structural adaptive resistance-based approaches in various contexts worldwide. Therefore, an integrated modelling framework is essential for flood risk management to incorporate the relationship between climate change, anthropogenic factors and hydrological responses.

15.4 Integrated Modelling Framework

The use of numerical modelling for flood risk management has become vital and indispensable, particularly under the uncertainty associated with predictions

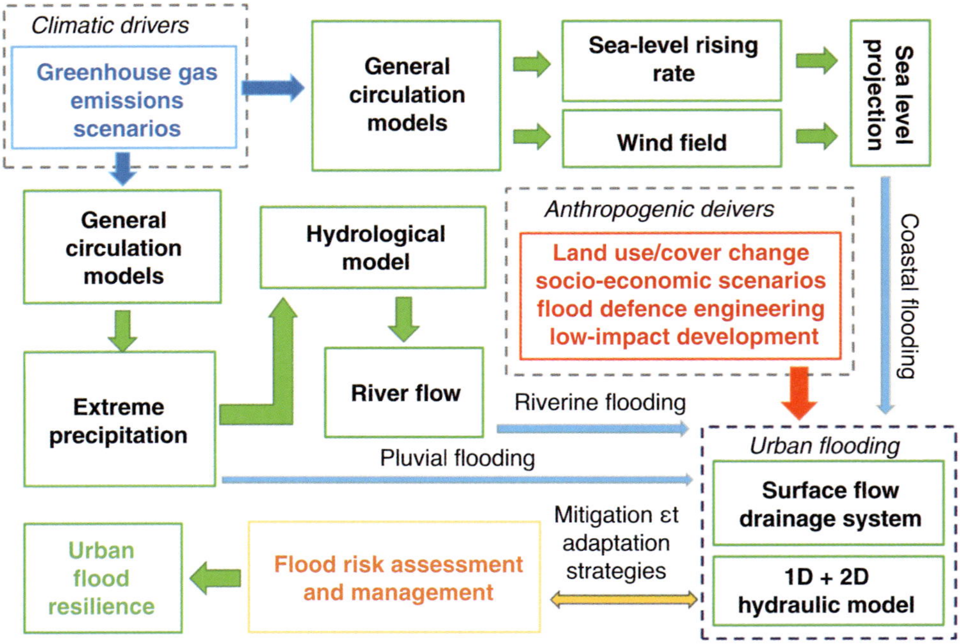

Figure 15.1 The integrated modelling framework diagram

derived from various climate change scenarios. An integrated modelling framework is necessary to encompass and quantify the effects of a wide range of natural and anthropogenic factors associated with flood risk. With an integrated model, urban flood adaptation strategies can be set out and developed to mitigate the effects of urban flooding. An integrated modelling framework requires interdisciplinary knowledge from climatology, geography, hydrology, hydraulic and socio-economic fields. Any framework must consider theoretical and technical issues associated with the coupling mechanisms and temporal and spatial scale-matching between different numerical models. Figure 15.1 highlights an integrated modelling framework for urban flooding. This framework focuses on delivering the climatic and anthropogenic drivers to urban flooding characteristics (i.e., flood extent, water depth, velocity and arrival time), in order to provide relevant information of flood hazard and vulnerability for flood risk assessment and management.

15.4.1 Integrated Framework

The proposed integrated modelling framework for flood risk management is based on the concept of drivers, hazard and vulnerability (Figure 15.1). The main driver, climate change, is the primary factor influencing the flooding hazard under various greenhouse gas emission scenarios. The induced extreme precipitation and sea-level rise are projected by applying General Circulation Models (GCMs) combined with downscaling methods to generate input data for the urban flooding model. The secondary driver, urbanization, is considered as an anthropogenic factor, having structural and non-structural aspects. Structural factors include land use/cover change, flood defence engineering, drainage system and sustainable development approaches and influence flood hazard and vulnerability in terms of drainage capacity, flooding pathways and urban resilience to flood. Non-structural factors, such as socio-economic development, mainly affect flood vulnerability in terms of the social, economic and political elements of individuals and society. The induced urban flood risk can be assessed based on hazard and vulnerability. According to the assessment results, the urban resilience to flood can be estimated and analysed. Based on this investigation, the mitigation and adaptation strategies of flood risk management can be improved to support the modification of national/local flood policy and urban planning design.

15.4.2 GCMs Projection

Extreme precipitation and sea-level rise, as the main drivers of urban flooding due to climate change, can be projected by GCMs under various greenhouse gas emission scenarios (e.g., Mehrotra et al., 2013; Chen et al., 2017; Yin et al., 2017; Reshmidevi et al., 2018), such as Special Report on Emission Scenarios (SRES) (IPCC-SRES, 2000) and Representative Concentration Pathways (RCP) scenarios (Meinshausen et al., 2011). The issue of uncertainty has been raised due to the discrepancies between model scale and physical-process scale in time and space. Uncertainty has been estimated by an ensemble approach and multi-model ensemble approach. The ensemble approach is implemented to minimize the impact of initial conditions on uncertainty, by averaging the outputs of a series of analogous models running based on the same historical and future scenarios. The multi-model ensemble approach is based on the intercomparison of GCMs. The World Climate Research Programme (WCRP) of the World Meteorological Organization (WMO) operates a project named the Coupled Model Intercomparison Project, which began in 1995 and is now in the sixth phase (CMIP6). The multi-model results provide more perspectives on uncertainty of GCMs simulations.

The implementation of GCMs at regional/local work is limited by the low spatial resolution. Therefore, statistical and dynamic downscaling approaches are implemented to achieve finer spatial resolution. A statistical downscaling method utilizes the empirical or statistical relationship between large-scale variables and local variables to estimate the corresponding regional/local climate projection

(Reynard et al., 2001; Mehrotra & Sharma, 2010). The dynamic downscaling approach is also called a Regional Climate Model (RCM) based on the same principle as GCMs with a higher spatial and temporal resolution, such as the UK Met Office Hadley Centre's regional climate model version 3 (HadRM3); US regional climate model version 3 (RegCM3) and European Centre Hamburg Model (ECHAM) (Trzaska & Schnarr, 2014). Downscaling techniques are generally calibrated to account for various objective conditions and requirements, such as the geographical location and observed climate data that may vary across regions.

15.4.3 Urban Flooding Model

As discussed in Section 15.2, urban flood risk can be viewed as a concept that includes both hazard and vulnerability. Urban flooding hazards driven by different sources present different hydrologic and hydrodynamic characteristics. To address climatic and anthropogenic impacts on urban flood, an urban flooding model is developed and implemented to simulate/predict the flooding hazard under various climatic and anthropogenic scenarios. The flooding characteristics produced by modelling, in terms of flood extent, water depth and flow velocity, are crucial for flood risk assessment and improvement strategies of flood mitigation and adaptation.

In this context, physically-based hydrologic and hydraulic models have been extensively studied and employed in urban flooding simulation. Hydrologic models are mostly adopted at basin scale to calculate the upstream runoff into a city (e.g., Wang et al., 2019) while hydraulic models are generally applied with high resolution (metres) to compute the dynamics of flooding in the urban area (e.g., Gallien et al., 2014; Smith & Liang, 2015; Glenis et al., 2018). Therefore, hydrologic and hydraulic models are jointly implemented in the integrated modelling framework presenting the complete flooding process (e.g., Wang et al., 2019). Moreover, over the last decade, models based on machine learning techniques have also been developed and applied to urban flood risk assessment, which are not informed by the physical mechanisms but learn the functional form between flooding conditioning factors and flooding occurrence from historical records (e.g., Chang et al., 2014; Darabi et al., 2019; Zhao et al., 2019).

15.4.4 Vulnerability Assessment

Vulnerability generally refers to the characteristics of hazard-bearing body, in terms of exposure, sensitivity and capacity to adapt, due to the flood hazard (Wisner et al., 2004). In case of urban flooding, vulnerability is induced by the dual impacts of geographic and anthropogenic factors and classified into two categories, that is, geophysical vulnerability and social vulnerability (Imran et al., 2019).

The assessment of geophysical vulnerability is based on geophysical condition such as topography, location, slope, elevation of the property and asset, directly related to the flooding processes. The corresponding assessment methods can be categorized into three types: (1) GIS-based method, (2) data-based method and (3) physically-based method. The GIS-based method is operated in the GIS platform to assess the flood vulnerability of local property and asset by evaluating the spatial data of hydrologic characteristics. The data-based method is developed to analyse the comprehensive data based on the statistical method or artificial intelligence approaches (e.g., Lee et al., 2017; Gebrehiwot et al., 2019). The physically-based method is implemented to assess the flood vulnerability based on the projected dynamic information of flooding process and features produced by the global/local hydrologic and hydraulic models (e.g., Grimaldi et al., 2013; Chini et al., 2014; van Dijk et al., 2014).

However, the assessment of social vulnerability is relatively complex and qualitative as it depends on a wide set of social, economic and political factors associated with individuals, society, stakeholders, local and national authorities. Therefore, the assessment of social vulnerability is a great challenge since it requires to collect, collate and analyse data from different disciplines, scales and viewpoints. The assessment method is mainly based on experts' experience and statistical methods, such as indicator-based methods with and without weighting (e.g., Balica & Wright, 2010; Khan, 2012; Koks et al., 2015), data-based method (e.g., Huang et al., 2012) and curve method (e.g., Aerts et al., 2018).

15.4.5 Summary

Urban flood risk is induced by the constant changes of a complex system driven by natural and anthropogenic factors in interaction. Climate change is projected to increase urban flooding risks that may affect people, properties, assets, economies and the environment by means of extreme precipitation, sea-level rises and storm surges (IPCC, 2014). Moreover, the socio-economic development and population growth during urbanization are also increasing flood vulnerability in urban areas. Under the threat of climate change and intense urbanization, strategies of mitigation and adaptation should benefit from flood risk assessments that consider both hazards and vulnerability. To achieve the sustainable development goals, flood risk management should be based on an integrated framework, to project the future urban flood risk, to evaluate the urban development strategies, to guide urban planning and construction, to assist and support the decision-making, and to promote urban resilience to flooding disaster.

15.5 Conclusion

Due to climate change and rapid urbanization that have affected many regions across the world in the most recent decades, increasing urban flooding hazards have been recorded by the Centre for Research on the Epidemiology of Disasters EM-DAT database. Moreover, extreme precipitation and sea-level rise are expected to significantly increase urban flood risk in the twenty-first century (IPCC, 2014). Numerous researchers, international organizations and governments have committed to mitigating and adapting to urban flooding risk and achieving urban sustainable development. This chapter has reviewed the role of the main drivers associated with flood risk, the progresses and advances of flood risk management in developing, emerging and developed economies, and the numerical modelling framework for assessment of flood hazard and vulnerability. Conclusive remarks are summarized as follows.

First, urban flood risk is composed of flood hazard and vulnerability caused by a combination of natural and anthropogenic factors that are interrelated and may interact. Climate change is considered as the main driver of flood risk and can be largely influenced by the anthropogenic factors both positively and negatively. For instance, the reduction of greenhouse gas emission and increase of green surface can reduce the extent of climate change while unplanned urbanization may increase greenhouse gas emissions and cut off the natural hydrologic cycle due to the increasing urban surface impermeability. In this case, a comprehensive understanding of flood-driving mechanisms should consider the entire system driven by climatological, geographical, hydrological, hydraulic, ecological and socio-economic factors. Based on a better understanding of the system, effective and active measures can be developed and taken to slow down the speed of climate change and minimize its effect on urban flooding.

Second, most countries have operated mitigation and adaption strategies of flood risk management to tackle increasing urban flooding. However, the flood policy and strategic executive capability may vary in developing, emerging and developed countries based on income level. An emerging economy such as China has implemented the ‘sponge city’ initiative to adopt resilient strategies based on nature-based hybrid systems, in order to utilize 70 per cent of rainwater. The situation in Africa is quite different. Sub-Saharan African governments have adopted large scale operations of infrastructure engineering and national response plans to tackle the threat of flood. However, the cost of maintenance of the infrastructure system and the participation of stakeholders at multiple levels are important issues for low-income countries. The governments of developed countries such as the United States and the United Kingdom have put significant investment into the evaluation and maintenance of flood defence structures. The non-structural strategies, such as insurance system, grants and local agencies, are widely utilized in developed countries. Further semi-structural strategies, such as green-infrastructures and natural defences, have been widely studied and implemented as a nature-based strategy of flood risk management. These kinds of semi- and non-structural strategies are preferred since they encompass natural, social, economic and ecological factors.

Third, an integrated modelling framework is essential, but complicated to achieve due to several issues:

1. The difficulty to carry out transdisciplinary research including numerous factors, such as climate change, land use/cover change, flood defence engineering, drainage system, urban flooding processes, socio-economic and cultural scenarios.
2. These factors are represented by multiple models and then combined to deliver an integrated modelling system. The temporal and spatial scales of different models have to be unified by implementing downscaling/upscaling approaches. This operation may produce error and uncertainty and influence the running of the integrated modelling system.
3. The GCMs projections based on various greenhouse gas emission scenarios present uncertainty associated with the predictions. The CMIP have compared the GCMs results to assess the uncertainty of the projections.
4. Urban flooding systems are dynamic over a timescale of hours or days and a spatial scale of metres. The predictions of flooding hazard with fine resolution under climatic scenarios can guide decision-making in urban planning and flood management strategies. However, the accuracy of fine-scale models is often limited due to difficulties in collecting detailed historical data.
5. Geophysical vulnerability can be assessed more easily than social vulnerability. Social vulnerability assessment mainly depends on the experts' experiences and statistical methods and is influenced by socio-economic and cultural factors.

Many researchers have devoted their efforts to the study of urban flooding risk and sustainable water management in the context of climate change. Great progress and advances have been made in the past decades aiming at achieving urban flood resilience in an uncertain future. Sustainable development goals can only be achieved by concerted efforts across nations, governments, international organizations, the local community and individuals. Systematic study, practice and sharing of mitigation success globally appears to be the most suitable approach to ensure a continuous improvement of flood risk management strategies.

References

Aerts, J. C., Botzen, W. J., Clarke, K. C., et al. (2018). Integrating human behaviour dynamics into flood disaster risk assessment. *Nature Climate Change* 8(3): 193.

Aitsi-Selmi, A., Murray, V., Wannous, C., et al. (2016). Reflections on a science and technology agenda for 21st century disaster risk reduction. *International Journal of Disaster Risk Science* 7(1): 1–29.

Andoh, R. Y. G., & Iwugo, K. O. (2012). Sustainable urban drainage systems: A UK perspective. *Global Solutions for Urban Drainage* 1–16. DOI:10.1061/40644(2002) 19.

Balica, S., & Wright, N. G. (2010). Reducing the complexity of the flood vulnerability index. *Environmental Hazards* 9(4): 321–339.

Behzadian, K., & Kapelan, Z. (2015) Modelling metabolism based performance of an urban water system using Water MET 2. *Resources, Conservation and Recycling* 99: 84–99.

Bennet, O., & Hartwell-Naquib, S. (2014). Flood defence spending in England. House of Commons library, standard note SN/SC/5755.

Bensona, D., Fritsch, O., Cook, H., & Schmidd, M. (2014). Evaluating participation in WFD river basin management in England and Wales: Processes, communities, outputs and outcomes. *Land Use Policy* 38: 213–222.

Berndtsson, R., Becker, P., Persson, A., et al. (2019). Drivers of changing urban flood risk: A framework for action. *Journal of Environmental Management* 240: 47–56.

Bertilsson, L., Wiklund, K., Tebaldi, I. D. M., et al. (2019). Urban flood resilience – A multi-criteria index to integrate flood resilience into urban planning, *Journal of Hydrology* 573: 970–982.

Beven, K. (2012). *Rainfall-Runoff Modelling: The Primer* (2nd ed., pp. i–xxix). Chichester: John Wiley & Sons.

Bottazzi, P., Winkler, M. S., & Speranza, C. I. (2019). Flood governance for resilience in cities: The historical policy transformations in Dakar's suburbs. *Environmental Science & Policy* 93: 172–180.

Brown, R. R., Keath, N., & Wong, T. H. F. (2009). Urban water management in cities: Historical, current and future regimes. *Water Science & Technology* 59(5): 847–855.

Bruwier, M., Maravat, C., Mustafa, A., et al. (2020). Influence of urban forms on surface flow in urban pluvial flooding. *Journal of Hydrology* 582: 124493.

Bubeck, P., Kreibich, H., Penning-Rowsell, E. C., Botzen, W. J. W., De Moel, H., & Klijn, F. (2015). Explaining differences in flood management approaches in Europe and in the USA – A comparative analysis. *Journal of Flood Risk Management* 10: 436–445.

Burby, R. J., Deyle, R. E., Godschalk, D. R., & Olshansky, R. B. (2000). Creating hazard resilient communities through land-use planning. *Natural Hazards Review* 1(2): 99–106.

Cabral, P., Augusto, G., Akande, A., et al. (2017). Assessing Mozambique's exposure to coastal climate hazards and erosion. *International Journal of Disaster Risk Reduction* 23: 45–52.

Cai, T., Li, X., Ding, X., Wang, J., & Zhan, J. (2019). Flood risk assessment based on hydrodynamic model and fuzzy comprehensive evaluation with GIS technique. *International Journal of Disaster Risk Reduction* 35: 101077.

Cariolet, J.-M., Vuillet, M. & Diab, Y. (2019). Mapping urban resilience to disasters – A review. *Sustainable Cities and Society* 51: 101746,

Chan, F. K. S., Griffiths, J. A., Higgitt, D., et al. (2018). 'Sponge City' in China – A breakthrough of planning and flood risk management in the urban context. *Land Use Policy* 76: 772–778.

Chang, F., Chen, P., Lu, Y., Huang, E., & Chang, K. (2014). Real-time multi-step-ahead water level forecasting by recurrent neural networks for urban flood control. *Journal of Hydrology* 517: 836–846.

Chelleri, L., Schuetze, T., & Salvati, L. (2015). Integrating resilience with urban sustainability in neglected neighbourhoods: Challenges and opportunities of transitioning to decentralized water management in Mexico City. *Habitat International* 48: 122–130.

Chen, J., Gao, C., Zeng, X., et al. (2017). Assessing changes of river discharge under global warming of 1.5° and 2° in the upper reaches of the Yangtze River Basin: Approach by using multiple-GCMs and hydrological models. *Quaternary International* 453: 63–73.

Chini, M., Giustarini, L., Matgen, P., Hostache, R., Pappenberger, F., & Bally, P. (2014). Flood hazard mapping combining high resolution multi-temporal SAR data and coarse resolution global hydrodynamic modelling. 2014 IEEE Geoscience and Remote Sensing Symposium, Quebec City, QC (pp. 2394–2396).

Cigler, B. (2017). U.S. floods: The necessity of mitigation. State and local government *Review* 49(2): 127–139.

Cobbinah, P. B., Asibey, M. O., Opoku-Gyamfi, M., & Peprah, C. (2019). Urban planning and climate change in Ghana. *Journal of Urban Management* 8(2): 261–271.

Cobbinah, P. B., Poku-Boansi, M., & Peprah, C. (2017). Urban environmental problems in Ghana. *Environmental Development* 23: 33–46.

Costabile, P., Costanzo, C., De Lorenzo, G., & Macchione, F. (2020). Is local flood hazard assessment in urban areas significantly influenced by the physical complexity of the hydrodynamic inundation model? *Journal of Hydrology* 580: 124231.

Costanza, R., Perez-Maqueo, O., Martinez, M. L., Sutton, P., Anderson, S. J., & Mulder, K. (2008). The value of coastal wetlands for hurricane protection. *AMBIO: A Journal of the Human Environment* 37(4): 241–248.

Darabi, H., Choubin, B., Rahmati, O., Haghighi, A. T., Pradhan, B., & Kløve, B. (2019). Urban flood risk mapping using the GARP and QUEST models: A comparative study of machine learning techniques. *Journal of Hydrology* 569: 142–154.

Dawson, R. J., Speight, L., Hall, J. W., Djordjevic, S., Savic, D., & Leandro, J. (2008). Attribution of flood risk in urban areas. *Journal of Hydroinformatics* 10(4): 275–288.

Department for Environment Food and Rural Affairs (DEFRA) (2017). *Evaluation of the Arrangements for Managing Local Flood Risk in England* (p. FD2680). Final Report. London: DEFRA.

Diagne, K., Ndiaye, A., Pelling, M., & Wisner, B. (2012). History, governance and the millennium development goals: Flood risk reduction in Saint-Louis, Senegal. In M. Pelling and B. Wisner (eds.), *Disaster Risk Reduction: Cases From Urban Africa*. (p. 147). London: Routledge.

van Dijk, E., van der Meulen, J., Kluck, J., & Straatman, J. H. M. (2014). Comparing modelling techniques for analysing urban pluvial flooding. *Water Science & Technology* 69(2): 305–311.

Dixon, L., Clancy, N., Seabury, S. A., & Overton, A. (2006). *The National Flood Insurance Program's Market Penetration Rate, Estimates and Policy Implications*. Arlington, VA: RAND.

FEMA (Federal Emergency Management) (1994). *A Unified National Program for Floodplain Management*. Washington, DC: FEMA. Available from www.fema.gov/media-library-data/20130726-1733-25045-0814/unp_floodplain_mgmt_1994.pdf (Last accessed 15 March 2020).

Fenner, R. A., O'Donnell, E., Ahilan, S., et al. (2019). Achieving urban flood resilience in an uncertain future. *Water* 11(5): 1082.

Field, C. B., Barros, V., Stocker, T. F., et al. (2012). *Managing the Risks of Extreme Events and Disasters to Advance Climate Change Adaptation. A Special Report of Working Groups I and II of the Intergovernmental Panel on Climate Change*. Cambridge: Cambridge University Press.

Filho, W. L., Balogun, A., Ayal, D. Y., et al. (2018). Strengthening climate change adaptation capacity in Africa – Case studies from six major African cities and policy implications. *Environmental Science & Policy* 86: 29–37.

Fletcher, T. D., Andrieu, H., & Hamel, P. (2013). Understanding, management and modelling of urban hydrology and its consequences for receiving waters: A state of the art. *Advances in Water Resources* 51: 261–279.

Foresight (2004). *Foresight: Future Flooding. Executive Summary*. London: Department of Trade and Industry, The Government Office for Science.

Fritsch, O., & Benson, D. (2013). Integrating the principles of IWRM? River basin planning in England and Wales. *International Journal of Water Governance* 1 (3–4): 265–284.

Gaitan, S., van de Giesen, N. C., & ten Veldhuis, J. A. E. (2016). Can urban pluvial flooding be predicted by open spatial data and weather data? *Environmental Modelling & Software: With Environment Data News* 85: 156–171.

Gallien, T. W., Sanders, B. F., & Flick, R. E. (2014). Urban coastal flood prediction: Integrating wave overtopping, flood defences and drainage. *Coastal Engineering* 91: 18–28.

Gebrehiwot, A., Hashemi-Beni, L., Thompson, G., Kordjamshidi, P., & Langan, T. E. (2019). Deep convolutional neural network for flood extent mapping using unmanned aerial vehicles data. *Sensors* 19(7): 1486.

Gersonius, B. (2012). *The Resilience Approach to Climate Adaptation Applied for Flood Risk*. PhD Thesis, UNESCO-IHE/TU Delft.

Gersonius, B., van Buuren, A., Zethof, M., & Kelder, E. (2016). Resilient flood risk strategies: Institutional preconditions for implementation. *Ecology and Society* 21(4): 28.

Glenis, V., Kutija, V., & Kilsby, C. G. (2018). A fully hydrodynamic urban flood modelling system representing buildings, green space and interventions. *Environmental Modelling & Software* 109: 272–292.

Griffiths, J. A., Zhu, F., Chan, F. K. S., & Higgitt, D. L. (2019). Modelling the impact of sea-level rise on urban flood probability in SE China. *Geoscience Frontiers* 10(2): 363–372.

Grimaldi, S., Petroselli, A., Arcangeletti, E., & Nardi, F. (2013). Flood mapping in ungauged basins using fully continuous hydrologic–hydraulic modeling. *Journal of Hydrology* 487: 39–47.

Hall, J. W., Meadowcroft, I. C., Sayers, P. B., & Bramley, M. E. (2003). Integrated flood risk management in England and Wales. *Natural Hazards Review* 4(3): 126–135.

Hegger, D., Alexander, M., Raadgever, T., Priest, S., & Bruzzone, S. (2020). Shaping flood risk governance through science-policy interfaces: Insights from England, France and the Netherlands. *Environmental Science & Policy* 106: 157–165.

Hirabayashi, Y., Mahendran, R., Koirala, S., et al. (2013). Global flood risk under climate change. *Nature Climate Change* 3: 816–821.

Huang, D., Zhang, R., Huo, Z., et al. (2012). An assessment of multidimensional flood vulnerability at the provincial scale in China based on the DEA method. *Natural Hazards* 64(2): 1575–1586.

van den Hurk, M., Mastenbroek, E., & Meijerink, S. (2014). Water safety and spatial development: An institutional comparison between the United Kingdom and the Netherlands. *Land Use Policy* 36: 416–426.

Hutter, G., & Schanze, J. (2008). Learning how to deal with uncertainty of flood risk in long-term planning. *Journal of River Basin Management* 6(2): 175–184.

Imran, M., Sumra, K., Mahmood, S. A., & Sajjad, S. F. (2019). Mapping flood vulnerability from socioeconomic classes and GI data: Linking socially resilient policies to geographically sustainable neighbourhoods using PLS-SEM. *International Journal of Disaster Risk Reduction* 41: 101288.

IPCC (2012). *Managing the Risks of Extreme Events and Disasters to Advance Climate Change Adaptation. A Special Report of Working Groups I and II of the Intergovernmental Panel on Climate Change*, 582 pp. [C. B. Field, V. Barros, T. F. Stocker, et al. (eds.)]. Cambridge: Cambridge University Press.

IPCC (2014). *Climate Change 2014: Synthesis Report. Contribution of Working Groups I, II and III to the Fifth Assessment Report of the Intergovernmental Panel on Climate Change*, 151 pp. [Core Writing Team, R. K. Pachauri & L. A. Meyer (eds.)]. Geneva: IPCC.

IPCC-SRES (2000). Emission scenarios. In N. Nakićenovic, & R. Swart (eds.), *Special Report on Emissions Scenarios: A Special Report of Working Group III of the Intergovernmental Panel on Climate Change* (pp. 239–292). Cambridge: Cambridge University Press.

Jenkins, K., Surminski, S., Hall, J., & Crick, F. (2017). Assessing surface water flood risk and management strategies under future climate change: Insights from an agent-based model. *Science of The Total Environment* 595: 159–168.

Ke, Y., Wang, S., Chan, A. P. C., & Lam, P. T. I. (2010). Preferred risk allocation in China's public–private partnership (PPP) projects. *International Journal of Project Management* 28(5): 482–492.

Khan, S. (2012). Vulnerability assessments and their planning implications: A case study of the Hutt Valley, New Zealand. *Natural Hazards* 64(2): 1587–1607.

Koks, E. E., Jongman, B., Husby, T. G., & Botzen, W. J. W. (2015). Combining hazard, exposure and social vulnerability to provide lessons for flood risk management. *Environmental Science & Policy* 47: 42–52.

Kolsky, P. (1999). *Storm Drainage: An Engineering Guide to the Low-Cost Evaluation of System Performance*. Rugby: Practical Action.

Kunreuther, H., & Roth, R. J. (1998). *Paying the Price: The Status and Role of Insurance against Natural Disasters in the United States*. Washington, DC: Joseph Henry.

Lee, S., Kim, J. C., Jung, H. S., Lee, M. J., & Lee, S. (2017). Spatial prediction of flood susceptibility using random-forest and boosted-tree models in Seoul metropolitan city, Korea. *Geomatics Natural Hazards & Risk* 8(2): 1185–1203.

Li, Y., Li, H. X., Huang, J., & Liu, C. (2020). An approximation method for evaluating flash flooding mitigation of sponge city strategies – A case study of Central Geelong. *Journal of Cleaner Production* 257: 120525.

Liao, K. (2012). A theory on urban resilience to floods – A basis for alternative planning practices. *Ecology and Society* 17(4): 48.

Löschner, L., & Nordbeck, R. (2019). Switzerland's transition from flood defence to flood-adapted land use: A policy coordination perspective. *Land Use Policy*, 95: 103873.

Lu, X., & Ran, L. (2011). China flood havoc highlights poor urban planning. *Natural Hazards* 56: 575–576.

Macchione, F., Costabile, P., Costanzo, C., & De Lorenzo, G. (2019). Extracting quantitative data from non-conventional information for the hydraulic reconstruction of past urban flood events. A case study. *Journal of Hydrology* 576: 443–465.

Maheu, A. (2012). Urbanization and flood vulnerability in a peri-urban neighbourhood of Dakar, Senegal: How can participatory GIS contribute to flood management? In W.

L. Filho (ed.), *Climate Change and the Sustainable Use of Water Resources. Climate Change Management* (pp. 185–207). Berlin: Springer.

Mai, T., Mushtaq, S., Reardon-Smith, K., et al. (2020). Defining flood risk management strategies: A systems approach. *International Journal of Disaster Risk Reduction* 47: 101550.

Meerow, S., Newell, J. P., & Stults, M. (2016). Defining urban resilience: A review. *Landscape and Urban Planning* 147: 38–49.

Mehrotra, R., & Sharma, A. (2010). Development and application of a multisite rainfall stochastic downscaling framework for climate change impact assessment. *Water Resources Research* 46(7): 759–768.

Mehrotra, R., Sharma, A., Nagesh Kumarb, D., & Reshmidevi, T. V. (2013). Assessing future rainfall projections using multiple GCMs and a multi-site stochastic downscaling model. *Journal of Hydrology* 488: 84–100.

Meinshausen, M., Smith, S. J., Calvin, K., et al. (2011). The RCP greenhouse gas concentrations and their extension from 1765 to 2300. *Climate Change* 109(1–2): 213–241.

Merz, B., Hall, J., Disse, M., & Schumann, A. (2010). Fluvial flood risk management in a changing world. *Natural Hazards and Earth System Sciences* 10(3): 509–727.

Metz, F., Angst, M., & Fischer, M. (2020). Policy integration: Do laws or actors integrate issues relevant to flood risk management in Switzerland? *Global Environmental Change* 61: 101945,

Miguez, M. G., & Veról, A. P. (2017). A catchment scale Integrated Flood Resilience Index to support decision making in urban flood control design. *Environment and Planning B: Planning and Design* 44(5): 925–946.

Miller, J. D., & Hutchins, M. (2017). The impacts of urbanisation and climate change on urban flooding and urban water quality: A review of the evidence concerning the United Kingdom. *Journal of Hydrology: Regional Studies* 12: 345–362.

Ministry of Water Resources of the People's Republic of China (2013). *China Flood and Drought Disaster Bulletin 2013*. Beijing: SinoMaps Press. Available from www.mwr.gov.cn/sj/tjgb/zgshzhgb/201612/t20161222_776091.html (Last accessed 9 March 2020).

Moftakhari, H. R., AghaKouchak, A., Sanders, B. F., Allaire, M., & Matthew, R. A. (2018). What is nuisance flooding? Defining and monitoring an emerging challenge. *Water Resource Research* 54(7): 4218–4227.

Mugume, S. N., Gomez, D. E., Fu, G., Farmani, R., & Butler, D. (2015). A global analysis approach for investigating structural resilience in urban drainage systems, *Water Research* 81: 15–26.

Müller, H., & Haberlandt, U. (2018). Temporal rainfall disaggregation using a multiplicative cascade model for spatial application in urban hydrology. *Journal of Hydrology* 556: 847–864,

NatCatSERVICE Database (2019). Available from www.munichre.com/en/risks/natural-disasters-losses-are-trending-upwards/floods-and-flash-floods-underestimated-natural-hazards.html (Last accessed 20 October 2019).

NOAA National Centers for Environmental Information (NCEI) (2020). *U.S. Billion-Dollar Weather and Climate Disasters*. Available from www.ncdc.noaa.gov/billions/ (Last accessed 22 August 2021).

Nkhonjera, G. K. (2017). Understanding the impact of climate change on the dwindling water resources of South Africa, focusing mainly on Olifants River basin: A review. *Environmental Science & Policy* 71: 19–29.

Nkwunonwo, U. C., Whitworth, M., & Baily, B. (2020). A review of the current status of flood modelling for urban flood risk management in the developing countries. *Scientific African* 7: e00269.

Noonan, D. S., & Sadiq, A. A. (2018). Flood risk management: Exploring the impacts of the community rating system program on poverty and income inequality. *Risk Analysis* 38(3): 489–503.
O'Donnell, E. C., Lamond, J. E., & Thorne, C. R. (2018). Learning and action alliance framework to facilitate stakeholder collaboration and social learning in urban flood risk management. *Environmental Science & Policy* 80(Suppl C): 1–8.
O'Donnell, E., Thorne, C., Ahilan, S., et al. (2019) The blue-green path to urban flood resilience. *Blue-Green Systems* 2(1): 28–45.
Pelling, M., & Wisner, B. (2012). *Disaster Risk Reduction: Cases from Urban Africa*. London: Routledge.
Petersen, M. S. (2001). Impact of flash floods. In E. Gruntfest, & J. Handmer (eds.), *Coping with Flash Floods* (pp. 11–13). Netherlands: Klumer Academic Publishers.
Qiao, X., Liao, K., & Randrup, T. B. (2020). Sustainable stormwater management: A qualitative case study of the Sponge Cities initiative in China. *Sustainable Cities and Society* 53: 101963.
Reshmidevi, T. V., Nagesh Kumar, D., Mehrotra, R., & Sharma, A. (2018). Estimation of the climate change impact on a catchment water balance using an ensemble of GCMs. *Journal of Hydrology* 556: 1192–1204.
Reynard, N. S., Prudhomme, C., & Crooks, S. M. (2001). The flood characteristics of large UK rivers: Potential effects of changing climate and land use. *Climate Change* 48: 343–359.
Rubinato, M., Nichols, A., Peng, Y., et al. (2019). Urban and river flooding: Comparison of flood risk management approaches in the UK and China and an assessment of future knowledge needs. *Water Science and Engineering* 12(4): 274–283.
Sadler, J. M., Goodall, J. L., Morsy, M. M., & Spencer, K. (2018). Modelling urban coastal flood severity from crowd-sourced flood reports using Poisson regression and Random Forest. *Journal of Hydrology* 559: 43–55.
Saul, A. J., Djordjevic, S., Maksimovic, C., & Blanksby J. (2011). Integrated urban flood modelling In G. Pender, & H. Faulkner (eds.), *Flood Risk Science and Management* (pp. 258–288). Hoboken, NJ: Wiley-Blackwell.
Sayers, P., Li, Y., Galloway, G., et al. (2013). *Flood Risk Management: A Strategic Approach*. Paris: UNESCO.
Schanze, J. (2006). Flood risk management – A basic framework. In J. Schanze, E. Zeman, & J. Marsalek (eds.), *Flood Risk Management – Hazards, Vulnerability and Mitigation Measures* (pp. 1–20). Berlin: Springer.
Scholz, M. (2013). Water quality improvement performance of geotextiles within permeable paving systems: A critical review. *Water* 5(2): 462–479.
Schwartz, S. S., & Smith, B. (2016). Restoring hydrologic function in urban landscapes with suburban subsoiling. *Journal of Hydrology* 543(Part B): 770–781.
Sendzimir, J., Magnuszewski, P., Flachner, Z., et al. (2007). Assessing the resilience of a river management regime: Informal learning in a shadow network in the Tisza River Basin. *Ecology and Society* 13(1): 11.
Smith, L. S., & Liang, Q. (2015). A high-performance integrated hydrodynamic modelling system for urban flood simulations. *Journal of Hydroinformatics* 17(4): 518–533.
Smits, A. J. M., Nienhuis, P. H., & Saeijs, H. L. F. (2006). Changing estuaries, changing views. *Hydrobiologia* 565: 339–355.
Sörensen, J., & Mobini, S. (2017). Pluvial, urban flood mechanisms and characteristics – Assessment based on insurance claims. *Journal of Hydrology* 555: 51–67.
Sörensen, J., Persson, A., Sternudd, C., et al. (2016). Re-thinking urban flood management – Time for a regime shift. *Water* 8(8): 332.

Tang, Q. (2020). Global change hydrology: Terrestrial water cycle and global change. *Science China Earth Sciences* 63: 459–462.
Tanoue, M., Hirabayashi, Y., & Ikeuchi, H. (2016). Global-scale river flood vulnerability in the last 50 years. *Scientific Reports* 6: 36021.
Tiepolo, M. (2014). Flood risk reduction and climate change in large cities south of the Sahara. In S. Macchi, & M. Tiepolo (eds.), *Climate Change Vulnerability in Southern African Cities* (pp. 19–36). Cham: Springer.
Trzaska, S., & Schnarr, E. (2014). *A review of downscaling methods for climate change projections*. United States Agency for International Development by Tetra Tech ARD (pp. 1–42).
UK Government (2010). *Flood and Water Management Act*. London: UK Government.
UN (2018). *68% of the world population projected to live in urban areas by 2050, says UN*. Available from www.un.org/development/desa/en/news/population/2018-revision-of-world-urbanization-prospects.html (Last accessed 17 October 2019).
UN (United Nations, Department of Economic and Social Affairs, Population Division) (2019). *World Urbanization Prospects 2018: Highlights* (ST/ESA/SER.A/421).
Ungaro, F., Calzolari, C., Pistocchi, A., & Malucelli, F. (2014). Modelling the impact of increasing soil sealing on runoff coefficients at regional scale: A hydrogeological approach. *Journal of Hydrology & Hydromechanics* 62(1): 33–42.
UNISDR (2004). *Living with risk. A Global Review of Disaster Reduction Initiatives*. Geneva: UNISDR. Available from www.unisdr.org/files/657_lwr21.pdf (Last accessed 21 March 2020).
Vahedifard, F., AghaKouchak, A., & Jafari, N. H. (2016). Compound hazards yield Louisiana flood. *Science* 353(6306): 1374.
ten Veldhuis, J. A. E., Harder, R. C., & Loog, M. (2010). Automatic classification of municipal call data to support quantitative risk analysis of urban drainage systems. *Structure and Infrastructure Engineering* 9(2): 1–10.
Vercruysse, K., Dawson, D., & Wright, N. (2019) Interoperability: A conceptual framework to bridge the gap between multi-functional and multi-system urban flood management. *Journal of Flood Risk Management* 12(S2): e12535.
van Vuuren, D. P., Edmonds, J., Kainuma, M., et al. (2011). The representative concentration pathways: An overview. *Climatic Change* 109: 5–31.
Waghwala, R. K., & Agnihotri, P. G. (2019). Flood risk assessment and resilience strategies for flood risk management: A case study of Surat City. *International Journal of Disaster Risk Reduction* 40: 101155.
Wang, Q., Xu, Y., Wang, Y., et al. (2020). Individual and combined impacts of future land-use and climate conditions on extreme hydrological events in a representative basin of the Yangtze River Delta, China. *Atmospheric Research* 236: 104805.
Wang, Y., Sun, M., & Song, B. (2017). Public perceptions of and willingness to pay for sponge city initiatives in China. *Resources, Conservation and Recycling* 122: 11–0.
Wang, Y., Zhang, X., Tang, Q., et al. (2019). Assessing flood risk in Baiyangdian Lake area in a changing climate using an integrated hydrological-hydrodynamic modelling. *Hydrological Sciences Journal* 64(16): 2006–2014.
Wang, Z., Lai, C., Chen, X., Yang, B., Zhao, S., & Bai, X. (2015). Flood hazard risk assessment model based on random forest. *Journal of Hydrology* 527: 1130–1141.
Ward, P. J., Jongman, B., Weiland, F. S., et al. (2013). Assessing flood risk at the global scale: Model setup, results, and sensitivity. *Environmental Research Letters* 8(4): 044019.
Watson, K. M., Harwell, G. R., Wallace, D. S., Welborn, T. L., Stengel, V. G., & McDowell, J. S. (2018). Characterization of peak streamflows and flood inundation

of selected areas in southeastern Texas and southwestern Louisiana from the August and September 2017 flood resulting from Hurricane Harvey: U.S. Geological Survey Scientific Investigations Report 2018-5070, 44 p., https://doi.org/10.3133/sir20185070.

Winsemius, H. C., Aerts, J. C. J. H., van Beek, L. P. H., et al. (2016). Global drivers of future river flood risk. *Nature Climate Change* 6: 381–385.

Wisner, B., Blaikie, P. M., Blaikie, P., Cannon, T., & Davis, I. (2004). *At Risk: Natural Hazards, People's Vulnerability and Disasters*. New York: Routledge.

Yao, L., Wei, W., & Chen, L. (2016). How does imperviousness impact the urban rainfall-runoff process under various storm cases? *Ecological Indicators* 60: 893–905.

Yin, Y., Tang, Q., Liu, X., & Zhang, X. (2017). Water scarcity under various socio-economic pathways and its potential effects on food production in the Yellow River Basin. *Hydrology and Earth System Sciences* 21: 791–804.

Zhang, L., Sun, X., & Xue, H. (2019). Identifying critical risks in Sponge City PPP projects using DEMATEL method: A case study of China. *Journal of Cleaner Production* 226: 949–958.

Zhao, G., Pang, B., Xu, Z., Peng, D., & Xu, L. (2019). Assessment of urban flood susceptibility using semi-supervised machine learning model. *Science of the Total Environment* 659: 940–949.

Zheng, Z., Qi, S., & Xu, Y. (2013). Questionable frequent occurrence of urban flood hazards in modern cities of China. *Natural Hazards* 65: 1009–1010.

16

Soft Computing Methods and Water Management

MOHAMMAD ZOUNEMAT-KERMANI AND MEYSAM ALIZAMIR

16.1 Introduction

Soft computing refers to a set of computing procedures applied to computer science, artificial intelligence, machine learning and other research fields. Despite their wide applications in water management, mathematical methods, such as analytical procedures, and hard computing techniques, such as numerical methods, are not always applicable. In the presence of non-linearity, soft computing methods offer a promising simulation framework for predictive purposes. These methods can be employed to construct data-driven and machine learning models and are easier to be applied in comparison to dynamic or physical models.

In principle, the development of a soft computing model can be summarized as a five-step process.

1. Data acquisition, pre-processing and preparation

The first step in preparing a soft computing model is the accurate determination of the model's inputs for capturing the non-linear and complex properties of the studied process. After identifying the problem, it is necessary to determine the database consisting of input and output data. Given the nature of the problem, one must ensure that the accuracy of the available data is sufficient. Then the following steps can be considered before using the data for modelling:

a. Assessing data and distinguishing the input and output data

After data collection, the type and usability of data must be determined for modelling. Given the type of the problem, one should distinguish the input from output data series.

b. Data cleansing and validation

In this step, it is essential to ensure the integrity and accuracy of data. To this end, if necessary, missing records can be completed through data imputation methods.

In some cases, the corrupt and inaccurate records cleansing from the database is necessary. Next, the data must be validated. For example, in some hydrological phenomena, a run test can be used as an effective method for testing the randomness of data.

c. Standardization and normalization

To achieve good predictive performance, the data should be standardized or normalized before being used in the soft computing model. Data normalization means adjusting values measured in different scales to a notionally common scale which is done often prior to averaging. Data standardization transforms the data so that the scale is identical between different variables and eases comparison between variables and computational procedures. For instance, one can standardize a dataset by using a simple formula such as (Max (yi) – (yi))/Max (yi) – Min (yi)), where yi represents a data value.

2. Feature selection and input data

In many soft computing modelling problems, considering all independent features as input vectors to the model is not only computationally more demanding but also reduces the accuracy of results. Therefore, it is more suitable to select only the effective features that will be used in the model. Different methods can be considered to select model features according to the nature of the problem. Common methods include the principal component analysis, factor analysis, subset analysis and Gamma Test. In the context of time-series data, autocorrelation function and average mutual information methods have been widely applied.

3. Model building

After determining the input vector and output parameters of the model, one should design the structure of the soft computing model. In this stage, two different steps can be taken, which may vary depending on the model type.

a. Structure and architecture of the model

Appropriate determination of the structure and architecture of the model is key to establish a soft computing model. In this phase, various approaches are applied by researchers depending on the type of model. For example, in artificial neural networks, determining the number of computational layers and neurons of each layer in the network can be achieved through trial-and-error and user experience.

b. Learning approach

Once the structure of the model is built, a learning method must be selected for training. In this stage, various mathematical techniques such as gradient-based or heuristic methods can be used.

4. Implementation and deployment of the model

The determination of training sets for model training and test sets for model evaluation and validation is a crucial step to assess the in-sample and out-of-sample performance of the model. Depending on the type of problem, various approaches, such as hold-out-method and k-fold cross-validation, can be used.

5. Post-processing and data visualization

After the model fitting processes, the results are examined to evaluate the accuracy and validity of the predictive performance of the model. Models are generally evaluated based on statistical indices, diagnostic analysis or statistical tests.

16.2 Soft Computing in Water Management: State of the Art

Reliability, accuracy and high predictive performance put the soft computing-based models among the most widely used and noteworthy models in the analysis of water management-related problems. Conventional soft computing methods such as artificial neural networks (ANNs) have proved to be effective and flexible enough to accommodate various water management problems (Zounemat-Kermani et al., 2013; Kisi et al., 2015). However, despite their prominent merits over traditional models and high capability to forecast hydrological events, soft computing-based models exhibit limitations in expressing and explaining the physical concepts associated with the investigated hydrological process. Besides, most of these methods suffer from slow training speed and are sometimes trapped in the local minima. This does not mean that modern soft computing methods necessarily overcome all the disadvantages of conventional soft computing procedures. Thus, modern soft computing methods are likely to exhibit higher accuracy than traditional approaches but this can often be achieved at the expense of higher computational costs. For instance, integrating metaheuristic algorithms with other soft computing methods improves the performance of the primitive model, but at the same time imposes a considerable extra computational time on the learning procedure. Therefore, while paying attention to the nature of the problem, the selection of the soft computing model should be based on user expectations from the point of view of computational accuracy, difficulty levels of model development, computational speed and available software.

The modelling procedure of water management problems (e.g., climatic and hydrological processes) can be classified into two main perspectives, either as knowledge-driven (based on physics) or data-driven (data derived from the governing

Table 16.1. A brief summary of the recent developments and modelling applications of soft computing methods in various fields of water management

Author(s), year	Water management problem	Soft computing approach	Remarks
Afzaal et al. (2020)	Groundwater level modelling	Deep learning	The applied deep learning models acted better in estimating groundwater fluctuations comparing to the traditional neural network models.
Robles-Velasco et al. (2020)	Water supply networks	Support vector machine	Pipe failure in water supply networks was predicted based on logistic regression (LR) and support vector classification methods (SVC). The SCV, as a soft computing method, provided promising results in the prediction phase.
Ashofteh et al. (2020)	Water resources management	Evolutionary algorithms	The study proposed the successful usage of a logical genetic algorithm, as an evolutionary soft computing technique, in modelling reservoir systems of dams.
Yue et al. (2020)	Water management in irrigation systems	Fuzzy logic	A full fuzzy-interval credibility-constrained nonlinear programming approach was applied to a case study for planning irrigation water allocation in China.
Sattar et al. (2019)	Water network management	Extreme learning machine	A failure rate prediction model for maintaining and rehabilitating of water pipe network was developed based on the extreme learning machine model. The developed extreme learning machine model was superior to traditional soft computing techniques.
Li et al. (2019)	Performance assessment of stormwater	Neural network	It was concluded that the neural networks are effective tools for evaluating the performance of stormwater.
Darabi et al. (2019)	Urban flood risk mapping	Fuzzy logic	A combination of expert knowledge method and a fuzzy analytical network process was applied to produce an urban flood water risk map.

system). Soft computing techniques offer resilient and reliable methods and can be considered as a reference model for modelling a variety of water management problems.

In Table 16.1, we review recent publications related to the use of different types of soft computing methods in modelling water resources problems. We highlight a wide variety of soft computing methods recently applied to water management, including groundwater modelling, water supply networks and flood mapping.

A review of the literature shows that almost all these methods exhibit better predictive performance than traditional methods (such as statistical methods). The main aim of this chapter is to provide an appropriate context for the introduction and scientific achievement of modern soft computing methods in water management. Many valuable published works and studies based on conventional and common soft computing models in water management are available. Further detail on the performance of common soft computing methods

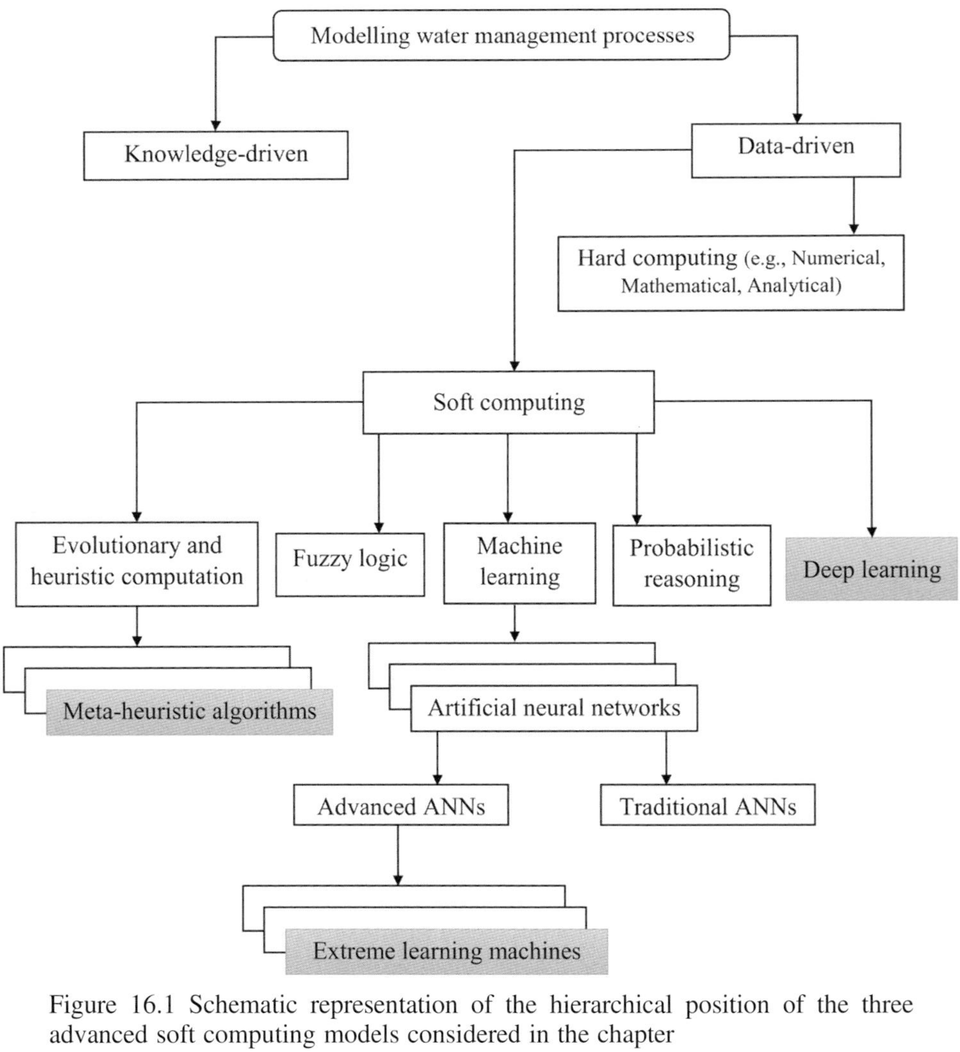

Figure 16.1 Schematic representation of the hierarchical position of the three advanced soft computing models considered in the chapter

in the field of water sciences can be found in Maier and Dandy (2000) and Nourani et al. (2014).

Given the high capabilities of metaheuristic methods in the optimization of complex problems, the use of these methods (alone or in combination with other soft computing methods) in the field of water management has increased over recent years. The extreme learning machine methods and deep learning networks, due to their unique ability to simulate nonlinear systems, have recently received more attention. Accordingly, this chapter focuses on describing the application of these types of modern soft computing models in water management problems. Figure 16.1 provides a diagram that represents the general procedure of the aforementioned models among other widely used soft computing methods in modelling water management problems.

In Section 16.3, the principles and theories of these models are introduced, followed by a description of the applicability of these models in modelling different water management systems.

16.3 Advanced Soft Computing Methods; Fundamentals and Theory

This section aims to describe variants of advanced soft computing models, including extreme learning machines, deep learning models and metaheuristic algorithms.

16.3.1 Extreme Learning Machines

16.3.1.1 Basic Extreme Learning Machine

Huang et al. (2004) suggested a novel training approach based on an extreme learning machine (ELM) that uses a single hidden layer artificial neural networks (SLFNs). The main objective of the ELM is to obtain better generalization performance via decreasing learning phase errors and output weights. Learning algorithms often encounter issues during the training phase associated with the time-consuming process of finding the optimum values of parameters of networks. The ELM method has gained popularity among researchers in different fields of studies by overcoming these problems commonly associated with gradient-based algorithms. Huang et al. (2004) compared the accuracy of ELM in regression and classification applications by carrying out different experiments using benchmark data sets. The findings of the study showed that the ELM may outperform classical gradient-based artificial neural networks in various domains. First, ELM can considerably improve the training speed. Second, ELM shows a better generalization performance in comparison with traditional learning models. In addition, ELM is less likely to encounter problems associated with local minima and overfitting. Finally,

ELM can apply non-differentiable activation functions for training SLFNs. Note that back-propagation feedforward neural networks can have more than one hidden layer; however, ELM is only applicable for single hidden layer feedforward networks. In general, ELM can model complex systems since they allow for more flexibility than classical methods in the functional relationship associated with the features and the label. The general SLFNN with *N* hidden nodes can be described as:

$$\sum_{i=1}^{N} \beta_i g(x_k, ; c_i, a_i) = y_k, k = 1, 2, \ldots, M \tag{16.1}$$

where $g(.)$, c_i and β_i denote the transfer function, bias and weight vector for linking the hidden nodes to output nodes. The sigmoid and Gaussian functions are among the most widely used non-linear mapping functions in ELM.

$$G(a, b, x) = \frac{1}{1 + \exp(-(a \cdot x + b))}, \tag{16.2}$$

$$G(a, b, x) = \exp(-b\|x - a\|^2). \tag{16.3}$$

Equation (16.1) can be rewritten as (Huang et al., 2004):

$$H\beta = Y \tag{16.4}$$

where

$$H = \begin{bmatrix} g(x_1; c_1, w_1) & \cdots & g(x_1; c_m, w_M) \\ g(x_N; c_1, w_1) & \cdots & g(x_N; c_M, w_M) \end{bmatrix}_{N*M}, \tag{16.5}$$

$$H\beta = \left(\beta_1^T \beta_2^T, \ldots, \beta_L^T\right)^T_{m*M}. \tag{16.6}$$

In this case, the weight of outputs can be calculated using the least square method (LSM) by applying the Moore–Penrose generalized inverse (H^+) of the matrix of the hidden layer. The schematic structure of the ELM model is shown in Figure 16.2.

$$\beta = H^+ Y \tag{16.7}$$

Over the past few years, the ELM methods have been extensively applied to solve different problems associated with water resources and hydrology (Abdullah et al., 2015; Deo et al., 2017; Kisi & Alizamir, 2018; Wen et al., 2019).

16.3.1.2 Fully Complex Extreme Learning Machine

Li et al. (2005) suggested the fully complex extreme learning machine (FCELM) by extending the basic ELM for the complex domain and applying non-linear channel equalization problems for SLFNs. Similar to the basic ELM, the weight of inputs and biases are randomly determined and the output weights are computed using a mathematical procedure. FCELMs are suitable approaches for modelling

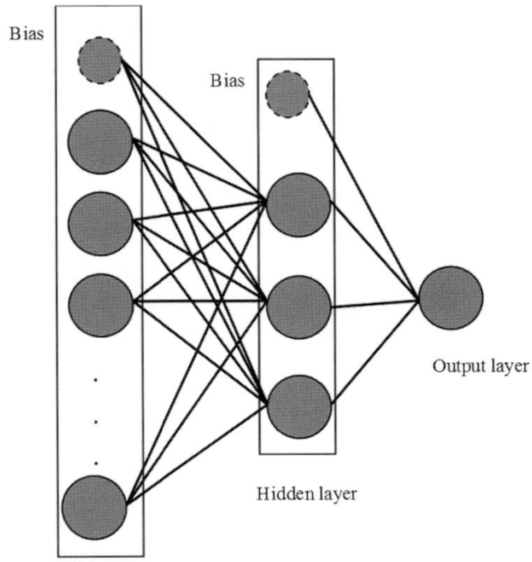

Figure 16.2 The general architecture of an ELM model

complex domains by providing parsimonious structures. The complex matrix H for the network can be written compactly as:

$$H = \begin{bmatrix} g_c(w_1 \cdot z_1 + b_1) & \cdots & g_c(w_{\tilde{N}} \cdot z_1 + b_{\tilde{N}}) \\ g_c(w_1 \cdot z_N + b_1) & \cdots & g_c(w_{\tilde{N}} \cdot z_N + b_{\tilde{N}}) \end{bmatrix}_{N*\tilde{N}} \quad (16.8)$$

$$\beta = \begin{bmatrix} \beta_1^T \\ \vdots \\ \beta_{\tilde{N}}^T \end{bmatrix}_{\tilde{N}*m}, \quad O = \begin{bmatrix} O_1^T \\ \vdots \\ O_N^T \end{bmatrix}_{N*m} \quad \text{and} \quad Y = \begin{bmatrix} y_1^T \\ \vdots \\ y_N^T \end{bmatrix}_{N*m}. \quad (16.9)$$

In FCELM, methods such as singular value decomposition (SVD), orthogonalization, iterative technique and orthogonal projection have been proposed to solve the Moore–Penrose generalized inverse of the elements.

16.3.1.3 Online Sequential Extreme Learning Machine

Liang et al. (2006) developed an online sequential learning ELM (OS-ELM) algorithm for SLFNs by applying additive or radial basis functions for hidden nodes. There are many limitations for the basic ELM to learn large scale data; therefore, optimization techniques have been applied for the basic ELM to bridge the gap. OS-ELM can learn one-by-one systems or chunk-by-chunk systems of different chunk sizes. By selecting hidden node parameters randomly and calculating the weight of outputs mathematically, this approach has shown better generalization performance and faster training processes than sequential methods (Liang et al., 2006). In the concept of OS-ELM, the output weight matrix can be written as:

$$H^{+} = (H^{T}H)^{-1}H^{T}. \tag{16.10}$$

β can be calculated as:

$$\beta = (H^{T}H)^{-1}H^{T}T. \tag{16.11}$$

In the next step, $\|H_0\beta - T_0\|$ should be minimized to solve the problem. By applying a recursive method for updating the LSM during the initialization phase and sequential learning phase, the matrix of the partial hidden layer can be defined as:

$$H_{k+1} = \begin{bmatrix} G\left(a_1, b_1, x_{\left(\sum_{j=0}^{k} N_j\right)+1}\right) & \cdots & G\left(a_{\tilde{N}}, b_{\tilde{N}}, x_{\left(\sum_{j=0}^{k} N_j\right)+1}\right) \\ \vdots & \cdots & \vdots \\ G\left(a_1, b_1, x_{\left(\sum_{j=0}^{k+1} N_j\right)}\right) & \cdots & G\left(a_{\tilde{N}}, b_{\tilde{N}}, x_{\left(\sum_{j=0}^{k+1} N_j\right)}\right) \end{bmatrix}_{N_{k+1*\tilde{N}}}. \tag{16.12}$$

Finally, the output weight ($\beta^{(k+1)}$) can be calculated as:

$$\beta^{(k+1)} = \beta^{(k)} + P_{k+1}H_{k+1}^{T}(T_{k+1} - H_{k+1}\beta^{(k)}) \tag{16.13}$$

Various researchers have applied the OS-ELM model in water management fields. Yadav et al. (2016) forecasted discharge in Neckar River, Germany. Kisi and Alizamir (2018) applied OS-ELM to model reference evapotranspiration. Ali et al. (2018) predicted rainfall using OS-ELM.

16.3.1.4 Self-Adaptive Evolutionary Extreme Learning Machine

Cao et al. (2012) suggested another scheme of ELM, called self-adaptive evolutionary extreme learning machine (SaE-ELM) for SLFNs that enhances the performance of the ELM training algorithm. In this approach, a powerful population-based stochastic searching method (self-adaptive differential evolution algorithm) optimizes parameters of hidden nodes by applying a trial vector generation strategy. Also, the weights of outputs are calculated by the Moore–Penrose generalized inverse. The SaE-ELM algorithm can be summarized as follows:

Step 1: Initialization: by generating random vectors of the parameters of network hidden node using populations of the first generation.

$$\theta_{k,G} = \left[a_{1,(k,G)}^{T}, \ldots, a_{L,(k,G)}^{T}, b_{1,(k,G)}, \ldots, b_{L,(k,G)}\right]. \tag{16.14}$$

Step 2: Finding output weights and root mean square error (RMSE): by considering each population vector, RMSE and output weights can be computed using the following equations.

$$\beta_{k,G} = H^{*}_{k,G}T, \tag{16.15}$$

$$\text{RMSE}_{k,G} = \sqrt{\frac{\sum_{j=1}^{L}\left\| \sum_{j=1}^{L} \beta_j g(a_j, (k,G), b_j, (k,G,x_i) - t_i) \right\|}{\text{m} * \text{n}}}, \tag{16.16}$$

where $H_{k,G}$ can be defined as:

$$H_{k,G} = \begin{bmatrix} g(a_{1(k,G)}, b_{1(k,G)}, x_1) & \cdots & g(a_{L(k,G)}, b_{L(k,G)}, x_1) \\ \vdots & \cdots & \vdots \\ g(a_{1(k,G)}, b_{1(k,G)}, x_N) & \cdots & g(a_{L(k,G)}, b_{L(k,G)}, x_N) \end{bmatrix}. \tag{16.17}$$

Step 3: Operators of mutation and crossover: in this phase, the trial vector generation strategy is applied and a specific number of iterations for the period of learning is employed.

Step 4: Evaluation: this step assesses the generated vectors. The generalization performance is enhanced by repeating steps (3) and (4) to obtain the optimum values for output weights. Cao et al. (2012) compared the accuracy of the proposed model with the evolutionary extreme learning machine (E-ELM), support vector regression, composite function wavelet neural network with ELM (CFWNN-ELM) and several evolutionary Levenberg–Marquardt methods in regression and classification problems. The results showed better generalization performance of SaE-ELM in both regression and classification simulations (Ghimire et al., 2018; Shabanlou, 2018).

16.3.1.5 Weighted Extreme Learning Machine

Zong et al. (2013) proposed the weighted extreme learning machine (WELM) to consider data with a skewed distribution. This approach has several advantages compared to the basic ELM, including: (1) easy implementation, (2) several kernel functions applied to generalize balanced data and (3) WELM applied to multiclass classification problems. The optimization problem procedure to find the minimum of the weighted cumulative error can be represented as:

$$\text{Minimize:} \quad \|H\beta - T\|^2 \tag{16.18}$$

and

$$\text{Minimize:} \quad L_{pELM} = \frac{1}{2}\|\beta\|^2 + CW\frac{1}{2}\sum_{i=1}^{N}\|\xi_i\|^2 \tag{16.19}$$

$$\text{Subject to:} \quad h(x_i)\beta = t_i^T - \xi_i^T, \quad i = 1, \ldots, N, \tag{16.20}$$

where $h(x_i)$ and β are the hidden layer feature mapping vector and output weight vector, respectively that link the output layer to hidden layer. Also, ξ_i indicates the training error of each sample. By applying Karush–Kuhn–Tucker (KKT) theorem, the problem can be written as:

$$L_{pELM} = \frac{1}{2}\|\beta\|^2 + CW\frac{1}{2}\sum_{i=1}^{N}\|\xi_i\|^2 - \sum_{i=1}^{N}\alpha_i(h(x_i)\beta - t_i + \xi_i) \tag{16.21}$$

where α_i is the Lagrange multiplier. Finally, the N*N kernel output function for WELM as a binary classifier can be represented as:

$$f(x)_{\text{kernel}} = \text{signh}(x)H^T\left(\frac{I}{C}+WHH^T\right)^{-1}WT = \text{sign}\begin{bmatrix} K(x,x_1) \\ \vdots \\ K(x,x_N) \end{bmatrix}^T \left(\frac{I}{C} + W\Omega_{ELM}\right)^{-1} WT. \tag{16.22}$$

Also, a similar process can be conducted to find the kernel output function for WELM as a multiclass classifier. Alternatively, we can have:

$$f(x) = h(x)H^T\left(\frac{I}{C} + HH^T\right)^{-1} T = \begin{bmatrix} K(x,x_1) \\ \vdots \\ K(x,x_N) \end{bmatrix}^T \left(\frac{I}{C} + \Omega_{ELM}\right)^{-1} T. \tag{16.23}$$

In WELM, several feature mapping functions can be applied.

16.3.1.6 Bidirectional Extreme Learning Machine

Yang et al. (2012) suggested a new version of ELM, called the bidirectional extreme learning machine (BELM), which optimizes the learning process by using a lower number of hidden nodes compared with the basic ELM. In the BELM, hidden nodes are not randomly chosen. This process leads to reducing the model output error at the early stages of the training phase. Note that the BELM is faster than the basic and conventional ELM models. The value of the obtained error feedback function can be defined as ($\|e_n\|$ indicates the sequence):

$$H_{2n}^e = e_{2n-1}\cdot(\beta_{2n-1})^{-1}. \tag{16.24}$$

By defining $\Delta = \|e_{2n-1}\|^2 - \|e_{2n}\|^2$, it can be rewritten as follows:

$$\begin{aligned} \Delta &= \|e_{2n-1}\|^2 - \|e_{2n-1} - H_{2n}^e\cdot\beta_{2n}\|^2 = 2\beta_{2n}\langle e_{2n-1}, H_{2n}^e\rangle - \|H_{2n}^e\|^2\cdot\beta_{2n}^2 \\ &= \|H_{2n}^e\|^2\left(\frac{2\beta_{2n}\langle e_{2n-1}, H_{2n}^e\rangle}{\|H_{2n}^e\|^2} - \beta_{2n}^2\right). \end{aligned} \tag{16.25}$$

By considering β_{2n}, Δ can be written as:

$$\Delta = \|H^e_{2n}\|^2(2\beta_{2n}\cdot\beta_{2n} - \beta^2_{2n}) = \|H^e_{2n}\|^2\cdot\beta^2_{2n} \geq 0. \tag{16.26}$$

Equation (16.25) can be rewritten as:

$$\|e^2_{2n-1}\| - \|e^2_{2n}\| = \|H^e_{2n}\|^2\cdot\beta^2_{2n} = \left(\frac{\beta_{2n}}{\beta_{2n-1}}\right)\cdot\|e^2_{2n-1}\|^2, \tag{16.27}$$

then

$$\left(\frac{\beta^2_{2n}}{\beta^2_{2n-1}}\right) + \frac{\|e_{2n}\|^2}{\|e_{2n-1}\|^2} = 1. \tag{16.28}$$

To assess the accuracy of the proposed model, nine benchmark applications have been used by the authors and compared by error-minimized LM (EM-ELM), convex incremental ELM (CI-ELM), enhanced incremental ELM (EI-ELM) and optimal pruned ELM (OP-ELM). The results of the simulation processes indicated that the new learning algorithm could enhance the training effectiveness by decreasing the number of hidden nodes for regression problems. Also, the learning time decreases by about two orders of magnitude.

16.3.2 Deep Learning Models

One of the most popular machine learning models is deep learning, which has gained success in different fields of study, including pattern recognition, image detection, natural language processing and image and video recognition. The main categories of these models are:

- Deep Feedforward Neural Networks
- Deep Belief Networks
- Deep Extreme Learning
- Convolutional Neural Networks
- Recurrent Neural Networks
- Pre-trained Unsupervised Networks
- Deep Echo State Network
- Deep Belief Echo State Network

Here, we describe the first three groups of deep learning models.

16.3.2.1 Deep Feedforward Neural Networks

Deep feedforward neural networks are multilayer perceptron-based models. A multi-layer deep learning model consists of several layers in the hidden part and rectified linear (ReLU) transfer function. Figure 16.3 shows the general

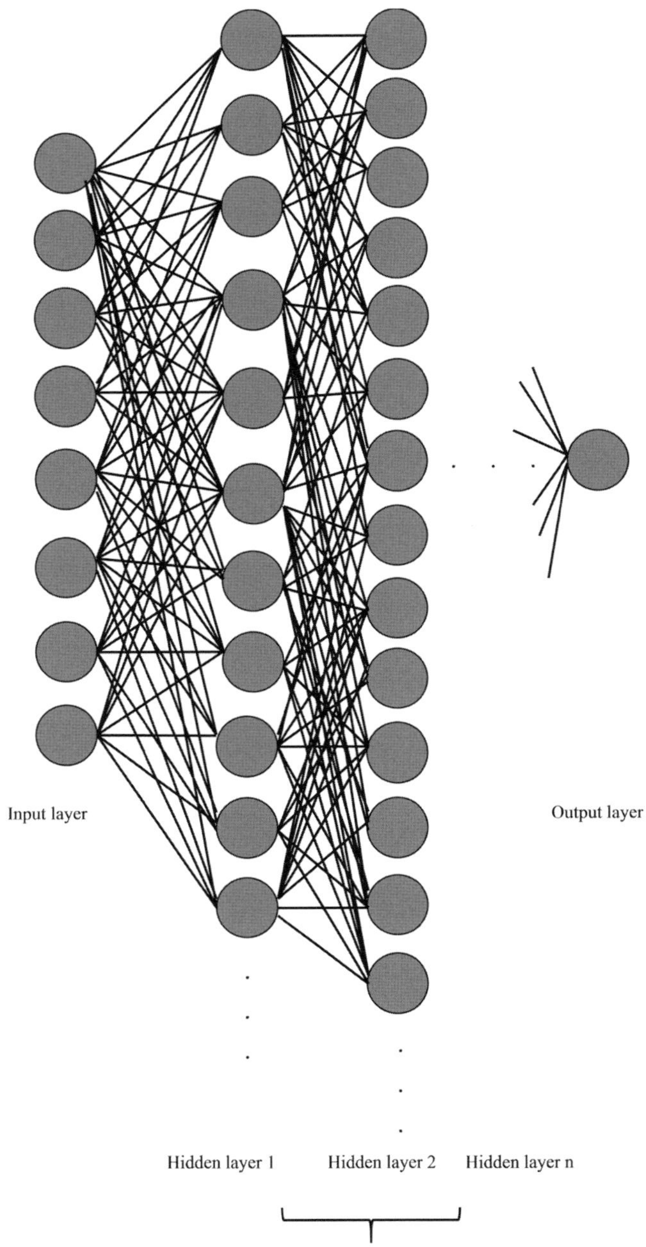

Figure 16.3 The general structure of a deep feedforward neural network

structure of the deep learning model. The main properties of a multi-layer deep learning model are rate annealing, rate of adaptive learning, momentum training, L_1 or L_2 regularization and dropout. The general form of this network is mathematically modelled as follows (Saggi & Jain, 2019):

$$Y(\text{net}) = \sum_{i=1}^{n} w_i * x_i + b. \tag{16.29}$$

The ReLU function is a computationally efficient and easy-to-implement method given by:

$$f(x) = \max(0, x). \tag{16.30}$$

By calculating the derivative of ReLU:

$$f'(x) = \begin{Bmatrix} 1, & \text{if} \quad x \succ 0 \\ 0, & \text{otherwise} \end{Bmatrix}. \tag{16.31}$$

As mentioned, the formulation of a multi-layer deep learning model is written as:

$$\text{net} = \sum_{i} (w_i * x_i + b) \tag{16.32}$$

$$Y = \text{activation}(\text{net}), \tag{16.33}$$

where Y and x are output and input parameters. Also, w and b indicate weight connection between different layers and the bias of hidden neurons. In the context of water management, Liu et al. (2017) developed a flood forecasting model based on deep learning methodology. He et al. (2019a) have applied deep neural networks for forecasting daily runoff.

16.3.2.2 Deep Belief Networks

Hinton et al. (2006) suggested deep belief networks (DBNs) as probabilistic models, which are composed of multiple hidden units. The main shortcomings of the earlier versions of deep learning models are: (a) the large number of samples required for the training process, (b) the considerable computational time needed to reach convergence and (c) encountering the local optima problem. In general, DBNs are composed of unsupervised networks like restricted Boltzmann machines (RBMs) or autoencoders (Bengio et al., 2007). RBMs can effectively identify the structure of predictor parameters. In this case, each visible layer is connected using the weight matrix to the hidden layer, and each hidden layer of the network plays the role of a visible layer for the next one. The general form of DBNs is mathematically modelled as follows:

$$P(\upsilon, h_1, h_2, \ldots, h_L) = \left(\prod_{k=0}^{L-2} P(h_k | h_{k+1}) \right) P(h_{L-1}, h_L) \tag{16.34}$$

where υ and h_i are predictors and multiple hidden units, respectively. The process of learning using weighted connections by an RBM model employs a Bayesian procedure that considers $P(\upsilon|h, W)$ and $P(h|W)$ so that:

$$P(v) = P(h|W)P(v|h, W) \tag{16.35}$$

where h is the vector of binary states and $P(h|W)$ indicates the prior distribution of hidden vectors. The training procedure of DBNs has two steps: (a) a greedy layer-wise learning approach based on unsupervised learning algorithms (pre-training) and (b) back-fitting with an up–down technique (fine-tuning). Practical instances for using DBNs in water management include drought forecasting. Chen et al. (2012) applied deep belief networks for the forecasting index of drought in the Huaihe river basin. Agana and Homaifar (2018) predicted drought using a deep belief network.

16.3.2.3 Deep Extreme Learning

Huang and Sun (2015) introduced a new soft computing model named the deep and stable extreme learning machine (DS-ELM) that is applicable in both regression and classification problems. The approach is based on integrating the extreme learning machine (ELM) and deep belief network (DBN) using several layers of restricted Boltzmann machines (RBMs). Generally, the DS-ELM model shares the advantages of ELM, including better generalization performance and less computational complexity, with the benefit of the ability of DBN to model highly non-linear and complex structures. In this approach, a DBN is applied to produce a stable input space H transferred to an ELM to obtain the weights of output parameters. The following steps briefly describe the suggested DS-ELM algorithm.

Step 1. The model is trained using the DBN structure using input vectors and is based on unsupervised pre-training and supervised fine-tuning at a limited number of epochs.

Step 2. After feeding ELM by outputs of DBN, the ELM model computes the weights of outputs.

Huang and Sun (2015) performed several experiments to evaluate the performance of the DS-ELM model. They compared the performance of the DS-ELM model with standalone ELM and DBN for both regression and classification applications. They reported that the DS-ELM model achieved better accuracy in the testing phase compared with the other two models in the context of complex high dimensional datasets. Besides, in binary and multi-class classification problems, the DS-ELM model solves the problems with smaller standard deviations.

16.3.3 Metaheuristic Algorithms

Metaheuristic algorithms or global optimization methods are derivative-free techniques to solve non-linear and non-convex problems by considering several non-linear constraints with some design variables. These models have gained the attention of researchers in the last decade as effective optimization approaches to solve complicated problems. Finding the global optimum of a complex optimization problem by considering constraints can be defined as:

$$\max f(x), \quad x = (x_1, \ldots, x_d) \in \Re^d \tag{16.36}$$

$$\text{subject to} \quad x \in X. \tag{16.37}$$

To solve this problem, a population of candidate solutions is initialized as the starting point and can be modified to a total number of generations. For each generation, a new solution population is produced by performing stochastic operators over the population. The objective function investigates the quality of each possible solution using fitness values. In recent years, many metaheuristic approaches have been suggested, which include: Artificial Bee Colony (ABC), Particle Swarm Optimization (PSO), Salp Swarm Algorithm (SSA), Harmony Search (HS), Gravitational Search Algorithm (GSA), Bat Algorithm (BA), Grey Wolf Optimization (GWO), Social Spider Optimization (SSO), FireFly Algorithm (FFA), Clonal Selection Algorithm (CSA), Cuckoo Search (CS), Locust Algorithm (LA), Genetic Algorithms (GA), and Differential Evolution (DE). In this section, we briefly describe the most relevant methods for our study.

16.3.3.1 Gravitational Search Algorithm

The gravitational search algorithm (GSA) was suggested by Rashedi et al. (2009) based on the gravitational forces among different masses. The situation of each mass can be evaluated using fitness values. Equation (16.38) defines a gravitational force from a mass *i* to a mass *j*:

$$\mathrm{F}_{ij}^{h}(t) = \mathrm{G}(t)\frac{\mathrm{Mp}_i(t) * \mathrm{Ma}_j(t)}{\mathrm{R}_{ij}(t) + \varepsilon}\left(x_j^h(t) - x_i^h(t)\right) \tag{16.38}$$

where Ma_j, Mp and G(t) are the active gravitational mass, the passive gravitational mass and the gravitational constant, respectively. Also, R_{ij} and ε denote the Euclidean distance and a constant, respectively. The total force on each mass can be computed as

$$\mathrm{F}_i^h(t) = \sum_{j=1, j\neq i}^{N} \mathrm{F}_{ij}^h(t). \tag{16.39}$$

The acceleration of each mass can be computed as

$$\mathrm{a}_i^h(t) = \frac{\mathrm{F}_i^h(t)}{\mathrm{Mn}_i(t)}. \tag{16.40}$$

Also, once velocity and position for each mass are computed, one can view heavier mass as a better solution. By evaluating the fitness function, updating inertia and gravitational masses is completed as follows:

$$\mathrm{m}_i(t) = \frac{f\left(x_i(t)\right) - \mathrm{worst}(t)}{\mathrm{best}(t) - \mathrm{worst}(t)} \tag{16.41}$$

$$\mathrm{M}_i(t) = \frac{\mathrm{m}_i(t)}{\sum_{j=1}^{N} \mathrm{m}_j(t)} \tag{16.42}$$

where $f(\cdot)$ stands for the fitness function. GSA has been applied by Ojugo et al. (2013), Moeini et al. (2017) and Niu et al. (2019) for solving water resources problems.

16.3.3.2 Grey Wolf Optimizer

The Grey Wolf Optimizer (GWO) algorithm was introduced by Mirjalili et al. (2014) to emulate the social life of a grey wolf pack. The hunting mechanism and hierarchy structure of grey wolves are the two main features of GWO. Each wolf pack is classified into four types by their responsibility, including α, β, δ and ω. Each potential solution is determined using the position of a grey wolf. For example, the position of an omega wolf is calculated as follows:

$$\vec{\mathrm{P}}_{\omega}^{N} = \frac{1}{3}\left(\vec{\mathrm{EP}}_{\alpha} + \vec{\mathrm{EP}}_{\beta} + \vec{\mathrm{EP}}_{\delta}\right) \tag{16.43}$$

where $\vec{\mathrm{EP}}_{\alpha}$, $\vec{\mathrm{EP}}_{\beta}$ and $\vec{\mathrm{EP}}_{\delta}$ denote the position of alpha, beta and omega grey, respectively. Some examples that illustrate the use of GWO can be found in Tikhamarine et al. (2019) and Zou et al. (2019).

16.3.3.3 Artificial Bee Colony

Karaboga (2005) introduced the artificial bee colony (ABC), which is inspired by the behaviour of honeybee swarm. This algorithm runs a series of iterations based on the population of food sources to assess the quality of food sources through a fitness function that simulates the production of a new source of food. This process can be described as:

$$v_i = s_i^k + \phi(s_i^k - sk_j), \quad i,j \in (1, 2, \ldots, N). \tag{16.44}$$

The fitness function can be written as:

$$\mathrm{fit}(s_i^k) = \left\{ \begin{array}{cc} \dfrac{1}{1+f(s_i^k)} & \text{if } f(s_i^k) \geq 0 \\ 1 + \mathrm{abs}(f(s_i^k)) & \text{if } f(s_i^k) < 0 \end{array} \right\}. \tag{16.45}$$

Research applications of the ABC algorithm can be found in Ahmad et al. (2016) and Yilmaz et al. (2019).

16.3.3.4 Harmony Search

Geem et al. (2001) introduced the harmony search evolutionary algorithm based on harmonies in jazz music for a population of N individuals. This algorithm evaluates each harmony using a fitness function. It finds new solutions using pitch adjustment, which leads to the generation of a new harmony. The procedure can be shown as:

$$H_{new} = \begin{cases} H_j & \in \{x_1, x_2, \ldots, x_{HMS,j}\} \text{ with probability } \quad HMCR \\ lower + (upper - lower) \cdot rand & \text{with probability } \ (1 - HMCR) \end{cases}. \tag{16.46}$$

A pitch-adjusting rated (PAR) can be applied to consider which harmony should be pitch rated for tuning a bandwidth factor (BW) to evaluate the local search of chosen harmonies. It can be described as:

$$H_{new} = \begin{cases} H_{new} \pm rand(0,1) \cdot \mathrm{BW} & \text{with probability} \quad \mathrm{PAR} \\ H_{new} & \text{with probability } (1 - \mathrm{HMCR}) \end{cases}. \tag{16.47}$$

PAR can change previous parameters such as the mutation operator by producing new potential samples.

16.3.3.5 Differential Evolution

One of the stochastic vector-based evolutionaries is the differential evolution (DE) algorithm, which was suggested by Storn and Price (1997). The DE uses the weighted difference between two vectors to produce a third new vector as shown:

$$v_i^{k+1} = x_p^k + F(x_q^k - x_r^k). \tag{16.48}$$

The crossover operator in the DE algorithm can be written as:

$$u_{j,i}^{k+1} = \begin{cases} v_{j,i} & \text{if } \ r_i \leq C_r \\ x_{j,i}^k & \text{otherwise} \end{cases}, \quad j = 1, 2, \ldots, d, \quad i = 1, 2, \ldots, n. \tag{16.49}$$

The DE algorithm has been applied by Al-Sudani et al. (2019) and He et al. (2019b).

16.3.3.6 Particle Swarm Optimization

Kennedy and Eberhart (1995) proposed particle swarm optimization (PSO) that is based on the behaviour of birds flocking. The PSO finds the best positions using a cost function. In this algorithm, values of velocity are calculated as:

$$v_{i,j}^{k+1} = \omega \cdot v_{i,j}^{k} + c_1 \cdot r_1 \cdot (p_{i,j}^{*} - p_{i,j}^{k}) + c_2 \cdot r_2 \cdot (g_{i,j}^{*} - p_{i,j}^{k}) \tag{16.50}$$

where ω is the inertia weight and c_1 and c_2 are the acceleration coefficients. The following equation is applied to compute the new position of each particle.

$$p_{i,j}^{k+1} = p_{i,j}^{k} + v_{i,j}^{k+1}. \tag{16.51}$$

An illustration of PSO in water management can be found in Meshram et al. (2019) and Rahgoshay et al. (2019).

16.3.3.7 Differential Search Algorithm

Civicioglu (2012) proposed the differential search algorithm (DSA), which is inspired by the Brownian movement of organisms. In the DSA, a Brownian movement is applied to generate a stopover vector after the initialization of organisms. This phase is computed as follows:

$$s_{i,N} = X_{i,N} + A \cdot (X_{ri,N} - X_{i,N}). \tag{16.52}$$

Also, the stopover position is described as follows:

$$s_{i,j}' = \left\{ \begin{array}{l} s_{i,j} \text{ if } r_{i,j} = 0 \\ X_{i,j} \text{ if } r_{i,j} = 1 \end{array} \right\}. \tag{16.53}$$

By selecting the potential solution, each organism is investigated by a cost function. The process of selection is based on the following criterion:

$$X_i^{k+1} = \left\{ \begin{array}{l} s_i^k \text{ if } f(s_i') \leq f(X_i^k) \\ X_i^k \text{ if } f(s_i') \geq f(X_i^k) \end{array} \right. . \tag{16.54}$$

16.3.3.8 Cuckoo Search Algorithm

Yang and Deb (2009) suggested the cuckoo search algorithm (CSA) based on the behaviour of cuckoos. In this algorithm, Lévy flights are applied to enhance the processes of the exploration in the solution space. A symmetric distribution is produced through a random step that is obtained using the following equation:

$$s_{i,} = \frac{u}{|v|^{\frac{1}{\beta}}} \tag{16.55}$$

where β is a constant and the vectors u and v are random variables that follow a normal distribution:

Table 16.2. General features of some of the well-known metaheuristic algorithms

Algorithm	Type	Origin
GSA	Physics-based algorithm	Newtonian physics principals of gravity and interaction between masses
GWO	Swarm-based algorithm	Hunting mechanism of grey wolves (Canis lupus)
ABC	Swarm-based algorithm	Searching for food by bees
HS	Evolutionary algorithm	Pitch adjustment in music
DE	Evolutionary algorithm	The natural phenomenon of evolution
PSO	Swarm-based algorithm	Flocking behaviour of birds
DSA	Evolutionary algorithm	Brownian-like random-walk movement used by an organism to migrate
CSA	Swarm-based algorithm	Intelligent behaviour of crows

Note: GSA, Gravitational Search Algorithm; GWO, Grey Wolf Optimizer; ABC, Artificial Bee Colony; HS, Harmony Search; DE, Differential Evolution; PSO, Particle Swarm Optimization; DSA, Differential Search Algorithm; CAS, Cuckoo Search Algorithm.

$$u \approx N(0, \sigma_u^2), \qquad v \approx N(0, \sigma_u^2) \tag{16.56}$$

$$\sigma_u = \left(\frac{\Gamma(1+\beta) \cdot \sin\left(\pi \cdot \beta/2\right)}{\Gamma(1+\beta/2) \cdot \beta . 2^{(\beta-1)/2}} \right), \tag{16.57}$$

where $\Gamma(\cdot)$ stands for the Gamma distribution. The new potential solution is obtained as follows:

$$e_i^{k+1} = e_i^k + c_i. \tag{16.58}$$

A fitness function evaluates this new solution; if this solution enhances the quality of the previous one, it is replaced, otherwise it remains. Meng et al. (2019) and Naidu et al. (2020) applied the CSA algorithm in their work. Table 16.2 provides a general comparison for the basic properties of the aforementioned metaheuristic algorithms.

16.4 Application of Advanced Soft Computing Methods in Water Management

Various problems and issues related to water management policies have gained attention from the research community. This section covers the application of three advanced soft computing methods, including deep learning networks, metaheuristic algorithms and extreme learning machines used to address three hydrological modelling problems. We focus on three main components of climatic conditions that have a considerable impact on water management: drought, rainfall-runoff and evapotranspiration processes. Here, we describe the use of modern soft computing methods in each of the domains of investigation.

16.4.1 Drought Modelling and Forecasting

The problem of drought modelling, as a critical water management factor, is of great importance to hydrologists. Getting reliable drought predictions is a key element in the management of hydrological and water resources: it can mitigate the consequent damage of rapid climate changes.

Obtaining accurate predicted values of the characteristics of drought, such as the beginning, severity, termination and frequency, enables the hydrologists to provide some precautionary water management and planning policies in reducing the drought impacts. Note that the concept of drought can be associated and measured through various approaches. In a forecasting context, drought can be specified through different climatological indices, availability and input data. Furthermore, drought can be defined in terms of meteorological, hydrological, agricultural and socio-economic perspectives.

The most common drought indices include the standardized precipitation index (SPI), the surface water supply index (SWSI), the effective drought index (EDI), the Palmer drought severity index (PDSI) and the moisture anomaly index (Z-index). Each of these indices emphasizes a specific concept and measure of drought. For instance, based on the data of precipitation, the SPI index categorizes the drought severity into five separate classes (D0–D4). Table 16.3 provides a general overview of drought severity based on the SPI index.

Methodologies of drought forecasting can be divided into two main categories: (1) traditional mathematical methods and (2) methods based on soft computing

Table 16.3. Different classes of drought severity based on the SPI index (US drought monitor)

Class	Description	SPI range
D0	Abnormally dry	–0.5 to –0.7
D1	Moderate drought	–0.8 to –1.2
D2	Severe drought	–1.3 to –1.5
D3	Extreme drought	–1.6 to –1.9
D4	Exceptional drought	$\leq$–2.0

techniques. One of the main advantages of applying soft computing techniques in modelling drought lies in its procedural simplicity associated with entering various types of useful indices to the models. After determining the optimal input vector using input selection methods, normalized or standardized data can be easily introduced to the soft computing models. Due to the differences in the nature of the problems and data availability, various types of meteorological and climatological input data can be applied. Precipitation data (e.g., rainfall) and air temperature variables (e.g., minimum and maximum air temperatures) are the impartible entering data for most models. In addition to these parameters, some models may also account for large-scale climate mode indices (Deo & Sahin, 2015).

Artificial neural networks (ANNs) are among the most frequently applied soft computing methods in drought modelling. ANNs are known to be powerful and versatile machine learning models for capturing, analysing and mapping complicated relationships between the input parameters and output targets. There are numerous reports on the successful usage of different types of ANNs in drought modelling; however, recent studies suggested that improved classes of neural networks such as extreme learning machines and other advanced soft computing methods perform better in the face of complex systems. Some researchers have exploited the capability of deep belief networks (DBN) to forecast drought and have claimed to obtain promising results. Chen et al. (2012) applied DBN for drought prediction based on different time-scale indices of SPI. The DBN model was employed as a promising tool to predict short-time drought in the Huaihe River Basin, located in China. Four hydrologic stations were chosen with available monthly precipitation data from 1958–2006. The forecasting results of the DBN model were evaluated and compared with the traditional multi-layer perceptron neural network (MLPNN). It was reported that the prediction accuracy of the DBN is superior to the conventional neural network for all of the stations. In a similar study, Agana and Homaifar (2018) developed a hybrid DBN model for drought forecasting in the Colorado River Basin, USA. The proposed hybrid DBN model showed better prediction accuracy in comparison with standard soft computing methods such as support vector regression models.

Other advanced soft computing techniques using extreme learning machine have been successfully applied to simulate drought characteristics in different regions across the world (Deo & Sahin, 2015; Deo et al., 2017). The ELM model can also be combined with signal processing techniques such as wavelet decomposition for the input parameters. This hybrid method can improve the results of the basic ELM model in drought forecasting. Overall, it is obvious that ELM gives excellent performance in drought forecasting and provides better results compared to traditional soft computing methods such as ANNs. Moreover, the ELM exhibits computational advantages over traditional approaches with regard to both the learning and the training processes.

Metaheuristic algorithms can be employed in multivariate modelling and forecasting drought. Their utility is illustrated in two different contexts. First, metaheuristic methods may be utilized to estimate parameters of copula models related to drought characteristics such as drought severity–duration–frequency (S-D-F) curves (Reddy & Singh, 2014). Second, they may be implemented as optimization tools to constructing soft computing models, including ANNs.

S-D-F curves aggregate comprehensive and concise information about drought characteristics in the region of interest. S-D-F curves can be used to design water management strategies in regions affected by drought. Metaheuristic algorithms used to estimate the parameters of probability models are new and therefore conventional approaches have been so far preferred by the research community. However, the usage of metaheuristic algorithms is not restricted to the estimation of S-D-F curves and its potential applications go beyond drought modelling.

As mentioned, metaheuristic algorithms have also been combined with machine learning and data-driven models to construct resilient and powerful soft computing drought forecasting models. Metaheuristic algorithms such as particle swarm optimization, genetic algorithm, ant colony and butterfly optimization algorithm have been used in drought modelling (Cao et al., 2009; Kisi et al., 2019; Malik et al., 2019). The predictive performance of combined soft computing models is superior to those obtained from single models. This brief review shows that metaheuristic algorithms can enhance the accuracy of drought characterization parameters in multivariate models such as copula and improve the efficiency of machine learning models such as ANNs in real-time forecasting of drought.

Advanced soft computing models show promising results in modelling and forecasting non-linear and complicated phenomena such as drought processes and provide a genuine alternative to the modelling of real-time hydrological phenomena.

16.4.2 Evapotranspiration Modelling

Evapotranspiration can be expressed as the combined process of water vaporizing to the air by both surface evaporation and vegetation transpiration. Given its important role in designing and operating agricultural and hydrological processes, modelling evapotranspiration (ET) is key to an effective management of water resources. Evapotranspiration and reference evapotranspiration can be directly calculated by field measurements and experimental techniques, including lysimeters and eddy covariance systems. Although these direct methods can accurately estimate actual evapotranspiration, they are usually difficult to implement and are costly. Besides, due to some operational, technical, and economic reasons, establishing direct devices for measuring ET is sometimes unpractical in some regions. As a result, the development of mathematical and indirect methods for precise estimating is more suitable in most cases. Indirect mathematical methods can be divided into two main categories, which include conventional mathematical methods (e.g., the FAO-56 Penman-Monteith model) and soft computing mathematical models (e.g., machine learning models).

The FAO-56 Penman-Monteith model has been successfully used in different climatic regions. This method requires several independent input variables such as maximum and minimum air temperature, relative humidity, solar radiation and wind speed. However, the FAO-56 Penman-Monteith is not applicable in the absence of meteorological records or when the reliability of the measured data is doubtful (e.g., in developing countries). When these difficulties are present, modelling and predicting ET should be ideally chosen based on minimal independent input variables, which can be achieved through soft computing techniques.

In the last decade, various soft computing models have successfully modelled ET using a limited meteorological input dataset (Kisi & Zounemat-Kermani, 2014; Wang et al., 2017). So far, it appears that ET modelling has been used by applying traditional and modern soft computational methods. Therefore, there is a promising research path that might investigate the use of deep learning in soft computational modelling of ET.

Several studies have reported the results of modelling reference ET based on limited input data in different regions using ELM models (Abdullah et al., 2015; Gocic et al., 2016; Feng et al., 2017). According to Gocic et al. (2016), the ELM models can be considered as a reliable method in eliminating the vagueness in input data variables and are capable of producing promising ET predictions. Although the overestimating tendency of the ELM models in ET prediction has been observed, the ELM model predictive performance remains better than its traditional neural-based soft computing counterparts (Abdullah et al., 2015). The literature suggests that the ELM models provide not only more accurate results than traditional neural-based soft computing models but also exhibit faster convergence in both the training and testing phases.

Metaheuristic algorithms have been used in both evapotranspiration and evaporation modelling (Abdullah et al., 2014; Petković et al., 2016; Yin et al., 2016; Zounemat-Kermani et al., 2019). Different types of metaheuristic algorithms, including particle swarm optimization (PSO), firefly algorithm (FA), continuous ant colony optimization (CACO), artificial bee colony (ABC) and genetic algorithm (GA), have been implemented for evaluating the efficiency of combined soft computing models. The results suggest that combining metaheuristic algorithms with other soft computing methods such as adaptive neuro-fuzzy inference systems, support vector machines and ANNs increases the potential of these models in predicting reference ET.

16.4.3 Rainfall-Runoff Modelling

Accurate forecasting of runoff and river flow rates plays a vital role in water resources management projects and related policies to watershed management. Rainfall-runoff processes in hydrology are complex and exhibit local variability and non-linearity, which makes their predictions particularly challenging.

Methods that account for general and local conditions of the investigated case such as data acquisition, land use and climatic issues can be employed for modelling rainfall-runoff processes. Without the availability of physiographical and geological parameters related to the watershed, developing models based on the available datasets (such as precipitation) provide a suitable alternative. In the last few decades, data-driven models have been developed and widely applied in this context. From a general perspective, data-driven models can be divided into two main categories: time-series models and soft computing models. The former includes deterministic and stochastic models such as auto-regressive (AR) and auto-regressive integrated moving average (ARIMA) models. Although these types of models have been widely used in numerous research works on runoff modelling, they have been gradually superseded by soft computing techniques and machine learning models. These emerging hydrological models have shown growth in recent years. One reason of their success is their ability to model non-linearity and chaos associated with complex systems like rainfall-runoff processes, which cannot be captured with traditional data-driven models restricted by the formulation of their mathematical and/or stochastic components.

Li et al. (2018) proposed novel neural computing techniques such as convolutional deep belief networks to predict the rainfall-runoff process. They claimed that DBN (based on its complex and multi-layer architecture) provided promising results while using a restricted set of data. He et al. (2019a) combined the decomposition techniques and deep neural networks (DNN) to construct a robust and reliable predictive model for predicting runoff. One of the main constraints in constructing deep neural networks is to determine a relevant number of hidden layers and neurons in each layer. A prevailing method for solving this problem is to use an ascending technique for the number of layers and neurons. Several architectures can be constructed for the DNNs and the best of them is selected and defined as the selected model. However, fitting these models is

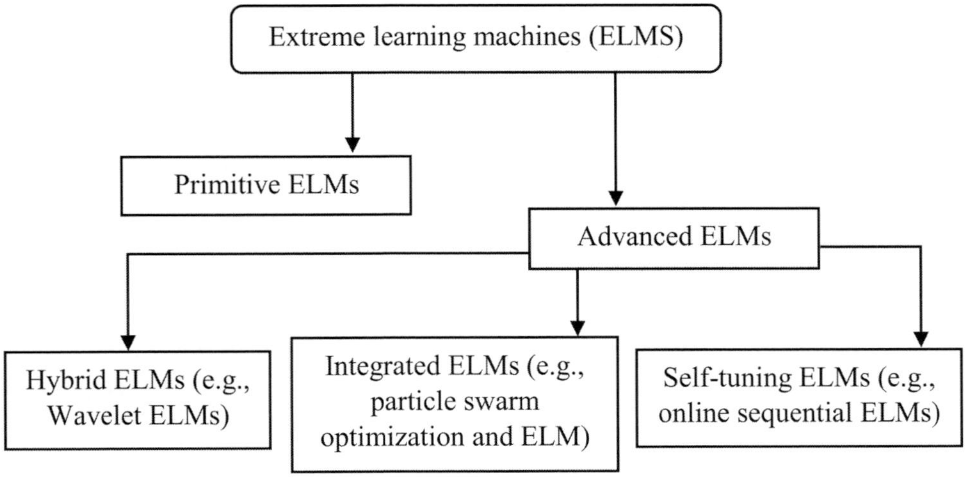

Figure 16.4 Diagram for the recent developments of ELM models in modelling rainfall-runoff processes

time-consuming and future studies might provide further guidelines for the use of the DNN architecture.

Published works in the field of runoff modelling using ELMs and metaheuristic methods are not limited as the DNNs and DBNs. Considering the ELM model, several researchers claimed that this model is a reliable and fast tool to estimate or predict runoff values (Dongwen, 2013; Li & Cheng, 2014; Yaseen et al., 2016). In addition, recent findings suggest that ELM models are better tools for capturing extreme events like the prediction of floods. Even though the usage of standard ELM models provided satisfactory results in modelling rainfall-runoff processes, an uprising trend for developing hybrid and integrated ELM models has occurred over the last years for modelling runoff and river flow. These advanced ELM models include the Online Sequential Extreme Learning Machine (OS-ELM, capable of updating the model equation at the time of entering new data) (Yadav et al., 2016), wavelet ELM models (Roushangar et al., 2017) and integrated ELMs (Taormina & Chau, 2015). Figure 16.4 illustrates recent trends in the evolution of ELM applications for runoff modelling.

Results of the metaheuristic methods applied in modelling runoff indicate that integrating these techniques with conventional soft computing models generates more consistent and accurate outcomes than their counterparts (Wang et al., 2013).

16.5 Discussion, Limitations and Future Directions

On the one hand, the availability of a wide range of water management data and, on the other hand, the necessity of using them in problems related to the design and management of water systems in recent years, have led the researchers to create simulation models that rely on data-driven methods.

Soft computing-based models represent one of the most effective approaches in this field. The intuition behind the soft computing techniques is different from knowledge-based, hard computing methods (e.g., numerical methods) and traditional data analysis. In most cases, soft computing water management models can process big data, learn the governing rules and handle complex dynamic systems, without accessing geographical, physiographical and geometric data. This modelling approach of water management is beneficial when information on the physical domain is limited. These new methods can potentially predict the rate of rainfall-runoff in a watershed merely based on the available data for precipitation and river flow, which would not be feasible with a traditional approach.

Despite the advantages provided by soft computing techniques over traditional, empirical, statistical and knowledge-based methods in water management engineering, they are not exempt from shortcomings and limitations. The main disadvantage of soft computing models in water management is their high dependency and sensitivity to the reliability of the data used in the model. Another drawback of the soft computing methods is that most of them have a black-box (e.g., traditional and novel neural networks) structure, which prevents the user from modifying and manipulating the models, if needed.

Given the current development trend of soft computing models in water management engineering, future research in this area can consider the following elements:

- Since soft computing models are more suitable than knowledge-based models for water management when data is lacking, not reliable or incomplete, it is expected that the application of these methods in the modelling of various case studies will spread.
- One of the disadvantages of computational methods, including soft computing models, is the high computational time required to simulate high-dimensional problems. Increased hardware capabilities of computers and the introduction of faster and more efficient models should increase the use of soft computing models in the future.
- Currently, most of the research is conducted on the evaluation and feasibility of soft computing methods for simulating and predicting hydrological phenomena based on academic studies. However, these methods have not reached the public yet. Management of functional systems such as flood and precipitation warning systems or drought forecasting systems could benefit from soft computing methods. We expect that these methods will be applied to real-world simulations in the near future.

References

Abdullah, S. S., Malek, M. A., Abdullah, N. S., Kisi, O., & Yap, K. S. (2015). Extreme learning machines: A new approach for prediction of reference evapotranspiration. *Journal of Hydrology* 527: 184–195.

Abdullah, S. S., Malek, M. A., Mustapha, A., & Aryanfar, A. (2014). Hybrid of artificial neural network-genetic algorithm for prediction of reference evapotranspiration (ET0) in arid and semiarid regions. *Journal of Agricultural Science* 6(3): 191.

Afzaal, H., Farooque, A. A., Abbas, F., & Acharya, B. (2020). Groundwater estimation from major physical hydrology components using artificial neural networks and deep learning. *Water* 12(1): 5.

Agana, N., & Homaifar, A. (2018). EMD-based predictive deep belief network for time series prediction: An application to drought forecasting. *Hydrology* 5(1): 18.

Ahmad, A., Razali, S. F. M., Mohamed, Z. S., & El-shafie, A. (2016). The application of artificial bee colony and gravitational search algorithm in reservoir optimization. *Water Resources Management* 30(7): 2497–2516.

Ali, M., Deo, R. C., Downs, N. J., & Maraseni, T. (2018). Multi-stage hybridized online sequential extreme learning machine integrated with Markov Chain Monte Carlo copula-Bat algorithm for rainfall forecasting. *Atmospheric Research* 213: 450–464.

Al-Sudani, Z. A., Salih, S. Q., & Yaseen, Z. M. (2019). Development of multivariate adaptive regression spline integrated with differential evolution model for streamflow simulation. *Journal of Hydrology* 573: 1–12.

Ashofteh, P. S., Bozorg-Haddad, O., & Loáiciga, H. A. (2020). Logical genetic programming (LGP) application to water resources management. *Environmental Monitoring and Assessment* 192(1): 34.

Bengio, Y., Lamblin, P., Popovici, D., & Larochelle, H. (2007). Greedy layer-wise training of deep networks. *Advances in Neural Information Processing Systems 19. Proceedings of the Twentieth Annual Conference on Neural Information Processing Systems*, Vancouver, BC, 4–7 December 2006 (pp. 153–160).

Cao, J., Chi, D., Liu, L., Li, S., & Yu, M. (2009). Application of BP network model based on PSO for the forecast of drought and flood. *Journal of Shenyang Agricultural University* 40(1): 118–121.

Cao, J., Lin, Z., & Huang, G. B. (2012). Self-adaptive evolutionary extreme learning machine. *Neural Processing Letters* 36(3): 285–305.

Chen, J., Jin, Q., & Chao, J. (2012). Design of deep belief networks for short-term prediction of drought index using data in the Huaihe river basin. *Mathematical Problems in Engineering* 2012(6): 235929.

Civicioglu, P. (2012). Transforming geocentric cartesian coordinates to geodetic coordinates by using differential search algorithm. *Computers & Geosciences* 46: 229–247.

Darabi, H., Choubin, B., Rahmati, O., Haghighi, A. T., Pradhan, B., & Kløve, B. (2019). Urban flood risk mapping using the GARP and QUEST models: A comparative study of machine learning techniques. *Journal of Hydrology* 569: 142–154.

Deo, R. C., & Şahin, M. (2015). Application of the extreme learning machine algorithm for the prediction of monthly Effective Drought Index in eastern Australia. *Atmospheric Research* 153: 512–525.

Deo, R. C., Tiwari, M. K., Adamowski, J. F., & Quilty, J. M. (2017). Forecasting effective drought index using a wavelet extreme learning machine (W-ELM) model. *Stochastic Environmental Research and Risk Assessment* 31(5): 1211–1240.

Dongwen, C. (2013). Comprehensive evaluation of water resources development and utilization in Yangtze River Basin based on extreme learning machine. *Advances in Science and Technology of Water Resources* 2: 005.

Feng, Y., Peng, Y., Cui, N., Gong, D., & Zhang, K. (2017). Modelling reference evapotranspiration using extreme learning machine and generalized regression neural network only with temperature data. *Computers and Electronics in Agriculture* 136: 71–78.

Geem, Z. W., Kim, J. H., & Loganathan, G. V. (2001). A new heuristic optimization algorithm: Harmony search. *Simulation* 76(2): 60–68.

Ghimire, S., Deo, R. C., Downs, N. J., & Raj, N. (2018). Self-adaptive differential evolutionary extreme learning machines for long-term solar radiation prediction with remotely-sensed MODIS satellite and reanalysis atmospheric products in solar-rich cities. *Remote Sensing of Environment* 212: 176–198.

Gocic, M., Petković, D., Shamshirband, S., & Kamsin, A. (2016). Comparative analysis of reference evapotranspiration equations modelling by extreme learning machine. *Computers and Electronics in Agriculture* 127: 56–63.

He, X., Luo, J., Zuo, G., & Xie, J. (2019a). Daily runoff forecasting using a hybrid model based on variational mode decomposition and deep neural networks. *Water Resources Management* 33(4): 1571–1590.

He, Z., Zhou, J., Qin, H., Jia, B., & Lu, C. (2019b). Long-term joint scheduling of hydropower station group in the upper reaches of the Yangtze River using partition parameter adaptation differential evolution. *Engineering Applications of Artificial Intelligence* 81: 1–13.

Hinton, G. E., Osindero, S., & Teh, Y. W. (2006). A fast learning algorithm for deep belief nets. *Neural Computation* 18(7): 1527–1554.

Huang, G. B., Zhu, Q. Y., & Siew, C. K. (2004). Extreme learning machine: A new learning scheme of feedforward neural networks. *Neural Networks* 2: 985–990.

Huang, W. B., & Sun, F. C. (2015). A deep and stable extreme learning approach for classification and regression. *Proceedings of ELM-2014*, Vol. 1 (pp. 141–150). Cham: Springer.

Karaboga, D. (2005). *An Idea Based on Honey Bee Swarm for Numerical Optimization* (Vol. 200, pp. 1–10). Technical report-tr06, Erciyes University, Engineering Faculty, Computer Engineering Department.

Kennedy, J., & Eberhart, R. (1995). Particle swarm optimization. *Neural Networks, Proceedings, IEEE International Conference*, vol. 4, 27 November–1 December, Perth, WA (pp. 1942–1948).

Kisi, O., & Alizamir, M. (2018). Modelling reference evapotranspiration using a new wavelet conjunction heuristic method: Wavelet extreme learning machine vs wavelet neural networks. *Agricultural and Forest Meteorology* 263: 41–48.

Kisi, O., Gorgij, A. D., Zounemat-Kermani, M., Mahdavi-Meymand, A., & Kim, S. (2019). Drought forecasting using novel heuristic methods in a semi-arid environment. *Journal of Hydrology* 578: 124053.

Kisi, O., Sanikhani, H., Zounemat-Kermani, M., & Niazi, F. (2015). Long-term monthly evapotranspiration modeling by several data-driven methods without climatic data. *Computers and Electronics in Agriculture* 115: 66–77.

Kisi, O., & Zounemat-Kermani, M. (2014). Comparison of two different adaptive neuro-fuzzy inference systems in modelling daily reference evapotranspiration. *Water Resources Management* 28(9): 2655–2675.

Li, B., & Cheng, C. (2014). Monthly discharge forecasting using wavelet neural networks with extreme learning machine. *Science China Technological Sciences* 57(12), 2441–2452.

Li, M. B., Huang, G. B., Saratchandran, P., & Sundararajan, N. (2005). Fully complex extreme learning machine. *Neurocomputing* 68: 306–314.

Li, S., Kazemi, H., & Rockaway, T. D. (2019). Performance assessment of stormwater GI practices using artificial neural networks. *Science of the Total Environment* 651: 2811–2819.

Li, X., Du, Z., & Song, G. (2018). A method of rainfall runoff forecasting based on deep convolution neural networks. *2018 Sixth International Conference on Advanced Cloud and Big Data (CBD)*, 12–15 August, Lanzhou, China (pp. 304–310). IEEE.

Liang, N. Y., Huang, G. B., Saratchandran, P., & Sundararajan, N. (2006). A fast and accurate online sequential learning algorithm for feedforward networks. *IEEE Transactions on Neural Networks* 17(6): 1411–1423.

Liu, F., Xu, F., & Yang, S. (2017). A flood forecasting model based on deep learning algorithm via integrating stacked autoencoders with BP neural network. *2017 IEEE Third International Conference on Multimedia Big Data (BigMM)*, 19–21 April, Laguna Hills, CA (pp. 58–61). IEEE.

Maier, H. R., & Dandy, G. C. (2000). Neural networks for the prediction and forecasting of water resources variables: A review of modelling issues and applications. *Environmental Modelling & Software* 15(1): 101–124.

Malik, A., Kumar, A., & Singh, R. P. (2019). Application of heuristic approaches for prediction of hydrological drought using multi-scalar streamflow drought index. *Water Resources Management* 33(8): 3985–4006.

Meng, X., Chang, J., Wang, X., & Wang, Y. (2019). Multi-objective hydropower station operation using an improved cuckoo search algorithm. *Energy* 168: 425–439.

Meshram, S. G., Ghorbani, M. A., Deo, R. C., Kashani, M. H., Meshram, C., & Karimi, V. (2019). New approach for sediment yield forecasting with a two-phase feedforward neuron network-particle swarm optimization model integrated with the gravitational search algorithm. *Water Resources Management* 33(7): 2335–2356.

Mirjalili, S., Mirjalili, S. M., & Lewis, A. (2014). Grey wolf optimizer. *Advances in Engineering Software* 69: 46–61.

Moeini, R., Soltani-Nezhad, M., & Daei, M. (2017). Constrained gravitational search algorithm for large scale reservoir operation optimization problem. *Engineering Applications of Artificial Intelligence* 62: 222–233.

Naidu, M. N., Boindala, P. S., Vasan, A., & Varma, M. R. (2020). Optimization of water distribution networks using cuckoo search algorithm. In R. Renkata Rao, & J. Taler (eds.), *Advanced Engineering Optimization Through Intelligent Techniques* (pp. 67–74). Singapore: Springer.

Niu, W. J., Feng, Z. K., Zeng, M., et al. (2019). Forecasting reservoir monthly runoff via ensemble empirical mode decomposition and extreme learning machine optimized by an improved gravitational search algorithm. *Applied Soft Computing* 82: 105589.

Nourani, V., Baghanam, A. H., Adamowski, J., & Kisi, O. (2014). Applications of hybrid wavelet–artificial intelligence models in hydrology: A review. *Journal of Hydrology* 514: 358–377.

Ojugo, A. A., Emudianughe, J., Yoro, R. E., Okonta, E. O., & Eboka, A. O. (2013). A hybrid artificial neural network gravitational search algorithm for rainfall runoffs modeling and simulation in hydrology. *Progress in Intelligent Computing and Applications* 2: 22–33.

Petković, D., Gocic, M., Shamshirband, S., Qasem, S. N., & Trajkovic, S. (2016). Particle swarm optimization-based radial basis function network for estimation of reference evapotranspiration. *Theoretical and Applied Climatology* 125(3–4): 555–563.

Rahgoshay, M., Feiznia, S., Arian, M., & Hashemi, S. A. A. (2019). Simulation of daily suspended sediment load using an improved model of support vector machine and genetic algorithms and particle swarm. *Arabian Journal of Geosciences* 12(9): 277.

Rashedi, E., Nezamabadi-Pour, H., & Saryazdi, S. (2009). GSA: A gravitational search algorithm. *Information Sciences* 179(13): 2232–2248.

Reddy, M. J., & Singh, V. P. (2014). Multivariate modelling of droughts using copulas and metaheuristic methods. *Stochastic Environmental Research and Risk Assessment* 28(3): 475–489.

Robles-Velasco, A., Cortés, P., Muñuzuri, J., & Onieva, L. (2020). Prediction of pipe failures in water supply networks using logistic regression and support vector classification. *Reliability Engineering & System Safety* 196: 106754.

Roushangar, K., Alizadeh, F., & Nourani, V. (2017). Improving capability of conceptual modelling of watershed rainfall–runoff using hybrid wavelet-extreme learning machine approach. *Journal of Hydroinformatics* 20(1): 69–87.

Saggi, M. K., & Jain, S. (2019). Reference evapotranspiration estimation and modeling of the Punjab Northern India using deep learning. *Computers and Electronics in Agriculture* 156: 387–398.

Sattar, A. M., Ertuğrul, Ö. F., Gharabaghi, B., McBean, E. A., & Cao, J. (2019). Extreme learning machine model for water network management. *Neural Computing and Applications* 31(1): 157–169.

Shabanlou, S. (2018). Improvement of extreme learning machine using self-adaptive evolutionary algorithm for estimating discharge capacity of sharp-crested weirs located on the end of circular channels. *Flow Measurement and Instrumentation* 59: 63–71.

Storn, R., & Price, K. (1997). Differential evolution – A simple and efficient heuristic for global optimization over continuous spaces. *Journal of Global Optimization* 11(4): 341–359.

Taormina, R., & Chau, K. W. (2015). Data-driven input variable selection for rainfall–runoff modelling using binary-coded particle swarm optimization and extreme learning machines. *Journal of Hydrology* 529: 1617–1632.

Tikhamarine, Y., Souag-Gamane, D., & Kisi, O. (2019). A new intelligent method for monthly streamflow prediction: Hybrid wavelet support vector regression based on grey wolf optimizer (WSVR–GWO). *Arabian Journal of Geosciences* 12(17): 540.

Wang, L., Kisi, O., Zounemat-Kermani, M., & Li, H. (2017). Pan evaporation modeling using six different heuristic computing methods in different climates of China. *Journal of Hydrology* 544: 407–427.

Wang, W. C., Xu, D. M., Chau, K. W., & Chen, S. (2013). Improved annual rainfall–runoff forecasting using PSO–SVM model based on EEMD. *Journal of Hydroinformatics* 15(4): 1377–1390.

Wen, X., Feng, Q., Deo, R. C., et al. (2019). Two-phase extreme learning machines integrated with the complete ensemble empirical mode decomposition with adaptive noise algorithm for multi-scale runoff prediction problems. *Journal of Hydrology* 570: 167–184.

Yadav, B., Ch, S., Mathur, S., & Adamowski, J. (2016). Discharge forecasting using an online sequential extreme learning machine (OS-ELM) model: A case study in Neckar River, Germany. *Measurement* 92: 433–445.

Yang, X. S., & Deb, S. (2009). Cuckoo search via Lévy flights. *2009 World Congress on Nature & Biologically Inspired Computing (NaBIC)*, 9–11 December, Coimbatore, India (pp. 210–214). IEEE.

Yang, Y., Wang, Y., & Yuan, X. (2012). Bidirectional extreme learning machine for regression problem and its learning effectiveness. *IEEE Transactions on Neural Networks and Learning Systems* 23(9): 1498–1505.

Yaseen, Z. M., Jaafar, O., Deo, R. C., et al. (2016). Stream-flow forecasting using extreme learning machines: A case study in a semi-arid region in Iraq. *Journal of Hydrology* 542: 603–614.

Yilmaz, B., Aras, E., Kankal, M., & Nacar, S. (2019). Prediction of suspended sediment loading by means of hybrid artificial intelligence approaches. *Acta Geophysica* 67(6): 1693–1705.

Yin, Z., Wen, X., Feng, Q., He, Z., Zou, S., & Yang, L. (2016). Integrating genetic algorithm and support vector machine for modelling daily reference evapotranspiration in a semi-arid mountain area. *Hydrology Research* 48(5): 1177–1191.

Yue, Q., Zhang, F., Zhang, C., Zhu, H., Tang, Y., & Guo, P. (2020). A full fuzzy-interval credibility-constrained nonlinear programming approach for irrigation water allocation under uncertainty. *Agricultural Water Management* 230: 105961.

Zong, W., Huang, G. B., & Chen, Y. (2013). Weighted extreme learning machine for imbalance learning. *Neurocomputing* 101: 229–242.

Zou, Q., Liao, L., Ding, Y., & Qin, H. (2019). Flood classification based on a fuzzy clustering iteration model with combined weight and an immune grey wolf optimizer algorithm. *Water* 11(1): 80.

Zounemat-Kermani, M., Kisi, O., & Rajaee, T. (2013). Performance of radial basis and LM-feed forward artificial neural networks for predicting daily watershed runoff. *Applied Soft Computing* 13(12): 4633–4644.

Zounemat-Kermani, M., Kisi, O., Piri, J., & Mahdavi-Meymand, A. (2019). Assessment of artificial intelligence-based models and metaheuristic algorithms in modelling evaporation. *Journal of Hydrologic Engineering* 24(10): 04019033.

17

Rainwater Harvesting for Sustainable Water Resource Management under Climate Change

RAM L. RAY, RAJENDRA P. SISHODIA AND TOLULOPE OLUTIMEHIN

17.1 Introduction

Climate change has a significant impact on ecosystems, food security, public health, water security and infrastructure development. Water resources are particularly vulnerable to climate change (Ayele, 2014). Water security issues are increasing across the world in developing and developed countries under climate change and growing populations. About 71 per cent of irrigated areas and 47 percent of large cities (population >0.5 million) have experienced periodic, annual, seasonal drought or water stress (Brauman et al., 2016; Ghimire et al., 2019). As climate change exerts further pressure on socio-economic and natural systems, water scarcity is likely to increase globally. If extreme events become more frequent under climate change, it would require the community to develop different means of adaptation.

While individual and community-based rainwater harvesting can be a good option to solve water security issues at the local level, rainwater harvesting is required to address water security at larger scales (e.g., national and continental levels). The harvesting culture for rainwater should be developed and coordinated from individual to regional and national levels. Moreover, a combined effort for rainwater/floodwater harvesting is necessary to enhance water security and reduce water stress. Rainwater harvesting can be the most suitable method to address water scarcity problems when rainfall is low; however, it is not adequate in the context of high water demand. Rainwater harvesting directly benefits the nation by reducing the demand on the public water supply, along with reducing runoff, erosion and contamination of surface water and increasing groundwater recharge (TALE, 2012).

This work was supported by the Evans-Allen funds of the United States Department of Agriculture (USDA), National Institute of Food and Agriculture.

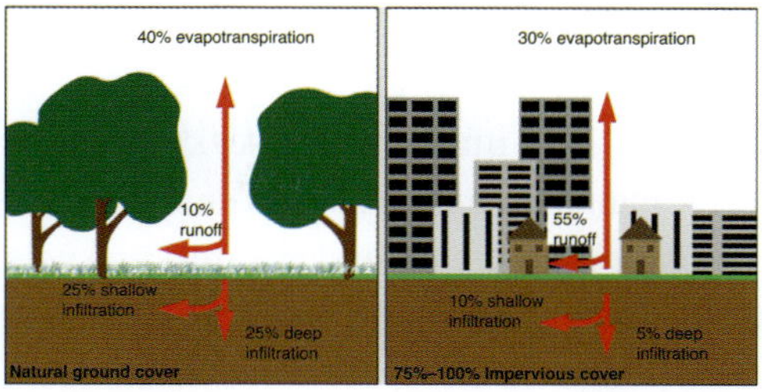

Figure 17.1 Relationship between impervious cover and runoff (Source: US EPA, TALE, 2012)

Rainwater harvesting has numerous benefits if the water is properly harvested, stored and effectively used. For example, rainwater harvesting promotes conservation of water and energy, reduces undesired stormwater runoff, improves water quality, supplies nutrients to plants, mitigates the effect of drought, reduces soil erosion, reduces flood hazards and improves groundwater storage (Gupta, 2006; TALE, 2012). Rainfall generates stormwater runoff and rainwater harvesting is one way to decrease the runoff, along with the issues associated with it, such as flood and/or flash flood (TALE, 2012). Rainwater harvesting enhances soil infiltration either through storage or through increased infiltration over the porous land surfaces (Figure 17.1). Increased infiltration through rainwater harvesting does not only decrease surface runoff but also supports/enhances groundwater recharge systems. Since the major parts of urban areas are impervious, the potential for harvesting rainwater is considerable in urban areas compared to rural areas because the latter has a more natural ground cover. Therefore, rainwater harvesting in urban areas is not only a source of generating non-potable water, but it also helps reduce urban flooding (Mehrabadi et al., 2013).

Rainwater harvesting is a common practice in various countries. However, it directly relies on rainfall and surface water of rivers, lakes and groundwater to satisfy the associated water demand requirements. Harvesting rainwater/floodwater poses challenges to the community on a large scale. However, an efficient management of rainwater/floodwater, which requires suitable water storage and utilization technologies, can contribute to managing water resources in a sustainable way (Tesfay, 2011). Rainwater harvesting is not exempt from water losses such as evaporation, spillage and infiltration, to name but three. However, rainwater harvesting is a promising alternative to reduce potable water demand or water scarcity, both in rural and urban areas and in locations with various precipitation regimes (Sturm et al., 2009; Matos et al., 2013; Guo & Guo, 2018).

The performance of the rainwater harvesting system is commonly estimated with historical rainfall data without considering the potential impacts of climate change on temporal and spatial distributions, frequency and the amount of rainfall in the investigated regions (Haque et al., 2016). The large uncertainty associated with climate change projections makes the design of future climate resilient water harvesting systems difficult. It is important to seek alternative water sources and develop improved water management strategies and methods, which can reduce pressure on water resources systems under stress (Ayele, 2014). Therefore, comprehensive knowledge of climate change and corresponding adaptation by human society is crucial to ensure water resource sustainability under a changing climate (Pandey et al., 2003). Moreover, innovative water resources management practices through emerging strategies and advanced technologies could address water resource sustainability issues (Ghimire et al., 2019).

This chapter provides an overview of traditional and more advanced methods and systems used for rainwater harvesting in rural and urban areas. It discusses the potential and challenges as well as the environmental benefits and water resources sustainability using rainwater harvesting under climate change.

17.2 Historical Development of Water Harvesting

Rainwater harvesting systems are traditional technologies that have met the needs of communities for many centuries (Bhattacharya, 2015). Moreover, this is one of the oldest and most commonly used water management systems practiced across the world. The world's major civilizations, such as Indus Valley Civilization, Ancient Egypt (the Nile) and Mesopotamia, developed along rivers to gather water for their domestic and agricultural use and develop their society. Over time, these civilizations have increased their techniques to efficiently store, save and distribute water. Rainwater harvesting has been practiced since at least 2000 BC in India, China, Mesopotamia and ancient Rome (UNEP 2001; Bhattacharya et al., 2011). Rainwater harvesting can be defined as the collection and storage of atmospheric precipitation (e.g., form of rain) for its later use (Antoniou et al., 2014). In ancient civilizations, rainwater was often collected in storage tanks or reservoirs such as cisterns, a form of artificial reservoir (Angelakis & Spyridakis, 1996; Antoniou et al., 2014; Angelakis, 2016).

In densely populated areas, supplying water to meet domestic water needs has always been challenging. Roman villas and cities were planned in a way to efficiently collect rainwater for drinking and air-cooling purposes (UNEP, 1979). The evidence of roof catchment dates back to early Roman times (UNEP, 2001). During the Hellenistic period (323–30 BC), rainwater harvesting became popular, but there are indications that Minoans had knowledge of designing and constructing rainwater harvesting structures (e.g., cisterns) earlier than this period (Mays et al., 2013; Antoniou et al., 2014). The early Minoan rainwater harvesting systems were based on collecting runoff from open yards and roofs, mainly in urban areas (Angelakis, 2016). In many urban areas, especially in arid and semi-arid regions, where good quality groundwater is lacking, rainwater becomes a vital water resource (UNEP, 2001). For example, in the western parts of China, rainwater harvesting has been in practice for at least a millennium, especially in controlling severe erosion (Falkenmark et al., 2001). Evidence was found that water harvesting systems were constructed in Jordan and Southern Mesopotamia over 9,000 and 6,500 years ago, respectively (Bruins et al., 1986).

In warmer climatic conditions, when rainfall events are irregular and inconsistent, rainwater collection is an important strategy to meet the essential water demands of the communities. There are numerous shreds of evidence that illustrate standard practices for rainwater harvesting during the past era. For example, Roman villas and cities engaged in rainwater harvesting to provide water to the citizens, especially under siege (Mays et al., 2013). For instance, on the hills of Mumbai, India, the Buddhist monastic cells had gutters and cisterns cut into the rock for collecting rainwater over the entire year (UNEP, 1979). The construction and use of cisterns to store rainwater can be traced back to the late Neolithic era in Mesopotamia (Yannopoulos et al., 2017). Falkenmark et al. (2001) state that rainwater harvesting techniques are believed to originate over 5,000 years ago in the Fertile Crescent of Iraq, the cradle of agriculture (Falkenmark et al., 2001).

Moreover, more than 4,000 years ago, water harvesting was also employed in India and China (Oweis et al., 2004). Collection and utilization of rainwater is an old practice used to provide a potable water supply in many parts of the developing world (Handia et al., 2003). Rainwater storage facilities have been constructed in the Mediterranean and the Near East since the third millennium BC where cisterns were in the forms of tanks dug out of sand or loose rocks lined with plaster to the sophisticated structures such as those built on Crete by the Minoans (Mays et al., 2013). Archaeological evidence shows that water storage and supply systems have been found for periods dating back to the Indus Valley Civilization (3,000–1,500 BC) to the present. Currently, rainwater is harvested in many ways both across small (bucket or small tank in the backyard) and large (large reservoirs constructing dams) scales.

Overall, rainwater harvesting played a crucial role in satisfying water demands in ancient civilizations, mainly in arid regions. In recent decades, it has been used in many regions across the world to address the increasing demands of water that result from climate change and the increase in population, urbanization, industrial development and irrigation (Amos et al., 2016; Campisano et al., 2017; Zhang et al., 2019). Figure 17.2 shows the area covered under spate irrigation in different countries worldwide (FAO, 2016). Spate irrigation is a popular rainwater harvesting method in arid and semi-arid regions where floodwater from upstream catchments is diverted from the stream or river channels for irrigation and other

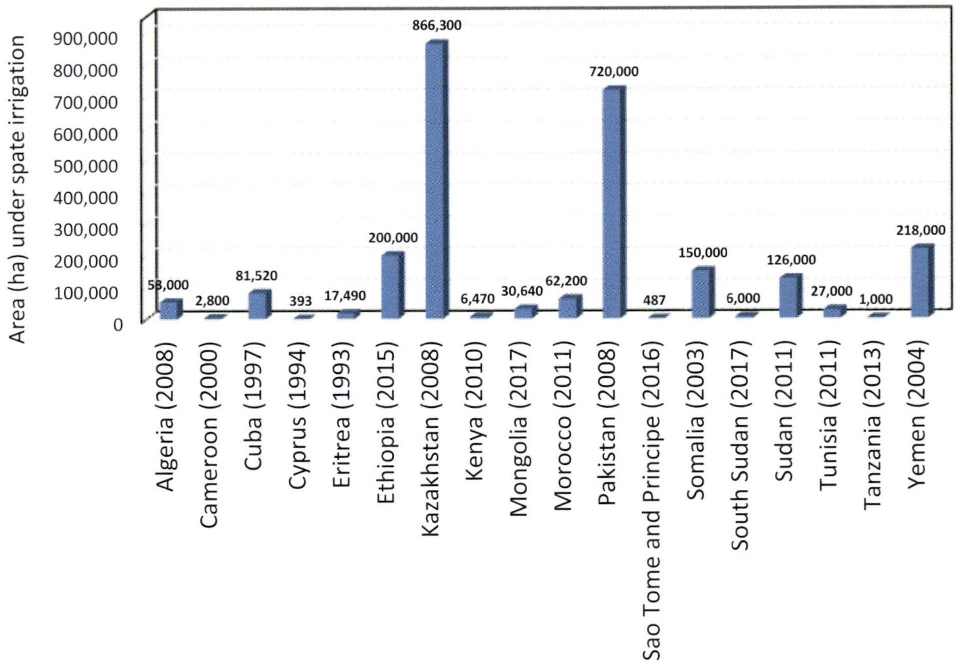

Figure 17.2 The area under spate irrigation, a popular rainwater harvesting method in arid and semi-arid regions, for different countries (FAO, 2016). On the x-axis, the year in parentheses following the country name indicates the period of data collection

uses. Various methods of rainwater harvesting, discussed in Section 17.3, are employed across the world to provide water supplies for multiple purposes.

17.3 Rainwater Harvesting Methods

Many methods of water harvesting have been developed in the past. Harvesting methods can be classified in multiple ways depending on where and how the water is generated and collected, for instance, how it is stored and how far it should be transported (Bunclark et al., 2018). A typical water harvesting system has three basic components, (1) a catchment area that collects rainwater falling on to it, (2) a storage facility, which could either be a surface reservoir, soil profile or aquifer and (3) a target area, where the harvested water is applied (Oweis et al., 2004). The three most common methods of classification found in the literature are based on: (1) the size of the catchment, (2) where and how the water is harvested and (3) the purposes of use, such as agricultural or domestic usage and type of land, such as rural or urban areas (Figure 17.3) (FAO, 1994; Ngigi, 2003; Oweis et al., 2004;

Figure 17.3 Some of the selected rainwater harvesting practices: (a) Pitting; (b) In situ RWH, ridging; (c) Micro-basins; (d) Rooftop RWH; (e) Traditional pond; (f) Shallow wells; (g) A runoff diversion system; (h) Micro-dam and (i) Underground well. Source: Ferrand and Cecunjanin (2014)

Rockström et al., 2010; Critchley et al., 2013). Table 17.1 provides a general classification of water harvesting methods based on the three criteria mentioned. There are overlaps in the classification of commonly used rainwater harvesting systems. For example, in-situ methods may belong to the micro-catchment category (Oweis et al., 2004). However, the micro-catchment category also includes ex-situ methods of water harvesting practiced in small catchment areas. Similarly, while most ex-situ methods are classified within the macro-catchment category, this same category also includes in-situ methods applied on large size catchments.

In-situ methods such as contour and graded bunds, runoff strips and conservation agriculture aim to reduce runoff velocity and increase flow path length, thereby increasing infiltration and soil water storage. Reduced runoff and velocity also reduce soil erosion and helps maintain soil fertility and crop production. In-situ water harvesting techniques are generally applied at a small field-scale and therefore are more affordable and easier to implement and manage

Table 17.1. Basic classification of water harvesting systems

Basis of classification	Types	Description and examples
Where and how the water is collected and used	In-situ	Water harvesting systems mainly aimed to enhance soil moisture regime for improved crop production (Rockström et al., 2010). Examples include bunds, ridges, furrows, runoff strips, small runoff basins, pits and other forms of conservation agriculture.
	Ex-situ Also, referred to as runoff harvesting or floodwater harvesting	It is designed to provide a supplemental water supply for irrigation, domestic use and livestock. Examples include check dams, large percolation tanks, farm ponds and storage reservoirs. It usually provides water supply in areas located downstream of the catchment from where the runoff is collected.
Size of the catchment	Micro-catchment	Water collected and used to enhance soil moisture and water supply for catchment areas are generally smaller than 1,000 m^2 (Oweis et al., 2004). It includes most in-situ and some ex-situ water harvesting methods and urban or rural rooftop water harvesting.
	Macro-catchment	Water collected and used to enhance soil moisture, and water supply for catchment areas are generally larger than 1,000 m^2 (Oweis et al., 2004). Includes most of the in-situ and some ex-situ water harvesting methods employed globally, such as water spreading and spate irrigation (Pina et al., 2018).
Purpose of rainwater harvesting	Agriculture	It includes methods aimed to enhance soil moisture and provide supplemental irrigation. It also includes bunds, ridges, furrows, runoff strips, small runoff basins, pits, check dams, large percolation tanks, farm ponds and storage reservoirs.
	Domestic/Urban	Systems aimed to store runoff water from rooftops and courtyards to enhance domestic water supply. A typical system has storage, conveyance and filtration components.

compared to the ex-situ methods. Ex-situ or macro-catchment methods such as check dams and percolation tanks require planning and management at a catchment scale as it may affect other water users and water rights in upstream and downstream areas. However, ex-situ methods can be very effective in providing critically needed supplemental irrigation, especially in areas where most of the runoff is generated from a few high intensity storms. At a watershed or catchment scale, a combination of in-situ and ex-situ practices is likely to be the most cost-effective in maximizing crop production (Garg et al., 2012).

The size of the micro-catchment defined in the literature varies. For example, Critchley et al. (2013) defined it as a catchment with length varying from 1 to 30 metres, whereas Oweis et al. (2004) put it as areas smaller than 1,000 m^2. Surface runoff from micro-catchment systems is stored in the root zone or soil profile of an adjacent infiltration basin (Prinz, 2002). While macro-catchment systems are larger systems that typically collect and store runoff water flowing in a channel or stream for its later use, they also include floodwater collection and uses such as spate irrigation (Pina et al., 2018). The amount of water captured per unit of catchment area (capture efficiency) is much lower in macro-catchment systems compared to the micro-catchment; it ranges from a few per cent to 50 per cent of annual rainfall (Oweis et al., 2004). A micro-catchment system is more efficient because captured water is evenly spread and evaporation losses are low (Mutunga, 2001).

Providing an adequate drinking and irrigation water supply is a major problem in water-scarce regions (Khatakho & Koju, 2017). Rainwater harvesting can help to increase the water supply for both agricultural and domestic purposes (Velasco-Muñoz et al., 2019). Therefore, water harvesting systems are very popular in both urban, rural and agricultural areas, especially in arid and semi-arid regions. Rainwater can be harvested from rooftops, courtyards, hill slopes, fields and ephemeral streams and channels for domestic and agricultural uses (Saleh et al., 2017). Rainwater harvesting systems have been used to handle water scarcity as far back as 4,500 BC (Frasier, 1985). In Minoan Crete, technologies were developed to collect and transport water to settlements. Water was conveyed into cisterns; the technology of rainwater storage was very well developed and has been used consistently up to modern times (Mays et al., 2013; Antoniou et al., 2014). The Minoan water cisterns were cylindrical and constructed with stones under the soil surface, with a diameter ranging from 1.5 to 7.0 metres and a depth from 2.5 to 5.0 metres (Mays et al., 2013; Yannopoulos et al., 2017). These early systems were used to collect rainwater from hillsides, open yards and rooftops mostly for domestic purposes (Mays et al., 2013).

17.3.1 Rainwater Harvesting Systems in Rural Areas

Access to clean water is a basic human need critical to human health. Without clean water, poverty reduction in third world countries is impossible. Also, rainwater harvesting is considered a good source of the drinking water in arsenic and salinity affected rural areas (Rana, 2005). Rainwater harvesting adds more benefits to households and individuals in rural areas (Velasco-Muñoz et al., 2019). If the generated runoff is low, domestic water harvesting systems in rural areas are generally low-cost and simpler in design. Most of the rain can be collected and stored, even though storing the first rain is avoided because it is not clean enough

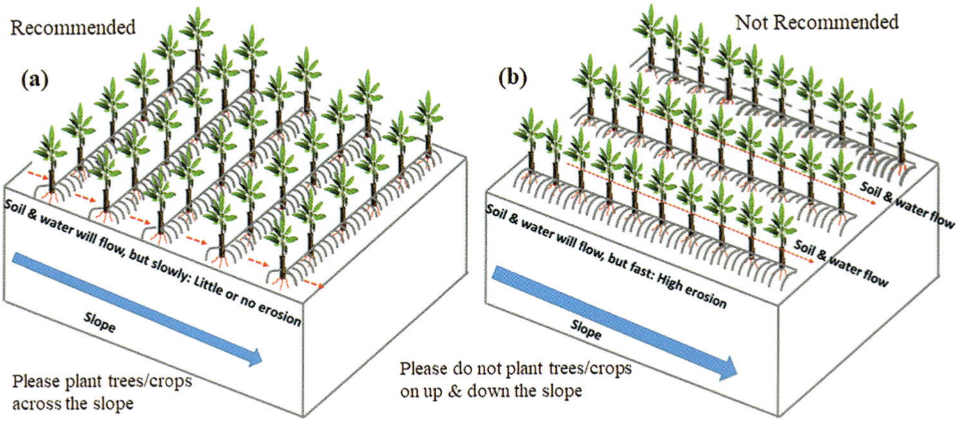

Figure 17.4 (a) A recommended agricultural practice, which can support rainwater harvesting and help improve soil moisture conditions, reduce soil erosion, retain soil nutrients and provide supplementary irrigation to enhance crop production; (b) An agricultural practice which is not recommended, mainly on the slope

for drinking (Oweis et al., 2004). It is a common practice in many remote rural areas in Africa and Asia to collect rainwater from roofs or caves and store it in local containers (UNEP, 2001). For example, in Zambia, rainwater is harvested using basic technologies, usually by placing buckets outside in open areas to collect rainwater (Handia et al., 2003). In rural areas, rainwater harvesting is employed mainly for domestic purposes when tap water is not accessible (Oweis et al., 2004). Also, water harvesting systems are used for drinking water and to sustain farm yields through supplementary irrigation because it is economical and, therefore, more suitable for rural dwellers (Falkenmark et al., 2001).

For agriculture, water harvesting can help improve soil moisture conditions and provide supplementary irrigation to enhance crop production (Figure 17.4). In-situ or micro-catchment methods such as bunds, contour strips, ridges, furrows and other forms of conservation agriculture help retain rainwater and improve infiltration (Rockström et al., 2010). Increased infiltration improves the conditions for soil moisture, growing and yields. Contour cropping and field bunding, among other watershed interventions, have increased crop production and improved farmers' livelihoods in a semi-arid watershed in India (Wani et al., 2003; Garg et al., 2012). Collection and storage of runoff into tanks, pits, check dams, ponds and other forms of reservoirs (Figure 17.5) can help increase groundwater recharge and provide a supplemental irrigation supply later in the season when rainfall is not sufficient. Collection of runoff water into large reservoirs called *tanks*, for later use in the dry season, is an old practice in semi-arid regions of India (Gunnell & Krishnamurthy, 2003; Bhattacharya et al., 2011). In semi-arid south India, *tanks* are generally constructed in series along the natural drainage pathway to store

Figure 17.5 Ex-situ water harvesting systems in south India. (a) Check dam and (b) Percolation tank. Source: Author

runoff for later use. In Africa and West Asia, these methods of runoff collection in reservoirs are named *wadi-bed*, off-*wadi* or *hafair* (Oweis et al., 2004). Water storage in such structures also increases groundwater recharge, thus increasing groundwater storage and supply during the dry season when other sources of surface water are not available (Sishodia et al., 2017). In-situ water harvesting techniques also reduce the risk of soil erosion and help maintain soil fertility and crop yields (Garg et al., 2012).

There are multiple challenges associated with water harvesting systems at field, watershed and catchment scales. Lack of financial and labour resources is probably the most limiting factor affecting the adoption of water harvesting systems by marginal farmers and communities. Many ex-situ methods, such as storage in reservoirs, involve high design and construction costs. Implementation of water harvesting at large watershed or catchment scale requires institutional engagement and a clear understanding of the water rights and regulations (Kahinda & Taigbenu, 2011). Catchment-scale benefits and trade-off analyses for users in upstream and downstream areas can help plan the number and storage capacity of water harvesting structures.

17.3.2 Rainwater Harvesting Systems in Urban Areas

Collecting rainwater from roof catchment is an ancient method to provide a potable water supply to urban dwellers in various regions of the developing world (Handia et al., 2003; Kumar, 2004; Angelakis, 2016). Providing an adequate water supply in urban areas is challenging in the face of the increasing population and water demands (Falkenmark et al., 2001; Handia et al., 2003; Antoniou et al., 2014). The number of people living in urban areas has risen from 750 million in 1950 to nearly 2.9 billion in 2001 (UNEP, 2001; UN-HABITAT, 2005). The world's population

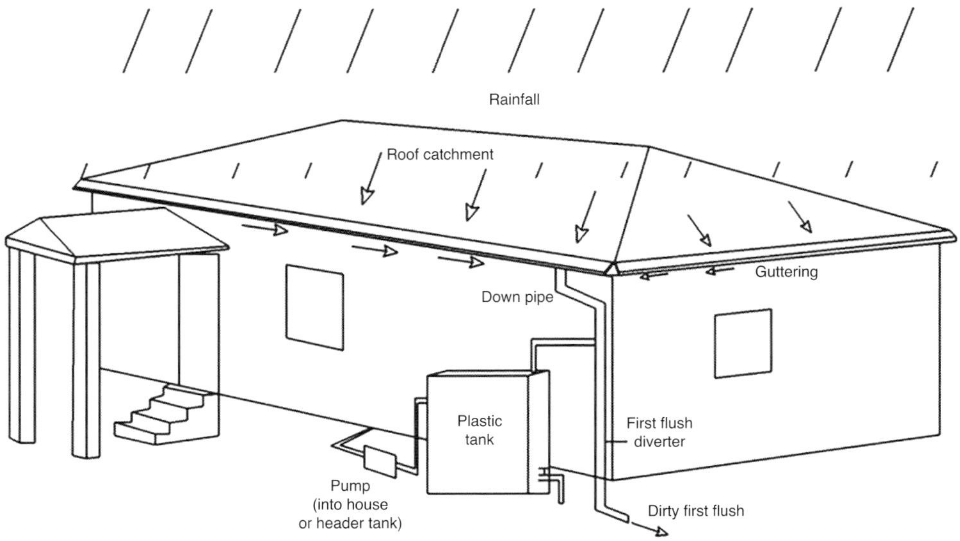

Figure 17.6 Major components of a rooftop water harvesting system. Adapted from Varad Shende, Ecosan Services Foundation

is currently about 7.2 billion, and more people live in urban areas than in rural areas (Wirtenberg et al., 2019). Urbanization affects flow regimes and eco-hydrology, which further affects natural hydrologic processes. Large dams have been constructed in many regions across the world to increase the urban water supply and potential flooding risks (Boelee et al., 2013).

A thorough understanding of rainfall amount, intensity and distribution is necessary to design a suitable water harvesting system (Prinz & Singh, 2000). A typical rainwater harvesting system has three basic components (Figure 17.6):

i. A collection system, such as roofs and/or gutters that direct water into the storage facility (Prinz, 2002; Handia et al., 2003);
ii. A conveyance system, with the connection of drain pipes and piping from the rooftop, rainwater from the roof catchment area is transferred through a filter system to the storage tanks (Khoury-Nolde, 2006) and
iii. A storage facility, where the water is stored to be used immediately or later (UNEP, 2001; Saleh et al., 2017; Pina et al., 2018). The urban water harvesting system also includes a water treatment or filtration system, especially if the collected water is intended to provide a potable water supply (GhaffarianHoseini et al., 2016).

Increasing the size of rainwater harvesting storage tanks can help maximize the water savings in urban multi-unit buildings. However, large capacity storage tanks are difficult to fit in limited urban spaces and effective planning and engineering solutions may be needed to increase the storage capacity. In addition, large tanks are associated with higher purchasing and installation costs. Due to the high initial investment costs, a life cycle analysis approach is often used to design and evaluate the profitability of rainwater harvesting systems in urban settings. In addition to high initial costs, the implementation of rainwater harvesting is often constrained by a lack of information on the effectiveness of these systems and the specifications needed to optimize their performance (GhaffarianHoseini et al., 2016).

Depending on the quality of the water, the harvested rainwater from the rooftop can be used for multiple purposes, including being a potable water supply, toilet flushing, clothes washing and gardening (Nawaz et al., 2013). The water quality of the harvested rainwater depends on many factors. Contamination occurs from the moment the raindrops leave the clouds. Abassi and Abassi (2011) identified three types of water contaminants: physical, chemical and microbiological. Contamination of roof catchment from atmospheric deposition and dissolution of roof material are important variables that can affect the quality of harvested rainwater (Cupido et al., 2012). Rainwater, harvested from concrete roofs, is better and of higher quality than those from the green roof and terrace catchment (Nawaz et al., 2013). In a metropolitan area, rainwater harvesting can be applied on a large scale by harvesting rainwater from many homes and buildings (Hofman & Paalman, 2014). Decentralized multi-purpose rainwater harvesting systems could be useful to combat water-related disasters such as flooding and fire outbreaks, especially in urban areas (Yannopoulos et al., 2017).

17.4 Rainwater Harvesting under Climate change: Potentials and Challenges

Climate change is expected to affect future water demand and the availability of water resources (Muchuru & Nhamo, 2019). The sustenance of countries with a growing population and increasing demand for freshwater, which increases water stress, depends heavily on rainfall (Boelee et al., 2013). Due to industrial development and urbanization, water security has become a major concern (Haque et al., 2016). The growing population has led to an increase in the demand for freshwater in many countries across the world. Sustainable water management is required to satisfy the increasing water demand and address the uncertainty of the availability and quality of water under climate change (Mays et al., 2013). Extreme climate events such as drought, stormy rainfall or hurricane are expected to have more impact on human society compared to gradual changes in the climate system (Pandey et al., 2003).

Since the twentieth century, changes in temperature, precipitation and other climatic variables have affected water resources globally (Willett et al., 2008; Wild, 2009; McVicar et al., 2012; Donat et al., 2013; Stocker et al., 2013; Mishra et al.,

2015). In the future, climate change is expected to further alter surface water and groundwater resources in many parts of the world (Doll, 2009; Crosbie et al., 2013; Van Vliet et al., 2013; Vano et al., 2015). The effects of climate change on water availability are likely to vary spatially depending on the direction and/or magnitude of changes in rainfall and temperature. Based on the results from 12 Global Climate Models (GCMs), Milly et al. (2005) have predicted 10–40 per cent increases in runoff in eastern Africa, the La Plata basin and high-latitude North America and Eurasia, and 10–30 per cent decreases in runoff in southern Africa, southern Europe, the Middle East and mid-latitude western North America by the year 2050. Similar future changes in regional streamflow have been predicted by Koirala et al. (2014) using GCMs from the fifth Coupled Model Intercomparison Project (CMIP5). Under climate change, runoff, streamflow, rainfall and temperature patterns are expected to change in several watersheds globally (Jha et al., 2004; Githui et al., 2009; Chien et al., 2013; Gizaw et al., 2017; Sishodia et al., 2018).

An increased runoff under climate change may provide additional opportunities for runoff capture and use. Increased water capture potential due to increased runoff may justify further investments and/or incentives for water harvesting structures, especially in water-scarce arid and semi-arid regions, including urban areas that are already experiencing water scarcity (e.g., Chennai, India; Krueger et al., 2019). However, studies have indicated that most increased rainfall may occur in the form of increased frequency and/or magnitude of high intensity rainfall events (Meehl et al., 2005; Kumar et al., 2011; O'Gorman, 2015). Future increases of high intensity rainfall will lead to a rise in peak flow events and runoff volumes (Taye et al., 2011; Zahmatkesh et al., 2014). Unless a majority of the increased runoff is captured, the potential benefits from increased runoff in the future may be limited. Capturing high flows may require significant infrastructure expansion and financial resources, which vary according to the magnitude and timing of increased flows. Wallace et al. (2015) found that, under climate change, significantly larger storage tanks would be needed to maintain the performance and reliability of urban rooftop water harvesting systems in Micronesia. Increased runoff capture during high intensity rainfall events will also reduce downstream flooding risks, which is an additional benefit of water harvesting. In many arid and semi-arid regions, where annual water demand is higher than renewable or replenishable water (e.g., Middle East and northwest India), flash floods or runoff from high rainfall enhance groundwater recharge. Construction or modification of rainwater harvesting structures in these naturally water-scarce regions could be justifiable, considering that most of the runoff will otherwise be lost as streamflow or evapotranspiration (ET) (Sen et al., 2013). A site or watershed-specific cost-benefit analysis may be needed to assess the feasibility of a water harvesting infrastructure expansion/construction under the changing climate.

On the other hand, reduced runoff, streamflow and groundwater recharge due to reduced rainfall and increased temperatures would make water harvesting more critical to satisfy future water demands. Even with a positive or no change in annual/seasonal rainfall in the future, an increased variability in rainfall and runoff is likely to diminish the reliability of surface water resources (e.g., rivers and reservoirs), thereby increasing the dependence on groundwater supply (Taylor et al., 2013). A future increase in temperature will also increase ET and irrigation water requirements for a large variety of crops (Kreins et al., 2015; Shrestha et al., 2017). Rainwater harvesting might prove to be a good adaptation strategy to cope with future changes in the climate, especially in arid and semi-arid regions. For example, the Intergovernmental Panel on Climate Change (IPCC) has listed rainwater harvesting as a key strategy for a planned adaptation to meet the future increasing water demands (Ayele, 2014). Water harvesting structures such as check dams and percolation tanks can help increase groundwater recharge and enhance groundwater availability for irrigation during the dry or post-rainy season (Stiefel et al., 2009). Thus, runoff collected by water harvesting structures (e.g., check dams and reservoirs) can help increase both surface and groundwater supply for irrigation and other purposes. Potential benefits of rainwater harvesting brought by changes in future climate will vary depending on the future changes in rainfall, temperatures, streamflow and other hydrologic processes. In addition, current optimum design parameters (e.g., size of storage) for water harvesting systems may also change in the future depending on the changes in the climate (Wallace et al., 2015). Water harvesting is not a panacea to water scarcity and it will likely need to be complemented with other management interventions to enhance water availability in the future. Studies have pointed out that water harvesting in upstream areas may negatively influence the water availability for downstream users (Garg et al., 2013; Dile et al., 2016). A basin-specific analysis can help understand the benefits or trade-offs of water harvesting for upstream and downstream users under changed climatic conditions.

Climate change projections, including rainfall, runoff and streamflow, are subject to the uncertainty associated with global climate modelling, greenhouse gases (GHGs) emission trajectories, climate downscaling and hydrologic modelling (Wilby & Harris, 2006; Schewe et al., 2014). A selection of GCMs, emission trajectories, downscaling methods and hydrologic models affect the uncertainty in predicting future changes in water availability and changes. Uncertainty in future climate and hydrologic predictions make it challenging to develop and devise management and policy plans aimed to enhance future water supply. However, many countries have already developed and/or modified water management plans and policies to mitigate or adapt to potential climate change effects (Pittock, 2011; Pietrapertosa et al., 2018). Development of climate-resilient water resources management and policy, including rainwater harvesting, should help enhance future water availability and sustainability.

17.5 Environmental Benefits of Rainwater Harvesting

Natural and agricultural landscapes provide valuable ecosystem services such as food, water, home for natural habitats, timber, air purification, soil formation, pollination and others (Millennium Ecosystem Assessment, 2005; Costanza et al., 2014). The level of ecosystem services provided depends on the availability of water, which is often a limiting factor, especially in arid and semi-arid regions. Increased water availability due to rainwater harvesting can improve ecosystem services. For example, rainwater harvesting for agriculture in Africa may improve the ecosystem services such as food and biomass production, groundwater recharge, aquatic ecosystems, animal diversity and nutrient cycling (Vohland & Barry, 2009). Similarly, in a semi-arid mountainous area of Egypt, Norfolk et al. (2012) found significantly higher biodiversity of arthropods in agricultural areas utilizing captured rainwater as compared to the surrounding natural habitats. Although conventional intensification of agriculture is considered a driving force causing a loss in terrestrial biodiversity, rainwater harvesting, as an alternative practice, may maintain or improve ecosystem services while improving the livelihood and food security (Garibaldi et al., 2017).

Rainwater harvesting structures (e.g., reservoirs, ponds) also function as wetlands, providing a range of ecosystem services such as flood control, nutrient retention and recycling, provision of avian habitats and increased biodiversity of plants and other organisms (Bassi et al., 2014). Capturing runoff generated from extreme rainfall events helps reduce peak flows and flooding risk in downstream areas (Welderufael et al., 2013; Singh et al., 2014). Runoff generated in urban landscapes is usually higher than rural and agricultural landscapes due to the higher proportion of impervious areas. Therefore, runoff capture in urban areas is likely to provide higher water quality, flood protection and water supply benefits. Water captured in rural water harvesting structures (e.g., check dams) also increases groundwater recharge, baseflow and downstream water availability, especially during the dry season (Welderufael et al., 2013; Meter et al., 2014; Singh et al., 2014). Increased baseflow in the dry season helps maintain flows over the year and levels in rivers and lakes to sustain aquatic ecosystems and water supply (Andersson et al., 2011).

Excess sediment and nutrients flowing into freshwater and coastal water bodies cause water quality impairment and eutrophication (Dodds & Smith, 2016; Smith et al., 1999). Rainwater harvesting structures such as check dams help capture and retain nutrients and sediments. Reduced sediments and nutrients into rivers, lakes and estuaries can help improve water quality and the provision of ecosystem services. Scrapings of nutrient-rich soil/silt accumulated in rainwater harvesting structures and its use in agricultural fields facilitates nutrient recycling and increases overall nutrient use efficiency (Gunnell & Krishnamurthy, 2003; Meter et al., 2014). Removal of the accumulated soil/sediment also helps maintain water storage and nutrient retention functions of rainwater harvesting structures. An improved soil moisture regime and vegetation cover in agricultural areas supplied with harvested rainwater also helps reduce wind and water erosion, maintains soil fertility and increases biomass and yields. Increased biomass and crop production also help sequester more carbon and reduce atmospheric CO_2 levels and mitigate climate change effects (Wang et al., 2010).

17.6 Natural Resources Conservation and Sustainability through Rainwater Harvesting

17.6.1 Natural Resources Conservation

Traditionally, rainwater harvesting has been viewed as a soil and water conservation strategy, especially in water-scarce arid and semi-arid regions. Retention of runoff water and its use for agriculture, domestic and industrial purposes helps conserve water resources that would otherwise be lost to non-beneficial uses (e.g., to the ocean). For example, the use of rainwater reduced potable home water use by 40 per cent in Victoria, Australia (Muthukumaran et al., 2011). Similarly, rainwater use led to potable water savings ranging from 6 to 92 per cent in Jordan and Brazil (Ghisi et al., 2006; Abdulla & Al-Shareef, 2009). Rainwater harvesting in rural areas helps reduce soil loss and facilitates nutrient cycling, thus reducing fertilizer use and negative environmental impacts (e.g., water quality impairment). In many arid and semi-arid regions, groundwater is the main source of water supply and the water demand often exceeds the renewable water supply (e.g., from recharge). This unsustainable use of groundwater – when water withdrawal is larger than groundwater recharge – has resulted in groundwater depletion in many countries, including India, China, the western United States, Middle East, Mexico and north Africa (Wada et al., 2010; Gleeson et al., 2012; Famiglietti, 2014). The use of rainwater harvesting to increase groundwater recharge can potentially help reduce groundwater depletion in such areas.

In many arid and semi-arid regions worldwide, the majority of annual groundwater recharge and runoff/streamflow are generated from few heavy rainfall events (Love et al., 2010; Van Wyk et al., 2012). In a basin/watershed, where runoff or streamflow generated from such events is unallocated (e.g., for irrigation or urban water supply), water harvesting can help reduce groundwater depletion and increase water availability at the basin or watershed scale. Reduced streamflow due to captured rainwater in upstream areas is not likely to affect the water supply and availability in downstream areas of the corresponding basins. In a large part of the sub-Saharan Africa region, surface or river flows are unallocated for water resources planning and, therefore, local water harvesting could improve water availability and food security and may lead to a more equitable distribution of resources (Ngigi, 2003; Meter et al., 2014). However, water harvesting may not increase the overall availability of water in the basins, where runoff generated streamflow is usually included in water resources planning and allocation in the downstream areas of the basin. This is true for many of the basins in India, such as 'closed' basins where no extra renewable water resources are available for diversions to satisfy the consumptive water demand (Kumar et al., 2008). In this case, increased water availability in upstream areas is likely to come at the expense of reduced water availability downstream. An examination of water demands, renewable water, water productivity and the status of water resources (e.g., groundwater depletion) should be conducted at basin-scale to determine the potential benefits and trade-offs of rainwater harvesting.

17.6.2 Water Resources Sustainability

Globally, about 1.7–4 billion people (up to two-thirds of the population) are facing water scarcity (Kummu et al., 2016; Mekonnen & Hoekstra, 2016). With a growing population and changing socio-economic and climatic patterns, more people will be vulnerable to water shortages in the future (Gosling & Arnell, 2016; Vörösmarty et al., 2000). Ensuring adequate water availability in the face of growing environmental concerns makes water management a challenging task for water managers and policymakers. The provision of adequate water to meet present and future demands while protecting the environment sets forth the concept of sustainable water resources management.

In general, the term sustainability has often been associated with the protection of ecosystems to maintain their functionality while meeting the current and projected needs of growing populations (Loucks, 2000). Mays (2007) defined water resources sustainability as 'the ability to use water in sufficient quantities and quality from the local to the global scale to meet the needs of humans and ecosystems for the present and the future to sustain life and to protect humans from the dangers brought about by natural and human-caused disasters that affect sustaining life' (p. 3). However, the increasing water demands emanating from the increasing population and changing socio-economic and climatic conditions and resulting unsustainable use of water resources have degraded the ecosystems services and caused groundwater depletion, reduced surface flows, saltwater intrusion, land subsidence, loss of springs and wetlands and water quality problems in many regions worldwide (Konikow & Kendy, 2005; Wen & Chen, 2006; Wada et al., 2010).

Development of sustainable water resources management and policy becomes further challenging, given the uncertainty in future climate projections (Clark et al., 2016). Currently, on the global scale, surface and groundwater are under enormous pressure because the demand for freshwater is increasing and the supply of freshwater (both groundwater and surface water) is decreasing. Therefore, developing robust strategies for sustainable water resources management, which includes planning, developing, conserving, harvesting, distributing and managing the efficient use of water resources, is of utmost importance. Planning the exploitation of rainfall through rainwater harvesting is not an easy task because rain falling on one portion of the watershed may be demanded by different users from other parts of the watershed (Hatibu & Mahoo, 2000). Since the global water consumption increases two-fold every 20 years – this is more than twice the rate of human population growth – sustainable water resources management is key to satisfying future water demands (Bhattacharya et al., 2011).

Rainwater harvesting, also called water harvesting, is considered a sustainable water management system. It can help improve current and future water availability while maintaining or enhancing ecosystem services (Vohland & Barry, 2009; Garibaldi et al., 2017). Water harvesting systems are traditional methods that contributed to satisfy the water needs of local communities for many centuries. These systems can, therefore, contribute to managing water resources in a sustainable way (Bhattacharya, 2015). Rainwater harvesting is an important adaptation strategy to achieve global water resources security and sustainable water resources management under climate change (Sharda, 2006). Rainwater harvesting is the most sustainable method of water resources management, which could easily be used for potable and non-potable purposes in industry, agriculture, rural/urban water supply, sanitation, etc. (Rahman et al., 2014).

Rainwater harvesting enhances agricultural production and food security. It also helps reduce the impact of water scarcity on agricultural production during a drought in a humid climate zone and water stress in semi-arid and arid climate zones. Furthermore, water harvesting also reduces the impact of floods during hurricane or severe storm events. Therefore, rainwater and floodwater harvesting may increase water and food security and reduce the impact of water stress on agriculture. In addition, it can reduce the risk of floods on agriculture production, infrastructure and human lives. Although there are institutional and regulatory challenges associated with rainwater/floodwater harvesting at large scales, sound management of rainwater/floodwater requires the development of appropriate water storage and utilization technologies, which help enhance water resources management in the long run (Tesfay, 2011). For example, Zhang et al. (2019) found that rainwater harvesting systems with larger storage capacity are more resilient to climate change under higher water demand scenarios in humid and semi-arid cities.

Water harvesting may alleviate water scarcity under the current and future climatic conditions. However, its effects vary spatially depending on the climate, water demands, rainfall patterns, institutional and regulatory framework pertaining to water resources management, people's perception regarding rainwater harvesting and future changes in the climate. Also, the amount of water harvested and the reliability of a water harvesting system will mainly vary according to design parameters associated with the catchment area and size of the storage unit. These parameters are affected by the availability of financial resources and relative cost-benefits. A large range of water savings (6–92 per cent) obtained from rooftop rainwater harvesting systems in Brazil, Jordan and Australia shows that the benefits of rainwater harvesting may vary in different settings (Ghisi et al., 2006; Abdulla & Al-Shareef, 2009; Muthukumaran et al., 2011).

Rainwater harvesting can improve the livelihoods of farmers and water resources sustainability in various biophysical and socio-economic settings (Pachpute et al., 2009). Overall, effective adaptations of rainwater harvesting to climate change may include: designing adaptation strategies in sequence, integrating adaptation and mitigation, effective systems for assessment and mapping of vulnerability and institutional efforts for rainwater harvesting under potential future climate change (Ayele, 2014). For example, location-specific adaptive adjustments (e.g., adjusting tank/reservoir sizes, catchment areas or water demands) can improve the efficiency of rainwater harvesting systems and help them satisfy current and future water demands and stormwater control requirements under the changing climate (Zhang et al., 2019).

17.7 Conclusion

For centuries, rainwater harvesting has been employed across the world. Rainwater harvesting is a promising technique that can be used to satisfy the current and future demands for water. Furthermore, it offers a framework for sustainable management of the water resources, which are under stress due to the changing climate and growing population that affect many countries worldwide. This chapter has reviewed the historical methods and uses of rainwater collection and distribution systems, methods used for rainwater harvesting in rural and urban landscapes, potential opportunities and challenges to harvest, store and use rainwater at small to large scales. In addition, this chapter has highlighted the potential environmental benefits of harvesting rainwater and floodwater in rural agricultural landscapes as well as impervious urban landscapes. This comprehensive review has discussed the contribution of rainwater harvesting to sustainable water resources management. The analysis carried out in this chapter suggests that adequate location-specific rainwater harvesting systems can be used to satisfy current and future water demands and stormwater control requirements under climate change. To improve the sustainability of rainwater harvesting systems, a site or watershed-specific cost-benefit analysis is required to assess the feasibility of a water harvesting infrastructure expansion and construction under climate change.

Overall, rainwater harvesting appears to be one of the most promising alternatives for sustainable water resources management to address the increase of water scarcity and water demands in agricultural, industrial, commercial, rural and urban water supply sectors. This review can be useful for policy and decision-makers, and water resource managers in their planning and management strategies for sustainable water resources management, making them resilient to climate change.

References

Abbasi, T., & Abbasi, S. A. (2011). Sources of pollution in rooftop rainwater harvesting systems and their control. *Critical Reviews in Environmental Science and Technology* 41(23): 2097–2167.

Abdulla, F. A., & Al-Shareef, A. W. (2009). Roof rainwater harvesting systems for household water supply in Jordan. *Desalination* 243(1–3): 195–207.

Amos, C. C., Rahman, A., & Gathenya, J. M. (2016). Economic analysis and feasibility of rainwater harvesting systems in urban and peri-urban environments: A review of the global situation with a special focus on Australia and Kenya. *Water* 8(4): 1–21.

Andersson, J. C., Zehnder, A. J., Rockström, J., & Yang, H. (2011). Potential impacts of water harvesting and ecological sanitation on crop yield, evaporation and river flow regimes in the Thukela River basin, South Africa. *Agricultural Water Management* 98(7): 1113–1124.

Angelakis, A. (2016). Evolution of rainwater harvesting and use in Crete, Hellas, through the millennia. *Water Science and Technology: Water Supply* 16(6): 1624–1638.

Angelakis, A. N., & Spyridakis, S. V. (1996). The status of water resources in Minoan times: A preliminary study. In A. N. Angelakis, & A. S. Issar (eds.), *Diachronic Climatic Impacts on Water Resources.* NATO ASI Series (Series I: Global Environmental Change, vol. 36, pp. 161–191). Berlin: Springer.

Antoniou, G., Kathijotes, N., Spyridakis, D. S., & Angelakis, A. N. (2014). Historical development of technologies for water resources management and rainwater harvesting in the Hellenic civilizations. *International Journal of Water Resources Development* 30(4): 680–693.

Ayele, Y. A. (2014). Rainwater harvesting for climate change adaptation in Ethiopia: Policy and institutional analysis. *V.R.F. Series 488*: 84.

Bassi, N., Kumar, M. D., Sharma, A., & Pardha-Saradhi, P. (2014). Status of wetlands in India: A review of extent, ecosystem benefits, threats, and management strategies. *Journal of Hydrology: Regional Studies* 2: 1–19.

Bhattacharya, S. (2015). Traditional water harvesting structures and sustainable water management in India: A socio-hydrological review. *International Letters of Natural Sciences* 37: 30–38.

Bhattacharya, S., Dasgupta, A., Mahansaria, R., Ghosh, S., Chattopadhyay, D., & Mukhopadhyay, A. (2011). Traditional water harvesting in India: Historical perspectives, present scenario, and future prospects. *Water Advisory Committee Meeting*, 612–623.

Boelee, E., Mekonnen, Y., Jean-Noel, P., et al. (2013). Options for water storage and rainwater harvesting to improve health and resilience against climate change in Africa. *Regional Environmental Change* 13: 509–519.

Brauman, K. A., Richter, B. D., Postel, S., Malsy, M., & Florke, M. (2016). Water depletion: An improved metric for incorporating seasonal and dry-year water scarcity into water risk assessments. *Elementa: Science of the Anthropocene* 4: 000083.

Bruins, H. J., Evenari, M., & Nessler, U. (1986). Rainwater-harvesting agriculture for food production in arid zones: The challenge of the African famine. *Applied Geography* 6 (1): 13–32.

Bunclark, L., Gowing, J., Oughton, E., Ouattara, K., Ouoba, S., & Benao, D. (2018). Understanding farmers' decisions on adaptation to climate change: Exploring adoption of water harvesting technologies in Burkina Faso. *Global Environmental Change* 48: 243–254.

Campisano, A., Butler, D., Ward, S., et al. (2017). Urban rainwater harvesting systems: Research, implementation and future perspectives. *Water Research* 115: 195–209.

Chien, H., Yeh, P. J. F., & Knouft, J. H. (2013). Modeling the potential impacts of climate change on streamflow in agricultural watersheds of the Midwestern United States. *Journal of Hydrology* 491(1): 73–88.

Clark, M. P., Wilby, R. L., Gutmann, E. D., et al. (2016). Characterizing uncertainty of the hydrologic impacts of climate change. *Current Climate Change Reports* 2(2): 55–64.

Costanza, R., de Groot, R., Sutton, P., et al. (2014). Changes in the global value of ecosystem services. *Global Environmental Change* 26(1): 152–158.

Critchley, W., Siegert, K., Chapman, C., & Finket, M. (2013). *Water Harvesting: A Manual for the Design and Construction of Water Harvesting Schemes for Plant Production*. Jodhpur: Scientific Publishers.

Crosbie, R. S., Scanlon, B. R., Mpelasoka, F. S., Reedy, R. C., Gates, J. B., & Zhang, L. (2013). Potential climate change effects on groundwater recharge in the High Plains Aquifer, USA. *Water Resources Research*, 49(7): 3936–3951.

Cupido, A., Brian, B., Guo, Y., & Robertson, A. (2012). An evaluation of rainwater runoff quality from selected white roof membranes. *Water Quality Research Journal* 47(1): 66–79.

Dile, Y. T., Karlberg, L., Daggupati, P., Srinivasan, R., Wiberg, D., & Rockström, J. (2016). Assessing the implications of water harvesting intensification on upstream–downstream ecosystem services: A case study in the Lake Tana basin. *The Science of the Total Environment* 542(Part A): 22–35.

Dodds, W. K., & Smith, V. H. (2016). Nitrogen, phosphorus, and eutrophication in streams. *Inland Waters* 6(2): 155–164.

Doll, P. (2009). Vulnerability to the impact of climate change on renewable groundwater resources: A global-scale assessment. *Environmental Research Letters* 4: 035006.

Donat, M. G., Alexander, L. V., Yang, H., et al. (2013). Updated analyses of temperature and precipitation extreme indices since the beginning of the twentieth century: The HadEX2 dataset. *Journal of Geophysical Research: Atmospheres* 118(5): 2098–2118.

Falkenmark, M., Fox, P., Persson, G., & Rockstrom, J. (2001). Water harvesting for upgrading of rainfed agriculture: Problems analysis and research needs. *SIWI Report 11*. Stockholm: Stockholm International Water Institute. 94 pp.

Famiglietti, J. S. (2014). The global groundwater crisis. *Nature Climate Change* 4(11): 945–948.

FAO (1994). Water harvesting for improved agricultural production: Proceedings of the FAO Expert Consultation, Cairo, Egypt. *Water Reports Series* 6: 424 pp.

FAO (2016). AQUASTAT Main Database – Food and Agriculture Organization of the United Nations (FAO). Available from https://ceowatermandate.org/resources/fao-aquastat-2016/ (Last accessed 14 January 2020).

Ferrand, E. A., & Cecunjanin, F. (2014). Potential of rainwater harvesting in a thirsty world: A survey of ancient and traditional rainwater harvesting applications. *Geography Compass* 8(6): 395–413.

Frasier, G. W. (1985). Technical, economic, and social considerations of water harvesting and runoff farming. *Paper presented at the Conference Arid Lands: Today and Tomorrow*, 20–25 October, University of Arizona, Tucson, AZ.

Garg, K. K., Karlberg, L., Barron, J., Wani, S. P., & Rockstrom, J. (2012). Assessing impacts of agricultural water interventions in the Kothapally watershed, Southern India. *Hydrological Processes* 26(3): 387–404.

Garg, K. K., Wani, S. P., Barron, J., Karlberg, L., & Rockstrom, J. (2013). Upscaling potential impacts on water flows from agricultural water interventions: Opportunities and tradeoffs in the Osman Sagar catchment, Musi subbasin, India. *Hydrological Processes* 27(26): 3905–3921.

Garibaldi, L. A., Gemmill-Herren, B., D'Annolfo, R., Graeub, B. E., Cunningham, S. A., & Breeze, T. D. (2017). Farming approaches for greater biodiversity, livelihoods, and food security. *Trends in Ecology and Evolution* 32(1): 68–80.

GhaffarianHoseini, A., Tookey, J., GhaffarianHoseini, A., Yusoff, S. M., & Hassan, N. B. (2016). State of the art of rainwater harvesting systems towards promoting green built environments: A review. *Desalination and Water Treatment* 57(1): 95–104.

Ghimire, S. R., Johnston, J. M., Garland, J., Edelen, A., Ma, X., & Jahne, M. (2019). Life cycle assessment of a rainwater harvesting system compared with an AC condensate harvesting system. *Resources, Conservation and Recycling* 146: 536–548.

Ghisi, E., Montibeller, A., & Schmidt, R. W. (2006). Potential for potable water savings by using rainwater: An analysis over 62 cities in southern Brazil. *Building and Environment* 41(2): 204–210.

Githui, F., Gitau, W., Mutua, F., & Bauwens, W. (2009). Climate change impact on SWAT simulated streamflow in western Kenya. *International Journal of Climatology* 29 (12): 1823–1834.

Gizaw, M. S., Biftu, G. F., Gan, T. Y., Moges, S. A., & Koivusalo, H. (2017). Potential impact of climate change on streamflow of major Ethiopian rivers. *Climatic Change* 143(3–4): 371–383.

Gleeson, T., Wada, Y., Bierkens, M. F., & van Beek, L. P. (2012). Water balance of global aquifers revealed by groundwater footprint. *Nature* 488(7410): 197–200.

Gosling, S. N., & Arnell, N. W. (2016). A global assessment of the impact of climate change on water scarcity. *Climatic Change* 134(3): 371–385.

Gunnell, Y., & Krishnamurthy, A. (2003). Past and present status of runoff harvesting systems in dryland peninsular India: A critical review. *AMBIO: A Journal of the Human Environment* 32(4): 320–324.

Guo, R., & Guo, Y. (2018). Stochastic modeling of the hydrologic operation of rainwater harvesting systems. *Journal of Hydrology* 562: 30–39.

Gupta, A. K. (2006). *Rainwater Harvesting*. Pune: Indian Railways Institute of Civil Engineering.

Handia, L., Tembo, J. M., & Mwiindwa, C. (2003). Potential of rainwater harvesting in urban Zambia. *Physics and Chemistry of the Earth* 28(20–27): 893–896.

Haque, M. M., Rahman, A., & Samali, B. (2016). Evaluation of climate change impacts on rainwater harvesting. *Journal of Cleaner Production* 137: 60–69.

Hatibu, N., & Mahoo, H. F. (2000). Rainwater harvesting for natural resources management: A planning guide for Tanzania. In *Sida's Regional Land Management Unit, Technical Handbook* (No. 22). Nairobi: Regional Land Management Unit, RELMA/Sida.

Hofman, J. A., & Paalman, M. (2014). Rainwater harvesting, a sustainable solution for urban climate adaptation? *KWR Watercycle Research Institute.* KFC report number KfC 142/2014, 66 p.

Jha, M., Pan, Z., Takle, E. S., & Gu, R. (2004). Impacts of climate change on streamflow in the Upper Mississippi River Basin: A regional climate model perspective. *Journal of Geophysical Research: Atmospheres* 109(D9105): 1–12.

Kahinda, J. M., & Taigbenu, A. E. (2011). Rainwater harvesting in South Africa: Challenges and opportunities. *Physics and Chemistry of the Earth, Parts A/B/C* 36 (14–15): 968–976.

Khatakho, R., & Koju, N. (2017). Rain Water Harvesting (RWH) technique: An alternative way to irrigation in Arid and Semi-Arid Regions (ASARs). Available from www .researchgate.net/publication/324803601 (Last accessed 20 December).

Khoury-Nolde, N. (2006). *Rainwater Harvesting*. Germany: Zero-M Organization. Available from https://pdfcoffee.com/rainwater-harvesting-5-pdf-free.html (Last accessed 20 December 2019).

Koirala, S., Hirabayashi, Y., Mahendran, R., & Kanae, S. (2014). Global assessment of agreement among streamflow projections using CMIP5 model outputs. *Environmental Research Letters* 9(6): 064017.

Konikow, L. F., & Kendy, E. (2005). Groundwater depletion: A global problem. *Hydrogeology Journal* 13(1): 317–320.

Kreins, P., Henseler, M., Anter, J., Herrmann, F., & Wendland, F. (2015). Quantification of climate change impact on regional agricultural irrigation and groundwater demand. *Water Resources Management* 29(10): 3585–3600.

Krueger, E., Rao, P. S. C., & Borchardt, D. (2019). Quantifying urban water supply security under global change. *Global Environmental Change* 56: 66–74.

Kumar, K. K., Patwardhan, S. K., Kulkarni, A., Kamala, K., Rao, K. K., & Jones, R. (2011). Simulated projections for summer monsoon climate over India by a high-resolution regional climate model (PRECIS). *Current Science* 101(3): 312–326.

Kumar, M. D. (2004). Roof water harvesting for domestic water security: Who gains and who loses? *Water International* 29(1): 43–53.

Kumar, M. D., Patel, A., Ravindranath, R., & Singh, O. P. (2008). Chasing a mirage: Water harvesting and artificial recharge in naturally water-scarce regions. *Economic and Political Weekly* 43(35): 61–71.

Kummu, M., Guillaume, J. H. A., de Moel, H., et al. (2016). The world's road to water scarcity: Shortage and stress in the 20th century and pathways towards sustainability. *Scientific Reports* 6(1): 38495.

Loucks, D. P. (2000). Sustainable water resources management. *Water international* 25(1): 3–10.

Love, D., Uhlenbrook, S., Corzo-Perez, G., Twomlow, S., & van der Zaag, P. (2010). Rainfall–interception–evaporation–runoff relationships in a semi-arid catchment, northern Limpopo basin, Zimbabwe. *Hydrological Sciences Journal* 55(5): 687–703.

Matos, C., Santos, C., Pereira, S., Bentes, I., & Imteaz, M. (2013). Rainwater storage tank sizing: A case study of a commercial building. *International Journal of Sustainable Built Environment* 2(2): 109–118.

Mays, L. W. (2007). *Water Resources Sustainability* (No. 363.6 M3.). New York: McGraw-Hill.

Mays, L. W., Antoniou, G. P., & Angelakis, A. N. (2013). History of water cisterns: Legacies and lessons. *Water* 5(4): 1916–1940.

McVicar, T. R., Roderick, M. L., Donohue, R. J., et al. (2012). Global review and synthesis of trends in observed terrestrial near-surface wind speeds: Implications for evaporation. *Journal of Hydrology* 416–417: 182–205.

Meehl, G. A., Arblaster, J. M., & Tebaldi, C. (2005). Understanding future patterns of increased precipitation intensity in climate model simulations. *Geophysical Research Letters* 32(18): 1–4.

Mehrabadi, M. H. R., Saghafian, B., & Fashi, F. H. (2013). Assessment of residential rainwater harvesting efficiency for meeting non-potable water demands in three climate conditions. *Resources, Conservation and Recycling* 73: 86–93.

Mekonnen, M. M., & Hoekstra, A. Y. (2016). Four billion people facing severe water scarcity. *Science Advances* 2(2): e1500323.

Meter, K. J. V., Basu, N. B., Tate, E., & Wyckoff, J. (2014). Monsoon harvests: The living legacies of rainwater harvesting systems in South India. *Environmental Science and Technology* 48(8): 4217–4225.

Millennium Ecosystem Assessment (2005). *Ecosystems and Human Well-Being: Synthesis.* Washington, DC: Island Press.

Milly, P. C., Dunne, K. A., & Vecchia, A. V. (2005). Global pattern of trends in streamflow and water availability in a changing climate. *Nature* 438(7066): 347–350.

Mishra, V., Ganguly, A. R., Nijssen, B., & Lettenmaier, D. P. (2015). Changes in observed climate extremes in global urban areas. *Environmental Research Letters*, 10(2), 024005.

Muchuru, S., & Nhamo, G. (2019). Sustaining African water resources under climate change: Emerging adaptation measures from UNFCCC national communications. *African Journal of Science, Technology, Innovation and Development* 11(2): 181–196.

Muthukumaran, S., Baskaran, K., & Sexton, N. (2011). Quantification of potable water savings by residential water conservation and reuse – A case study. *Resources, Conservation and Recycling* 55(11): 945–952.

Mutunga, K. (2001). Water conservation, harvesting and management (WCHM) – Kenyan experience. In *Sustaining the Global Farm. 10th International Soil Conservation Organization Meeting*. Purdue University and the USDA-ARS National Soil Erosion Research Laboratory (pp. 1139–1143).

Nawaz, M., Amin, M. T., Han, M., Alazba, A. A., Manzoor, U., & Amin, M. N. (2013). Variation of pseudomonas aeruginosa in rainwater harvesting systems: Effects of seasons, catchments and storage conditions. *Clean – Soil Air Water* 42(7): 893–900.

Ngigi, S. N. (2003). What is the limit of up-scaling rainwater harvesting in a river basin? *Physics and Chemistry of the Earth, Parts A/B/C* 28(20–27): 943–956.

Norfolk, O., Abdel-Dayem, M., & Gilbert, F. (2012). Rainwater harvesting and arthropod biodiversity within an arid agro-ecosystem. *Agriculture, Ecosystems and Environment* 162: 8–14.

O'Gorman, P. A. (2015). Precipitation extremes under climate change. *Current Climate Change Reports* 1(2): 49–59.

Oweis, T., Hachum, A., & Bruggeman, A. (2004). *The Role of Indigenous Knowledge in Improving Present Water Harvesting Practices*. Aleppo, Syria: ICARDA.

Pachpute, J. S., Tumbo, S. D., Sally, H., & Mul, M. L. (2009). Sustainability of rainwater harvesting systems in rural catchment of Sub-Saharan Africa. *Water Resources Management* 23(13): 2815–2839.

Pandey, D. N., Gupta, A. K., & Anderson, D. M. (2003). Rainwater harvesting as an adaption to climate change. *Current Science* 85(1): 46–59.

Pietrapertosa, F., Khokhlov, V., Salvia, M., & Cosmi, C. (2018). Climate change adaptation policies and plans: A survey in 11 southeast European countries. *Renewable and Sustainable Energy Reviews* 81: 3041–3050.

Pina, C. L., Kassaye, R. B., & Schaldach, R. (2018). *Working Paper: Rainwater Harvesting Methods*. Available from https://ruvival.de/rainwater-harvesting (Last accessed 20 December 2019).

Pittock, J. (2011). National climate change policies and sustainable water management: Conflicts and synergies. *Ecology and Society* 16(2): 25.

Prinz, D. (2002). The role of water harvesting in alleviating water scarcity in arid areas. *Keynote Lecture, Proceedings, International Conference on Water Resources Management in Arid Regions*. Kuwait Institute for Scientific Research, Kuwait, 3, 107–122.

Prinz, D., & Singh, A. (2000). Technological potential for improvements of water harvesting. *Contribution Paper to the World Commission on Dams*, Cape Town, South Africa. Technical Report, 12 p.

Rahman, S., Khan, M. T. R., Akib, S., Din, N. B. C., Biswas, S. K., & Shirazi, M. (2014). Sustainability of rainwater harvesting system in terms of water quality. *The Scientific World Journal* 2014: 721357.

Rana, S. (2005). *Rainwater harvesting for drinking in rural area: A case study on three villages of Paikgacha Thana in Khulna District*. Available from www.ctahr.hawaii .edu/hawaiirain/Library/papers/Rana_Md.Sohel.pdf (Last accessed 20 December 2019).

Rockström, J., Karlberg, L., Wani, S. P., et al. (2010). Managing water in rainfed agriculture-The need for a paradigm shift. *Agricultural Water Management* 97(4): 543–550.

Saleh, S. A., Taher, T., & Noaman, A. (2017). *Manual for Rooftop Rainwater Harvesting Systems in the Republic of Yemen*. Available from http://spate-irrigation.org/wp-content/uploads/2018/02/Rain-Water-Harvesting-Manual-WEC-FINAL2-Small.pdf (Last accessed 19 December 2019).

Schewe, J., Heinke, J., Gerten, D., et al. (2014). Multimodel assessment of water scarcity under climate change. *Proceedings of the National Academy of Sciences (USA)* 111 (9): 3245–3250.

Şen, Z., Al Alsheikh, A., Al-Turbak, A. S., Al-Bassam, A. M., & Al-Dakheel, A. M. (2013). Climate change impact and runoff harvesting in arid regions. *Arabian Journal of Geosciences* 6(1): 287–295.

Sharda, V. N. (2006). *Watershed Management and Water Harvesting as Strategic Tools for Groundwater Augmentation*. Colombo, Sri Lanka: International Water Management Institute, IWMI Books, Reports H039314.

Shrestha, S., Chapagain, R., & Babel, M. S. (2017). Quantifying the impact of climate change on crop yield and water footprint of rice in the Nam Oon Irrigation Project, Thailand. *Science of the Total Environment* 599: 689–699.

Singh, R., Garg, K. K., Wani, S. P., Tewari, R. K., & Dhyani, S. K. (2014). Impact of water management interventions on hydrology and ecosystem services in Garhkundar-Dabar watershed of Bundelkhand region, Central India. *Journal of Hydrology* 509: 132–149.

Sishodia, R. P., Shukla, S., Graham, W. D., Wani, S. P., Jones, J. W., & Heaney, J. (2017). Current and future groundwater withdrawals: Effects, management and energy policy options for a semi-arid Indian watershed. *Advances in Water Resources* 110: 459–475.

Sishodia, R. P., Shukla, S., Wani, S. P., Graham, W. D., & Jones, J. W. (2018). Future irrigation expansion outweigh groundwater recharge gains from climate change in semi-arid India. *Science of the Total Environment* 635(3): 725–740.

Smith, V. H., Tilman, G. D., & Nekola, J. C. (1999). Eutrophication: Impacts of excess nutrient inputs on freshwater, marine, and terrestrial ecosystems. *Environmental Pollution* 100(1–3): 179–196.

Stiefel, J. M., Melesse, A. M., McClain, M. E., Price, R. M., Anderson, E. P., & Chauhan, N. K. (2009). Effects of rainwater-harvesting-induced artificial recharge on the groundwater of wells in Rajasthan, India. *Hydrogeology Journal* 17(8): 2061–2073.

Stocker, T. F., Qin, D., Plattner, G.-K., et al. (eds.) (2013). *IPCC, 2013: Climate Change 2013: The Physical Science Basis. Contribution of Working Group I to the Fifth Assessment Report of the Intergovernmental Panel on Climate Change*. Cambridge: Cambridge University Press.

Sturm, M., Zimmermann, M., Schütz, K., Urban, W., & Hartung, H. (2009). Rainwater harvesting as an alternative water resource in rural sites in central northern Namibia. *Physics and Chemistry of the Earth, Parts A/B/C* 34(13): 776–785.

Texas AgriLife Extension (TALE) (2012). Rainwater Harvesting: System Planning, report, November 2012; College Station, TX: University of North Texas Libraries, The Portal to Texas History. Available from https://texashistory.unt.edu/ark:/67531/metapth639744/ (Last accessed 23 August 2021).

Taye, M. T., Ntegeka, V., Ogiramoi, N. P., & Willems, P. (2011). Assessment of climate change impact on hydrological extremes in two source regions of the Nile River Basin. *Hydrology and Earth System Sciences* 15(1): 209–222.

Taylor, R. G., Scanlon, B., Döll, P., et al. (2013). Groundwater and climate change. *Nature Climate Change* 3(4): 322–329.

Tesfay, G. (2011). On-Farm Water Harvesting for Rainfed Agriculture Development and Food Security in Tigray, Northern Ethiopia. Investigation of Technical and Socioeconomic Issues, DCG Report No. 61.

UNEP. (1979). Rain and Storm Water Harvesting for Additional Water Supply in Rural Areas. Nairobi, Kenya. *Expert Group Meeting*, 30 October to 2 November 1979. 28 p.

UNEP International Environmental Technology Centre -Osaka/Shiga, JP, IETC. (2001). *Rainwater Harvesting and Utilization: An Environmentally Sound Approach for Sustainable Urban Water Management: An Introductory Guide for Decision-makers*. UNEP International Environmental Technology Centre, 2, 12 p.

UN-HABITAT. (2005). *Rainwater Harvesting and Utilisation; Blue Drop Series, Book 2: Beneficiaries & Capacity Builders*. Mtwapa, Kenya: UN-HABITAT.

Van Vliet, M. T. H., Franssen, W. H. P., Yearsley, J. R., et al. (2013). Global river discharge and water temperature under climate change. *Global Environmental Change* 23(2): 450–464.

Van Wyk, E., Van Tonder, G. J., & Vermeulen, D. (2012). Characteristics of local groundwater recharge cycles in South African semi-arid hard rock terrains: Rainfall–groundwater interaction. *Water SA* 38(5): 747–754.

Vano, J. A., Nijssen, B., & Lettenmaier, D. P. (2015). Seasonal hydrologic responses to climate change in the Pacific Northwest. *Water Resources Research*, 51(4): 1959–1976.

Velasco-Muñoz, J. F., Aznar-Sánchez, J. A., Batlles-delaFuente, A., & Fidelibus, M. D. (2019). Rainwater harvesting for agricultural irrigation: An analysis of global research. *Water* 11(1320): 1–18.

Vohland, K., & Barry, B. (2009). A review of in situ rainwater harvesting (RWH) practices modifying landscape functions in African drylands. *Agriculture, Ecosystems and Environment* 131(3–4): 119–127.

Vörösmarty, C. J., Green, P., Salisbury, J., & Lammers, R. B. (2000). Global water resources: Vulnerability from climate change and population growth. *Science* 289 (5477): 284–288.

Wada, Y., Van Beek, L. P., Van Kempen, C. M., et al. (2010). Global depletion of groundwater resources. *Geophysical Research Letters*, 37(20): L20402.

Wallace, C. D., Bailey, R. T., & Arabi, M. (2015). Rainwater catchment system design using simulated future climate data. *Journal of Hydrology* 529: 1798–1809.

Wang, Q., Li, Y., & Alva, A. (2010). Cropping systems to improve carbon sequestration for mitigation of climate change. *Journal of Environmental Protection*, 1(3): 207–215.

Wani, S. P., Singh, H. P., Sreedevi, T. K., et al. (2003). *Farmer-Participatory Integrated Watershed Management: Adarsha Watershed, Kothapally India*. Case 7.

Welderufael, W. A., Woyessa, Y. E., & Edossa, D. C. (2013). Impact of rainwater harvesting on water resources of the Modder river basin, central region of South Africa. *Agricultural Water Management* 116: 218–227.

Wen, F., & Chen, X. (2006). Evaluation of the impact of groundwater irrigation on streamflow in Nebraska. *Journal of Hydrology* 327(3–4): 603–617.

Wilby, R. L., & Harris, I. (2006). A framework for assessing uncertainties in climate change impacts: Low-flow scenarios for the River Thames, UK. *Water Resources Research*, 42(2).

Wild, M. (2009). Global dimming and brightening: A review. *Journal of Geophysical Research-Atmospheres* 114: D00D16.

Willett, K. M., Jones, P. D., Gillett, N. P., & Thorne, P. W. (2008). Recent changes in surface humidity: Development of the HadCRUH dataset. *Journal of Climate* 21(20): 5364–5383.

Wirtenberg, J., Kelley, L.M., Lipsky, D., & Russell, W.G. (2019). *The Sustainable Enterprise Fieldbook: Building New Bridges*. New York: Routledge.

Yannopoulos. S., Antoniou, G., Kaiafa-Saropoulou, M., & Angelakis, A. N. (2017). Historical development of rainwater harvesting and use in Hellas: A preliminary review. *Water Science & Technology: Water Supply* 17(4): 1022–1034.

Zahmatkesh, Z., Karamouz, M., Goharian, E., & Burian, S. J. (2014). Analysis of the effects of climate change on urban stormwater runoff using statistically downscaled precipitation data and a change factor approach. *Journal of Hydrologic Engineering* 20(7): 05014022.

Zhang, S., Zhang, J., Yue, T., & Jing, X. (2019). Impacts of climate change on urban rainwater harvesting systems. *Science of the Total Environment* 665(10): 262–274.

18

Variability of Runoff Coefficient and Precipitation Elasticity at Watersheds across China

YIN TANG, QIUHONG TANG AND ZHONGGEN WANG

18.1 Introduction

Water is a vital resource for human well-being and the functioning of ecosystems. Recent warnings worldwide have highlighted the growing challenges posed by water scarcity. Since the 1970s, reports have brought to light a global decreasing of freshwater discharge and an increase in global aridity that can be mainly attributed to climate change. Water resource planning and management increasingly need to account for the impacts of global climate change in order to accurately predict future water supplies. The changing pattern of global precipitation induced by climate change deserves urgent and systematic attention. Previous studies showed the emergence of a changing pattern in precipitation: (a) an increase of precipitation in high latitudes (Northern Hemisphere); (b) a reduction of precipitation in China, Australia and the Small Island States in the Pacific and (c) large uncertainty in the changes of precipitation in equatorial regions. Precipitation is one of the most important sources for freshwater. Annual changes in precipitation are the most important driver of runoff. Thus, future changes in precipitation will affect the regional availability of water resources. In the long-term, changes in precipitation might affect runoff. In the short-term, more intense and intermittent precipitation might increase the risk of floods and droughts. Consequently, hydrologists and geoscientists have attempted to better understand the sensitivity of streamflow to precipitation changes.

Several methods have been proposed to assess the sensitivity of streamflow to climate change for river basins across the world. Most hydrologic climate sensitivity studies involve calibrating a conceptual deterministic watershed model, and varying the model's meteorological forcing, to observe the resulting changes in streamflow. Another approach is to analytically derive the sensitivity of streamflow in terms of model parameters. A third approach is to fit multivariate

This work was supported by the National Natural Science Foundation of China (41730645 and 41901042).

regional hydrologic models using climate and streamflow data for many basins in a region. A fourth approach is to empirically estimate changes in streamflow which resulted from historical changes in climate. A fifth approach is to use multivariate statistical methods to estimate the relationship between climate and streamflow on a single site. Of all these approaches the use of conceptual deterministic watershed models is by far the most common because such models can model the complex spatial and temporal variations associated with evapotranspiration, soil moisture, groundwater and streamflow processes. In this chapter, we provide a more detailed discussion of the advantages of conceptual watershed models for modelling climate change impacts. However, this method also presents fundamental challenges. Climate sensitivity analyses performed on the same basin using different conceptual watershed models can lead to significant differences in the results. More problematically, climate sensitivity analyses performed on the same basin using identical conceptual watershed models can lead to considerable differences in the results.

Recently, Sankarasubramanian and Vogel (2003) noticed a concept defined as the climate elasticity of streamflow by Schaake (1990). Based on this concept, they successfully developed a metric method to assess the variations of the response of watershed hydrology to precipitation changes. The climate elasticity of streamflow can be defined as the proportional change in streamflow divided by the proportional change in a climatic variable such as precipitation. Therefore, the precipitation elasticity of streamflow can be defined as:

$$\varepsilon = \frac{\delta Q/Q}{\delta P/P}, \tag{18.1}$$

where Q is the streamflow, P is the precipitation and ε is the precipitation elasticity of streamflow. Similarly, the climate elasticity of a quantity different from precipitation (e.g., potential evapotranspiration) can also be defined using Eqn. (18.1) by replacing the precipitation with other climatic variables. The estimation of elasticity defined in Eqn. (18.1) is problematic since this definition considers a discrete range of the variables of interest but the functions of P and Q vary across time and space. To solve this paradox, Sankarasubramanian et al. (2001) introduced a specific case of Eqn. (18.1) at the mean value of the climatic variable:

$$\varepsilon_P(\mu_P, \mu_Q) = \frac{dQ}{dP}\Big|_{P=\mu_P} \frac{\mu_P}{\mu_Q}, \tag{18.2}$$

where μ_P, μ_Q are mean values of precipitation and streamflow, respectively. The definition in Eqn. (18.2) can provide a relatively reliable estimation of precipitation elasticity in investigated watersheds. Sankarasubramanian et al. (2001) further developed a non-parametric estimator associated with the definition in

Eqn. (18.2) and verified that the precipitation elasticity is robust for a wide class of hydrological models – the estimator does not depend on the form of the hydrological model. Additionally, several methods to estimate the elasticity of streamflow defined as Eqn. (18.2) have been proposed. The analytical derivation of elasticity is often based on the Budyko hypothesis under which several simple functions for modelling streamflow through the aridity index (i.e., ratio of precipitation and potential evapotranspiration) and one parameter are considered. Empirical approaches are possible alternatives. They compute the ratio of annual average runoff and climate variables and are often illustrated by contours of runoff change over annual departures from normal precipitation and temperature.

The precipitation elasticity is an efficient index used to quantify the sensitivity of streamflow to precipitation changes. It provides us with a tool to understand the historical and predicted changes in streamflow in investigated watersheds which is key to assess the impact of climate change on water resources. In this chapter, we will provide the details of two types of frequently used methods for estimating precipitation elasticity of streamflow and a case study on the temporal and spatial variations in streamflow and the estimated precipitation elasticity by a non-parametric estimator during three periods of interest.

18.2 Estimators of Precipitation Elasticity

In this section, we introduce two types of methods on evaluating precipitation elasticity of streamflow. The first type is a parametric estimation model which requires knowing the function that links precipitation with streamflow. The second type is a non-parametric model which requires long-term observations of precipitation and streamflow.

18.2.1 Model-based Estimates of Precipitation Elasticity

Usually, a simple and robust function is considered in the model because it can simply approximate observed climate and streamflow data and easily derive the theoretical expressions for precipitation elasticity. Several models have been proposed to describe the relationship between long-term annual precipitation and streamflow. These models vary in physical basis from black statistical formulae to grey conceptual formulae. To elaborate the model-based estimates of precipitation elasticity, a black statistic model and a grey conceptual model are taken as examples.

Within a statistical model framework, we consider a trivariate linear model relating Q to P and E_P:

Table 18.1. One-parameter Budyko-type equations

Budyko equation	Parameter	References
$\frac{E}{P} = \left[1 + \left(\frac{E_P}{P}\right)^{-n}\right]^{-1/n}$	n	Choudhury (1999); Mezentsev (1955); Yang et al. (2008)
$\frac{E}{P} = 1 + \frac{E_p}{P} - \left[1 + \left(\frac{E_p}{P}\right)^{\omega}\right]^{1/\omega}$	ω	Fu (1981); Zhang et al. (2004)
$\frac{E}{P} = \frac{1 + w\frac{E_p}{P}}{1 + w\frac{E_p}{P} + \left(\frac{E_p}{P}\right)^{-1}}$	w	Zhang et al. (2001)
$\frac{E}{P} = \frac{1 + E_p/P - \sqrt{(1 + E_p/P)^2 - \frac{4\alpha(2-\alpha)E_p}{P}}}{2\alpha(2-\alpha)}$	α	Wang and Tang (2014)

$$Q = \alpha + \beta P + \delta E_P + \eta, \tag{18.3}$$

where α, β and δ are model parameters and η are independent and identically distributed model errors with mean zero and constant variance. For the trivariate model, the precipitation elasticity can be expressed as follows:

$$\varepsilon_P = \rho_{Q,P}(C_Q/C_P), \tag{18.4}$$

where $\rho_{Q,P}$ is the cross-correlation of Q and P and C_Q and C_P are the coefficient of variation of Q and P, respectively. Previous studies show that the value of C_Q is often significantly larger than the value of C_P and that $\rho_{Q,P}$ is usually near unity. Hence, the value of ε_P is expected to be larger than 1 in this linear model.

The Budyko framework is a common conceptual model that provides a simple but effective tool to estimate long-term evaporation and streamflow and to evaluate their responses to climate change. Based on a large number of observations, the climate aridity index, defined as the ratio between mean annual potential evaporation (E_p) and P (i.e., $\emptyset = \frac{E_p}{P}$), predominantly controls the partitioning of mean annual precipitation. The other impact factors (e.g., climate variability, vegetation, soil and topography) are lumped into a single parameter under the Budyko framework. Several one-parameter Budyko equations have been proposed or derived as illustrated in Table 18.1. The functional forms of the Budyko-type equations are relatively simple and have been used to derive climate elasticity to quantify the sensitivity of streamflow to changes in climatic variables. Generally, the Budyko-type equations can be expressed as follows:

$$E = f(E_p, P, x), \tag{18.5}$$

where E represents the actual evapotranspiration, x represents the parameter of different models (e.g., ω for Fu's equation) and $f()$ represents the Budyko-type equations as listed in Table 18.1. On a long-term scale, the total water storage change is negligible (i.e., $\Delta S \approx 0$) and the mean annual precipitation equals the sum of the mean annual actual evapotranspiration and the mean annual total runoff. The corresponding water balance equation is:

$$P = E + Q. \tag{18.6}$$

According to Eqns. (18.5) and (18.6), we can express the total differential of actual evapotranspiration:

$$dE = \frac{\partial f}{\partial P} dP + \frac{\partial f}{\partial E_p} dE_p + \frac{\partial f}{\partial x} dx \tag{18.7}$$

$$dQ = dP - dE. \tag{18.8}$$

Substituting Eqn. (18.7) into Eqn. (18.8), we obtain:

$$dQ = \left(1 - \frac{\partial f}{\partial P}\right) dP - \frac{\partial f}{\partial E_p} dE_p - \frac{\partial f}{\partial x} dx. \tag{18.9}$$

Given that the changes in watershed properties are negligible (i.e., $dx = 0$), Eqn. (18.9) can be simplified as:

$$dQ = \left(1 - \frac{\partial f}{\partial P}\right) dP - \frac{\partial f}{\partial E_p} dE_p. \tag{18.10}$$

Dividing by Q on both sides of Eqn. (8.10) and rearranging the terms, we obtain:

$$\frac{dQ}{Q} = \varepsilon_{Q,P} \frac{dP}{P} - \varepsilon_{Q,E_p} \frac{dE_p}{E_p}, \tag{18.11}$$

where the elasticity of streamflow to precipitation and potential evapotranspiration are:

$$\varepsilon_P = \left(1 - \frac{\partial f}{\partial P}\right) \frac{P}{P - E}, \quad \varepsilon_{E_p} = \frac{\partial f}{\partial E_p} \frac{E_p}{P - E}. \tag{18.12}$$

To obtain precipitation elasticity and potential evaporation elasticity, partial derivatives expressed in the Budyko equations need to be found. Previous studies substituted the function of potential evapotranspiration into Eqn. (18.12) and found the partial derivatives of potential evapotranspiration with respect to the climatic variables such as air temperature, net solar radiation, relative humidity, and wind speed, to derive the climate elasticity other than precipitation. Recently, this method has been widely used to estimate the climate elasticity of streamflow and attribute the proportional change in streamflow to the proportional changes in climatic variables.

18.2.2 Non-parametric Elasticity

To develop a robust and approximately unbiased estimator of the sensitivity of streamflow to climate which can perform well under different model assumptions, Sankarasubramanian et al. (2001) proposed a non-parametric estimator. This estimator is a statistical description (median over time) of the observed annual elasticity:

$$\varepsilon_P = \text{median}\left[\frac{(Q_i - \bar{Q})/\bar{Q}}{(P_i - \bar{P})/\bar{P}}\right], \tag{18.13}$$

where i is the ith year, $\bar{Q}$ is the mean annual runoff and $\bar{P}$ is the mean annual precipitation.

This estimator is simple and robust for a wide class of hydrologic models. The uncertainty associated with this estimator and other alternative methods (i.e., statistical model and hydrological model) have been evaluated using climate and streamflow data generated by Monte Carlo experiments (Sankarasubramanian et al., 2001). Results indicated that Eqn. (18.13) is reliable for general usage. Several studies have used Eqn. (18.13) to investigate the impacts of climate change on annual streamflow. Sankarasubramanian and Vogel (2003) used it to document the precipitation elasticity of streamflow for 1,337 basins in the United States, showing that a 1 per cent change in precipitation results in a 1.5–2.5 per cent change in basin runoff, depending upon the degree of buffering by storage processes and other factors. Chiew (2006) used it within a hydrological modelling approach to estimate the precipitation elasticity in 219 watersheds across Australia, suggesting that a 1 per cent change in mean annual rainfall results in a 2.0–3.5 per cent change in mean annual streamflow. Tang et al. (2019) used it to examine the assumption that there is the same streamflow sensitivity for both an increase and a decrease in precipitation in 353 watersheds across the United States. Their results suggest that the assumption is valid in humid and sub-humid regions (aridity index less than 1.0) while in arid and semi-arid regions (aridity index greater than 1.0), the sensitivity of runoff to precipitation is higher with a decrease in precipitation compared to that with an increase in precipitation.

18.3 Variability of Streamflow and Precipitation Elasticity across China

In this section, the robust non-parametric estimator of precipitation elasticity, (i.e., Eqn. (18.13)) is employed for estimations in watersheds across China. China is the world's most populous country. Climate change and its impacts on water resources and crop production is a major issue for this country. To develop the economy, China initiated the economic reform and opening-up policy in 1978. The inconsistency between rare economic supply and rising social demands for natural resources (e.g., water resources) in the eastern coastal area of China has gradually emerged under the pressures of rapid industrialization and urbanization. In China, despite the growing importance of industry, agriculture has a central role in ensuring the food security and welfare of 1.3 billion people and put growing pressure on water supply. The spatial distribution of water resources is uneven between the south where water is abundant and the north where water is scarce. Many regions lie in areas where water resources could be affected positively or negatively by changes in climate. Over the past several decades, China has already experienced some devastating climate extremes. For instance, the great flood of 1998 inundated 2,131 km^2 of land and destroyed about 5,000,000 houses in the Yangtze basin, causing an economic loss of over US$20 billion. Additionally, a large-scale drought disaster from November 2009 to March 2010 occurred in Yunnan province, which is one of the most important provinces in western China. The disaster threatened 9,640,000 residents with a shortage of drinking water and about 28,300 km^2 bared farmland, causing an economic loss of over 9.27 per cent of GDP in Yunnan province. It is therefore crucial to better understand the interannual variability of streamflow and its sensitivity to climate change (e.g., changes in precipitation).

In this case study, we will first use historical observation data to quantify interannual variabilities of precipitation and streamflow in investigated watersheds. Then, precipitation elasticity will be estimated to understand the sensitivity of streamflow to changes in precipitation. We chose a total of 164 study watersheds across China that cover large geographical areas and climatic diversity (Figure 18.1). The sizes of the 164 investigated watersheds range from 128 km^2 to 129,361 km^2. The median size is about 5974 km^2. About 10 per cent of watersheds are greater than 43,000 km^2 while about 10 per cent are smaller than 597 km^2. The mean annual precipitations range from 262 to 1712 mm. Daily meteorological data from 1951 to 2010 at national meteorological stations, which include precipitation, air temperature, wind speed at 10 metres and relative humidity were obtained from the National Meteorological Information Centre of the China Meteorological Administration. The daily or monthly streamflow data from 1957 to 2000 at the stream gauge of each watershed were provided by the Ministry of Water Resources of China. This case study focuses on the high-quality and overlapped period of the data sets from 1957 to 2000, which counts a total of 44 years. To ensure a good representation of hydrometeorological data for various hydrologic events, the extent of data records should exceed 25 years. All

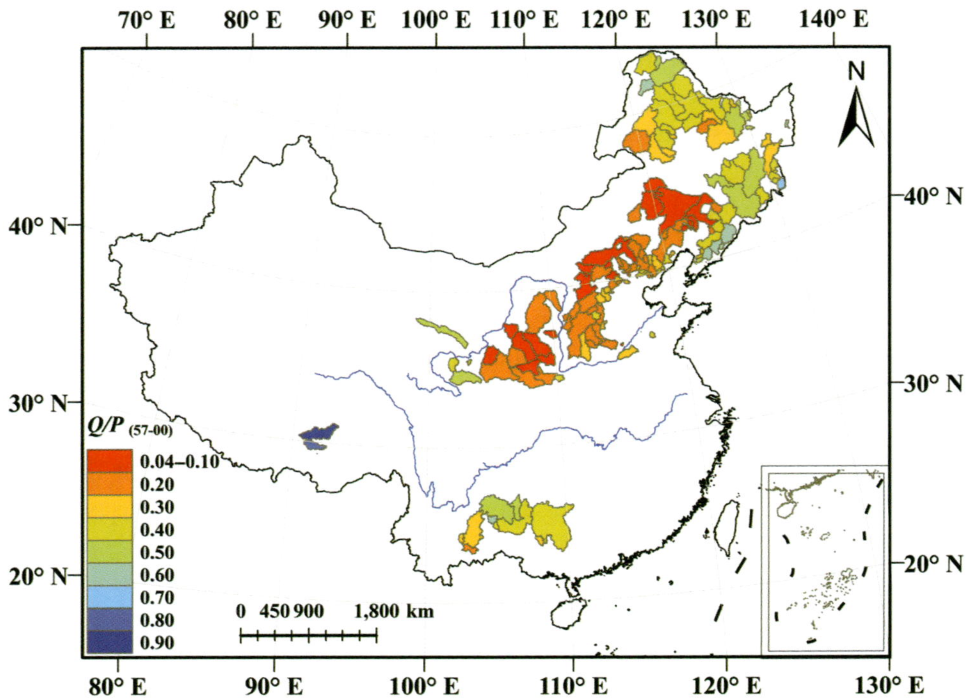

Figure 18.1 Spatial distribution of runoff coefficients during 1957–2000 in 164 study watersheds across China

meteorological gauges have at least 41 years of records. More than 80 percent of the stream gauges have more than 30 years of records, and the average duration of records is about 42 years. In order to detect the influence of anthropogenic activities (e.g., economic development), the analyses will be conducted in three periods of interest including (1) 1957–2000: the entire historic period that covers a total of 44 years; (2) 1957–1978: a relatively low-speed development period covering a total of 22 years and (3) 1979–2000: a relatively high-speed development period, covering a total of 22 years.

The watershed-averaged precipitation was obtained by calculating the arithmetic mean values of records of all gauges at each investigated watershed. The quality of precipitation data is critically important to elasticity estimation. Previously, MOPEX watersheds have established a minimum density requirement for rain gauges based on the basin size. We tried to satisfy this criterion for calculating the watershed-averaged precipitation. However, due to the relatively sparse density of meteorological stations in some watersheds (e.g., western China), the boundaries of these watersheds are extended by creating features within a 20–50 kilometres buffered distance to include additional meteorological observation stations. The original unit (i.e., m^3 s^{-1}) of daily or monthly streamflow is converted to millimetres per day or millimetres per month by dividing the area of each watershed and multiplying the total seconds of a day or a month. The annual precipitation and streamflow for each watershed were obtained by summarizing daily watershed-averaged precipitation and daily or monthly unit converted streamflow, respectively. The linear trends of annual precipitation and streamflow were calculated for each study period, respectively. The statistical significance of each trend was tested by the two-tailed Student's t test. The precipitation elasticity of streamflow for each watershed during each study period was estimated by the non-parametric estimator (i.e., Eqn. (18.13)). Previous studies have shown that the uncertainty associated with the non-parametric estimator increases with the reduction in sample size. Therefore, we computed the significance of the estimated precipitation elasticity with a focus on the two short periods with the 22-year observations.

At first glance, a map of runoff coefficients during 1957–2000 illustrated at the locations of the investigated watersheds across China (Figure 18.1) suggests a relationship between precipitation and streamflow. The runoff coefficient is a dimensionless factor equal to the ratio of mean annual streamflow and precipitation (i.e., Q/P) that is used to convert the rainfall amounts to runoff. This coefficient represents the effect of catchment losses and hence depends upon the nature of land surface, slope and rainfall intensity. It is also affected by the proximity to a water table, degree of soil compaction, porosity of soil, vegetation and depression storage. Some of its plausible values are associated with different land uses. For instance, it takes the value 0.9 for metal or tile roofs and could take values up to 0.3 for green roofs. The runoff coefficients during 1957–2000 in the investigated watersheds across China range from about 0.04 to 0.85, with an average value of 0.27. More than half (about 84) of the watersheds have a value of runoff coefficients larger than 0.27. The coefficient of variation (C_V) of runoff coefficients across the watersheds is 0.57, which indicates a large variability of runoff coefficients. This result brings evidence of spatial heterogeneity in the relationship between precipitation and streamflow. The watersheds in central and mid-eastern China exhibit small values of runoff coefficients while watersheds in northeastern and southwestern China show large values of runoff coefficients. Two watersheds in the lower Yarlung Zangbo river basin show runoff coefficient values larger than 0.72. Both patterns of runoff coefficients during 1957–1978 and 1979–2000 are very similar with the spatial distribution during 1957–2000 (i.e., Figure 18.1). However, we observe differences in the runoff coefficients between the three investigated periods. The runoff coefficients during 1957–1978 range from about 0.03 to 0.94 with an average value of 0.30, while the runoff coefficients during 1979–2000 range from about 0.04 to 0.83 with an average value of 0.25. As shown in Figure 18.2, the values of runoff coefficients in the investigated

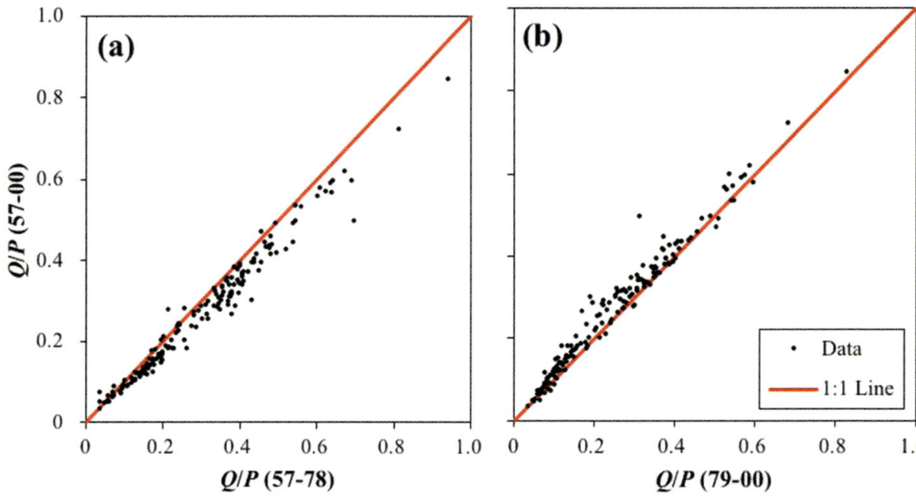

Figure 18.2 (a) Runoff coefficient values in 1957–2000 (y-axis) and 1957–1978 (x-axis); (b) Runoff coefficient values in 1957–2000 (y-axis) and 1979–2000 (x-axis) in the investigated watersheds across China

watersheds during 1957–1978 are generally larger than that during the whole period (Figure 18.2(a)) which contrasts with the observed runoff coefficient values during 1979–2000 (Figure 18.2(b)). The values of runoff coefficients in the investigated watersheds have the largest, smallest and middle-range values observed in 1957–1978, 1979–2000 and 1957–2000, respectively.

18.3.1 Temporal and Spatial Variations in Precipitation and Streamflow

For each variable, we consider a range of 44 years, 22 years and 22 years, which corresponds to their respective study period. Due to the data availability, about 26 per cent of watersheds have 44 years of data, about 91 per cent of studied watersheds have more than 40 years of data and about 98 pe cent of study watersheds have more than 30 years of data for 1957–2000.

Figure 18.3 shows the spatial distributions of trends in annual precipitation and streamflow during 1957–2000. Trends of annual precipitation during 1957–2000 at watersheds located in northeastern and southwestern China are generally positive, as illustrated by the green triangles in Figure 18.3(a), which suggests an increase of the annual precipitation over time throughout the study period. Twenty-three out of 92 trends are positive and statistically significant. The increasing rates range from 0 to 7 mm yr^{-1}, the average rate is about 1.7 mm yr^{-1}, the median is about 1.3 mm yr^{-1} and the *Cv* is about 0.79, indicating a relatively large spatial variability across study watersheds. Watersheds in eastern coastal region and the

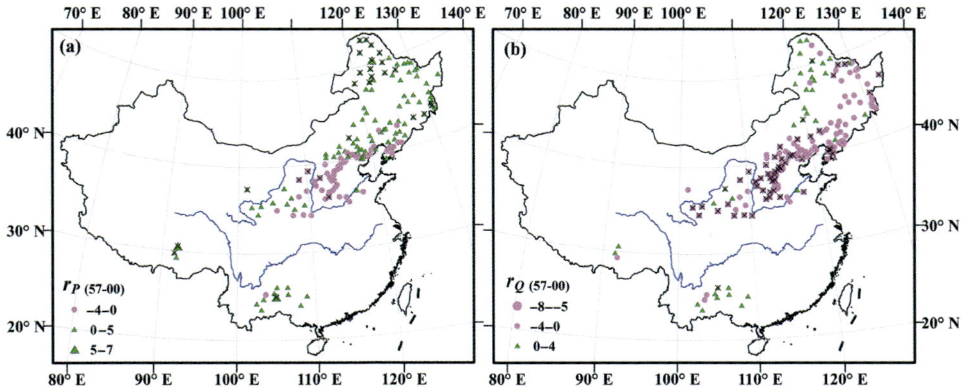

Figure 18.3 Spatial distributions of the trends of annual precipitation (a) and streamflow (b) during 1957–2000 in the investigated watersheds across China. Upper pointing green triangles and purple dots represent positive and negative trends, respectively. The cross indicates a significant trend with a p values smaller than 0.05

Loess Plateau tend to exhibit negative treads, which suggests a reduction in the annual precipitation throughout the investigated period, as illustrated by purple dots in Figure 18.3(a). Only 5 out of 72 negative trends are statistically significant. The decreasing rates range from –4 to 0 mm yr^{-1}, the average rate is about –1.2 mm yr^{-1}, the median is about –1.1 mm yr^{-1} and the Cv is about 0.77. Trends of annual streamflow during 1957–2000 in a few watersheds located in northeastern and southwestern China are generally positive, as illustrated by the green triangles in Figure 18.3(b). Only 4 out of 38 positive trends are statistically significant. The increasing rates range from 0 to 4 mm yr^{-1}, the average value is about 0.9 mm yr^{-1}, the median is about 0.5 mm yr^{-1} and the Cv is about 1.11, which indicates a large spatial heterogeneity in the annual streamflow. Watersheds in the eastern region and in the Loess Plateau tend to exhibit negative trends throughout the study period, as illustrated by the purple dots in Figure 18.3(b). A total of 49 out of 126 negative trends are statistically significant. The decreasing rates range from –8 to 0 mm yr^{-1}, the average rate is about –1.6 mm yr^{-1}, the median is about –1.1 mm yr^{-1} and the Cv is about 0.93. Figure 18.3(a) and (b) shows that the spatial patterns of precipitation and streamflow differ in northeastern China but are similar in central and southern China. The number of watersheds with negative streamflow trends is much larger than that for precipitation (Figure 18.3(b)). Additionally, the absolute decreasing rate of annual streamflow seems larger than that of annual precipitation in the same watersheds in central China where the patterns of streamflow and precipitation look similar.

Figure 18.4 shows the spatial distribution of trends of annual precipitation and streamflow during 1957–1978. In this period, trends of annual precipitation are

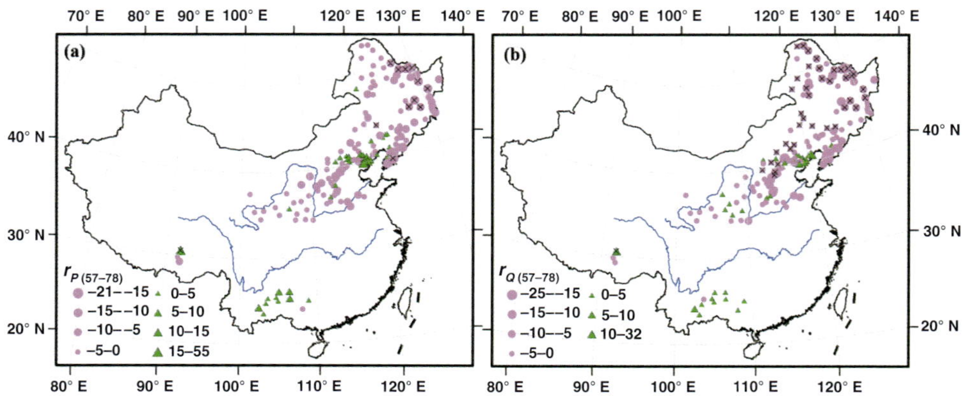

Figure 18.4 Spatial distributions of the trends of annual precipitation (a) and streamflow (b) during 1957–1978 in the investigated watersheds across China. Upper pointing green triangles and purple dots represent positive and negative trends, respectively. The cross indicates a significant trend with a p values smaller than 0.05

negative for most watersheds (Figure 18.4(a)). A total of 13 out of 125 negative trends are statistically significant. The decreasing rates range from –21 to 0 mm yr^{-1} and the average rate is about –4.7 mm yr^{-1}, the median is about –4.1 mm yr^{-1} and the Cv is about 0.75. Watersheds in east-central coastal region and southern China tend to exhibit positive treads of the annual precipitation during 1957–1978, as illustrated by the green triangles in Figure 18.4(a). Only 1 out of 39 positive trends is statistically significant. The rates range from 0 to 55 mm yr^{-1}, the average rate is about 5.4 mm yr^{-1}, the median is about 3.2 mm yr^{-1} and the C_v is about 1.63, which indicates a very large spatial heterogeneity of the trends in the annual precipitation. Similarly, trends of annual streamflow are negative in most watersheds during 1957–1978, as illustrated in Figure 18.4(b). A total of 39 out of 128 negative trends are statistically significant. The decreasing rates range from –25 to 0 mm yr^{-1}, the average rate is about –4.4 mm yr^{-1}, the median is about –3.8 mm yr^{-1} and the Cv is about 0.87. Watersheds in the east-central coastal region, the central Loess Plateau and southern China also exhibit positive trends, as shown in Figure 18.4(b). The increasing rates range from 0 to 32 mm yr^{-1}, the average rate is about 3.7 mm yr^{-1}, the median is about 2.8 mm yr^{-1} and the C_v is about 1.48.

The spatial distribution of trends during 1979–2000 are displayed in Figure 18.5. Trends of annual precipitation during 1979–2000 in watersheds located in central and southwestern China are generally positive, as illustrated by the green triangles in Figure 18.5(a). Five out of 86 positive trends are statistically significant. The increasing rates range from 0 to 13 mm yr^{-1}, the average rate is about 3.2 mm yr^{-1}, the median is about 2.2 mm yr^{-1} and the Cv is about 0.87. Watersheds in northeastern China and in the Loess Plateau tend to exhibit negative treads of

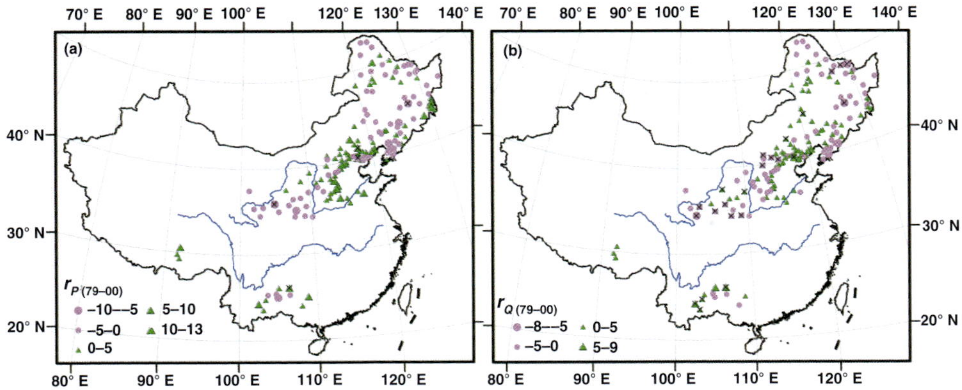

Figure 18.5 Spatial distributions of the trends of annual precipitation (a) and streamflow (b) during 1979–2000 in the investigated watersheds in China. Upper pointing green triangles and purple dots represent positive and negative trends, respectively. The cross indicates a significant trend with a p values smaller than 0.05.

annual precipitation during 1979–2000, as illustrated by the purple dots in Figure 18.5(a). A total of 2 out of 78 negative trends are statistically significant. The decreasing rates range from –10 to 0 mm yr^{-1}, the average rate is about –2.3 mm yr^{-1}, the median is about –2.4 mm yr^{-1} and the Cv is about 0.74. Trends of annual streamflow during 1979–2000 in watersheds located in parts of northeastern and southwestern China are generally positive, as illustrated by the green triangles in Figure 18.5(b). A total of 10 out of 84 positive trends are statistically significant. The increasing rates range from 0 to 9 mm yr^{-1}, the average rate is about 2.1 mm yr^{-1}, the median is about 1.6 mm yr^{-1} and the Cv is about 1.03. Watersheds in the eastern region and in the Loess Plateau tend to exhibit negative treads of annual streamflow during 1979–2000, as illustrated by purple dots in Figure 18.5(b). A total of 14 out of 81 negative trends are statistically significant. The decreasing rates range from –8 to 0 mm yr^{-1}, the average rate is about –1.9 mm yr^{-1}, the median is about –1.2 mm yr^{-1} and the Cv is about 0.98. Figure 18.5(a) and (b) shows similar patterns of precipitation and streamflow in this period, however, the trends of precipitation and streamflow are opposite in some watersheds located in central and eastern China.

Changes in streamflow during 1957–1978 – this period corresponds to a relatively slow development period – are generally in line with the changes in precipitation in most watersheds in China, as illustrated in Figure 18.4. Moreover, this matching holds for the trends observed in most watersheds during 1979–2000, as illustrated in Figure 18.5. These results indicate that precipitation change induced by climate change is a dominant factor of the annual streamflow changes. This conclusion is in line with Liu et al. (2019). However, the out-phase changes in streamflow in some watersheds in central and eastern China during 1979–2000 – this period corresponds to a relatively fast development period – may also imply that the anthropogenic activities also play a role in the changes in streamflow. After the opening-up policy conducted since the late 1970s, the eastern coastal region has experienced metaphase industrialization and rapid urbanization and the urban–rural development has recently entered a new period of transformation. In this process, losses of water resources are aggravated and conflicts between large populations for scarce water resources will strengthen as the needs for water for construction continue to increase. In this scenario, local water resources will be used for feeding a huge population, guaranteeing a growing urban and rural development and supporting ecological rehabilitation. Anthropogenic activities aimed at addressing these issues, such as reservoir construction and operation, and land use change will eventually affect the watershed hydrologic cycle.

18.3.2 Precipitation Elasticity of Streamflow

Figure 18.6 shows the precipitation elasticity (ε_{57-00}) using data from 1957 to 2000 in 164 investigated watersheds across China. The values of ε_{57-00} range from

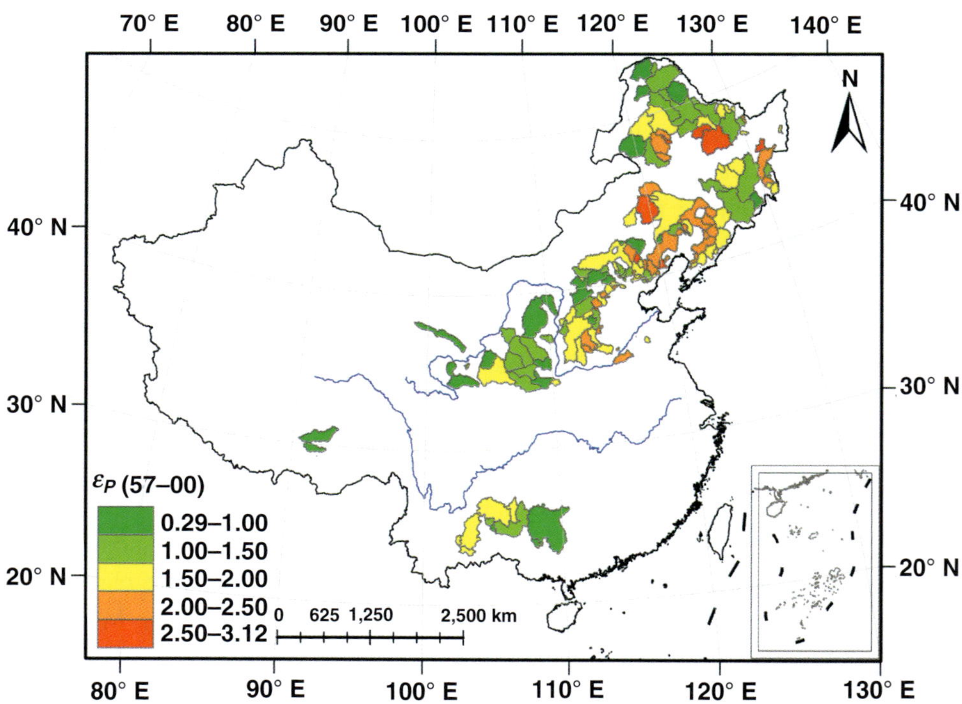

Figure 18.6 Spatial distribution of precipitation elasticity during 1957–2000 in the investigated watersheds across China

0.29 to 3.12, with values at about 75 per cent of the watersheds (about 124) falling between 1.0 and 2.4 (the 18th to 92th percentile range). The average value of ε_{57-00} is 1.63, while the median value is 1.65. The *Cv* is 0.35, indicating a relatively small variability across the watersheds. A significance test was conducted to estimate precipitation elasticity. The results show that all *p* values associated with the estimated elasticity are smaller than 0.05. Values of ε_{57-00} greater than 1.0 indicate that a 1 per cent change in the mean annual precipitation will result in more than 1 per cent change in the mean annual streamflow, depending on the degree of buffering by storage processes and other factors. The values of ε_{57-00} for watersheds in the northeastern coastal region are generally larger than that for watersheds in the central and southern part of China. In watersheds with a large runoff coefficient (i.e., Q/P larger than the average value of 0.27 during 1957–2000), the values of ε_{57-00} range from 0.29 to 2.74 (the smallest to the 99th percentile value), the average value is 1.59, the median is 1.65 and the *Cv* is about 0.35. At watersheds with a small runoff coefficient (i.e., $Q/P < 0.27$), the values of ε_{57-00} range from 0.29 to 3.12 (the smallest value to the largest value), the average value is 1.68, the median is 1.66 and the *Cv* is about 0.36. The precipitation elasticity during 1957–2000 in watersheds with a small runoff coefficient slightly larger than that in watersheds with a large runoff coefficient.

Figure 18.7 shows the precipitation elasticity (ε_{57-78}) using data from 1957 to 1978 in the investigated watersheds. The values of ε_{57-78} range from 0.29 to 3.58, with values at about 71 per cent of the watersheds (about 117) falling between 1.0 and 2.4 (the 17th to 87th percentile range). The average value of ε_{57-78} is 1.66, the median is 1.57 and the *Cv* is about 0.41. Most watersheds exhibit significant precipitation elasticity (*p* values smaller than 0.05) during 1957–1978 and only 11 out of 164 (about 7 per cent) watersheds show no significant precipitation elasticity (*p* values larger than 0.05). Compared to ε_{57-00}, the value of ε_{57-78} is relatively large in watersheds in northeastern and central China, while the value of ε_{57-78} is relatively small in watersheds in northern China (see Figures 18.6 and 18.7). In those watersheds with large runoff coefficients (i.e., Q/P larger than the average value of 0.30 during 1957–1978), the values of ε_{57-78} are varied from 0.29 to 3.31 (the smallest to the 98th percentile value). The average value is 1.63, the median value is 1.62, and the *Cv* is about 0.36. In watersheds with a small runoff coefficient (i.e., $Q/P < 0.30$), the values of ε_{57-78} range from 0.44 to 3.58 (the 3rd percentile value to the largest value). The average value is 1.68, the median value is 1.48 and the *Cv* is about 0.48. Similarly, the precipitation elasticity during 1957–1978 in watersheds with a small runoff coefficient is also larger than that in watersheds with a large runoff coefficient.

Figure 18.8 shows the precipitation elasticity (ε_{79-00}) using data from 1979 to 2000. The values of ε_{79-00} range from 0.05 to 3.75, with values in about 65 per

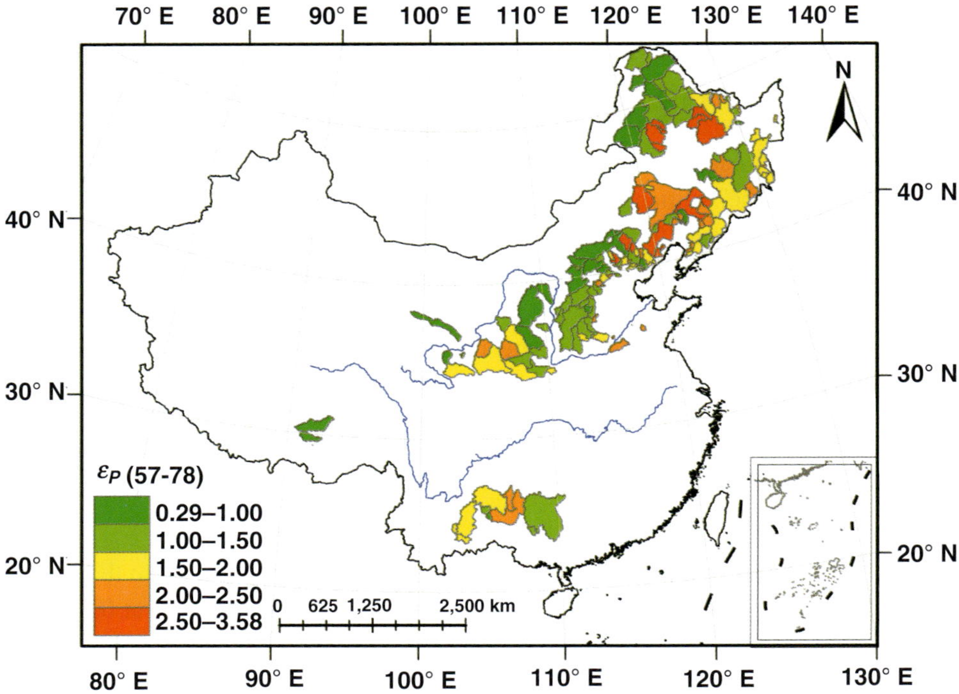

Figure 18.7 Spatial distribution of precipitation elasticity during 1957–1978 in the investigated watersheds across China

cent of the watersheds (about 106) falling between 1.0 and 2.4 (the 17th to 82th percentile range). The average value of ε_{79-00} across study watersheds is 1.76, the median value is 1.80 and the *Cv* is about 0.41. During 1979–2000, most watersheds show *p* values associated with the estimated elasticity smaller than 0.05 and only 8 out of 164 (about 5 per cent) watersheds exhibit *p* values larger than 0.05. Compared to ε_{57-00} (see Figure 18.6), the value of ε_{79-00} is generally larger for most watersheds. Compared to ε_{57-78} (see Figure 18.7), the value of ε_{79-00} is relatively large in watersheds in the northern, eastern coastal, central and western regions of China, while it is relatively small in watersheds in the southern part of China and in the Loess Plateau. In watersheds with a large runoff coefficient (i.e., *Q*/*P* larger than the average value of 0.25 during 1979–2000), the values of ε_{79-00} range from 0.43 to 3.07 (the 3rd to the 98th percentile value), the average value is 1.74, the median value is 1.80 and the *Cv* is about 0.38. In watersheds with a small runoff coefficient (i.e., $Q/P < 0.25$), the values of ε_{79-00} range from 0.05 to 3.75 (the smallest value to the largest value), the average value of ε_{79-00} is 1.78, the median is 1.84 and the *Cv* is about 0.44. Similarly, the precipitation elasticity during 1979–2000 at watersheds with a small runoff coefficient is also larger than that in watersheds with a large runoff coefficient.

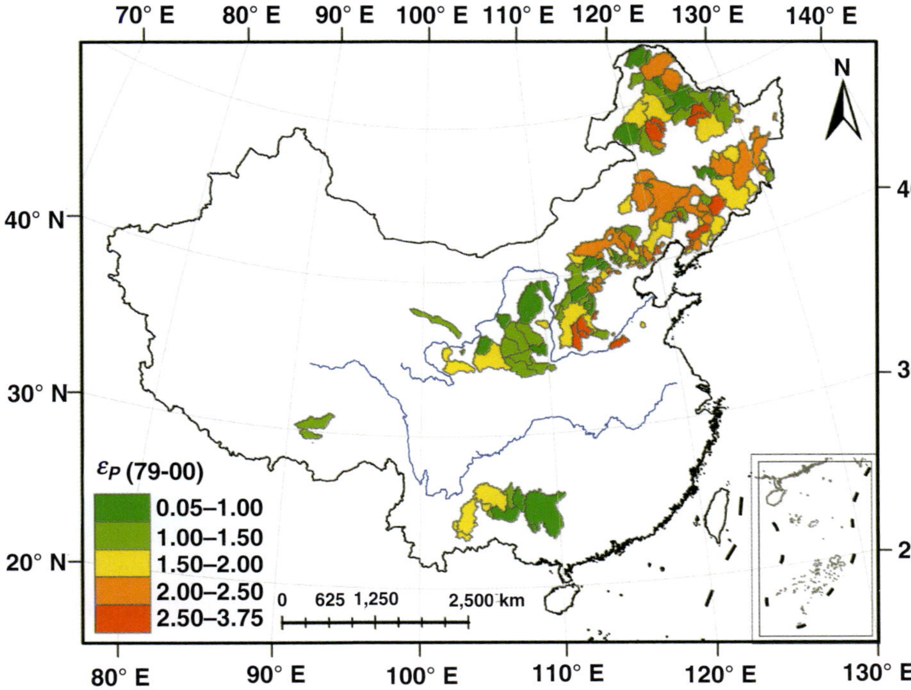

Figure 18.8 Spatial distribution of precipitation elasticity during 1979–2000 in investigated watersheds across China

18.4 Summary and Conclusion

The precipitation elasticity is an efficient index to quantify the sensitivity of streamflow to precipitation changes in watersheds. It provides us with a tool to understand the historical and predicted changes in streamflow, which is key to assess the impact of climate change on water resources. In this chapter, we have provided the details of three frequently used methods for estimating precipitation elasticity of streamflow and a case study on the temporal and spatial variability of runoff coefficient and precipitation elasticity estimated by a non-parametric estimator in three time periods in a sample of watersheds in China. Results suggest that the spatiotemporal patterns of streamflow are generally in line with those of precipitation, which indicates that climatic change is a dominant factor of annual streamflow changes. This conclusion agrees with recent findings (e.g., Liu et al., 2019). However, the intensification of anthropogenic activity in a context of economic development may also play an important role in the changes in streamflow. In most investigated watersheds, the precipitation elasticity of streamflow was larger than 1, indicating that a 1 per cent change in annual precipitation results in a more than 1 per cent change in the annual streamflow. The precipitation elasticity in watersheds with a small runoff coefficient was generally larger than that in watersheds with a large runoff coefficient.

Additionally, non-parametric estimators used to estimate precipitation elasticity have shortcomings. In this framework, the sensitivity of streamflow is only a function of precipitation, but does not account for the role of evapotranspiration, which is mainly associated with temperatures, humidity, wind speed and radiation. This may limit the applicability of this method when climatic variables other than precipitation are accounted for to assess their role on streamflow under future global warming scenarios. Limbrunner (1998) and Chiew (2006) estimated separately the precipitation elasticity of streamflow and the potential evaporation elasticity of streamflow in the United States and Australia, respectively. However, Fu et al. (2007) indicate these elasticities should not be considered separately, but rather as a joint precipitation and evapotranspiration (temperature) elasticity of streamflow. Arora (2002) introduced the use of the aridity index – a function of both precipitation and potential evapotranspiration – to assess climate change effects on annual streamflow. Notably, Sankarasubramanian and Vogel (2003) reported that the aridity index is not an adequate predictor, by itself, of either the long-term water balance or the variability of annual streamflow.

References

Arora, V. K. (2002). The use of the aridity index to assess climate change effect on annual runoff. *Journal of Hydrology* 265(1–4): 164–177.

Chiew, F. H. S. (2006). Estimation of rainfall elasticity of streamflow in Australia. *Hydrological Sciences Journal* 51(4): 613–625.

Choudhury, B. J. (1999). Evaluation of an empirical equation for annual evaporation using field observations and results from a biophysical model. *Journal of Hydrology* 216 (1–2): 99–110.

Fu, B. P. (1981). On the calculation of the evaporation from land surface. *Scientia Atmospherica Sinica* 5(1): 23–31 (in Chinese).

Fu, G., Charles, S. P., & Chiew, F. H. S. (2007). A two-parameter climate elasticity of streamflow index to assess climate change effects on annual streamflow. *Water Resources Research* 43(11): W11419.

Limbrunner, J. F. (1998). *Climatic elasticity of streamflow in the United States*, MS thesis, 39 pp., Tufts University, Medford, MA.

Liu, X., Liu, W., Yang, H., et al. (2019). Multimodel assessments of human and climate impacts on mean annual streamflow in China. *Hydrology and Earth System Sciences* 23(3): 1245–1261.

Mezentsev, V. S. (1955). More on the calculation of average total evaporation. *Meteorologija i gidrologija* 5: 24–26.

Sankarasubramanian, A., & Vogel, R. M. (2003). Hydroclimatology of the continental United States. *Geophysical Research Letters* 30(7): 1363.

Sankarasubramanian, A., Vogel, R. M., & Limbrunner, J. F. (2001). Climate elasticity of streamflow in the United States. *Water Resources Research* 37(6): 1771–1781.

Schaake, J. C. (1990), From climate to flow. In P. E. Waggoner (ed.), *Climate Change and U.S. Water Resources* (pp. 177–206). New York: Wiley.

Tang, Y., Tang, Q., Wang, Z., et al. (2019). Different precipitation elasticity of runoff for precipitation increase and decrease at watershed scale. *Journal of Geophysical Research: Atmospheres* 124(22): 11932–11943.

Wang, D., & Tang, Y. (2014), A one-parameter Budyko model for water balance captures emergent behavior in Darwinian hydrologic models, *Geophysical Research Letters* 41(13): 4569–4577.

Yang, H., Yang, D., Lei, Z., & Sun, F. (2008), New analytical derivation of the mean annual water-energy balance equation, *Water Resources Research* 44(3): W03410.

Zhang, L., Dawes, W. R., & Walker, G. R. (2001). Response of mean annual evapotranspiration to vegetation changes at catchment scale. *Water Resources Research* 37(3): 701–708.

Zhang, L., Hickel, K., Dawes, W. R., Chiew, F. H. S., Western, A. W., & Briggs, P. R. (2004). A rational function approach for estimating mean annual evapotranspiration. *Water Resources Research* 40(2): W02502.

19

Contribution of Hydrological Model Calibration Uncertainty to Future Hydrological Projections over Various Temporal Scales

A Case Study in the Boulder Creek Watershed

QINGHUAN ZHANG, QIUHONG TANG, GUOYONG LENG
AND SEYED-MOHAMMAD HOSSEINI-MOGHARI

19.1 Introduction

Streamflow variability in the mountains is influenced by the frequency, magnitude and type of precipitation, land cover characteristics, snow accumulation and ablation, as well as the timing and rate of snowmelt (Harding et al., 2012; Musselman et al., 2017; Zhang et al., 2018). Future climate change is predicted to impact all these factors, thus affecting regional water availability for municipal, industrial, mining and irrigation consumers, and hydropower generation (Barnett et al., 2005; Wagener et al., 2010; Liu et al., 2016; Yin et al., 2017). Furthermore, hydrological impacts due to warming are accentuated at high elevations because water yields are larger and the processes governing snow accumulation and melt are sensitive to changes in air temperature (Bales et al., 2006; Gao et al., 2010). Notwithstanding, given accurate information, land and water managers can develop targeted watershed management plans to adapt to and potentially mitigate the effects of climate change on streamflow and ecosystem function (Poff et al., 1997; Watts et al., 2016).

Downscaled climate data from General Circulation Models (GCMs) are frequently used as input data in hydrologic models to predict how changes in climate would affect regional and global water balances (Wilby et al., 2009; Liu et al., 2017). In particular, the magnitude of projected climate perturbation represents a major source of uncertainty associated with hydrologic variables obtained from hydrologic models (Raje & Krishnan, 2012; Schewe et al., 2014; Mendoza et al., 2015; Guo et al., 2018). In addition, different GCMs, emission scenarios and time periods over which the climate perturbation was obtained can affect the magnitude and direction of hydrologic projections (Kay et al., 2009; Addor et al., 2014). Previous studies have considered hydrological models in the

Funding for this research is provided by the International Partnership Program of the Chinese Academy of Sciences (131A11KYSB20170113) and the National Natural Science Foundation of China (41730645).

context of a "cascade of uncertainty" to compare the contribution of various elements of the model chain to total uncertainty. This approach provides valuable insights into the development of hydrological modelling in changing conditions (Addor et al., 2014), but few studies have focused specifically on the role of parametric uncertainty in the context of future climate change projections.

The effect of hydrologic modelling decisions, including model selection, model structure and choice of parameter sets, on future hydrological projections is currently an open question (e.g. Addor et al., 2014; Mendoza et al., 2016; Seiller et al., 2017; Yuan et al., 2017; Joseph et al., 2018). Recent works have suggested that calibration of hydrological models is fundamentally important for projecting important hydrological variables, including changes in water quality, soil moisture and groundwater (Her et al., 2019). However, it is unclear whether the impact of model calibration resultant parameter set selection alone is large enough to impact the direction and magnitude of projected hydrological changes across different temporal scales.

This work focuses on the contribution of model calibration resultant parameter uncertainty to future hydrological projections at different time scales (annual, monthly and daily) and for different hydrological variables. Specifically, we applied the distributed, physically based Variable Infiltration Capacity (VIC) hydrologic model to a regionally important (Boulder Creek) watershed that provides drinking water, agricultural irrigation, aquatic habitat and recreation (Murphy et al., 2003). The principal objectives of the study were to: (1) quantify the relative proportions of uncertainty resulting from calibrated model parameters to projections of hydrologic flux and state variables and (2) assess the temporal scale-dependency of this uncertainty within the context of hydrological change in the Boulder Creek Watershed. As human activity is increasing and human-induced climate warming is occurring, appropriate management strategies must be based on reliable predictions of freshwater distribution and changes under a perturbed climate, as provided by this work.

19.2 Study Area

The Boulder Creek Watershed at Orodell is located at 40.01° N and 105.33° W in Colorado, USA (USGS stream gauge 6727000), and is generally representative of mesoscale catchments in the southern Rocky Mountains (Figure 19.1). The watershed area is 264 km^2 and the elevation ranges from 1,779 to 4,117 metres, with a mean elevation of 3,139 metres. The mean annual precipitation is 840 mm. In general, low-flow occurs from October to March, while high-flow occurs from May to July and peaks in June, depending on snowpack depth and air temperature (Murphy et al., 2003). Soil water is maintained through the sustained input of

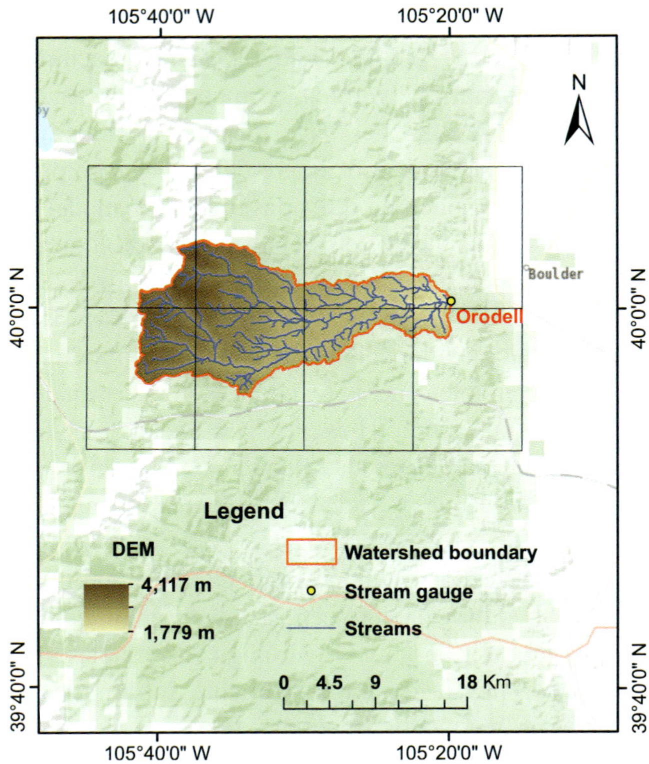

Figure 19.1 The geographical location of the study area along with the grid cells in the VIC model corresponding to the Boulder Creek Watershed

snowmelt water during the spring, which promotes significant runoff during wet periods (Hamlet & Lettenmaier, 2007). Otherwise, streamflow is relatively independent of soil moisture conditions due to thin soil cover (Rauscher et al., 2008).

19.3 Data and Methods

19.3.1 The Variable Infiltration Capacity (VIC) Model

19.3.1.1 Model Mechanism

The VIC model is a physically-based, distributed, macro-scale hydrologic model that is capable of simulating the water balance from distributed grid cells through calculations of runoff, baseflow, evapotranspiration (ET), snow water equivalent and other variables (Liang et al., 1994). It has been applied to assess the impact of climate change in many river basins throughout the western United States (Harding

Table 19.1 Summary of VIC parameters selected for calibration at Boulder Creek

			Calibration range	
Parameter	Description	Units	Min	Max
b	Variable infiltration curve parameter	Fraction	0.001	0.8
Ds	Fraction of Dsmax where non-linear baseflow begins	Fraction	0.001	1
Ds_max	Maximum velocity of baseflow	mm day^{-1}	1	20
Ws	Fraction of maximum soil moisture where non-linear baseflow occurs	Fraction	0.001	1
c	Exponent used in baseflow curve	–	1	4
Depth 1	Thickness of soil layer 1	m	0.01	0.5
Depth 2	Thickness of soil layer 2	m	0.1	2
Depth 3	Thickness of soil layer 3	m	0.1	2.9

et al., 2012). Each grid cell is modelled as a flat surface and sub-grid heterogeneity is handled via statistical distributions. No horizontal routing of surface overland flow, subsurface flow or channel flow is performed (Mendoza et al., 2015), although streamflow routing can be post-processed through a routing model (Lohmann et al., 1996). At each time step, surface runoff is defined as precipitation that exceeds the storage capacity of the soil at the previous time step. Baseflow is defined as a function of the soil moisture in the third soil layer (Arno formulation) from the surface, which is linear below a threshold and non-linear above that threshold (Gao et al., 2010). Total runoff is the sum of surface runoff and baseflow.

19.3.1.2 Historical Data Input and Streamflow

The VIC model simulations were driven by inputs of daily gridded precipitation, maximum and minimum air temperature and wind speed and were calibrated against streamflow observations between water years (1 October–30 September) 1981 to 1990. In total, eight parameters were calibrated at 1/8° spatial resolution (Table 19.1). Each parameter set was applied over the entire basin. The parameter ranges were selected following Demaria et al. (2007) for the contiguous United States. It is assumed that these parameters are constant over time and representative of the inherent properties of this watershed. The climate forcing data were obtained from Maurer et al. (2002).[1]

Naturalized daily streamflow data from 1906 to 2011 were combined from the US Geological Survey at station number 6727000 and the Colorado Division Support System at gauge BOCOROCO. Both of the streamflow gauges are located at Orodell. Naturalized flow represents a flow that would have occurred at the stream gauge had historical depletions and reservoir regulation not been present (Harding et al., 2012). Simulated streamflow from 1981 to 2010 was applied as the baseline level for subsequent analyses. Since both model parameters and climate forcing inputs could influence hydrologic simulations (Ren et al., 2016), a time series including wet and dry climatic conditions is necessary to calibrate the models.

19.3.1.3 Land Cover and Soil Parameters

Land cover types were obtained from the US Geological Survey National Gap Analysis Program and were reclassified into 11 classes according to the 1-km Global Land Cover Classification at the University of Maryland (Hansen et al., 2000). Leaf area index (LAI), in particular, has been shown to influence interception, evaporation and runoff generation (Chen et al., 2005; Dietz et al., 2006). In our study, the annual LAI was held constant over time but was allowed to vary on a monthly basis. For each vegetation type, monthly LAIs were applied from the VIC vegetation library file.

Soil parameters including saturated hydrologic conductivity, mineral type, fractional soil moisture at the critical point and at the wilting point and soil bulk density were compiled from the STATSGO database produced by the United States Department of Agriculture (Soil Survey Staff, 2015). In STATSGO, Map Unit Delineation consists of one or more closed polygons that are generally geographic mixtures of groups of soils or soils and non-soil areas. Each map unit can have as many as 21 components. A component is a phase of a soil series (which is the lowest category of the national soil classification system) representing the most homogenous classes in the taxonomy system. Each component can have multiple soil layers and the size of each component is provided by the areal percentage of the map unit (Huang, 2005). The dominant soil class in each grid cell was applied to calculate soil parameters for that grid cell.

19.3.1.4 Model Calibration and Performance Evaluation

Model parameters were calibrated by comparing observed streamflow with simulated streamflow from 1981 to 1990; the two years prior to 1981 was used for model spin-up (Shi et al., 2008). The period 1991 to 2010 was used for model validation. The use of different length calibration and validation time periods follows the approach of Bastola et al. (2011). We calibrated the model by comparing simulated and observed streamflow data using the Borg Multi-Objective Evolutionary Algorithm (MOEA), which is an evolutionary search algorithm for multi-objective optimization problems (Hadka & Reed, 2013). The Borg MOEA uses the epsilon non-dominance operator, which has the advantages of convergence and diversity with respect to approximating the true Pareto-optimal front over other MOEAs. The epsilons represent the resolutions of the objective functions. Specifically, the epsilon-box dominance archive divides the objective space into hyper-boxes with side-length epsilon, so called epsilon-boxes (Hadka & Reed, 2013, 2015). Parameter sets with similar performance but located in different regions of the parameter space will be chosen, that is, the best parameter sets from Borg represent epsilon-box non-dominated solutions that are sampled from the full Pareto-optimal front such that the results are expected to be robust.

We selected parameter sets with similar performance but located in different regions of the parameter space following the approaches used by Moriasi et al. (2007) and Demaria et al. (2007). Performance measures were selected for the following reasons: Nash-Sutcliffe efficiency (NSE) is the most commonly used metric in hydrologic modelling and quantifies the overall performance, putting an emphasis on the seasonal cycle. In contrast, root mean square error (RMSE) measures the error in the squared units of the simulated versus observed datasets and emphasizes high flows and outliers. Per cent bias (PBIAS) measures the average of the simulations compared to the observed datasets, evaluating the quality of the overall water balance. The coefficient of determination (R^2) measures the proportion of shared variance between observed and simulated datasets. The ratio of the standard deviation of simulations to observations (RSD) exclusively evaluates the variability of the model relative to the observation and thus represents a valuable quality control (Table 19.2). The use of multiple objective functions to calibrate model parameters can help to partition total runoff between direct runoff and baseflow (Liu et al., 2018) and to overcome the problem of lacking detail in conceptual models (Gupta et al., 2008).

19.3.2 Global Climate Models and Future Scenarios

The selection of a particular climate model simulation affects water yield predictions under climate change (Stone et al., 2003) and may give biased results. Hence, the use of multiple climate models constrains the possible range of hydrologic change. The Coupled Model Intercomparison Project Phase 5 (CMIP5) dataset compares multiple global climate models under four future emission scenarios. The four emission scenarios, also called representative concentration pathways (RCPs), are based on concentration pathways that approximate various level of radiative forcing (W m^{-2}) at the end of the twenty-first century (Meinshausen et al., 2011). Specifically, the greenhouse gas concentration trajectories RCP2.6 (Vuuren et al., 2011), 4.5 (Thomson et al, 2011), 6 and 8.5 (Riahi et al, 2011) are based on radiative forcing values of 2.6, 4.5, 6 and 8.5 W m^{-2}, respectively, in the year 2100. In particular, the RCP8.5 scenario assumes high

Table 19.2. Evaluating functions of streamflow simulations at Boulder Creek

Metric	Formula	Perfect value
Nash-Sutcliffe efficiency	$\text{NSE} = 1 - \frac{\sum (Q_{ob,i} - Q_{si,i})^2}{\sum (Q_{ob,i} - \overline{Q}_{ob})^2}$	1
Per cent bias	$\text{PBIAS} = \left\lvert \frac{\sum (Q_{ob,i} - Q_{si,i})}{\sum Q_{ob,i}} \right\rvert \times 100$	0
Root mean square error	$\text{RMSE} = \sqrt{\frac{\sum_{i=1}^{n} (Q_{ob} - Q_{si})^2}{n}}$	0
Coefficient of determination	$R^2 = \frac{\sum \left((Q_{ob,i} - \overline{Q}_{ob})(Q_{si,i} - \overline{Q}_{si}) \right)}{\sqrt{\sum (Q_{ob,i} - \overline{Q}_{ob})^2 \sum (Q_{si,i} - \overline{Q}_{si})^2}}$	1
Ratio of standard deviations	$\text{RSD} = \frac{\sqrt{\frac{1}{n} \sum (Q_{si,i} - \overline{Q}_{si})^2}}{\sqrt{\frac{1}{n} \sum (Q_{ob,i} - \overline{Q}_{ob})^2}}$	1

Note: $Q_{ob,i}$, observed streamflow at time step i (mm/time scale); $Q_{si,i}$, simulated streamflow at time step i (mm/time scale); $\overline{Q}_{ob}$, mean observed streamflow (mm/time scale); $\overline{Q}_{si}$, mean simulated streamflow (mm/time scale). Time scale is chosen as daily in the context of this study.

population growth and high-energy demand without implementation of policy to address climate change. In this study, eighteen CMIP5 GCMs were applied in RCP8.5 (Table 19.3) for the purpose of simulating hydrologic responses to future climate projections using the VIC model. Hydrologic predictions (including streamflow, evapotranspiration as ET and soil moisture) under future climate change were compared with baseline simulations in the historical period. The climate datasets were statistically downscaled using the Bias Corrected Spatial Downscaling (BCSD) method and were obtained from the US Bureau of Reclamation (Reclamation, 2013).

19.3.3 Climate Data

We applied the existing downscaled GCM data, namely the bias-correction and spatial disaggregation (BCSD) climate data, as input forcing for future hydrologic projections. The BCSD climate data from USBR used a quantile mapping approach on location-specific datasets between gridded observations (Maurer et al., 2002) and the GCM historical data. For each grid cell and the corresponding

Table 19.3. General circulation models used in this study

Model ID	Name	Institution
CCSM4	CCSM4	National Center for Atmospheric Research, USA
CNRM	CNRM	National Centre for Meteorological Research, France
CSIRO	CSIRO	Commonwealth Scientific and Industrial Research Organization, Queensland Climate Change Centre of Excellence, Australia
CanESM2	CESM2	Canadian Centre for Climate Modeling and Analysis, Canada
CSM1	CSM1	Beijing Climate Center, China Meteorological Administration
CSM1m	CSM1m	Beijing Climate Center, China Meteorological Administration
GFDLG	GFDLG	NOAA Geophysical Fluid Dynamics Laboratory, USA
GFDLM	GFDLM	NOAA Geophysical Fluid Dynamics Laboratory, USA
HadGEM2C	HGEMC	Met Office Hadley Centre, UK
HadGEM2E	HGEME	Met Office Hadley Centre, UK
MIROC-ESM	MIROC	Japan Agency for Marine-Earth Science and Technology, Atmosphere and Ocean Research Institute (The University of Tokyo) and National Institute for Environmental Studies, Japan
MIROC5	MIROC5	Atmosphere and Ocean Research Institute (The University of Tokyo), National Institute for Environmental Studies and Japan Agency for Marine-Earth Science and Technology, Japan
MIROC-CHEM	MIROCC	Japan Agency for Marine-Earth Science and Technology, Atmosphere and Ocean Research Institute (The University of Tokyo) and National Institute for Environmental Studies, Japan
INM-CM4	INMCM	Institute for Numerical Mathematics, Russia
IPSL-CM5A-MR	IPSLA	Institut Pierre-Simon Laplace, France
IPSL-CM5B-LR	IPSLB	Institut Pierre-Simon Laplace, France
MRI-CGCM3	MRI	Meteorological Research Institute, Japan
NorEMS1-M	NorESM	Norwegian Climate Center, Norway

climate variable, cumulative distribution functions (CDFs) of conditions from both observed and GCM historical datasets were constructed. At each percentile rank in the GCM historical CDFs, observed values in the same rank were identified and applied so that the historical GCM and observed CDFs match. Quantile mapping was performed (by USBR) by populating the sample distribution using a 15-day moving window centred on each calendar day. Quantile mapping adjustments from historical GCM and observed datasets were then transferred to future time slices (Abatzoglou & Brown, 2012; USBR, 2013). Two future time periods in RCP8.5 were considered: Phase 1 (2040–2069) and Phase 2 (2070–2099), which are representative of the middle and the end of the twenty-first century, respectively.

19.3.4 Percentage Change and Uncertainty Analysis

Percentage changes in future precipitation, streamflow and soil moisture were calculated as

$$\text{Delta X}\ (\%) = \frac{(X_{\text{future}} - X_{\text{historical}})}{X_{\text{historical}}} \times 100\% \tag{19.1}$$

where X_{future} represents precipitation, streamflow or soil moisture for a future time period and $X_{historical}$ represents precipitation, streamflow or soil moisture in the historical time period.

We applied two-way ANOVA to quantify relative proportions of uncertainties from calibrated parameter sets and climate models following Hattermann et al. (2018). The two groups (factors) that contribute to the variation are climate scenarios and calibrated parameter sets. There are 18 future scenarios and 16 parameter sets in our case, which represent 18 and 16 levels for the two factors, respectively. In ANOVA, the total sum of squares (SS_T) represents the total variation from different factors. Specifically, SS_T is decomposed into three parts, $SS_{CLIMATE}$, SS_{PARAM} and the interaction term $SS_{CLIMATE*PARAM}$.

$$SS_T = SS_{CLIMATE} + SS_{PARAM} + SS_{CLIMATE*PARAM} \tag{19.2}$$

thus, $SS_{CLIMATE}$ represents the squared deviation from climate scenarios, SS_{PARAM} represents the squared deviation from calibrated parameter sets and $SS_{CLIMATE*PARAM}$ represents the squared deviation from the interaction term of calibrated parameter sets and climate scenarios.

19.4 Results

19.4.1 Model Calibration Results

Figure 19.2 shows the Pareto plot for the multi-objective calibration (i.e., NSE against other calibration metrics). The Pareto front shows that there is a relatively slight trade-off between objective functions. The RMSE was inversely correlated with NSE, indicating autocorrelation and thus redundancy of the RMSE metric, whereby the minimization of RMSE is equivalent to the maximization of NSE (Mendoza et al., 2016). The R^2 values were directly proportional to NSE, whereas the PBIAS and RSD values demonstrated a non-linear relationship to NSE. The optimal parameters of the RSDs and NSEs form a Pareto Front where the best performing parameters along this front have no preference over one another (Hadka & Reed, 2013). As expected, the NSEs were lower during the validation time period than during the calibration time period. However, the NSEs were correlated during the validation and calibration time periods, which suggests that the performance of a given parameter set may be transferrable across time periods. Among the optimal parameter sets, the NSE values varied from 0.745 to 0.779. The ratio of standard deviations (RSD) ranged from 0.689 to 0.994, per cent bias (PBIAS) ranged from 0.0008 to 6.594 per cent, R^2 ranged from 0.76 to 0.79 and

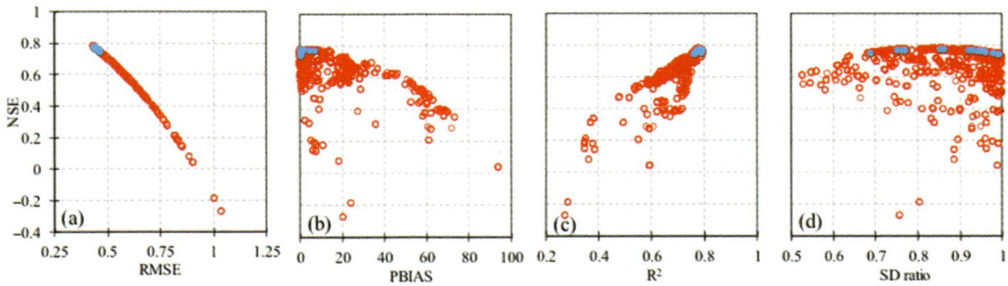

Figure 19.2 Pareto plot for the multi-objective calibration of the VIC model. Each dot represents one simulation of the VIC model in the final run of the Borg algorithm (10,000 simulations) and blue dots are the optimal parameter sets (i.e., the set of non-dominated simulations)

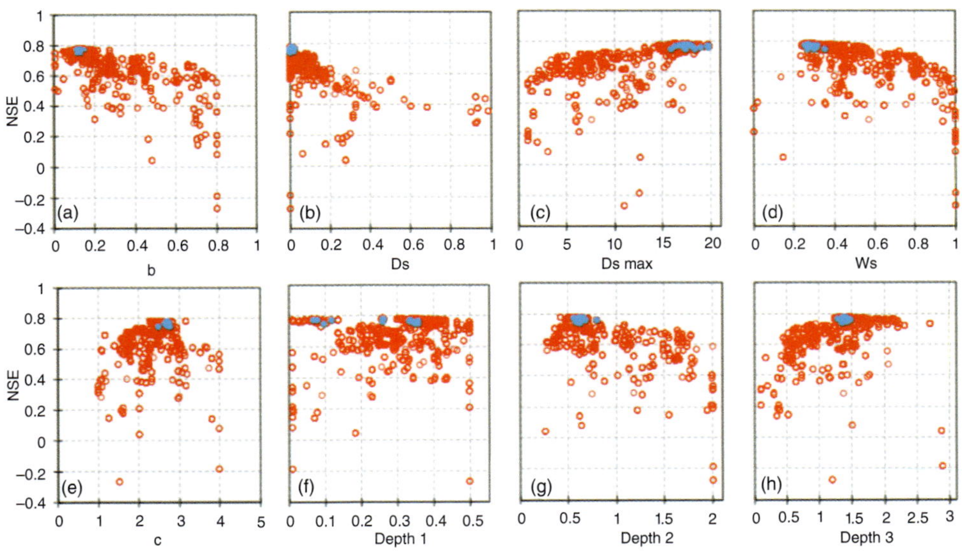

Figure 19.3 Parameter sets versus NSE at Boulder Creek. The x-axis represents parameters b, Ds, Ds_max, Ws, c, soil depth 1, soil depth 2 and soil depth 3, which are described in Table 19.1. Red dots represent all simulations of the VIC model in the final run of the Borg algorithm (10,000 population) and blue dots represent the 16 optimal parameter sets

RMSE ranged from 0.43 to 0.46 mm day^{-1}. Figure 19.3 shows the parameter sets against NSEs on a daily scale. This figure illustrates that larger values for Ds_max and smaller values for b, Ds, Ws and soil depth 2 lead to decreasing error. All parameters except soil depth 1 were sensitive to streamflow simulations, as evidenced by decreasing scatter with increasing NSE.

The selected parameters were subsequently used to simulate streamflow (Figure 19.4). During the validation period (1991–2010), daily NSEs ranged

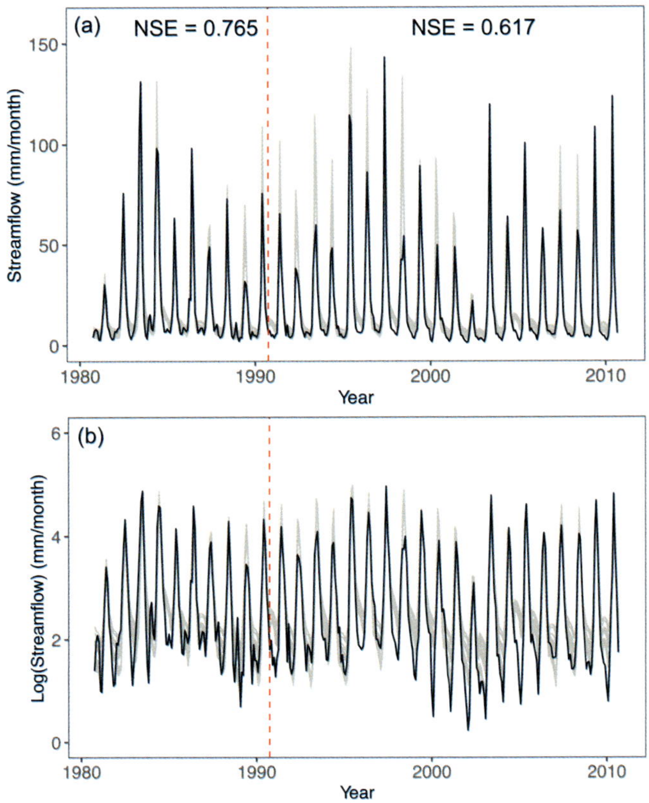

Figure 19.4 (a) Monthly simulated (grey lines) and observed (black line) streamflow at the Boulder Creek Orodell station between 1981 and 2010. (b) Same data shown on a log-scale. The red dashed line separates the calibration (1981–1990) and validation periods (1991–2010)

from 0.549 to 0.687 and averaged 0.617. Although there were discrepancies between flow volumes at peak flow, the timing of observed streamflow was well predicted by the model (Figure 19.4(a)). Figure 19.4(b) presents the daily simulated and observed streamflow on a log scale to emphasize the low flow period when different parameter sets introduced a large uncertainty range.

19.4.2 Precipitation and Air Temperature Changes under Perturbed Climate

Daily precipitation and maximum and minimum air temperature of BCSD datasets simulated between 2040–2069 (Phase 1) and 2070–2099 (Phase 2) were compared with the historical data from 1981 to 2010 (Figure 19.5). Despite the divergences on the average annual precipitation, the investigated climate models agree on the trends associated with the maximum and minimum temperatures. Most models predicted higher precipitation in Phase 2 relative to Phase 1. In Phase 1, changes in annual precipitation varied from –7.49 to +31.44 per cent and averaged +8.11 per cent compared to the historical period. In Phase 2, changes in average annual precipitation varied from –7.97 to +32.41 per cent and averaged +12.13 per cent compared to the historical period. In contrast, all models showed increasing trends in maximum and minimum air temperatures. Changes in daily maximum temperature varied from +2.02 to +4.58°C with an average of +3.5°C in Phase 1, and from +3.37 to +7.31°C with an average of +5.55°C in Phase 2. Changes in daily minimum temperature varied from +1.87 to +4.04°C with an average of

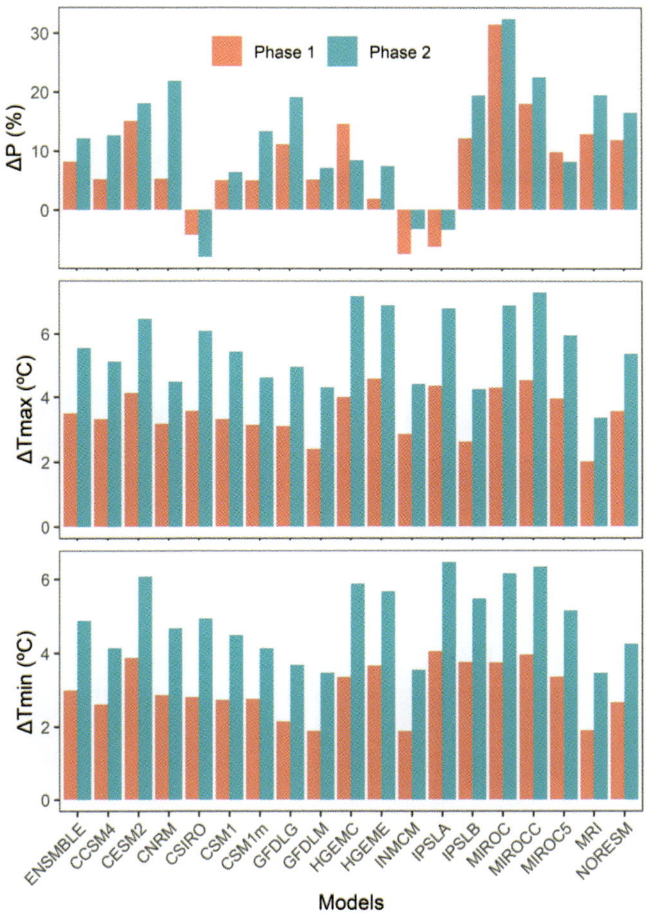

Figure 19.5 Changes in average annual precipitation (a), daily maximum air temperature (b) and daily minimum air temperature (c) in Phase 1 (2040–2069) and Phase 2 (2070–2099) compared to the historical period at Boulder Creek. Coloured bars correspond to individual climate models and boxplots summarize the variability

+2.99°C in Phase 1, and from +3.45 to +6.45°C with an average of +4.88°C in Phase 2.

19.4.3 Annual Changes in Flux and State Variables

Changes in average annual streamflow, ET and soil moisture in Phases 1 and 2 were calculated with respect to the historical period. Figure 19.6 presents the median values of percentage changes in average annual streamflow, ET and soil moisture using each parameter set and all climate data inputs and thus corresponds to parameter set uncertainty; the dashed lines represent the median values using all

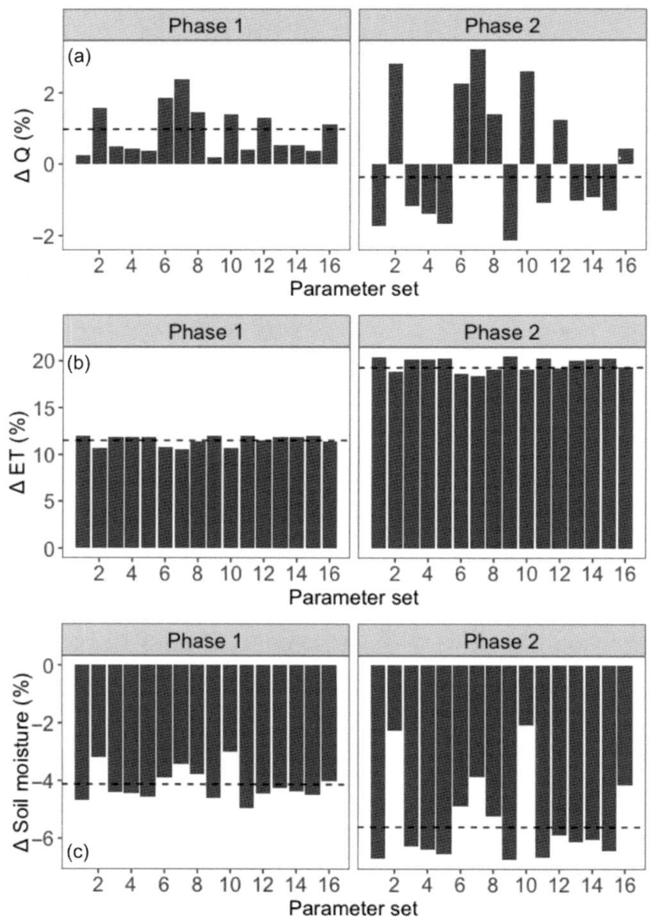

Figure 19.6 Median value of changes in average annual streamflow (a), ET (b) and soil moisture (c) using each parameter set in Phase 1 (left column) and Phase 2 (right column). The dashed lines represent median values in annual changes using all parameter sets

parameter sets and climate data. The median values show that annual streamflow increased by 0.9 per cent (0.2–2.4 per cent) and 0.1 per cent (–2.1 to 3.2 per cent), annual ET increased by 11.5 per cent (10.6–12 per cent) and 19.7 per cent (18.4–20.5 per cent) and soil moisture decreased by 4.1 per cent (–5.0 to –3.0 per cent) and 5.4 per cent (–6.8 to 2.0 per cent) in Phases 1 and 2, respectively (Figure 19.6). The relative uncertainties from the calibrated parameter sets were 0.2 and 0.8 per cent for annual streamflow, 0.8 and 2.0 per cent for ET and 3.1 and 4.3 per cent for soil moisture with respect to the total sum of squares in Phases 1 and 2, respectively (Table 19.4(a)). For climate data uncertainty, the relative proportions were 99.5 and 98.6 per cent for annual streamflow, 98.1 and 97.1 per cent for ET and 83.29 and 87.4 per cent for soil moisture.

We used q10 and q90 to represent the 10th and 90th percentile streamflow in each year, which can highlight extreme low and high flows. Figure 19.7 shows the per cent changes in average low and high flows by parameter and climate model choice. By taking into account uncertainty in the estimation of the parameters, our results show that q10 (low flows) will decrease by 5.1 per cent in Phase 1, and will decrease by 7.9 per cent in Phase 2, whereas q90 (high flows) will increase by 3.9 per cent in Phase 1 and 0.8 per cent in Phase 2.

19.4.4 Monthly and Daily Changes in Flux and State Variables

Figure 19.8 shows boxplots of per cent changes in monthly streamflow, ET and soil moisture using median values of per cent change relative to the historical period. In general, streamflow showed an increasing trend between January and May and a decreasing trend from June to November in both Phases 1 and 2 (Figure 19.8(a)). ET showed an increasing trend in all months, but the magnitude of increase was lower between June and September compared to the other months (Figure 19.8(b)). Soil moisture showed an increasing trend between February and May, and a decreasing trend in the other months (Figure 19.8(c)). For all variables, the magnitude and variability of changes were generally higher in Phase 2 than in Phase 1.

Monthly changes in streamflow uncertainty were higher in the spring and summer months and peaked in May in Phase 1 and in April in Phase 2 (Figure 19.8 (d) and (g)). ET uncertainty was highest in summer and early fall (Figure 19.8(e) and (h)); soil moisture uncertainty was highest between January and May (Figure 19.8(f) and (i)). The proportions of uncertainty from calibrated parameter sets were higher in February, March, August to October for changes in streamflow, in July to October for changes in ET, and in January to May for changes in soil moisture than in the other months, respectively (Table 19.4(a)). The relative proportions of uncertainty from calibrated parameter sets ranged from 0.6 to 15.1

Table 19.4. Percentage of the median interquartile range from (a) parameter uncertainty and (b) climate data uncertainty relative to the total interquartile range for average annual and monthly streamflow (Q), evapotranspiration (ET) and soil moisture (SM)

(a) Uncertainty from calibrated parameter sets

Variable	Time period	Annual	Jan	Feb	Mar	Apr	May	Jun	Jul	Aug	Sep	Oct	Nov	Dec
ΔQ	2040–2069	0.2	1.6	2.0	2.7	0.8	0.6	2.1	1.5	5.6	5.3	3.8	2.0	1.8
	2070–2099	0.8	2.3	3.4	5.4	2.1	2.7	2.2	4.7	15.1	12.3	7.3	5.0	3.7
ΔET	2040–2069	0.8	0.1	0.3	0.1	0.2	0.1	0.6	1.8	2.0	8.1	4.5	0.4	0.2
	2070–2099	2.0	0.1	0.4	0.2	0.4	0.9	2.5	3.5	6.3	11.8	6.5	0.9	0.1
ΔSoil moisture	2040–2069	3.1	7.8	12.2	9.6	5.6	8.9	1.0	1.7	2.9	2.4	1.6	1.3	3.5
	2070–2099	4.3	12.1	19.6	17.0	16.7	14.8	0.5	5.7	6.7	3.5	1.1	0.6	3.8

(b) Uncertainty from climate data

Variable	Time period	Annual	Jan	Feb	Mar	Apr	May	Jun	Jul	Aug	Sep	Oct	Nov	Dec
ΔQ	2040–2069	99.5	82.7	82.8	88.7	96.4	98.7	97.4	98.2	92.6	89.6	88.0	88.5	85.5
	2070–2099	98.6	86.5	87.2	87.7	95.8	96.2	97.7	94.7	82.8	83.2	87.1	88.9	87.5
ΔET	2040–2069	98.1	99.9	99.6	99.9	99.7	99.6	97.1	92.6	95.0	87.7	91.9	98.9	99.7
	2070–2099	97.1	99.8	99.5	99.8	99.6	98.6	93.9	92.3	90.8	85.0	90.8	98.5	99.8
ΔSoil moisture	2040–2069	83.2	69.4	68.4	77.8	87.1	84.5	95.1	91.2	85.8	84.0	81.4	79.6	73.7
	2070–2099	87.4	76.3	70.0	74.9	77.0	81.6	97.0	89.5	86.8	87.9	90.7	91.4	86.6

(c) Uncertainty from the interaction term of parameter sets and climate data

Variable	Time period	Annual	Jan	Feb	Mar	Apr	May	Jun	Jul	Aug	Sep	Oct	Nov	Dec
ΔQ	2040–2069	0.3	15.7	15.2	8.5	2.8	0.7	0.5	0.3	1.8	5.1	8.3	9.6	12.6
	2070–2099	0.5	11.2	9.4	6.9	2.0	1.1	0.1	0.6	2.1	4.5	5.6	6.2	8.8
ΔET	2040–2069	1.0	0.0	0.1	0.1	0.1	0.3	2.3	5.6	3.0	4.2	3.6	0.7	0.1
	2070–2099	1.0	0.1	0.1	0.0	0.1	0.5	3.6	4.2	2.9	3.2	2.7	0.6	0.1
ΔSoil moisture	2040–2069	13.7	22.8	19.4	12.7	7.2	6.6	3.9	7.1	11.3	13.6	17.0	19.2	22.8
	2070–2099	8.3	11.6	10.4	8.1	6.3	3.5	2.5	4.9	6.5	8.6	8.2	8.1	9.7

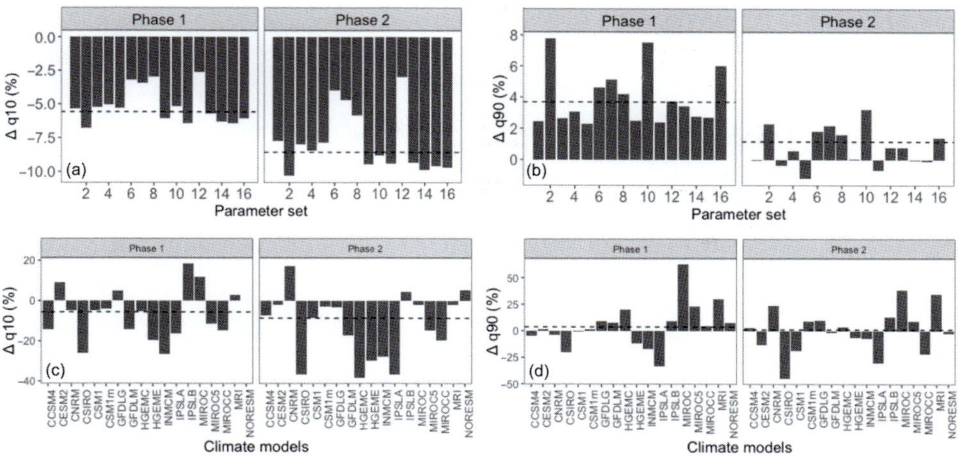

Figure 19.7 Median value of changes in average annual low flow and high flow using each parameter set (a and b) in Phases 1 and 2 and using each climate model data (c and d) in Phases 1 and 2. The dashed lines represent median values in annual changes using all parameter sets (top panel) or all climate model data (bottom panel). Annual low flow q10 is defined as the 10 per cent streamflow, while annual high flow q90 is defined as the 90 per cent streamflow

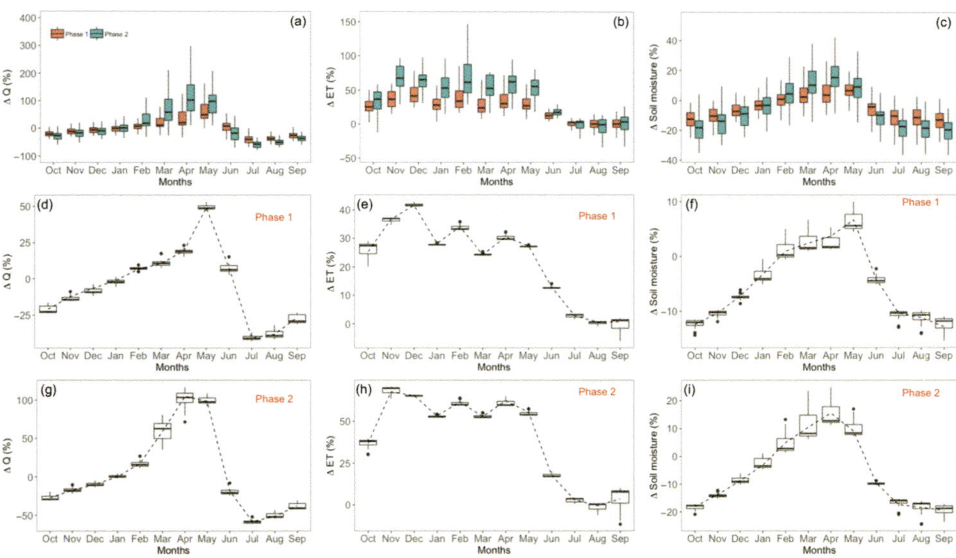

Figure 19.8 Boxplots of per cent changes using all climate data and parameter sets (a–c), and boxplots of median per cent changes using all parameter sets in Phase 1 (d–f) and Phase 2 (g–i) for basin-averaged monthly streamflow, ET and soil moisture. The dashed lines in the middle and bottom rows represent median changes from all climate data and parameter sets

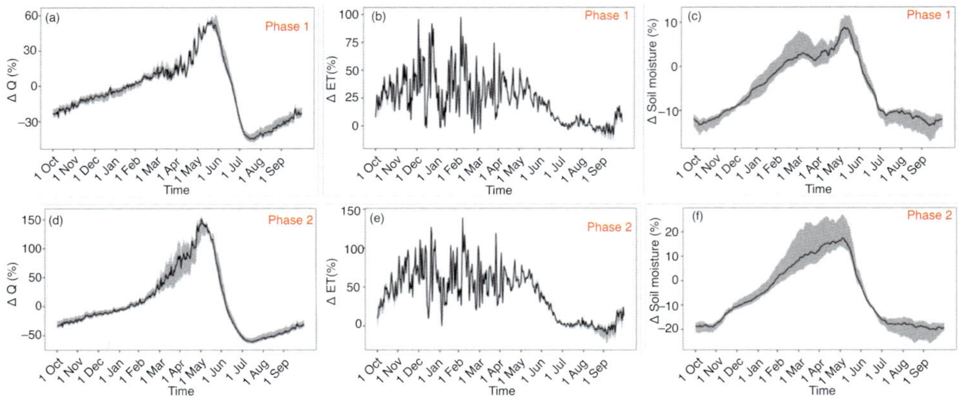

Figure 19.9 Median per cent changes using all parameter sets in Phase 1 (a–c) and Phase 2 (d–f) for basin-averaged daily streamflow, ET and soil moisture

per cent for streamflow, 0.1–11.8 per cent for ET and 0.5–19.6 per cent for soil moisture on a monthly basis. The relative proportions of uncertainty from climate data ranged from 82.7 to 98.7 per cent for streamflow, 85–99.9 per cent for ET and 68.4–97.0 per cent for soil moisture on a monthly basis. The relative proportions of uncertainty from the interaction term were below 23 per cent for the three variables. The per cent change from historical to future daily conditions was higher in Phase 2 compared to Phase 1 for all variables (Figure 19.9). Seasonally, relative changes in streamflow and soil moisture increased from January to May and then decreased through the summer. Median changes in daily ET were temporally inconsistent and were more variable in October to April during the low flow season. These results demonstrate the effects of both parametric and climatic uncertainty on hydrologic projections at fine time scales and demonstrate that these effects are magnified during the snowmelt season.

19.5 Discussion

Robust characterization of streamflow conditions under future climate scenarios is essential for assessing the sustainability of the long-term supply of water to natural and managed systems downstream (Pradhanang et al., 2013). Accordingly, we applied the VIC model to the Boulder Creek Watershed in Colorado, USA, to evaluate the impact of parameter uncertainty on projected hydrologic flux and state variables under future climate regimes. The selected model parameters were calibrated by comparing simulated streamflow against observed streamflow and climate forcing data applied BCSD downscaled data from eighteen GCM models. Between 2040 and 2069 (Phase 1), our results suggest 0.9 and 11.5 per cent increases in annual streamflow and ET, respectively, but a 4.1 per cent decrease in soil moisture relative to historical conditions. Between 2070 and 2099 (Phase 2), streamflow (+0.1 per cent) is predicted to return to near historical levels, whereas ET (+19.7 per cent) and soil moisture (–5.4 per cent) changes are exacerbated over time. Uncertainty resultant from different climate scenarios was systematically higher than uncertainty associated with the estimation of the parameter sets.

19.5.1 Multiple Objective Functions and Climate Data Processing

All selected parameter sets showed NSE values greater than 0.7, which indicates good model performance (Moriasi et al., 2007). Additionally, the choice of many parameter sets constrains uncertainty in streamflow and facilitates the transfer of model parameters from data-rich to data-sparse areas (e.g., Huang, 2005). In this case, all calibrated parameters were accurately estimated except soil depth 1. The parameter b and soil depth 1 may thus be important for high flows, while the other calibrated parameters may be important for low flows or baseflow. Over the long-term, the simulated monthly trends in streamflow represented observed values during the historical period.

Although the delta change method removes large variations in climate data from the GCMs, it also assumes that the magnitude of changes in precipitation and temperature remains stationary. Hewitson and Crane (2006) asserted that the degree of non-stationarity in projected climate change is relatively small, and that circulation dynamics may be more robust to non-stationarity. Gutmann et al. (2014) used climate data to compare several statistical downscaling methods and found that precipitation from the BCSD method is biased toward low values for individual months, but relatively unbiased on an annual scale. In this study, BCSD downscaling showed that average annual precipitation and maximum and minimum air temperature are expected to increase in both Phases 1 and 2, and that the increase in magnitude will be higher in Phase 2 than in Phase 1 (Figure 19.5).

19.5.2 Uncertainty in Projected Annual Hydrologic Fluxes

Although average annual precipitation was projected to increase by 4 per cent between Phases 1 and 2, the decrease in average annual streamflow between Phases 1 and 2 was not significantly different from the observed data in the historical period during Phase 2. This result indicates that the impact of increased air temperature on water resources availability may supersede that of precipitation changes in the future, as the increased temperature was translated to increased ET (+8.2 per cent between Phases 1 and 2). However, for each parameter set, simulation uncertainties were higher in Phase 2 than in Phase 1 (Figure 19.6).

Furthermore, differences in both the magnitude and the direction of the changes in streamflow and soil moisture associated with different parameter sets indicate that the use of only one parameter set might yield a biased result. Although different parameter sets resulted in a more consistent average annual ET trend, that is, an increase in Phases 1 and 2, changes in average annual streamflow were uncertain in Phase 2 (Figure 19.6). The relative uncertainties from calibrated parameter sets were up to 0.8 per cent for changes in annual streamflow, 2.0 per cent for changes in ET and 4.3 per cent for changes in soil moisture (Table 19.4). Although annual streamflow showed no significant change from the historical period to Phase 2, the range of values based on different parameter sets indicates a potential increase between 0.2 and 2.4 per cent. By incorporating calibrated parameter uncertainty, annual extreme low flows ranged from –10.3 to –3.0 per cent in Phase 2, and annual extreme high flows changed from –1.2 to 3.2 per cent in Phase 2. Although the uncertainty associated with the use of different climate models is larger than those that stem from the calibrated parameter sets, these results demonstrate that model calibration uncertainty must be accounted for in order to accurately model changes in hydrologic and state fluxes.

19.5.3 Monthly and Daily Projections

In snowmelt dominated watersheds, reservoirs are replenished in spring and water is released for agricultural, industrial and urban water use during summer. Previous work that used three different modelling approaches to project streamflow under the A2 emission scenario in the Upper Colorado River Basin suggested that spring and summer streamflow is likely to decline by the end of the twenty-first century (Ficklin et al., 2013). At a smaller spatial scale, the current results in the Boulder Creek Watershed refine this estimate temporally. This suggests that, on average, streamflow is expected to increase from March to May, but should decrease from June to August (Figure 19.8(a)). However, the use of different parameter sets led to large uncertainty in monthly streamflow and soil moisture projections during the spring snowmelt season and large uncertainty in ET projections during the summer months (Figure 19.8). For all variables, the seasonal characteristics of the uncertainty ranges were consistent with daily simulations (Figure 19.9). The relative uncertainty from calibrated parameter sets can account for 15.1 per cent of the changes in streamflow, 11.8 per cent of the changes in ET and 19.6 per cent of the changes in soil moisture (Table 19.4). The uncertainty from parameter sets was higher in February, March, August and September than in the other months for streamflow, in July to September than in the other months for ET and in January to May than in the other months for soil moisture. In the context of Köplin et al. (2012) that used a cluster analysis to reduce 186 catchments to 7 response types in

Switzerland, this study from a snowmelt dominated watershed provides additional insight into other snowmelt dominated basins where the use of multiple parameter sets lead to uncertainty in the results that are more likely to reflect the range of the possible directions of snowmelt processes under the most likely climate change scenarios (Seiller et al., 2017). Monthly changes based on multiple parameter sets result in a range of projected variables that allows for characterization of the uncertainty associated with predictions of both hydrologic flux and state variables to the benefit of effective water resources management. In addition, we chose historically optimal performing parameter sets from the Borg calibration algorithm, which led to expected smaller relative uncertainties than those from climate model data.

19.5.4 Implications for Decision-Making

The magnitude of parameter uncertainty relative to total uncertainty varies over time and space (Mendoza et al., 2016; Seiller et al., 2017). In larger (global) domains, parameter uncertainty may have a smaller impact on future projections due to the potential for compensating errors to partially offset at large spatial scales (Elsner et al., 2014). However, at smaller basins, the impact of parameter uncertainty has the potential to be significant. Here, we use a similar set of parameters which highlighted significant differences between hydrological responses to future projections, in spite of similar historical performance. Although previous work has considered the effects of model structure, objective function, multiple local optima, and forcing calibration data on projections of flux and state variables (see e.g., Mendoza et al., 2016), this current study extends the analysis to account for climate uncertainty. We used downscaled climate data from 18 GCMs, which enabled a broad characterization of relative uncertainty under multiple scenarios. In this suggested framework, we evaluated the effects of uncertainty across multiple time scales (annual, monthly and daily), which allowed us to characterize the impacts of hydroclimatic extremes which may occur at various fine time scales.

In the case study focused on Boulder Creek Watershed, the use of different parameter sets resulted in streamflow changes of opposite sign during Phase 2 between 2070 and 2099. Accordingly, this work highlights the importance of context when reporting data with ramifications for management decisions, since mean or median changes may change sign under different scenarios. To constrain this effect, we selected a small river basin, which prevents us from compensating errors, and ensured that calibration was not limited by computational power. We did not examine whether an increase in the number of parameter sets would reduce uncertainty since it would be difficult to interpret the ensuing ensemble mean value within the context of management policy, even if the ensemble mean values were to stabilize when the number of parameter sets was large. This raises a fundamental question: is it acceptable to treat all calibrated parameter sets equally? While we acknowledge that proposing a minimum number of parameter sets would be useful for decision-makers, we also posit that the requisite number may be catchment-dependent, and more importantly, that it would depend on the timescale of interest (e.g., flooding versus drought), which is a broader question and worthy of future inquiry.

Additional sources of uncertainty that may influence the accuracy of the results remain despite our efforts to consider a large number of GCM models and to carry out multi-objective calibration of a comprehensive hydrological model. We only consider the RCP8.5 scenarios, therefore the estimated hydrological model uncertainty does not capture the range of possible uncertainty values associated with other emission scenarios. Moreover, similar to many other studies, the shape and area of the basin in the VIC model do not accurately describe the actual one and provide an additional omitted source of uncertainty. Despite these short-comings, we have compared the future and historical period under the same conditions, which should reduce the effects of the ignored uncertainties. Therefore, we believe that these additional sources of uncertainty should not change the overall results.

19.6 Conclusion

Models are useful tools to evaluate the potential impacts of climate change on hydrologic variables. While climate data inputs are critical to model results, calibrated parameter sets provide uncertainty in the predictions under different climate scenarios. In this study, we selected a topographically complex watershed to investigate the contribution of model calibration uncertainty relative to climate model data uncertainty. We specifically evaluated the role of uncertainty at annual, monthly and daily time scales with respect to multiple hydrological variables. Relative to historical conditions in the Boulder Creek Watershed, annual streamflow increased by 0.9 and 0.1 per cent, annual ET increased by 11.5 and 19.7 per cent and soil moisture decreased by 4.1 and 5.4 per cent between 2040 and 2069 (Phase 1) and 2070 and 2099 (Phase 2), respectively. Projected changes in flux and state variables were subsequently interpreted within the context of model parameter uncertainty to identify changes that were robust. Overall, parameter uncertainty was a significant contributor to future annual hydrologic projections both annually (up to 4.3 per cent) and seasonally (up to 19.6 per cent) and was especially relevant to streamflow and soil moisture projections. These results demonstrate the need to consider multiple optimal parameter sets in order to provide a more complete range of future projections of water resource availability from snowmelt dominated montane domains in the future.

Note

1 Available from www.engr.scu.edu/~emaurer/gridded_obs/index_gridded_obs.html (Last accessed 1 March 2018).

References

Abatzoglou, J. T., & Brown, T. J. (2012). A comparison of statistically downscaling methods suited for wildfire applications. *International Journal of Climatology* 32(5): 772–780.

Addor, N., Rössler, O., Köplin, N., et al. (2014). Robust changes and sources of uncertainty in the projected hydrological regimes of Swiss catchments. *Water Resources Research* 50(10): 7541–7562.

Bales, R. C., Molotch, N. P., Painter, T. H., et al. (2006). Mountain hydrology of the western United States. *Water Resources Research* 42(8): W08432.

Barnett, T. P., Adam, J. C., & Lettenmaier, D. P. (2005). Potential impacts of a warming climate on water availability in snow-dominated regions. *Nature* 438: 303–309.

Bastola, S., Murphy, C., & Sweeney, J. (2011). The role of hydrological modelling uncertainties in climate change impact assessments of Irish river catchments. *Advances in Water Resources* 34(5): 562–576.

Chen, J. M., Chen, X., Ju, W., & Geng, X. (2005). Distributed hydrological model for mapping evapotranspiration using remote sensing inputs. *Journal of Hydrology* 305 (1–4): 15–39.

Demaria, E. M., Nijssen, B., & Wagener, T. (2007). Monte Carlo sensitivity analysis of land surface parameters using the Variable Infiltration Capacity model. *Journal of Geophysical Research* 112(D11): D11113.

Dietz, J., Hölscher, D., Leuschner, C., & Hendrayanto, H. (2006). Rainfall partitioning in relation to forest structure in differently managed montane forest stands in Central Sulawesi, Indonesia. *Forest Ecology and Management* 237(1–3): 170–178.

Elsner, M. M., Gangopadhyay, S., Pruitt, T., et al. (2014). How does the choice of distributed meteorological data affect hydrologic model calibration and streamflow simulations? *Journal of Hydrometeorology* 15(4): 1384–1403.

Ficklin, D. L., Stewart, I. T., & Maurer, E. P. (2013). Climate change impacts on streamflow and subbasin-scale hydrology in the upper Colorado River basin. *PLoS One* 8 (8): e71297.

Gao, H., Tang, Q., Shi, X., et al. (2010). Water budget record from Variable Infiltration Capacity (VIC) model. In Algorithm Theoretical Basis Document for Terrestrial Water Cycle Data Records (pp. 120–173). Available from: www.research.lancs.ac.uk/portal/en/publications/water-budget-record-from-variable-infiltration-capacity-vic-model(1e8618dd-212f-4c3b-bee3-65862ea5a4b9)/export.html (Last accessed 3 September 2021).

Guo, D., Johnson, F., & Marshall, L. (2018). Assessing the potential robustness of conceptual rainfall-runoff models under a changing climate. *Water Resources Research* 54(7): 5030–5049.

Gupta, H. V., Wagener, T., & Liu, Y. (2008). Reconciling theory with observations: Elements of a diagnostic approach to model evaluation. *Hydrological Processes* 22(18): 3802–3813.
Gutmann, E., Pruitt, T., Clark, M. P., et al. (2014). An intercomparison of statistical downscaling methods used for water resource assessments in the United States. *Water Resources Research* 50(9): 7167–7186.
Hadka, D., & Reed, P. (2013). Borg: An auto-adaptive many-objective evolutionary computing framework. *Evolutionary Computation* 21(2): 231–259.
Hadka, D., & Reed, P. (2015). Large-scale parallelization of the Borg multiobjective evolutionary algorithm to enhance the management of complex environmental systems. *Environmental Modelling & Software* 69(C): 353–369.
Hamlet, A. F., & Lettenmaier, D. P. (2007). Effects of 20th century warming and climate variability on flood risk in the western U.S. *Water Resources Research* 43(6): W06427.
Hansen, M. C., Defries, R. S., Townshend, J. R. G., & Sohlberg, R. (2000). Global land cover classification at 1 km spatial resolution using a classification tree approach. *International Journal of Remote Sensing* 21(6–7): 1331–1364.
Harding, B. L., Wood, A. W., & Prairie, J. R. (2012). The implications of climate change scenario selection for future streamflow projection in the Upper Colorado River Basin. *Hydrology and Earth System Sciences* 16(11): 3989–4007.
Hattermann, F. F., Vetter, T., Breuer, L., et al. (2018). Sources of uncertainty in hydrological climate impact assessment: A cross-scale study. *Environmental Research Letters* 13(1): 015006.
Her, Y., Yoo, S. H., Cho, J., et al. (2019). Uncertainty in hydrological analysis of climate change: Multi-parameter vs. multi-GCM ensemble predictions. *Scientific Reports* 9: 4974.
Hewitson, B. C., & Crane, R. G. (2006). Consensus between GCM climate change projections with empirical downscaling: Precipitation downscaling over South Africa. *International Journal of Climatology* 26(10): 1315–1337.
Huang, M. (2005). *Surface and Groundwater Interactions and their Impacts on Water and Energy Budgets at the Land Surface*. Berkeley, CA: University of California Press.
Joseph, J., Ghosh, S., Pathak, A., & Sahai, A. K. (2018). Hydrologic impacts of climate change: Comparisons between hydrological parameter uncertainty and climate model uncertainty. *Journal of Hydrology* 566: 1–22.
Kay, A. L., Davies, H. N., Bell, V. A., & Jones, R. G. (2009). Comparison of uncertainty sources for climate change impacts: Flood frequency in England. *Climatic Change* 92 (1–2): 41–63.
Köplin, N., Schädler, B., Viviroli, D., & Weingartner, R. (2012). Relating climate change signals and physiographic catchment properties to clustered hydrological response types. *Hydrology and Earth System Sciences* 16: 2267–2283.
Liang, X., Lettenmaier, D. P., Wood, E. F., & Burges, S. J. (1994). A simple hydrologically based model of land surface water and energy fluxes for general circulation models. *Journal of Geophysical Research* 99(D7): 14415–14428.
Liu, X., Tang, Q., Cui, H., et al. (2017). Multimodel uncertainty changes in simulated river flows induced by human impact parameterizations. *Environmental Research Letters* 12(2): 025009.
Liu, X., Tang, Q., Voisin, N., & Cui, H. (2016). Projected impacts of climate change on hydropower potential in China. *Hydrology and Earth System Sciences* 20: 3343–3359.
Liu, Y., Hejazi, M., Li, H., Zhang, X., & Leng, G. (2018). A hydrological emulator for global applications – HE v1.0.0. *Geoscientific Model Development* 11(3): 1077–1092.
Lohmann, D., Nolte-Holube, R., & Raschke, E. (1996). A large-scale horizontal routing model to be coupled to land surface parametrization schemes. *Tellus A: Dynamic Meteorology and Oceanography*, 48(5): 708–721.

Maurer, E. P., Wood, A. W., Adam, J. C., & Lettenmaier, D. P. (2002). A long-term hydrologically based dataset of land surface fluxes and states for the conterminous United States. *Journal of Climate* 15(22): 3237–3251.

Meinshausen, M., Smith, S. J., Calvin, K., et al. (2011). The RCP greenhouse gas concentrations and their extensions from 1765 to 2300. *Climatic Change* 109(1–2): 213–241.

Mendoza, P. A., Clark, M. P., Mizukami, N., et al. (2015). Effects of hydrologic model choice and calibration on the portrayal of climate change impacts. *Journal of Hydrometeorology* 16(2): 762–780.

Mendoza, P. A., Clark, M. P., Mizukami, N., et al. (2016). How do hydrologic modeling decisions affect the portrayal of climate change impacts? *Hydrological Processes* 30 (7): 1071–1095.

Moriasi, D. N., Arnold, J. G., Van Liew, M. W., et al. (2007). Model evaluation guidelines for systematic quantification of accuracy in watershed simulations. *ASABE* 50(3): 885–900.

Murphy, S. F., Barber, L. B., Verplanck, P. L., & Kinner, D. A. (2003). Environmental setting and hydrology of the Boulder Creek Watershed, Colorado. In S. F. Murphy, P. L. Verplanck, & L. B. Barber (eds.), *Comprehensive Water Quality of the Boulder Creek Watershed, Colorado, during High-Flow and Low-Flow Conditions. 2000. Water Resources Investigation Report 03-4045*. Denver, CO: US Geological Survey.

Musselman, K. N., Clark, M. P., Liu, C., Ikeda, K., & Rasmussen, R. (2017). Slower snowmelt in a warmer world. *Nature Climate Change* 7: 214–219.

Poff, N. L., Allan, J. D., Bain, M. B., et al. (1997). The natural flow regime – A paradigm for river conservation and restoration. *BioScience* 47(11): 769–784.

Pradhanang, S. M., Mukundan, R., Schneiderman, E. M., et al. (2013). Streamflow responses to climate change: Analysis of hydrologic indicators in a New York City water supply watershed. *Journal of American Water Resources Association* 49(6): 1308–1326.

Raje, D., & Krishnan, R. (2012). Bayesian parameter uncertainty modeling in a macroscale hydrologic model and its impact on Indian river basin hydrology under climate change. *Water Resources Research* 48(8): W08522.

Rauscher, S. A., Pal, J. S., Diffenbaugh, N. S., & Benedetti, M. M. (2008). Future changes in snowmelt-driven runoff timing over the western US. *Geophysical Research Letters* 35(16): L16703.

Reclamation (2013). *Downscaled CMIP3 and CMIP5 Climate Projections: Release of Downscaled CMIP5 Climate Projections, Comparison with Preceding Information, and Summary of User Needs*. U.S. Department of the Interior, Bureau of Reclamation, 104 pp. Available from http://gdo-dcp.ucllnl.org/downscaled_cmip_projections/techmemo/downscaled_climate.pdf (Last accessed 1 March 2018).

Ren, H., Hou, Z., Huang, M., et al. (2016). Classification of hydrological parameter sensitivity and evaluation of parameter transferability across 431 US MOPEX basins. *Journal of Hydrology* 536: 92–108.

Riahi, K., Rao, S., Krey, V., et al. (2011). RCP8.5-A scenario of comparatively high greenhouse gas emissions. *Climatic Change* 109(1): 33–57.

Schewe, J., Heinke, J., Gerten, D., et al. (2014). Multimodel assessment of water scarcity under climate change. *Proceedings of the National Academy of Sciences (USA)* 111 (9): 3245–3250.

Seiller, G., Roy, R., & Anctil, F. (2017). Influence of three common calibration metrics on the diagnosis of climate change impacts on water resources. *Journal of Hydrology* 547: 280–295.

Shi, X., Wood, A. W., & Lettenmaier, D. P. (2008). How essential is hydrologic model calibration to seasonal streamflow forecasting? *Journal of Hydrometeorology* 9(6): 1350–1363.

Soil Survey Staff (2015). Natural Resources Conservation Service. United States Department of Agriculture, Web Soil Survey. Available from http://websoilsurvey.nrcs.usda.gov/ (Last accessed 1 March 2018).

Stone, M. C., Hotchkiss, R. H., & Mearns, L. O. (2003). Water yield responses to high and low spatial resolution climate change scenarios in the Missouri River Basin. *Geophysical Research Letters* 30(4): 1186.

Thomson, A. M., Calvin, K. V., Smith, S. J., et al. (2011). RCP4.5: A pathway for stabilization of radiative forcing by 2100. *Climatic Change* 109(1–2): 77–94.

USBR (2013). *Downscaled CMIP3 and CMIP5 Climate and Hydrology Projections: Release of Downscaled CMIP5 Climate Projections, Comparison with Preceding Information, and Summary of User Needs* (47 pp.). Denver, CO: U.S. Department of the Interior, Bureau of Reclamation, Technical Services Center.

Vuuren, D. P. V., Stehfest, E., Elzen, M. G. J. D., et al. (2011). RCP2.6: Exploring the possibility to keep global mean temperature increase below 2°C. *Climatic Change* 109(1–2): 95–116.

Wagener, T., Sivapalan, M., Troch, P. A., et al. (2010). The future of hydrology: An evolving science for a changing world. *Water Resources Research* 46(5): W05301.

Watts, A., Grant, G., & Safeeq, M. (2016). *Flows of the Future – How Will Climate Change Affect Streamflows in the Pacific Northwest? Science Findings, 187.* Portland, OR: U.S. Department of Agriculture, Forest Service, Pacific Northwest Research Station.

Wilby, R. L., Troni, J., Biot, Y., et al. (2009). A review of climate risk information for adaptation and development planning. *International Journal of Climatology* 29(9): 1193–1215.

Yin, Y., Tang, Q., Liu, X., & Zhang, X. (2017). Water scarcity under various socio-economic pathways and its potential effects on food production in the Yellow River Basin. *Hydrology and Earth System Sciences* 21: 791–804.

Yuan, F., Zhao, C., Jiang, Y., et al. (2017). Evaluation on uncertainty sources in projecting hydrological changes over the Xijiang River basin in South China. *Journal of Hydrology* 554: 434–450.

Zhang, Q., Knowles, J. F., Barnes, R. T., et al. (2018). Surface and subsurface water contributions to streamflow from a mesoscale watershed in complex mountain terrain. *Hydrological Processes* 32(7): 954–967.

20

Future Water Scarcity over the Yellow River Basin and the Effects of Adaptive Measures

ZHONGWEI HUANG, QIUHONG TANG AND YUANYUAN YIN

20.1 Introduction

Water resources are essential to human life, socioeconomic development and ecology. Global water demand has increased about four times during past several decades as a result of rapid growth of population and increasing living standard (Shiklomanov, 2000; Döll et al., 2009; Wada & Bierkens, 2014; Huang et al., 2018). Furthermore, global change affects the terrestrial water cycle (Tang & Oki, 2016; Tang, 2020) and the amount and distribution of available fresh water resources (Stohlgren et al., 1998; Tang et al., 2008; Kustu et al., 2011; Wang & Hejazi, 2011; Taylor et al., 2013; Döll et al., 2014; Haddeland et al., 2014). As water demand approaches or even exceeds the total renewable freshwater resource, water scarcity has become one of the major risks to sustainable development in many areas of the world, threatening food security, human health, socioeconomic development as well as ecosystem services (Cheng et al., 2009; Rodell et al., 2009; Wada et al., 2011b; Taylor et al., 2013; Kummu et al., 2016; Yin et al., 2017). Furthermore, future global water scarcity is expected to intensify along with climate change and socioeconomic development, especially in the semi-arid and arid areas, such as the Middle East, western US and northern China (Hanasaki et al., 2013; Hejazi et al., 2014; Schewe et al., 2014; Mekonnen & Hoekstra, 2016). Accurate projection of future water scarcity conditions should consider both climate change, population growth, socioeconomic development land use change and other factors that may affect water demand and water availability. An interdisciplinary approach that accounts for the diversity and complexity of processes that lead to water scarcity is key to ensure the applicability of future water management within a sustainable development framework.

The Yellow River (YR) is the second-longest river in China and represents an important part of the Chinese culture. In 2010, the basin counted about 114 million

This research is supported by the National Natural Science Foundation of China (41730645, 41790424).

people and covered 12.6 million hectares of cropland (Yellow River Conservancy Commission, 2013). Increasing population and socio-economic development has led to significant increases in human water demand in the YR basin and have put pressure on the supply of freshwater resources. As reported in previous studies, YR has faced severe water scarcity conditions during past decades, which have brought great challenges for water managers to meet the human and environmental water demands (Wada et al., 2011a; Veldkamp et al., 2015, 2017; Yin et al., 2017; Liu et al., 2019).

Water scarcity assessment should take both water demand and water availability into consideration. Various studies have simulated the changes in water availability and demand in the YR basin. For example, streamflow of the YR has decreased partly because of the decrease in precipitation and increase in temperature since 1950 (Tang et al., 2008; Xu, 2011; Wang et al., 2012). Water availability is likely to decrease due to both precipitation decrease and temperature increase over the next few decades in YR (Li et al., 2012; Davie et al., 2013; Haddeland et al., 2014). In addition, changes in water demand by sectors have also been estimated and results agreed that population growth and socioeconomic development greatly led to increase the water demand (Yin et al., 2017; Huang et al., 2018) and improving techniques may reduce the pressure on the water supply (Zhou et al., 2020). Therefore, an explicit assessment of future water scarcity in YR considering both effects of climate change and socioeconomic development is necessary. However, many previous water scarcity projections have only focused on the impacts of climate change and have not considered the impacts from a socioeconomic development perspective. Yin et al. (2017) conducted a quantitative assessment of water scarcity in the YR basin under pressure from both climate change and socioeconomic development, but they ignored the impacts of future land use change on irrigation water demand (Huang et al., 2019). Based on the model simulations provided by Yin et al. (2017), this chapter analyses future water scarcity in the YR basin at sub-basin level while taking into account the impacts of climate change, socioeconomic change, population growth and land use change.

To mitigate regional water scarcity, various adaptive measures have been taken, such as improving water-saving techniques, inter-basin water transfer projects and virtual water trade (Zhao et al., 2015; Hanasaki et al., 2016; Flörke et al., 2018). Actually, the water managers have implemented a water flow regulation rule in the YR basin since the 1980s, which enforced an upper limit on water withdrawals for the eight provinces that rely on water supply from the river (Cai & Rosegrant, 2004). This inner-basin water allocation measure is expected to increase economic prosperity and alleviate water scarcity in the lower sub-basins where the major human water supply relies on upstream inflow (Yin et al., 2017). Furthermore, some efficient irrigation technologies and energy cooling technologies have also been applied in the YR basin to reduce sectoral water demand (Zhou et al., 2020). In addition, the South-to-North Water Diversion (SNWD) project, which is the world's largest and most expensive inter-basin water diversion project, will transfer water from the Yangtze River to the YR and increase its water availability (Barnett et al., 2015). However, the efficiency of these adaptive measures to mitigate future water scarcity in the YR basin remains unknown.

To address the issues discussed here, this chapter presents an explicit assessment about future water scarcity in the YR basin by considering the effects of both climate change and socioeconomic development. The aims of this chapter are to: (1) assess the effects of future water scarcity in YR at sub-basin scale while accounting for the impacts of climate change and socioeconomic development; and (2) further analyse the effects of inter-basin water transfer projects and water-saving techniques improvements on future water scarcity mitigation.

20.2 The Yellow River Basin

The YR originates from the northern foothills of the Tibetan Plateau and flows through nine provinces and autonomous regions of China before finally reaching the Bohai Gulf (Figure 20.1). The total area of the YR basin is 75,200 km^2, with most parts of the basin belonging to arid or semi-arid regions. The YR basin lies in a temperate continental climate zone and the mean annual temperature varies spatially from 5 to 15°C across the basin – in this region, temperature variability is mainly determined by the latitude. Precipitation also varies spatially across the entire river basin, with a range of mean annual precipitation from 60 to 900 mm and shows an increasing trend from the northwest to southeast. The YR basin is dominated by three major land cover types: grasslands (47.6 per cent), croplands (26.1 per cent) and forest and shrublands (13.4 per cent) (Cong et al., 2009). In addition, the croplands are mainly distributed in the middle and lower reach of the

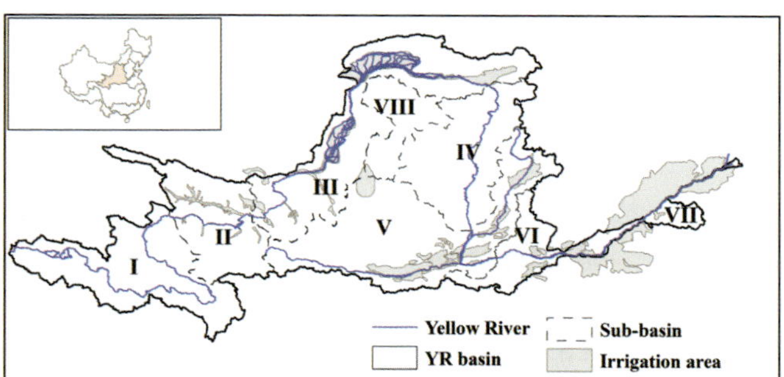

Figure 20.1 The Yellow River (YR) basin and its eight sub-basins

Table 20.1 Basic information of the eight sub-basins in the YR basin

Sub-basin	Areas (10^3 km^2)	Population (10^6)	Irrigated areas (km^2)	Rainfed areas (km^2)	Proportion of water availability	Reaches
I	127	0.65	219	27	0.6%	Upper
II	87	9.17	2,680	1,706	8.2%	
III	157	16.05	23,692	2,106	37.4%	
IV	107	8.71	1,591	3,940	3.4%	Middle
V	185	51.19	25,422	12,311	16.5%	
VI	40	13.40	5,717	2,956	6.1%	
VII	51	13.91	42,824	2,430	27.2%	Lower
VIII	42	0.57	446	225	0.6%	Endorheic

YR as there are two large irrigation districts and the urban cites and built-up lands are concentrated in the middle and lower reaches, which may greatly affect local water availability (Tang et al., 2008). The population in the YR basin is about 114 million, accounting for about 9 per cent of China's total population, and the gross domestic product (GDP) represents about 8 per cent of China's total GDP (as shown in Table 20.1). The population and high GDP areas are mostly concentrated downstream along the river (Yin et al., 2017).

20.3 Data and Methods

This study targets a middle of the road development scenario associated with climate change mitigation and adaption scenarios, by considering a representative concentration pathway 6.0 (RCP6.0) and shared socioeconomic pathway 2 (SSP2). The RCP6.0–SSP2 scenario is a suitable combination of the scenario matrix architecture (O'Neill et al., 2014; van Vuuren et al., 2014; Fricko et al., 2017). This combination depicts a moderate evolution of climate change and socioeconomics in the future, avoiding extreme pathways that are less likely to happen with the ongoing international efforts dedicated to mitigating climate change and seeking sustainable development goals (SDGs).

20.3.1 Water Demand Simulation

Agricultural water demand (*WD_agr*) was estimated by the Global Change Assessment Model (GCAM) and two of its ancillary modelling tools, namely Demeter and Xanthos, which can well represent the effects of future climate and land use changes, especially future changes in cropland area driven by population growth and socioeconomic development (Edmonds et al., 1997; Kim et al., 2006; Calvin et al., 2014, 2018; Huang et al., 2019). *WD_agr* was calculated by multiplying the crop water requirements per unit of irrigated area with the area of irrigated land and the irrigation efficiency. Here, crop-specific water requirements are closely related to climatic and hydrological conditions in croplands, such as temperature, precipitation, soil moisture and crop calendar; while irrigation efficiency and irrigated cropland area are determined by socioeconomic development, technological improvement and governmental policies. In this study, climate forcing from five global climate models (GCMs) was served as climate inputs for GCAM framework, namely HadGEM2-ES, NorES1-M, IPSL-CM5A-LR, MIROC-ESM-CHEM and GFDL-ESM2M under the RCP6.0 scenario. Future land use (especially irrigated cropland) was projected by GCAM at the regional level under SSP2 scenario, and Demeter, a disaggregation tool, was then used to disaggregate crop information from the regional to grid scale (West et al., 2014; Le Page et al., 2016; Vernon et al., 2018). Irrigation efficiency across global major regions under SSP2 scenario was obtained from Chaturvedi et al. (2015) in which irrigation efficiency is projected to rise from 0.35 in 2005 to 0.42 in 2050 under SSP2 scenario in the YR basin. In general, *WD_agr* was generated at a spatial resolution of $0.5^\circ \times 0.5^\circ$ at monthly time scale, the monthly gridded *WD_agr* was then aggregated to annual scale at sub-basin level. We used the 31-year moving average value of *WD_agr* to exclude the impact of inter-annual variability on the changing signal. However, there were considerable biases between simulated *WD_agr* and reported data in the YR basin during the historical period. By applying the correction factors between model estimates and reported data to the annual average agricultural water demand during 1996–2005 at sub-basin level in YR, annual agricultural water demand during 2010–2050 were bias-corrected.

The industrial water demand only included water demand for the manufacturing sector following Yin et al. (2017). The manufacturing water demand was positively correlated with the economic metric manufacturing gross value added and was calculated following Flörke et al. (2013) and Wada et al. (2016). Annual industrial water demand (WD_{ind}) was estimated by the value added of manufacturing sectors and the industrial water use intensities:

$$WD_{ind} = GDP_{manu} \times I_{ind,t0} \times (1 - r_{manu})^{(t-t0)}, \tag{20.1}$$

where GDP_{manu} is the value added to manufacturing sector; $I_{ind,t0}$ is water intensity for manufacturing (m^3 per 10,000 RMB) for the year 2010 and is set as 85.7; and r_{manu} is the change of the manufacturing water intensity (%), set as 0.6 per cent for YR (Wada et al., 2016; Yin et al., 2017).

The domestic water demand was mostly affected by population and human living standard (Alcamo et al., 2007). Following Hanasaki et al. (2013), the annual domestic water demand (WD_{dom}) was calculated by the following equation:

$$WD_{dom} = Pop \times (I_{dom,t0} + r_{dom} \times (t - t0)) \times 0.365, \quad (20.2)$$

where Pop is the population in the YR basin; $I_{dom,t0}$ is the domestic water intensity (L day^{-1} person^{-1} yr^{-1}) and set as 85.7 for the base year 2010, r_{dom} is the domestic water intensity change rate following Hanasaki et al. (2013) (L day^{-1} person^{-1} yr^{-1}) and the multiplier 0.365 is applied for unit conversion.

20.3.2 Water Availability Simulation

The annual runoff, which was treated as available surface water resources and obtained from the Inter-Sectoral Impact Model Inter-comparison Project Phase Fast-track,[1] was derived for the period 1995–2065 at a $0.5° \times 0.5°$ spatial resolution from an ensemble of five global hydrological models (GHMs), that is, H08 (Hanasaki et al., 2008a, 2008b), VIC (Liang et al., 1994), MPI-HM (Stacke & Hagemann, 2012), PCR-BLOBWB (van Beek et al., 2011; Wada et al., 2011b) and WBMplus (Wisser et al., 2010). These GHMs were forced by climate projections from five general circulation models (GCMs), namely HadGEM2-ES, NorES1-M, IPSL-CM5A-LR, MIROC-ESM-CHEM and GFDL-ESM2M under the RCP6.0 scenario. Due to the biases between GCM outputs and observations, climate datasets from these GCMs were bias-corrected using the WATCH climate data for the overlapping period (Piani et al., 2010; Hempel et al., 2013). Similar to agricultural water demand, water availability was also biased in the whole YR basin due to the reporting issues associated with the availability of water resources data from YRCC. We used the ensemble average for water scarcity assessment. Furthermore, to maintain the river in a desired environmental condition, we assumed that only a part of the renewable water resources could be appropriated by humans. The net water withdrawal, that is, water withdrawal minus return flow, accounted for 53 per cent of renewable water resources in the 1980s and 72 per cent in the 2010s (YRCC, 2013). In this study, we assume that only 40 per cent of natural water availability can be used for humans, similar to that in Schewe et al. (2014). In addition, the water flow regulation rule implemented by YRCC sets the upper limit on water withdrawals for each sub-basin. According to the rule, the proportion of water availability is prescribed for each sub-basin as shown in Table 20.1. Thus, water availability for each sub-basin is calculated by the total water availability in YR and the proportion of water availability in each sub-basin.

20.3.3 Water Scarcity Assessment

The water stress index (WSI), defined as the ratio of water demand to water availability, was used to quantify water scarcity condition in this study. This indicator can well represent the imbalance between water availability and total water demand for humans over a specific time period and for a specific region and has been applied in a wealth of studies to calculate water scarcity (Vörösmarty et al., 2000; Wada et al., 2011b; Mekonnen & Hoekstra, 2016; Veldkamp et al., 2017). The larger the WSI value, the more severe the water scarcity conditions are. In this study, WSI was calculated at annual scale at sub-basin spatial scale in the YR basin:

$$WSI_{i,y} = \frac{WD_{i,y}}{WA_{i,y}}, \tag{20.3}$$

where $WSI_{i,y}$represents the water scarcity condition for sub-basin i in the year y, $WD_{i,y}$ is the total water demand (the sum of agriculture, domestic and industrial) and $WA_{i,y}$ is the water availability for sub-basin i in the year y. In this study, we used the conventional WSI threshold of 1 to define the severe water scarcity which assumes that water scarcity occurs when water demand exceeds the total water availability for humans.

Furthermore, to further quantify water scarcity conditions at sub-basin level, we used water deficit as an indicator to evaluate water scarcity conditions (Bijl et al., 2018; Flörke et al., 2018). The indicator was defined as the difference between sub-basin water demand and water availability and was calculated as follows:

$$DEF_i = \max(WD_i - WA_i, 0), \tag{20.4}$$

where WD_i and WA_i are total water demand and water availability at sub-basin level, respectively (m^3). A 0 value of water deficit indicates that water demand is fully satisfied and a higher water deficit value means more severe water scarcity conditions.

20.3.4 Adaptive Measures

As water scarcity has become a significant issue in the YR basin, various adaptive measures have been taken from both water demand and water availability sides. This chapter chooses improvements in water-saving techniques and inter-basin water transfer as two typical measures to quantify the impacts of adaptive measures on future water scarcity mitigation in the YR basin (Table 20.2). In terms of agricultural water demand, irrigation efficiency (IE) is projected to rise from 0.35 in 2005 to 0.42 in 2050 under the SSP2 scenario, and we assume that IE will reach

Table 20.2 Adaptive measures during 2010–2050: irrigation efficiency without (IE_NOMEA) and with (IE_MEA_DE) adaptive measures; industrial water intensity change rate without (IND_NOMEA) and with (IND_MEA _DE) adaptive measures; water transfer volume by two water transfer projects (billion m^3), that is, Water transfer from Hanjiang River to Weihe River project (WTP_1) and Water transfer from the upstream of Yangtze River to the upstream of the YR (WTP_2)

	2010	2020	2030	2040	2050
IE_NOMEA	0.358	0.373	0.389	0.404	0.42
IE_MEA_DE	0.378	0.433	0.489	0.544	0.6
IND_NOMEA	0.60%				
IND_MEA_DE	1.20%				
WTP_1	0	1	1.5	1.5	1.5
WTP_2	0	0	8	8	8

0.6 by 2050 which considers a potential future improvement of the efficiency of water saving techniques. The industrial water intensity change rate (shown in Eq. (20.2)) is defined as 0.6 per cent under SSP2 scenarios, and we assume that the industrial water intensity change rate will increase by 1.2 per cent assuming high technical improvement for industrial water demand (Hanasaki et al., 2013; Yin et al., 2017). In this framework, we assume that adaptive measures for industrial and irrigation water demand do not vary across all sub-basins.

The inter-basin water transfer project is an efficient solution to alleviate severe water scarcity. As planned by the Chinese government, two inter-basin water transfer projects are considered in this study, that is, the Water transfer from Hanjiang River to Weihe River project from 2020, which will supply 1 billion m^3 water from 2020 and 1.5 billion m^3 from 2030 to YR, as well as Water transfer from the upstream of the Yangtze River to the upstream of the YR, which will also transfer 8 billion m^3 water from 2030 (YRCC, 2013). Water transferred to the Yellow River basin was allocated to eight sub-basins based on water allocation rules as described in Table 20.2. Thus, we compared water scarcity conditions when considering adaptive water allocation or not; the effects of adaptive measures on water scarcity can be analysed.

20.4 Results

20.4.1 Changes in Future Water Availability and Water Demand

Figure 20.2 shows the estimated water availability in the YR basin and eight sub-basins during 2010–2050. The total water availability in YR is projected to increase significantly from 32.5 billion m^3 in 2010 to 34.5 billion m^3 in 2030, and then is projected to decrease to 32.5 billion m^3 in 2050. Water availability in each sub-basin level – 60 per cent of local runoff in each sub-basin without inflow from

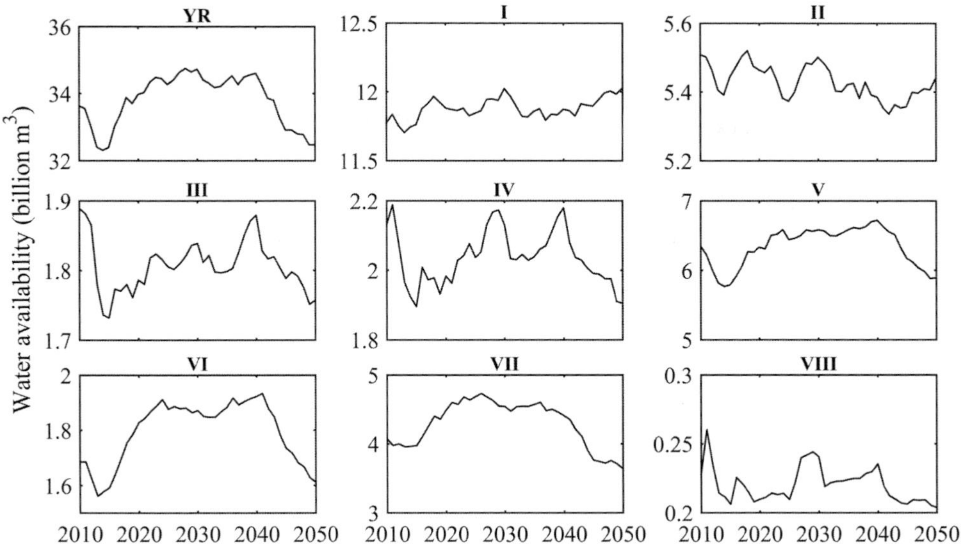

Figure 20.2 Local water availability (i.e., local runoff) during 2010–2050 in the Yellow River basin and its sub-basins

upstream areas – shows different patterns in 2010–2050. Sub-basins III–VII show similar changing trends with local water availability, first showing an increasing trend and then decreasing, in line with the patterns of water availability in the whole basin. However, local water availability in sub-basin I shows a significant increasing trend from 11.7 billion m^3 in 2010 to 12.1 billion m^3 in 2050, which accounts for one third of the water availability in the whole basin. Sub-basin II shows a slight decrease in water availability during 2010–2050.

Figure 20.3 shows the simulated total and sectoral (agricultural, domestic and industrial) water demand in the YR basin and eight sub-basins under RCP6.0–SSP2 during 2010–2050. In the YR basin, the total water demand is projected to increase from 35.4 billion m^3 in 2010 to 52.3 billion m^3 in 2050. This increase is primarily driven by a growth in the industrial water demand, which is expected to increase by 346 per cent in 2010–2050 as the result of a high increase in the value added from the manufacturing sectors. As the dominant water use sector, agricultural water demand in YR is expected to increase slightly by 19 per cent from 28.1 billion m^3 in 2010 to 33.5 billion m^3 in 2050, which indicates that the effects of land use change and warming climate is larger than that of improvement in irrigation efficiency. As both population and domestic water use intensities will increase and then decrease due to a decrease in population, domestic water demand varies very little. At sub-basin level, sub-basin III counts for the largest proportion due to a large agricultural water demand in the Hetao irrigation district. In sub-basins I–V and VIII, although the industrial water demand increases rapidly,

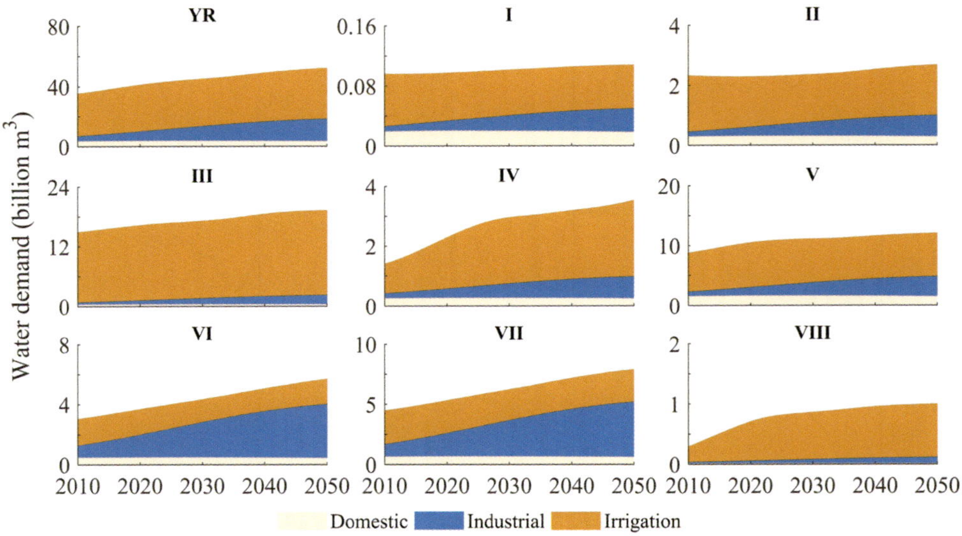

Figure 20.3 Changes in water demand for domestic, industrial and irrigation during 2010–2050 in the Yellow River basin and its sub-basins under RCP6.0–SSP2 scenario

agricultural remains the dominant water demand sector in 2010–2050. In sub-basins VI and VII, the agricultural water demand is the dominant water demand sector at the beginning of the period, while industrial might become the dominant water demand sector after the 2030s due to a significant predicted increase in the value added from the manufacturing sector in the downstream of the YR.

20.4.2 Future Water Scarcity in the Yellow River and its Changing Patterns

Figure 20.4 shows the annual WSI for the YR basin and eight sub-basins during 2010–2050 under RCP6.0–SSP2 scenario. The WSI for YR is projected to increase by 57 per cent from 1.05 in 2010 to 1.65 in 2050 due to an increase in water demand and changes in water availability. The upper reaches of the YR basin (sub-basins I and II) are projected to have a WSI less than 1, meaning that water availability would be larger than water demand. WSI in sub-basin II first shows a decreasing trend before 2030 because of a significant increase in water availability but then increases and reaches 1.01, indicating that sub-basin II is likely to face water scarcity after 2050. In the endorheic basin (sub-basin VIII) and the middle reaches (sub-basins III, IV, V and VI) the WSI is always larger than 1, representing that the water would be scarce in 2010–2050 in these areas. The endorheic basin (sub-basin VIII) is expected to suffer the most severe water scarcity, with WSI increasing from 1.45 in 2010 to 5.14 in 2050. WSI in sub-basins III, IV, V and VI

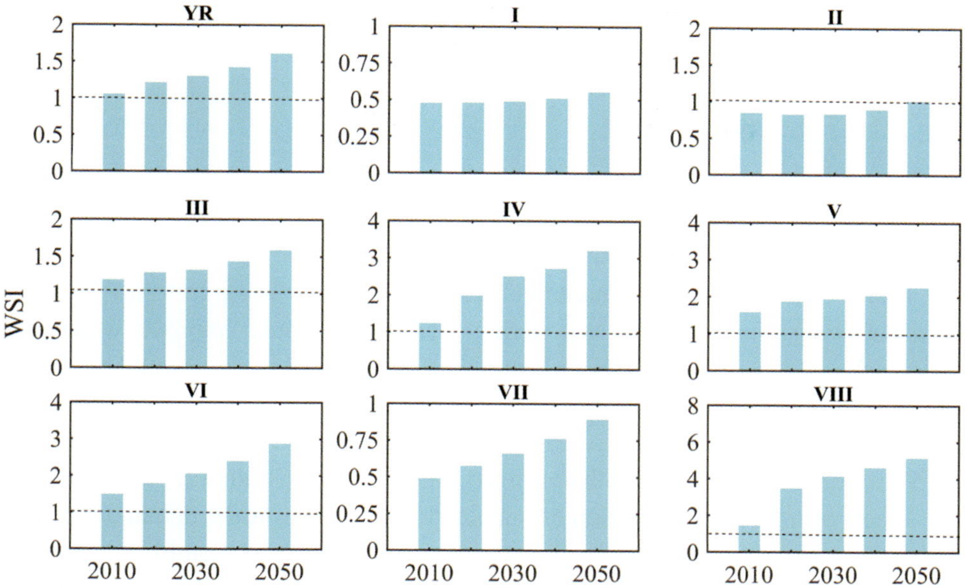

Figure 20.4 Water stress index (WSI) during 2010–2050 in the Yellow River basin and its sub-basins under RCP6.0–SSP2 scenario

will increase by 32, 159, 42 and 94 per cent, respectively, during 2010–2050, indicating that intensifying water scarcity conditions occur in the middle reaches of YR. In terms of the lower reach of YR (sub-basin VII), the WSI is expected to increase from 0.49 to 0.89, which appears an underestimated increase compared to a previous prediction (Yin et al., 2017). This discrepancy may be attributed to the fact that we partly considered the agricultural water demand outside the YR basin, which might have led to underestimation of the water demand in sub-basin VII.

The water deficit of the YR basin in 2010 was 6.93 billion m^3, which accounts for 20 per cent of total water demand. This deficit is likely to increase by 200 percent to 20.81 billion m^3 in 2050, which indicates that water scarcity in the YR will increase significantly in the future. At sub-basin level, water deficits in sub-basins I and II are 0 during the whole 2010–2050 period, where abundant water resources can fully satisfy the local water demand. In the endorheic basin (sub-basin VIII) the water deficit is also tiny but shows a significant increasing trend, that is, from 0.09 billion m^3 in 2010 to 0.81 billion m^3 in 2050. Furthermore, large water deficits mainly occur in sub-basins III and V, which occupy most of the total water demand in the YR basin. Therein, water deficits in sub-basins III and V should increase by 204 per cent and 103 per cent during 2010–2050, respectively. In addition, the water deficit is expected to increase from 0.27 billion m^3 to 2.43 billion m^3 in sub-basin IV and from 0.99 billion m^3 to 3.71 billion m^3 in sub-basin VI during 2010–2050. Despite the water deficit values in these two sub-basins (IV and VI) not being comparable to those of sub-basins III and V, the significant increase of water deficit values signal that the region might have to face severe water scarcity conditions in the future.

20.4.3 The Effects of Adaptive Measures on Water Scarcity Mitigation

This chapter provides two adaptive measures to cope with future water scarcity in the YR basin from both water demand and water availability sides. In terms of water demand, we assumed that water saving techniques would improve water use efficiency and water demand for industrial and agricultural sectors would decrease, which should help reduce water scarcity. As shown in Figure 20.5, irrigation water demand in YR is expected to decrease by about 30 per cent in 2050 (i.e., 33.54 and 23.57 billion m^3 without and with improved water saving technique applications). As the effects of improving irrigation efficiency on agricultural water demand increase along the period, the effects of adaptive measures on water demand became more efficient in the future. Improvement in the industrial water intensity change rate might also lead to a reduction in industrial water demand, that could reach 24 per cent in 2050. Thus, a reduction in the total water demand results in significant alleviation in water scarcity and WSI in the YR is expected to decrease by 35.3 per cent in 2050 with WSI values expected to reach 1.61 and follow a decrease to 1.19 if adaptive measures are applied. At sub-basin level (shown in Table 20.3) the WSI in

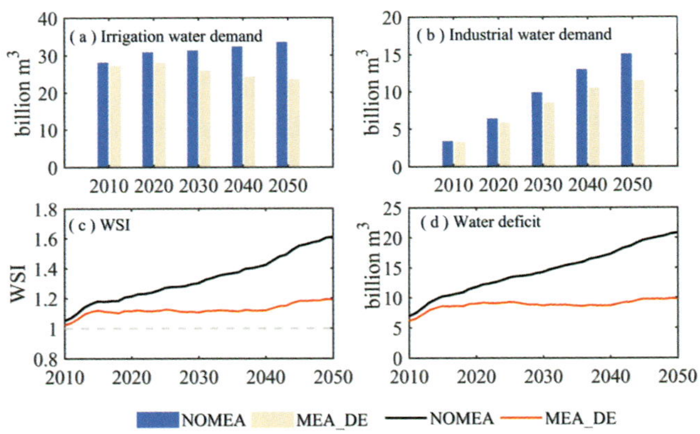

Figure 20.5 The effects of water-saving techniques improvements on water demand and water scarcity in the YR basin: (a) changes in irrigation water demand with consideration of improvement in irrigation efficiency; (b) changes in industrial water demand when considering high technical improvement; (c) and (d) the effects of water-saving techniques improvements on WSI and water deficits in YR during 2010–2050, respectively. NOMEA means without water-saving techniques improvements and MEA_DE means considering the effects of water-saving techniques improvements

Table 20.3 Water stress index (WSI) for the years 2010 and 2050 in the Yellow River basin and its eight sub-basins under three different scenarios: without adaptive measures (NOMEA), with water-saving technique applied in irrigation and industrial sectors (MEA_DE) and with inter-basin water transfer (MEA_AV)

	2010			2050		
	NOMEA	MEA_DE	MEA_AV	NOMEA	MEA_DE	MEA_AV
YR	1.05	1.02	1.05	1.61	1.19	1.35
I	0.48	0.47	0.48	0.56	0.43	0.46
II	0.85	0.83	0.85	1.01	0.76	0.84
III	1.19	1.15	1.19	1.59	1.13	1.33
IV	1.24	1.20	1.24	3.20	2.35	2.68
V	1.58	1.54	1.58	2.25	1.70	1.89
VI	1.48	1.45	1.48	2.88	2.20	2.41
VII	0.49	0.48	0.49	0.89	0.68	0.75
VII	1.45	1.44	1.45	5.14	3.86	4.30

2050 should decrease by 23–29 per cent for eight sub-basins when considering improved water saving techniques, and sub-basin II should be exempt of water scarcity in 2050 as the WSI in sub-basin II for 2050 decreases from 1.01 to 0.76. Water deficit should reduce by about 11 billion m^3 (53 per cent) in the YR basin, assuming that water saving techniques will improve. The reduction in water deficit at sub-basin level should vary from 0.02 to 5.5 billion m^3 in 2050 (shown in Table 20.3). The most significant mitigation of water scarcity occurs in sub-basin III, where improvement in irrigation efficiency leads to a large decrease in agricultural water demand. The water deficit of sub-basin V should decrease by about 2.94 billon m^3 in 2050 as water saving techniques reducing water demand for both agricultural and industrial sectors are expected to improve.

In this study, we considered two inter-basin water transfer projects: (1) water transfer from Hanjiang River to Weihe River project from 2020 and 2030, and (2) water transfer from the upstream of the Yangtze River to the upstream of the Yellow River from 2030. As shown in Figure 20.6, water availability in the YR is expected to increase by 0.6 billion m^3 in 2020 and by 6.3 billion m^3 in 2050, assuming the realization of the inter-basin water transfer projects. The significant increase in water availability in 2050 could help to mitigate water scarcity in the YR with an expected WSI decreasing by 16 per cent and water deficit decreasing by 19 per cent from 20.81 billion m^3 to 16.76 billion m^3, respectively. At sub-basin level, the WSI for the eight sub-basins should decrease and sub-basin II should become exempt from water scarcity in 2050 due to an expected increase in water availability (shown in Table 20.2). Water deficits for sub-basins III to VI should decrease by 33, 9, 15 and 10 per cent in 2050, respectively (shown in Table 20.4). In general, inter-basin water transfer projects are expected to greatly mitigate water scarcity conditions in the YR basin.

Table 20.4 Water deficit for the years 2010 and 2050 in the Yellow River basin and its eight sub-basins under three different experiments: without adaptive measures (NOMEA), with water-saving technique applied in irrigation and industrial sectors (MEA_DE) and with inter-basin water transfer (MEA_AV), unit: billion m^3

	2010			2050		
	NOMEA	MEA_DE	MEA_AV	NOMEA	MEA_DE	MEA_AV
YR	6.93	6.15	6.93	20.81	9.81	16.76
I	0.00	0.00	0.00	0.00	0.00	0.00
II	0.00	0.00	0.00	0.02	0.00	0.00
III	2.34	1.90	2.34	7.13	1.63	4.77
IV	0.27	0.23	0.27	2.43	1.49	2.22
V	3.24	3.01	3.24	6.71	3.77	5.67
VI	0.99	0.92	0.99	3.72	2.37	3.33
VII	0.00	0.00	0.00	0.00	0.00	0.00
VIII	0.09	0.09	0.09	0.81	0.56	0.77

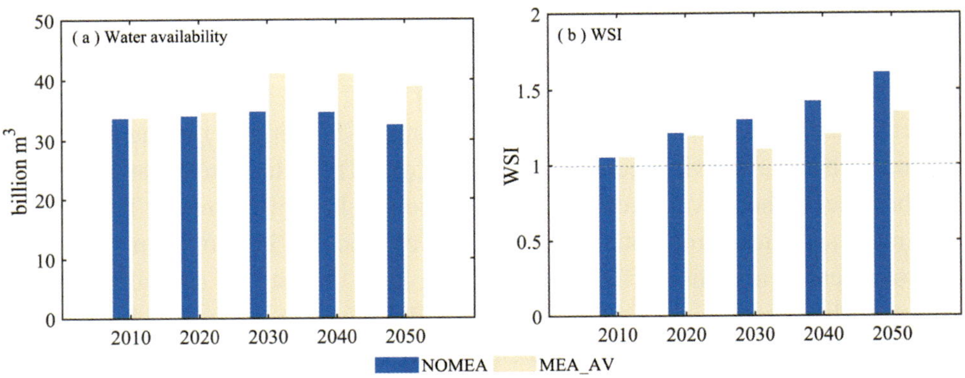

Figure 20.6 The effects of inter-basin water transfer projects on water availability (a) and water scarcity (b) in the YR basin during 2010–2050. NOMEA and MEA_AV mean with and without the consideration of inter-basin water transfer projects

20.5 Discussion

The quantitative projections of future water scarcity in YR under the RCP6.0–SSP2 scenario remain uncertain. Uncertainties stem from potential fluctuations in the future changes in the sectoral water demand, water availability and methodology in assessing water scarcity. Greve et al. (2018) reported three major sources of uncertainty in water scarcity assessment (i.e., GHMs, GCMs and future scenarios) and their results showed that the uncertainties caused by these three factors in water scarcity assessment are increasing worldwide. Among them, uncertainties from GHMs have been proven to be the largest source as a result of the differences in structure, parameters, formulations and assumptions in GHMs

(Wada et al., 2013; Piontek et al., 2014; Liu et al., 2017; Greve et al., 2018). In this chapter, future agricultural water demand was simulated by integrated assessment models that consider both future climate and land use effects and projections of future climate and land use change would contribute to uncertainties in future agricultural water demand simulation (Huang et al., 2019). This chapter used the historical reported data to correct the bias of future projections and applied an ensemble mean and moving-average to reduce the uncertainties of their inter-annual variability. In terms of domestic and industrial water demand, as water demand is influenced by many social, economic and political factors, this study used the methodology from Hanasaki et al. (2013) and Yin et al. (2017) to calculate domestic and water demand which may introduce large uncertainties as they only use some typical impact factors such as population, manufacturing gross value and technology improvement. In addition, future runoff simulated by an ensemble of five GHMs forced by five GCMs that could be treated as renewable surface water resources would reduce the uncertainties. Due to the large variability in future climate projections arising from different GCMs, the uncertainties in water scarcity projection caused by GCMs cannot be neglected in spite of bias-correction (Müller Schmied et al., 2014). Furthermore, similar to agricultural water demand, bias-correction and moving average, as well as ensemble average, were used to account for uncertainties in simulated water availability. In addition, this study only focuses on water scarcity in the YR under a sensible middle-range scenario (RCP6.0–SSP2). However, additional climate change and socioeconomic development scenarios need to be considered to better comprehend the future risk of water scarcity and its uncertainties in the YR region.

Besides the uncertainties from water demand and water availability simulation, a few limitations in the methods of water scarcity assessment, which may also introduce uncertainties in the simulations, need to be taken into account in future studies. First, this study assumed that the environmental flow requirement would cover 40 per cent of the total surface of water availability. An improved model could account for fine-scale spatial and temporal fluctuations of the environmental flow requirement that vary regionally due to hydrological regimes and ecosystems variations across regions worldwide (Pastor et al., 2014; Veldkamp et al., 2017). Furthermore, the upstream human water uses would often affect water quality and available water resources in the downstream might be reduced due to pollution in the upstream. Neglecting water quality aspects may result in underestimation of downstream water scarcity (Liu et al., 2016; Ma et al., 2020). Moreover, groundwater, which is an important available water resource, particularly in arid regions, is ignored in water scarcity assessment, leading to an overestimation of the actual water scarcity conditions (Wada et al., 2011a).

20.6 Summary

Based on future projections of sectoral water demand and water availability in the YR basin, this chapter assessed future water scarcity in the YR basin while accounting for the impacts of both climate change and socioeconomic development. Furthermore, it assessed the role of inter-basin water transfer projects and water-saving techniques improvements on future water scarcity mitigation. Results show that total water availability in the YR basin is projected to first increase significantly in 2010–2030 and then is projected to decrease to 32.5 billion m^3 in 2050. Total water demand is projected to increase by 48 per cent in 2010–2050, which is primarily driven by the industrial water demand. In terms of water scarcity, the WSI for the YR basin is projected to increase by 57 per cent during 2010–2050 under the RCP6.0–SSP2 scenario and the most significant increase should occur in the middle and lower reaches. Water deficit also shows a significant increase from 6.93 billion m^3 in 2010 to 20.81 billion m^3 in 2050. Assuming improvements in water-saving techniques, agricultural and industrial water demand in the YR basin is expected to decrease by 30 and 24 per cent in 2050, respectively. Changes in water demand should contribute to about 35 and 53 per cent mitigation in the WSI and water deficit in 2050, respectively. Furthermore, inter-basin water transfer projects are expected to increase water availability by 0.6 billion m^3 in 2020 and 6.3 billion m^3 in 2050 in the entire YR basin, leading to mitigation in water scarcity in the YR basin with a WSI decrease by 16 per cent and water deficit decrease by 19 per cent. Despite the uncertainties in water scarcity assessments, this chapter illustrated future plausible scenarios that might lead to water scarcity in the YR basin and assessed the effects of adaptive measures (i.e., improvements in water saving techniques and inter-basin water transfer projects) on water scarcity mitigation, which is of great significance for regional water management and sustainable development.

Note

1 ISI-MIP, www.isimip.org/ (Last accessed 3 September 2021).

References

Alcamo, J., Flörke, M., & Märker, M. (2007). Future long-term changes in global water resources driven by socio-economic and climatic changes. *Hydrological Sciences Journal* 52(2): 247–275.

Barnett, J., Rogers, S., Webber, M., Finlayson, B., & Wang, M. (2015). Sustainability: Transfer project cannot meet China's water needs. *Nature* 527(7578): 295.

van Beek, L., Wada, Y., & Bierkens, M. F. (2011). Global monthly water stress: 1. Water balance and water availability. *Water Resources Research* 47: W07517.

Bijl, D. L., Biemans, H., Bogaart, P. W., et al. (2018). A global analysis of future water deficit based on different allocation mechanisms. *Water Resources Research* 54(8): 5803–5824.

Cai, X., & Rosegrant, M. W. (2004). Optional water development strategies for the Yellow River Basin: Balancing agricultural and ecological water demands. *Water Resources Research* 40: W08S04.

Calvin, K., Patel, P., Clarke, L., et al. (2018). GCAM v5.1: Representing the linkages between energy, water, land, climate, and economic systems. *Geoscientific Model Development* 12(2): 677–698.

Calvin, K., Wise, M., Kyle, P., et al. (2014). Trade-offs of different land and bioenergy policies on the path to achieving climate targets. *Climatic Change* 123(3–4): 691–704.

Chaturvedi, V., Hejazi, M., Edmonds, J., et al. (2015). Climate mitigation policy implications for global irrigation water demand. *Mitigation and Adaptation Strategies for Global Change* 20(3): 389–407.

Cheng, H., Hu, Y., & Zhao, J. (2009). Meeting China's water shortage crisis: Current practices and challenges. *Environmental Science & Technology* 43(2): 240–244.

Cong, Z., Yang, D., Gao, B., Yang, H., & Hu, H. (2009). Hydrological trend analysis in the Yellow River basin using a distributed hydrological model. *Water Resources Research* 45: W00A13.

Davie, J. C. S., Falloon, P. D., Kahana, R., et al. (2013). Comparing projections of future changes in runoff from hydrological and biome models in ISI-MIP, *Earth System Dynamics* 4(2): 359–374.

Döll, P., Fiedler, K., & Zhang, J. (2009). Global-scale analysis of river flow alterations due to water withdrawals and reservoirs. *Hydrology and Earth System Sciences* 13(12): 2413–2432.

Döll, P., Mueller Schmied, H., Schuh, C., Portmann, F. T., & Eicker, A. (2014). Global-scale assessment of groundwater depletion and related groundwater abstractions: Combining hydrological modeling with information from well observations and GRACE satellites. *Water Resources Research* 50(7): 5698–5720.

Edmonds, J., Wise, M., Pitcher, H., et al. (1997). An integrated assessment of climate change and the accelerated introduction of advanced energy technologies – An application of MiniCAM 1.0. *Mitigation and Adaptation Strategies for Global Change* 1(4): 311–339.

Flörke, M., Kynast, E., Bärlund, I., et al. (2013). Domestic and industrial water uses of the past 60 years as a mirror of socio-economic development: A global simulation study. *Global Environmental Change* 23(1): 144–156.

Flörke, M., Schneider, C., & McDonald, R. I. (2018). Water competition between cities and agriculture driven by climate change and urban growth. *Nature Sustainability* 1(1): 51–58.

Fricko, O., Havlik, P., Rogelj, J., et al. (2017). The marker quantification of the shared socioeconomic pathway 2: A middle-of-the-road scenario for the 21st century. *Global Environmental Change: Human and Policy Dimensions* 42(1): 251–267.

Greve, P., Kahil, T., Mochizuki, J., et al. (2018). Global assessment of water challenges under uncertainty in water scarcity projections. *Nature Sustainability* 1(9): 486–494.

Haddeland, I., Heinke, J., Biemans, H., et al. (2014). Global water resources affected by human interventions and climate change. *Proceedings of the National Academy of Sciences (USA)* 111(9): 3251–3256.

Hanasaki, N., Fujimori, S., Yamamoto, T., et al. (2013). A global water scarcity assessment under shared socio-economic pathways – Part 2: Water availability and scarcity. *Hydrology and Earth System Sciences* 17(7): 2393–2413.

Hanasaki, N., Kanae, S., Oki, T., et al. (2008a). An integrated model for the assessment of global water resources – Part 1: Model description and input meteorological forcing. *Hydrology and Earth System Sciences* 12(4): 1007–1025.

Hanasaki, N., Kanae, S., Oki, T., et al. (2008b). An integrated model for the assessment of global water resources – Part 2: Applications and assessments. *Hydrology and Earth System Sciences* 12(4): 1027–1037.

Hanasaki, N., Yoshikawa, S., Kakinuma, K., & Kanae, S. (2016). A seawater desalination scheme for global hydrological models. *Hydrology and Earth System Sciences* 20 (10): 1–25.

Hejazi, M. I., Edmonds, J., Clarke, L., et al. (2014). Integrated assessment of global water scarcity over the 21st century under multiple climate change mitigation policies. *Hydrology and Earth System Sciences* 18(8): 2859–2883.

Hempel, S., Frieler, K., Warszawski, L., Schewe, J., & Piontek, F. (2013). A trend-preserving bias correction – The ISI-MIP approach. *Earth System Dynamics* 4(2): 219–236.

Huang, Z., Hejazi, M., Li, X., et al. (2018). Reconstruction of global gridded monthly sectoral water withdrawals for 1971–2010 and analysis of their spatiotemporal patterns. *Hydrology and Earth System Sciences* 22(4): 2117–2133.

Huang, Z., Tang, Q., Lo, M.-H., et al. (2019). The influence of groundwater representation on hydrological simulation and its assessment using satellite-based water storage variation. *Hydrological Processes* 33(8): 1218–1230.

Kim, S. H., Edmonds, J., Lurz, J., Smith, S. J., & Wise, M. (2006). The ObjECTS framework for integrated assessment: Hybrid modeling of transportation. *The Energy Journal* (Special Issue #2): 63–91.

Kummu, M., Guillaume, J. H. A., de Moel, H., et al. (2016). The world's road to water scarcity: Shortage and stress in the 20th century and pathways towards sustainability. *Scientific Reports* 6(1): 1–16.

Kustu, M. D., Fan, Y., & Rodell, M. (2011). Possible link between irrigation in the US High Plains and increased summer streamflow in the Midwest. *Water Resources Research* 47: W03522.

Le Page, Y., West, T. O., Link, R., & Patel, P. (2016). Downscaling land use and land cover from the Global Change Assessment Model for coupling with Earth system models. *Geoscientific Model Development* 9(9): 3055–3069.

Li, L., Shen, H. Y., Dai, S., Xiao, J. S., & Shi, X. H. (2012). Response of runoff to climate change and its future tendency in the source region of Yellow River. *Journal of Geographical Sciences* 23(3): 431–440.

Liang, X., Lettenmaier, D. P., Wood, E. F., & Burges, S. J. (1994). A simple hydrologically based model of land surface water and energy fluxes for general circulation models. *Journal of Geophysical Research: Atmospheres* 99(D7): 14415–14428.

Liu, J. G., Liu, Q. Y., & Yang, H. (2016). Assessing water scarcity by simultaneously considering environmental flow requirements, water quantity, and water quality. *Ecological Indicators* 60(1): 434–441.

Liu, J. G., Yang, H., Gosling, S. N., et al. (2017). Water scarcity assessments in the past, present, and future. *Earth's Future* 5(6): 545–559.

Liu, X., Tang, Q., Liu, W., et al. (2019). A spatially explicit assessment of growing water stress in China from the past to the future. *Earths Future* 7(9): 1027–1043.

Ma, T., Sun, S., Fu, G., et al. (2020). Pollution exacerbates China's water scarcity and its regional inequality. *Nature Communications* 11(1): 650.

Mekonnen, M. M., & Hoekstra, A. Y. (2016). Four billion people facing severe water scarcity. *Science Advances* 2(2): e1500323.

Müller Schmied, H., Eisner, S., Franz, D., et al. (2014). Sensitivity of simulated global-scale freshwater fluxes and storages to input data, hydrological model structure, human water use and calibration. *Hydrology and Earth System Sciences* 18(9): 3511–3538.

O'Neill, B. C., Kriegler, E., Riahi, K., et al. (2014). A new scenario framework for climate change research: The concept of shared socioeconomic pathways. *Climate Change* 122(3): 387–400.

Pastor, A. V., Ludwig, F., Biemans, H., Hoff, H., & Kabat, P. (2014). Accounting for environmental flow requirements in global water assessments. *Hydrology and Earth System Sciences* 18(12): 5041–5059.

Piani, C., Weedon, G. P., Best, M., et al. (2010). Statistical bias correction of global simulated daily precipitation and temperature for the application of hydrological models. *Journal of Hydrology* 395(3): 199–215.

Piontek, F., Müller, C., Pugh, T. A., et al. (2014). Multisectoral climate impact hotspots in a warming world. *Proceedings of the National Academy of Sciences (USA)* 111(9): 3233–3238.

Rodell, M., Velicogna, I., & Famiglietti, J. S. (2009). Satellite-based estimates of groundwater depletion in India. *Nature* 460(7258): 999–1002.

Schewe, J., Heinke, J., Gerten, D., et al. (2014). Multimodel assessment of water scarcity under climate change. *Proceedings of the National Academy of Sciences (USA)* 111 (9): 3245–3250.

Shiklomanov, I. A. (2000). Appraisal and assessment of world water resources. *Water International* 25(1): 11–32.

Stacke, T., & Hagemann, S. (2012). Development and evaluation of a global dynamical wetlands extent scheme. *Hydrology and Earth System Sciences* 16(8): 2915–2933.

Stohlgren, T. J., Chase, T. N., Pielke, R. A., Kittel, T. G., & Baron, J. (1998). Evidence that local land use practices influence regional climate, vegetation, and stream flow patterns in adjacent natural areas. *Global Change Biology* 4(5): 495–504.

Tang, Q. (2020). Global change hydrology: Terrestrial water cycle and global change. *Science China Earth Sciences* 63(3): 459–462.

Tang, Q., & Oki, T. (2016). *Terrestrial Water Cycle and Climate Change: Natural and Human-Induced Impacts, American Geophysical Union (AGU) Geophysical Monograph Series* (Vol. 221). Hoboken, NJ: John Wiley & Sons.

Tang, Q., Oki, T., Kanae, S., & Hu, H. (2008). Hydrological cycles change in the Yellow River basin during the last half of the twentieth century. *Journal of Climate* 21(8): 1790–1806.

Taylor, R. G., Scanlon, B., Döll, P., et al. (2013). Ground water and climate change. *Nature Climate Change* 3(4): 322–329.

Veldkamp, T. I. E., Wada, Y., Aerts, J., et al. (2017). Water scarcity hotspots travel downstream due to human interventions in the 20th and 21st century. *Nature communications* 8(1): 1–12.

Veldkamp, T. I. E., Wada, Y., de Moel, H., et al. (2015). Changing mechanism of global water scarcity events: Impacts of socioeconomic changes and inter-annual hydro-climatic variability. *Global Environmental Change* 32(1): 18–29.

Vernon, C. R., Le Page, Y., Chen, M., et al. (2018). Demeter – A land use and land cover change disaggregation model. *Journal of Open Research Software* 6(1): 15.

Vörösmarty, C. J., Green, P., Salisbury, J., & Lammers, R. B. (2000). Global water resources: Vulnerability from climate change and population growth. *Science* 289 (5477): 284–288.

van Vuuren, D. P., Kriegler, E., O'Neill, B. C., et al. (2014). A new scenario framework for climate change research: Scenario matrix architecture. *Climate Change* 122(3): 373–386.
Wada, Y., Beek, L. P. H. V., & Bierkens, M. F. P. (2011a). Modelling global water stress of the recent past: On the relative importance of trends in water demand and climate variability. *Hydrology and Earth System Sciences* 8(4): 3785–3805.
Wada, Y., Beek, L. P. H. V., Viviroli, D., et al. (2011b). Global monthly water stress: 2. Water demand and severity of water stress. *Water Resources Research* 47(7): 197–203.
Wada, Y., & Bierkens, M. F. (2014). Sustainability of global water use: Past reconstruction and future projections. *Environmental Research Letters* 9(10): 104003.
Wada, Y., Flörke, M., Hanasaki, N., et al. (2016). Modeling global water use for the 21st century: Water Futures and Solutions (WFaS) initiative and its approaches. *Geoscientific Model Development* 9(1): 175–222.
Wada, Y., Wisser, D., Eisner, S., et al. (2013). Multimodel projections and uncertainties of irrigation water demand under climate change. *Geophysical Research Letters* 40(17): 4626–4632.
Wang, D., & Hejazi, M. (2011). Quantifying the relative contribution of the climate and direct human impacts on mean annual streamflow in the contiguous United States. *Water Resources Research* 47: W00J12.
Wang, S. J., Yan, M., Yan, Y. X., Shi, C. X., & He, L. (2012). Contributions of climate change and human activities to the changes in runoff increment in different sections of the Yellow River. *Quaternary International* 282(1): 66–77.
West, T. O., Le Page, Y., Huang, M., Wolf, J., & Thomson, A. M. (2014). Downscaling global land cover projections from an integrated assessment model for use in regional analyses: results and evaluation for the US from 2005 to 2095. *Environmental Research Letters* 9(6): 064004.
Wisser, D., Frolking, S., Douglas, E. M., et al. (2010). The significance of local water resources captured in small reservoirs for crop production – A global-scale analysis. *Journal of Hydrology* 384(3): 264–275.
Xu, J. (2011). Variation in annual runoff of the Wudinghe River as influenced by climate change and human activity. *Quaternary International* 244(1): 230–237.
Yin, Y., Tang, Q., Liu, X., & Zhang, X. (2017). Water scarcity under various socio-economic pathways and its potential effects on food production in the Yellow River basin. *Hydrology and Earth System Sciences* 21(2): 791–804.
YRCC (Yellow River Conservancy Commission) (2013). *Comprehensive Planning of Yellow River Basin (2012–2030)*. Zhengzhou, China: The Yellow River Water Conservancy Press (in Chinese).
Zhao, X., Liu, J., Liu, Q., et al. (2015). Physical and virtual water transfers for regional water stress alleviation in China. *Proceedings of the National Academy of Sciences (USA)* 112(4): 1031–1035.
Zhou, F., Bo, Y., Ciais, P., et al. (2020). Deceleration of China's human water use and its key drivers. *Proceedings of the National Academy of Sciences (USA)* 117(14): 7702.

21

Shrinking Lake Urmia

Causes, Future Projection and Implications

SEYED-MOHAMMAD HOSSEINI-MOGHARI, QIUHONG TANG,
GUOYONG LENG AND XIMENG XU

21.1 Introduction

It is crucial to get a thorough understanding of the system to manage it efficiently. However, the multifaceted nature of water resource systems makes their management problematic. Water systems are inherently dynamic. External factors such as climatic variability along with changes in the water demand are difficult to predict, therefore a major challenge for the management of water resource systems. Additional constraints are added when naturally available water is scarce in a region (i.e., arid and semi-arid regions). Also, climate variability can affect renewable water resources, water demand and the generated streamflow (Wang et al., 2016; Azarnivand et al., 2020), which in arid and semi-arid regions usually leads to a decrease in renewable water resources and an increase in demand (Herrera-Pantoja & Hiscock, 2015).

Iran is a good example of a complex water resource management system. Although the climatic conditions in Iran vary from very wet to extra arid, the majority of the country (approximately 90 per cent) has arid to semi-arid climatic conditions (Rahimi et al., 2013). Iran receives 250 mm precipitation annually, which is less than one-third of the global average. The government follows the policy of food self-sufficiency and the majority of job opportunities are associated with agricultural activities. On the other hand, in Iran, with other arid and semi-arid regions, crop production is dependent on irrigation, which leads to the overexploitation of surface water and groundwater resources. When exploitation from water resources exceeds a certain level, adverse impacts on the environment begin (Zahedi et al., 2017). This impact can be more severe under climatic

Funding for this research was provided by the National Natural Science Foundation of China (41790424) and the International Partnership Program of the Chinese Academy of Sciences (131A11KYSB20170113). The first author is sponsored by the Chinese Academy of Sciences President's International Fellowship Initiative (2019VEA0019).

extremes, that is, drought (Hosseini-Moghari et al., 2017). In addition, Iran's population increased by more than 200 per cent between 1956 and 1996, from 18.9 to 59.5 million (Hoodfar & Assadpour, 2000) and reached 81 million in 2017. Over the last decades, an increase in temperature and decrease in precipitation across Iran were reported by Saboohi et al. (2012) and Khalili et al. (2016), respectively. In the Iranian context, water management is therefore particularly challenging.

With use of more than 70 per cent of the renewable freshwater resources, the water resources in Iran are beyond the stress level, established at 40 per cent (Madani, 2014). Extensive human water use has led to many wetlands and lakes drying up as well as declining the groundwater level in Iran over the last decades. About 50 per cent of Iran's plains are in critical condition (Madani et al., 2016). The shrinkage of the Lake Urmia basin located in the northwest of Iran – Lake Urmia was once the largest lake in the Middle East – is a tragic illustration of the depletion of water resources in the region. In this chapter, we explain the shrinkage of Lake Urmia through the analysis of several datasets and highlight the role of hydro-climate factors and their uncertainty on water resources in future scenarios.

21.2 Data

We assessed two types of data: *in-situ* and global datasets. In-situ data encompass the annual time series of water inflow, volume and area of the lake, along with human water use and the irrigated area in the basin. The global datasets include CRU (Climatic Research Unit) precipitation and temperature, GRACE TWSA (Gravity Recovery and Climate Experiment Total Water Storage Anomaly) and the outputs of 17 CORDEX (Coordinated Regional Climate Downscaling Experiment) models. The in-situ data were collected from the ULRP (Urmia Lake Restoration Programme)[1] and global data was obtained from the corresponding repository.

We used CRU TS 4.03 precipitation and temperature data to assess climate variability across the basin and over time, based on a monthly and annual time frame.[2] GRACE TWSA represents all water storage components, which include the storage of surface water, groundwater, soil moisture, snow water and canopy. We used the monthly time series of CSR GRACE RL06 mascon solutions (Save et al., 2016) between 2002 and 2017.[3]

To analyse future changes in precipitation and temperature across the basin, we considered the outputs of all available General Circulation Models (GCMs) in the CORDEX project for south Asia, that is, 17 models (Table 21.1). Among these models, only six models were run under the optimistic scenario (Representative

Table 21.1. List of the CORDEX models used to project future precipitation and temperature across the Lake Urmia basin

					Future period		
No.	Driving model (GCM)	RCM model	Institute	Historical period	RCP2.6	RCP4.5	RCP8.5
1	CCCma-CanESM2	RegCM4–4	IITM	1951–2005	–	2006–2099	2006–2099
2	CNRM-CERFACS-CNRM-CM5	RegCM4–4	IITM	1951–2005	–	2006–2099	2006–2085
3	CSIRO-QCCCE-CSIRO-Mk3–6-0	RegCM4–4	IITM	1951–2005	–	2006–2099	2006–2099
4	IPSL-IPSL-CM5A-LR	RegCM4–4	IITM	1951–2005	–	2006–2099	2006–2099
5	MPI-M-MPI-ESM-MR	RegCM4–4	IITM	1951–2005	–	2006–2099	2006–2099
6	NOAA-GFDL-GFDL-ESM2M	RegCM4–4	IITM	1951–2005	–	2006–2099	2006–2099
7	MPI-M-MPI-ESM-LR	REMO2009	MPI-CSC	1961–2005	2006–2100	2006–2100	2006–2100
8	CCCma-CanESM2	RCA4	SMHI	1951–2005	–	2006–2100	2006–2100
9	CNRM-CERFACS-CNRM-CM5	RCA4	SMHI	1951–2005	–	2006–2100	2006–2100
10	CSIRO-QCCCE-CSIRO-Mk3–6-0	RCA4	SMHI	1951–2005	–	2006–2100	2006–2100
11	ICHEC-EC-EARTH	RCA4	SMHI	1951–2005	2006–2100	2006–2100	2006–2100
12	IPSL-IPSL-CM5A-MR	RCA4	SMHI	1951–2005	–	2006–2100	2006–2100
13	MIROC-MIROC5	RCA4	SMHI	1951–2005	2006–2100	2006–2100	2006–2100
14	MOHC-HadGEM2-ES	RCA4	SMHI	1951–2005	2006–2099	2006–2099	2006–2100
15	MPI-M-MPI-ESM-LR	RCA4	SMHI	1951–2005	2006–2100	2006–2100	2006–2100
16	NCC-NorESM1-M	RCA4	SMHI	1951–2005	2006–2100	2006–2100	2006–2100
17	NOAA-GFDL-GFDL-ESM2M	RCA4	SMHI	1951–2005	–	2006–2100	2006–2100

Abbreviations: CanESM, Canadian Earth System Model; CCCma: Canadian Centre for Climate Modelling and Analysis; CERFACS, Centre Européen de Recherche et de Formation Avancée en Calcul Scientifique; CM5, Climate Model version 5; CNRM, Centre National de Recherches Météorologiques; CSIRO, Commonwealth Scientific and Industrial Research Organisation; ESM, Earth System model; ICHEC, Irish Centre for High-End Computing; IITM, Indian Institute of Technology Madras; IPSL, Institut Pierre-Simon Laplace; MIROC, Model for Interdisciplinary Research on Climate; MOHC, Met Office Hadley Centre; MPI-CSC: Max Planck Institute for Meteorology, Climate Service Center; MPI-M, Max-Planck-Institut for Meteorology; NCC, Norwegian Climate Centre; NOAA-GFDL, National Oceanic and Atmospheric Administration Geophysical Fluid Dynamics Laboratory; NorESM, Norwegian Earth System Model; QCCCE, Queensland Climate Change Centre of Excellence; RCA, Rossby Centres regionala atmosfärsmodell; RCM, Regional Climate Model; REMO, REgional Model; SMHI, Swedish Meteorological and Hydrological Institute.

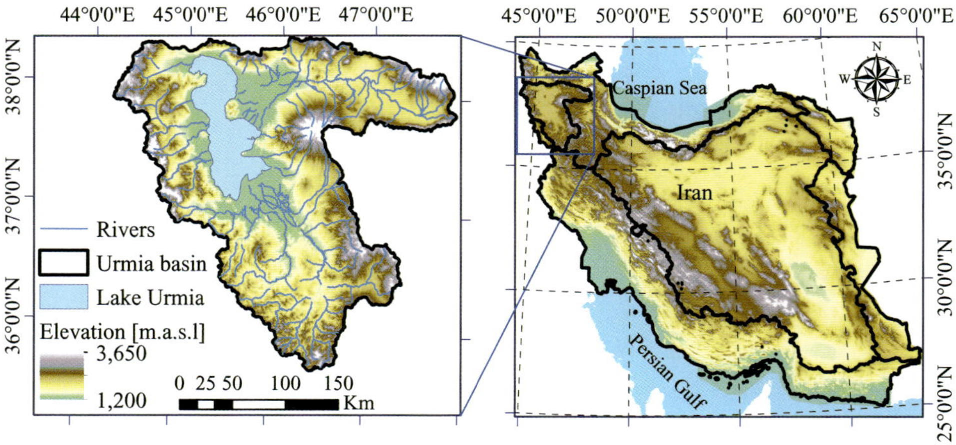

Figure 21.1 Lake Urmia basin location

Concentration Pathway 2.6; RCP2.6), since it represents the less likely scenario.[4] All models were run from 1951 to 2005 as historical runs and for RCP4.5 (medium scenario) and 8.5 (pessimistic scenario) as future projections from 2006 to the end of the century (see Table 21.1). For detailed information about the RCP, please refer to Van Vuuren et al. (2011).[5]

21.3 Lake Urmia Story

Lake Urmia basin, with an area of 52,000 km^2, is located in the northwest of Iran. Downstream in the basin, a discharge of 11 sub-basins is drained through 29 permanent and seasonal rivers to create Lake Urmia (Figure 21.1). The lake has unique ecological and natural features that made it a UNESCO (The United Nations Educational, Scientific and Cultural Organization) Biosphere Reserve and a Ramsar Site (Henareh Khalyani et al., 2014). Annually, the rivers transfer 2.2 million tons of salt to the lake. As a result, Lake Urmia exhibit very high levels of salinity – the lake used to be the second largest hypersaline lake in the world. Hence, the water of the lake is unsuitable for almost all human water use purposes. Also, there is no significant interaction between the lake and groundwater. Therefore, evaporation is the unique natural factor responsible for the water loss that has affected Lake Urmia, whose annual rate was estimated at about 3×10^9 m^3 yr^{-1} (Abbaspour & Nazaridoust, 2007).

The Urmia basin covers 3.15 per cent of Iran and receives 7 per cent of renewable surface water resources of the country. On the one hand, the potential for agricultural development in this basin is higher than most basins in Iran. On the other hand, the policy implemented by the government to secure food and employment coincide with a relatively humid period. The agricultural development in the basin associated with the policy seems to be a win–win game. In this context, farmers are allowed to use more renewable water to produce more food before the degradation of the water quality that occurs while flowing toward the lake. To this end, several dams were built to regulate river discharge for irrigation seasons and the number of drilled wells reached 85,000 (50 per cent are illegal) for withdrawal from groundwater resources (Danesh-Yazdi & Ataie-Ashtiani, 2019). As a result, the irrigated area of the basin increased from $3{,}186 \times 10^6$ m^2 in 1976 to approximately $4{,}337 \times 10^6$ m^2 in 2013 (Kamali & Youneszadeh Jalili, 2015).

Although the presence of a few wet years increased the volume of the lake to 35×10^9 m^3 (in 1995), the downward trend in precipitation and the increase (upward trend) in temperature (see Figure 21.2) associated with human-induced water use have caused the lake to lose about 30×10^9 m^3 of its maximum historical volume (ULRP, 2015a). Increasing temperature (and evaporation rate) and decreasing inflow into the lake caused the shrinkage rate to exceed 66 per cent from 1995 to 2018 (see Figure 21.2). From 1999 onwards, the inflow into the lake was less than 3×10^9 m^3 yr^{-1} in most years, which did not suffice to compensate for the lake evaporation (Abbaspour & Nazaridoust, 2007). Also, temperature and precipitation variability during this period might have increased the environmental inflow requirements of Lake Urmia up to 5.4×10^9 m^3 yr^{-1} (Alborzi et al., 2018), which resulted in a drastic decline in the lake water level from about 1,278 in 1995 to about 1,270 m.a.s.l (metres above sea level) in 2005. Therefore, the basin was rapidly depleted, which is confirmed by the GRACE TWSA time series between 2002–2017 (Figure 21.2(d)), which indicates that the basin lost $1{,}014 \times 10^6$ m^3 of its water storage annually. The contribution of lake shrinkage is estimated to account for about 80 per cent of the total water loss in the basin (Tourian et al., 2015).

Many studies have been carried out to identify the reasons for the depletion of the Lake Urmia basin. The causes of a decreasing inflow into the lake and its shrinkage include agricultural development, rising temperatures, reduction of precipitation, changes in precipitation frequency, building dams, construction of an east–west causeway across the lake and an increase in the number of wells (Zeinoddini et al., 2009; Delju et al., 2013; AghaKouchak et al., 2015; Fathian et al., 2015; Amirataee & Zeinalzadeh, 2016; Hesami & Amini, 2016; Shadkam et al., 2016; Azarnivand & Banihabib, 2017; Bavil et al., 2018; Chaudhari et al., 2018; Fazel et al., 2018; Ghale et al., 2018; Hosseini-Moghari et al., 2018, 2020).

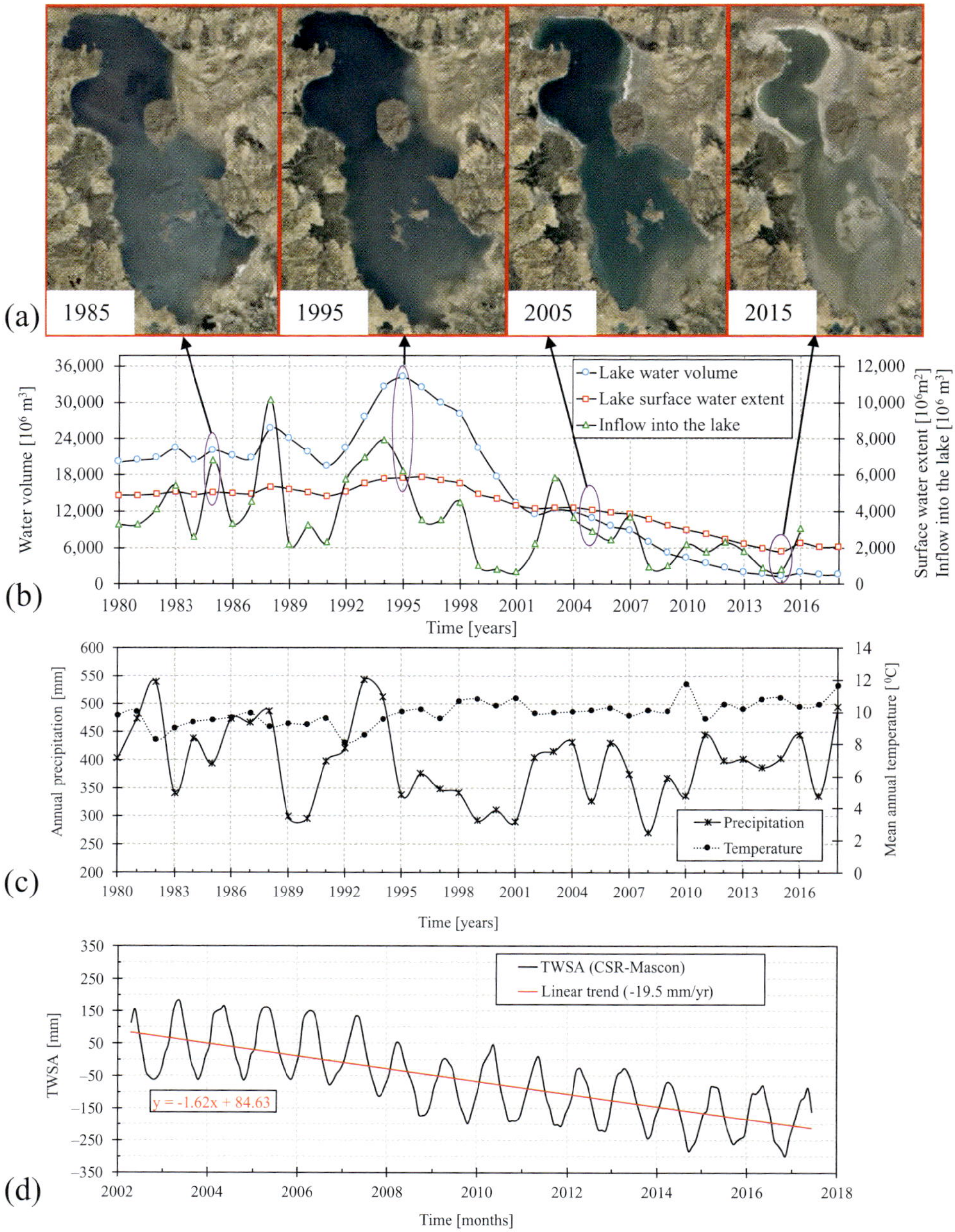

Figure 21.2 Google Earth Timelapse of Lake Urmia's surface area during the last four decades (a), the annual times series of lake volume and surface water extent as well as inflow into the lake (data from ULRP) (b), the annual times series of precipitation and temperature over the Lake Urmia basin (data from CRU) (c) and the monthly times series of TWSA over the Lake Urmia basin (data from GRACE CSR mascon solution) (d)

The lake shrinkage has generated an estimated 8 billion tons of salt over the lake bed, which may represent the primary source of local saline dust storms (Pengra, 2012). Evidence shows that more than 57 per cent of the total suspended particles in dust storms across the region stem from the lake bed (Gholampour et al., 2015). As a result, desertification of a large area of the lake threatens the health of about 7.1 million habitants living in basin area and affects 80 million people who live within a radius of 500 km from the lake (Mardi et al., 2018). The lake plays an important role in protecting the aquatic environment. An increase in salt concentration might cause damage to the aquatic life that depends on the resources from the lake (Karbassi et al., 2010).

Following the lake shrinkage and its consequences on health, society and the economy as well as national and international pressures, Iran's government committed to prevent the lake from drying up and made promises to restore it. To this end, the government launched a 10-year programme termed "Urmia Lake Restoration Program (ULRP)" in June 2013. The government pledged to allocate the equivalent of $ 5 billion (Guardian, 2015) to achieve the objectives of the programme. The ULRP (2014–2023) is composed of three main phases: (I) stabilization phase (2014–2016); (II) restoration phase (2016–2023); and (III) sustainable restoration phase (2023–2024). The stabilization phase was aimed at stopping the decreasing trend of the lake level and reducing the environmental consequences of the dried part of the lake. The goal of the restoration phase is to raise the lake water level to 1,273.7 m.a.s.l by 2023. The sustainable restoration phase is planned to achieve at least the minimum required water level to revive the biota life, which requires that the ecological water level of the lake reaches 1,274.1 m.a.s.l (Danesh-Yazdi & Ataie-Ashtiani, 2019).

ULRP considers six approaches including: (I) Reduction of water consumption in the agricultural sector by the increase of water efficiency; (II) Control and decrease of exploitation of surface and groundwater resources by prevention of new water resources development projects and implementation of the regulations for unauthorized water use and illegal wells; (III) Start initiatives on mitigation of adverse consequences, for example, stabilizing dust sources and reducing the likelihood risk; effects and increasing alternative employment for those affected by lake shrinkage or ULPR policies; (IV) Software measures, for example, holding comprehensive training programmes to increase awareness and getting local and public community contributions; (V) Hardware projects for example, helping water transfer from rivers to the lake through dredging; and (VI) Transferring water from new water resources into the lake, for example, treated wastewater or the study of transferring water from the Caspian Sea (Shadkam, 2017).

According to water withdrawal data in the Lake Urmia basin in 2009, the total withdrawals from surface water and groundwater were equal to $4{,}825 \times 10^6$ m^3

(ULRP, 2015b). According to the ULRP programme, a 40 per cent reduction in water consumption in the agricultural sector should increase the water inflow into the lake by $1{,}340 \times 10^6$ m^3 yr^{-1} (Danesh-Yazdi & Ataie-Ashtiani, 2019). This measure was considered as the key project for the lake restoration. Due to budget limitations, ULRP has not been able to fully implement its plans. However, some plans, e.g., blocking illegal wells or preventing unauthorized harvesting of surface water, pose a threat to farmers who often have no viable alternative to agriculture. This policy has led to various social problems, which include an increase in the unemployment rate, and has therefore not received wide political support.

Currently, we are in the middle of the restoration phase. We observe that the restoration measures have not met the targets regarding the water level of the lake, as stated in the ULRP roadmap. Along with the aforementioned barriers to the implementation of the policy, further issues might have added additional constraints on the realization of the project. These include: (i) planning without a proper understanding of the contribution of each factor associated with the shrinkage of the lake; (ii) insufficient understanding of the interaction of the water components in the basin; and (ii) difficulty to predict water availability due to uncertainty in the climate factors. ULRP works under the assumption that human activity is the main reason for the lake shrinkage and that there is no direct relationship between the lake and groundwater. Although most studies on Lake Urmia acknowledged human pressures as the primary contributor to this catastrophe, their magnitude has been under debate. While Hosseini-Moghari et al. (2020) report that human pressures have caused 40 per cent of the lake shrinkage, Ghale et al. (2018) suggest that, in effect, this might have explained 80 per cent of the shrinkage. To investigate the reason for this difference, Hosseini-Moghari et al. (2018) analysed the trend of daily precipitation and inflow into the main reservoirs in the Lake Urmia basin. They suggested that a decreasing trend in the annual precipitation along with a significant increasing trend in precipitation of less than 5 mm day^{-1} (negligible from a hydrological perspective) might have substantially contributed to the reduction of the river flow. Shadkam et al. (2016) and Farokhnia et al. (2018), who used daily data to model the role of anthropogenic and climatic factors in the basin, reported that climatic factors played a more important role than human activity in the lake shrinkage process. For a detailed discussion about this issue, please refer to Hosseini-Moghari et al. (2020). Most studies have shown that there is no direct relationship between lake and groundwater in the basin, however the debate about a potential link is still ongoing. For instance, Vaheddoost and Aksoy (2018) stated that there is strong evidence of interactions between the lake and groundwater, while Amiri et al. (2016), by using chemical tracer and isotope analyses, did not find a significant relationship between them in the basin.

Failure to reach the target in the restoration phase resulted in a revisit in the roadmap and the definition of a new target associated with health threshold. In this phase, the goal is to reach a water level of 1,271.72 m.a.s.l by 2021, which is far from the ecological level, defined as 1,274.1 m.a.s.l (Danesh-Yazdi & Ataie-Ashtiani, 2019). Fortunately, due to the heavy rainfall along with the flood that happened in spring 2019, the lake water level raised by 1.15 metres and reached 1,271.52 m.a.s.l. However, the actions carried out through ULRP did not compensate for the massive evaporation of the lake that occurs during summer. As a result, the water level declined and reached 1271.25 m.a.s.l in December 2019. This illustrates that extensive actions funded by a large budget may not necessarily suffice to offset the effects of unsustainable water resources management. Over-capacity utilization of water resources may be profitable in the short-term but offsetting its environmental effects would entail more cost.

21.4 Future Climate Change

In Section 21.3, we discussed the devastating effects of the unsustainable planning of water resources over Lake Urmia basin. In this section, we describe and analyse the results of possible future changes in the main climatic variables that is, precipitation and temperature across the basin, based on the outputs of 17 CORDEX models. Figure 21.3 shows the annual precipitation changes (%) and annual temperature anomalies (°C) in historical and future periods. It should be noted that to calculate precipitation change and temperature anomaly, each model was compared to itself relative to the mean for the period of 1951–2005.

Despite that the uncertainty of precipitation from the simulation models is high (Figure 21.3(a)), its bound encompasses observational data. However, the mean values of historical runs exhibit less fluctuations compared to the future values estimated in all RCP scenarios, implying that the basin is likely to face more extreme events in the future. The uncertainty associated with precipitation is larger in the future compared to the variability observed in historical periods (except for RCP2.6, since only six models were used in this scenario, while the other scenarios include more models). The highest values of uncertainty are related to RCP8.5, where annual precipitation changes fluctuate between –60 and +100 per cent for the end of the century compared to the average of historical runs.

Figure 21.3(b) illustrates the annual temperature anomaly for historical and future periods. The models show good fit with regard to the trend from the historical temperature data in comparison with the CRU time series. However, the estimation of the amplitude of temperature is less accurate. In most cases, the difference between the mean temperature values of 17 models and observation values is less than 1°C. All RCP scenarios suggest an increase in temperature over

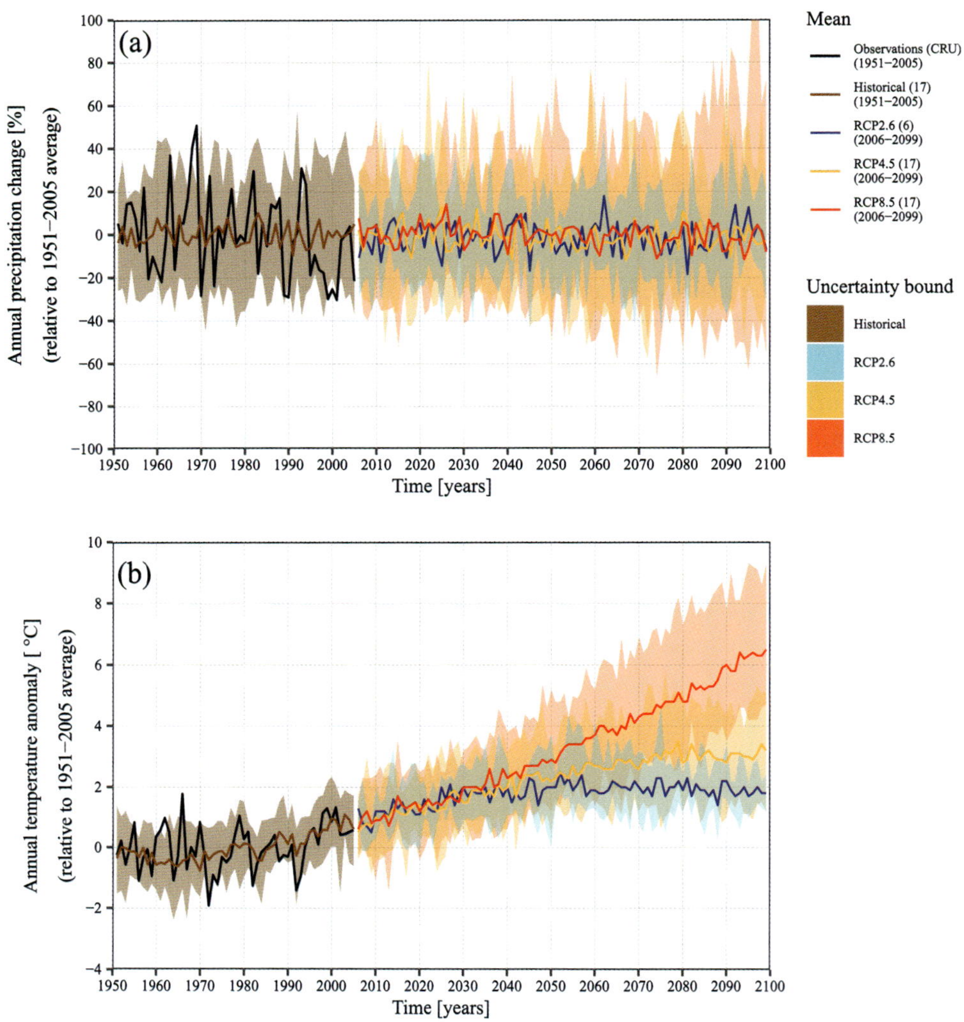

Figure 21.3 Times series of annual precipitation changes (a) and temperature anomalies (b) for CRU datasets and models' historical runs as well as models' future runs under different RCP scenarios along with their uncertainty bound over the Lake Urmia basin. The numbers in the parentheses represent the number of models

the basin in the future. On average, the temperature in future periods is expected to be 1.6, 2.3 and 3.3°C higher than historical mean, according to RCP2.6, 4.5 and 8.5, respectively. The mean of temperature anomaly at the end of the century obtained using different models under RCP2.6, 4.5 and 8.5 scenarios is about 1.8, 3.2 and 6.5°C greater than the historical mean (Figure 21.3(b)).

We provide a more detailed assessment of the changes in precipitation and temperature for the Lake Urmia basin over time. We divide the predictive temporal window into three sub-periods: near future (2010–2039), mid-future (2040–2069) and far future (2070–2099). Using a delta-change method, we estimate future precipitation and temperature values from the predictive models and historical data. For more details about the delta-change method, please refer to Ahmadi et al. (2015). Figure 21.4 shows the climatology of precipitation for different scenarios and different future periods. Based on Figure 21.4, uncertainty increases over time in all scenarios, in particular under the RCP8.5 scenario. Mean annual precipitation using the observed data equals 403 mm yr^{-1} while under RCP2.6, 4.5 and 8.5, this value is equal to 399, 405 and 410 mm yr^{-1} for near future, 409, 394 and 400 mm yr^{-1} for mid future and 402, 396 and 387 mm yr^{-1} for far future, respectively. Therefore, based on the ensemble mean values, the only significant decrease in precipitation happens in far future and under RCP8.5. Moreover, the seasonal characteristics of precipitation in far future under RCP8.5 are different from observation in the historical period; somehow, they have less amplitude between January to April (10 per cent less than observations). As a result, the rainfed areas that need rainwater between February and May (Kamali & Youneszadeh Jalili, 2015) are affected under this scenario. Also, with an expected decreasing precipitation in these months, the inflow into the dams should be reduced. According to the RCP8.5 scenario, the irrigated farms that need large volumes of water between April and October (Saemian et al., 2015) might face water shortages in the far future.

In all scenarios the model simulations suggest an increase in the temperatures in all investigated sub-periods. Figure 21.5 shows the regime of temperature over the Lake Urmia basin for historical periods and under future climate scenarios. The uncertainly associated with temperature increases from near future to far future. Since the number of models in RCP2.6 is smaller than the models used in the other scenarios, the temperature uncertainty also increases from RCP2.6 to RCP8.5. As expected, RCP8.5 suggests the highest temperature increase. The mean annual temperature in historical period equals 9.7°C. Under RCP2.6, 4.5 and 8.5, the average temperature should reach 10.9, 10.9 and 11°C in the near future, 11.3, 11.9 and 12.7°C in the mid future and 11.3, 12.4 and 14.8°C in the far future, respectively. By the end of the century, the mean annual temperature might increase between 1.6 and 5.1°C across the basin.

Figure 21.6 shows spatial changes of precipitation across the basin for different RCPs in near, mid and far futures. It should be noted that since the temperature changes are assumed to be spatially homogenous, spatial changes of temperature are not considered here. According to Figure 21.6, precipitation should decrease in the south part of the basin and increase in the northern part. The range of spatial changes of the annual precipitation varies from less than ±5 per cent (for RCP2.6 for all future sub-periods) to a maximum of +10 and –11 per cent (for RCP8.5 in

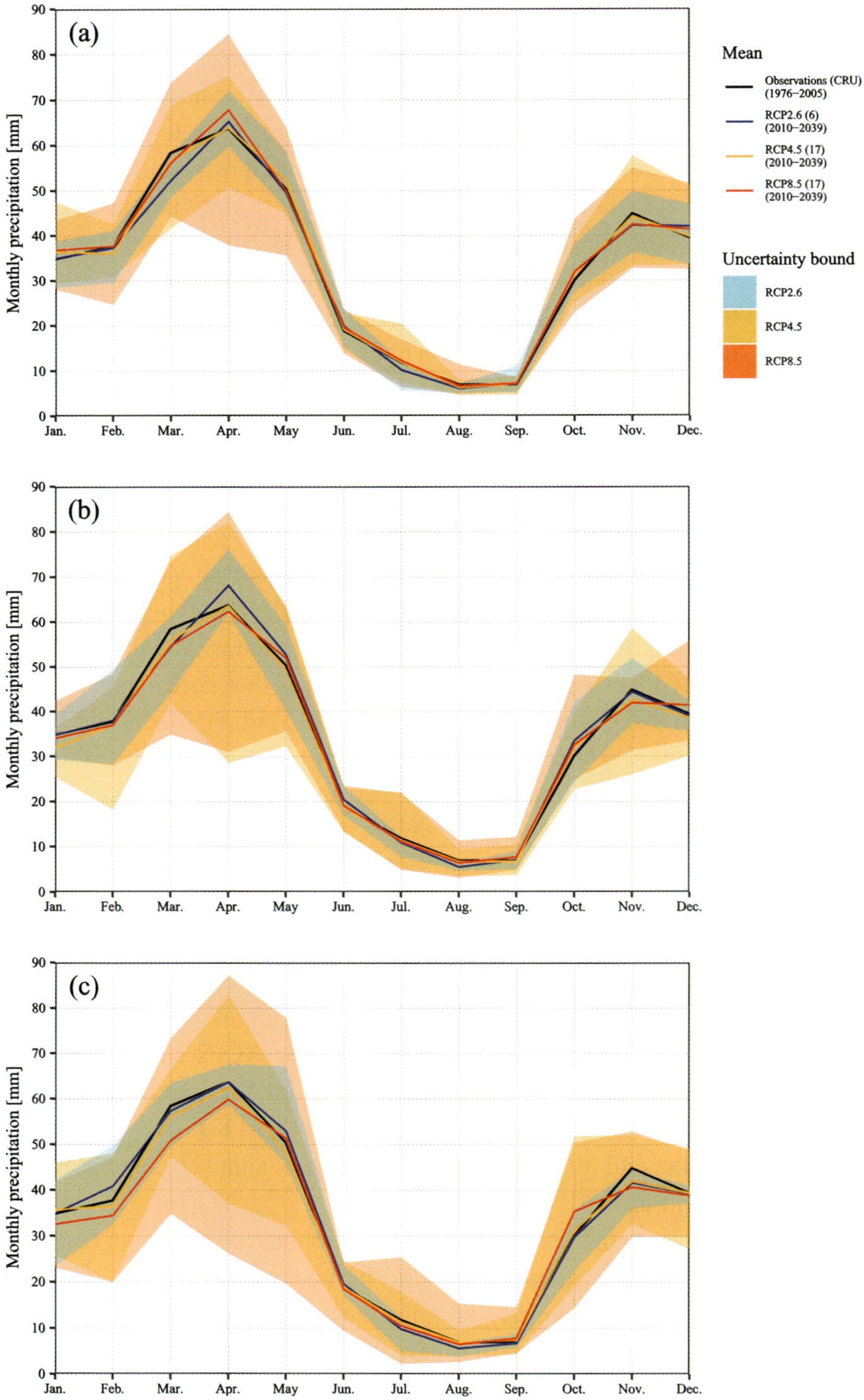

Figure 21.4 Regimes of precipitation over the Lake Urmia basin in the historical period and the near future (2010–2039) (a), mid future (2040–2069) (b) and far future (2070–2099) (c) along with their uncertainty bounds. The numbers in the parentheses represent the number of models

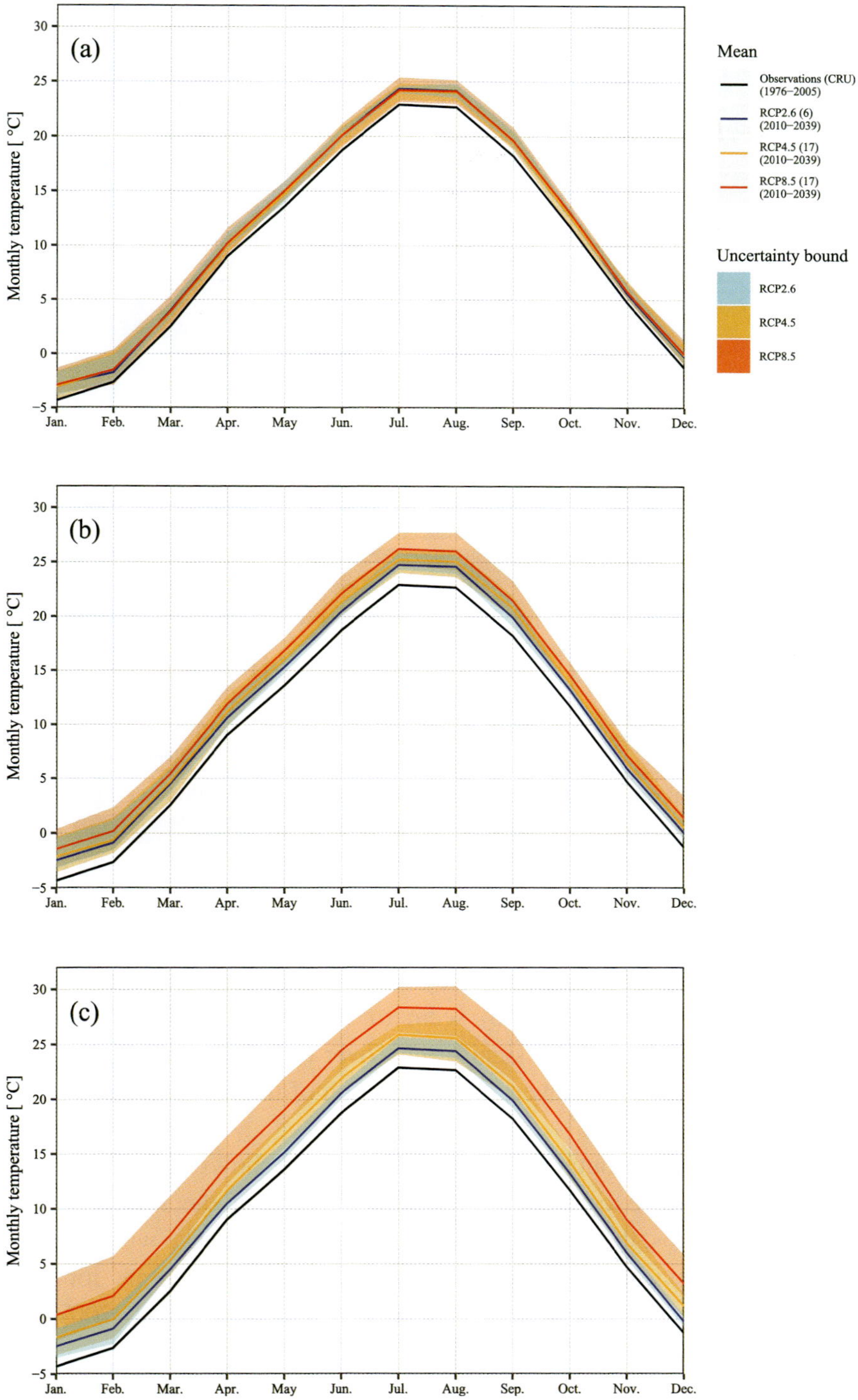

Figure 21.5 Regimes of temperature over the Lake Urmia basin in the historical period and near future (2010–2039) (a), mid future (2040–2069) (b) and far future (2070–2099) (c) along with their uncertainty bounds. The numbers in the parentheses represent the number of models

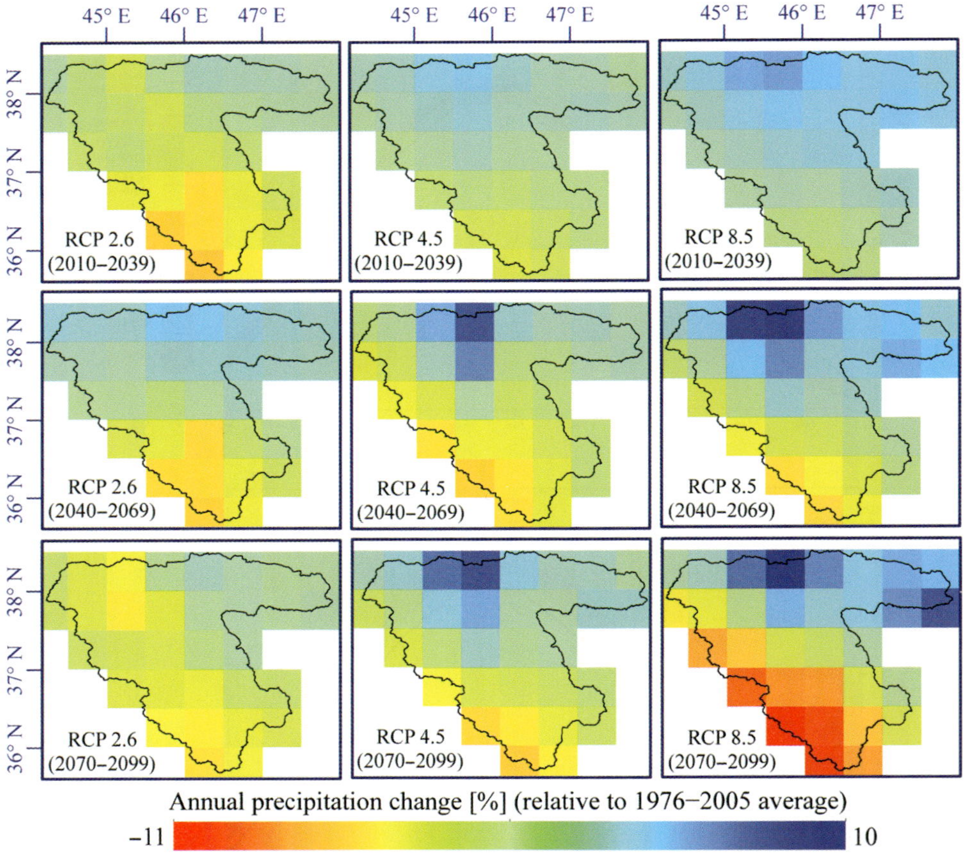

Figure 21.6 Spatial pattern of annual precipitation changes over the Lake Urmia basin in near, mid and far future periods under different RCP scenarios

the period 2070–2099). All scenarios (at the exception of the predictions for the far future under the RCP8.5 scenario) seem to suggest that the total amount of precipitation should not change significantly over time, this independently of the sub-period investigated. However, the distribution of the precipitation is expected to change across space.

The evaluation of the outputs of CORDEX models shows that the Lake Urmia basin is expected to experience higher temperatures in the future. Under the investigated scenarios, this means that evaporation from the lake will increase and, as a result, the environmental water requirements to compensate for the lake evaporation, that is, 3×10^9 m^3 yr^{-1} (Abbaspour & Nazaridoust, 2007) should increase to 5.4×10^9 m^3 yr^{-1} for dry years, as reported by Alborzi et al. (2018). Besides, precipitation is expected to decrease in the south part of the basin, where the rivers provide 65 per cent of the total inflow into the lake. Fathian et al. (2015) indicated that the inflow to the Lake Urmia basin is significantly related to precipitation and temperature variability and the inflow into the lake is more sensitive to changes in the temperature than precipitation variability. Therefore, it is expected that the inflow into the lake will naturally decrease in the future, while the lake environmental water requirements will increase.

Currently, the irrigation months in the basin occur between April and October (Saemian et al., 2015). The predicted precipitation is not expected to increase during the cultivation period, the increase in temperature makes the evapotranspiration rate go up, which increases the agricultural water demand. Therefore, the cultivation period might shift over time. Also, the efficiency of the reservoirs is reduced with increasing temperatures. Those reservoirs that could receive less inflow in the future compared to the historical period might act as an evaporation pan and lose much of their water through evaporation. Therefore, the results of the predictive models suggest that the basin situation might worsen in the future.

While the uncertainty of the predictions accounted for by the CORDEX models remains high, the renewable water resources in the basin are expected to decrease in the near, mid and far future. Thus, the rate of water allocation to water users in the basin should be revisited. According to the results of the model simulations, it is unlikely that the lake would reach levels that are sufficient to maintain the ecosystem and human activity in a sustainable manner. Therefore, it seems more realistic to aim at reaching the health threshold of the water level (1271.72 m.a.s.l.), which can contribute to controlling dust storms.

21.5 Conclusions

In this chapter, the story of the Lake Urmia basin, as an example of unsustainable water management, has been presented. We have shown how water resources development based on a wet period, without considering climate variability, can lead to a social, economic and environmental catastrophe. This example illustrates that an overuse of renewable water may temporarily be beneficial but can provoke undesirable side-effects in the longer term. Agricultural development in the Lake Urmia basin was a relatively quick solution to economic development and job creation in the region. However, in the absence of sufficient water due to climatic variability (change) or excessive pressure on water resources, this solution has escalated into environmental, social and economic problems. Assessment of the outputs of CORDEX models indicates that a decrease in the average precipitation across the basin is not likely to occur, but its spatial pattern might change. All models agree on a future increase in temperature, which is expected to increase the environmental, agricultural and urban water demands. Under such circumstances, achieving the targets planned by the ULRP with the water resources from the basin seems unattainable. Water transfer plans from other regions might offer a short-term solution to compensate for local water resources shortage, but might generate further social, economic and environmental problems in those regions. While reaching sustainable lake restoration and water management in the Lake Urmia basin remains challenging, implementing a management of water demand in the basin that accounts for climatic uncertainty and setting sensible targets on the restoration of the lake are key measures to mitigate the effects of the lake shrinkage in the long-term.

Notes

1 Available from www.ulrp.ir/en/ (Last accessed 14 December 2019).
2 CRU data are available through www.cru.uea.ac.uk/data (Last accessed 5 December 2019).
3 This data can be downloaded from www2.csr.utexas.edu/grace/RL06.html (Last accessed 19 October 2019).
4 Information about the CORDEX project can be found at www.cordex.org/ (Last accessed 14 December 2019).
5 The outputs of the CORDEX model are freely available from https://esgf-data.dkrz.de/search/cordex-dkrz/ (Last accessed 14 December 2019).

References

Abbaspour, M., & Nazaridoust, A. (2007). Determination of environmental water requirements of Lake Urmia, Iran: An ecological approach. *International Journal of Environmental Studies* 64(2): 161–169.

AghaKouchak, A., Norouzi, H., Madani, K., et al. (2015). Aral Sea syndrome desiccates Lake Urmia: Call for action. *Journal of Great Lakes Research* 41(1): 307–311.

Ahmadi, M., Haddad, O. B., & Loáiciga, H. A. (2015). Adaptive reservoir operation rules under climatic change. *Water Resources Management* 29(4): 1247–1266.

Alborzi, A., Mirchi, A., Moftakhari, H., et al. (2018). Climate-informed environmental inflows to revive a drying lake facing meteorological and anthropogenic droughts. *Environmental Research Letters* 13(8): 084010.

Amirataee, B., & Zeinalzadeh, K. (2016). Trends analysis of quantitative and qualitative changes in groundwater with considering the autocorrelation coefficients in west of Lake Urmia, Iran. *Environmental Earth Sciences* 75(5): 371.

Amiri, V., Nakhaei, M., Lak, R., & Kholghi, M. (2016). Geophysical, isotopic, and hydrogeochemical tools to identify potential impacts on coastal groundwater resources from Urmia hypersaline Lake, NW Iran. *Environmental Science and Pollution Research* 23(16): 16738–16760.

Azarnivand, A., & Banihabib, M. E. (2017). A multi-level strategic group decision making for understanding and analysis of sustainable watershed planning in response to environmental perplexities. *Group Decision and Negotiation* 26(3): 629–648.

Azarnivand, A., Camporese, M., Alaghmand, S., & Daly, E. (2020). Simulated response of an intermittent stream to rainfall frequency patterns. *Hydrological Processes* 34(3): 615–632,

Bavil, S. S., Zeinalzadeh, K., & Hessari, B. (2018). The changes in the frequency of daily precipitation in Urmia Lake basin, Iran. *Theoretical and Applied Climatology* 133 (1–2): 205–214.

Chaudhari, S., Felfelani, F., Shin, S., & Pokhrel, Y. (2018). Climate and anthropogenic contributions to the desiccation of the second largest saline lake in the twentieth century. *Journal of Hydrology* 560: 342–353.

Danesh-Yazdi, M., & Ataie-Ashtiani, B. (2019). Lake Urmia crisis and restoration plan: Planning without appropriate data and model is gambling. *Journal of Hydrology* 576: 639–651.

Delju, A. H., Ceylan, A., Piguet, E., & Rebetez, M. (2013). Observed climate variability and change in Urmia Lake Basin, Iran. *Theoretical and Applied Climatology* 111 (1–2): 285–296.

Farokhnia, A., Morid, S., Abbaspour, K., & Delavar, M. (2018). Development of SWAT-LU model for simulation of Urmia lake water level decrease and assessment of the proposed actions for its restoration (role of anthropogenic and climatic factors on hydrological change of the basin and lake). *Iranian Journal of Irrigation and Drainage* 12(5): 1041–1058.

Fathian, F., Morid, S., & Kahya, E. (2015). Identification of trends in hydrological and climatic variables in Urmia Lake basin, Iran. *Theoretical and Applied Climatology* 119(3–4): 443–464.

Fazel, N., Berndtsson, R., Uvo, C. B., Madani, K., & Kløve, B. (2018). Regionalization of precipitation characteristics in Iran's Lake Urmia basin. *Theoretical and Applied Climatology* 132(1–2): 363–373.

Ghale, Y. A. G., Altunkaynak, A., & Unal, A. (2018). Investigation anthropogenic impacts and climate factors on drying up of Urmia Lake using water budget and drought analysis. *Water Resources Management* 32(1): 325–337.

Gholampour, A., Nabizadeh, R., Hassanvand, M. S., et al. (2015). Characterization of saline dust emission resulted from Urmia Lake drying. *Journal of Environmental Health Science and Engineering* 13(1): 82.

Guardian (2015). Lake Urmia: How Iran's most famous lake is disappearing. Available from www.theguardian.com/world/iran-blog/2015/jan/23/iran-lake-urmia-drying-up-new-research-scientists-urge-actionv (Last accessed 14 December 2019).

Henareh Khalyani, A., Mayer, A. L., & Norman, E. S. (2014). Water flows toward power: Socioecological degradation of Lake Urmia, Iran. *Society & Natural Resources* 27 (7): 759–767.

Herrera-Pantoja, M., & Hiscock, K. M. (2015). Projected impacts of climate change on water availability indicators in a semi-arid region of central Mexico. *Environmental Science & Policy* 54: 81–89.

Hesami, A., & Amini, A. (2016). Changes in irrigated land and agricultural water use in the Lake Urmia basin. *Lake and Reservoir Management* 32(3): 288–296.

Hoodfar, H., & Assadpour, S. (2000). The politics of population policy in the Islamic Republic of Iran. *Studies in Family Planning* 31(1): 19–34.

Hosseini-Moghari, S. M., Araghinejad, S., & Azarnivand, A. (2017). Drought forecasting using data-driven methods and an evolutionary algorithm. *Modeling Earth Systems and Environment* 3(4): 1675–1689.

Hosseini-Moghari, S. M., Araghinejad, S. & Ebrahimi, K. (2018). Monthly precipitation assessment: A misleading tool for understanding the effects of climate change. *8th Global FRIEND-Water Conference*, Beijing, China, 6–9 November 2018.

Hosseini-Moghari, S. M., Araghinejad, S., Tourian, M. J., Ebrahimi, K., & Döll, P. (2020). Quantifying the impacts of human water use and climate variations on recent drying of Lake Urmia basin: The value of different sets of spaceborne and in situ data for calibrating a global hydrological model. *Hydrology and Earth System Sciences* 24(4): 1939–1956.

Kamali, M., & Youneszadeh Jalili, S. (2015). Investigation of land use changes in Lake Urmia Basin using remotely sensed images. *Report of Urmia Lake Restoration Program (ULRP)* (in Persian).

Karbassi, A., Bidhendi, G. N., Pejman, A., & Bidhendi, M. E. (2010). Environmental impacts of desalination on the ecology of Lake Urmia. *Journal of Great Lakes Research* 36(3): 419–424.
Khalili, K., Tahoudi, M. N., Mirabbasi, R., & Ahmadi, F. (2016). Investigation of spatial and temporal variability of precipitation in Iran over the last half century. *Stochastic Environmental Research and Risk Assessment* 30(4): 1205–1221.
Madani, K. (2014). Water management in Iran: What is causing the looming crisis? *Journal of Environmental Studies and Sciences* 4(4): 315–328.
Madani, K., AghaKouchak, A., & Mirchi, A. (2016). Iran's socio-economic drought: Challenges of a water-bankrupt nation. *Iranian Studies* 49(6): 997–1016.
Mardi, A. H., Khaghani, A., MacDonald, A. B., et al. (2018). The Lake Urmia environmental disaster in Iran: A look at aerosol pollution. *Science of The Total Environment* 633: 42–9.
Pengra, B. (2012). The drying of Iran's Lake Urmia and its environmental consequences. *UNEP-GRID, Sioux Falls, UNEP Global Environmental Alert Service (GEAS).*
Rahimi, J., Ebrahimpour, M., & Khalili, A. (2013). Spatial changes of extended De Martonne climatic zones affected by climate change in Iran. *Theoretical and Applied Climatology* 112(3–4): 409–418.
Saboohi, R., Soltani, S., & Khodagholi, M. (2012). Trend analysis of temperature parameters in Iran. *Theoretical and Applied Climatology* 109(3–4): 529–547.
Saemian, P., Faridzad, M., Youneszadeh, S., et al. (2015). Estimation of evapotranspiration, water saving and agricultural water consumption at Urmia Lake Basin, 2010. *Report of ULRP* (in Persian).
Save, H., Bettadpur, S., & Tapley, B. D. (2016). High-resolution CSR GRACE RL05 mascons. *Journal of Geophysical Research: Solid Earth* 121(10): 7547–7569.
Shadkam, S. (2017). *Preserving Urmia Lake in a Changing World: Reconciling Anthropogenic and Climate Drivers by Hydrological Modelling and Policy Assessment*. Wageningen, Netherlands: Wageningen University.
Shadkam, S., Ludwig, F., van Oel, P., Kirmit, Ç., & Kabat, P. (2016). Impacts of climate change and water resources development on the declining inflow into Iran's Urmia Lake. *Journal of Great Lakes Research* 42(5): 942–952.
Tourian, M. J., Elmi, O., Chen, Q., et al. (2015). A spaceborne multisensor approach to monitor the desiccation of Lake Urmia in Iran. *Remote Sensing of Environment* 156: 349–360.
ULRP (2015a). The trend of Lake Urmia changes. Available from http://ulrp.sharif.ir/fa (Last accessed 20 July 2020) (in Persian).
ULRP (2015b). Status of water resources and water consumption in the basin. Available from http://ulrp.sharif.ir (Last accessed 20 July 2020) (in Persian).
Vaheddoost, B., & Aksoy, H. (2018). Interaction of groundwater with Lake Urmia in Iran. *Hydrological Processes* 32(21): 3283–3295.
Van Vuuren, D. P., Edmonds, J., Kainuma, M., et al. (2011). The representative concentration pathways: An overview. *Climatic Change* 109(1–2): 5.
Wang, X. J., Zhang, J. Y., Shahid, S., et al. (2016). Adaptation to climate change impacts on water demand. *Mitigation and Adaptation Strategies for Global Change* 21(1): 81–99.
Zahedi, S., Azarnivand, A., & Chitsaz, N. (2017). Groundwater quality classification derivation using multi-criteria-decision-making techniques. *Ecological Indicators* 78: 243–252.
Zeinoddini, M., Tofighi, M. A., & Vafaee, F. (2009). Evaluation of dike-type causeway impacts on the flow and salinity regimes in Urmia Lake, Iran. *Journal of Great Lakes Research* 35(1): 13–22.

Index